Control and Optimization with Differential-Algebraic Constraints

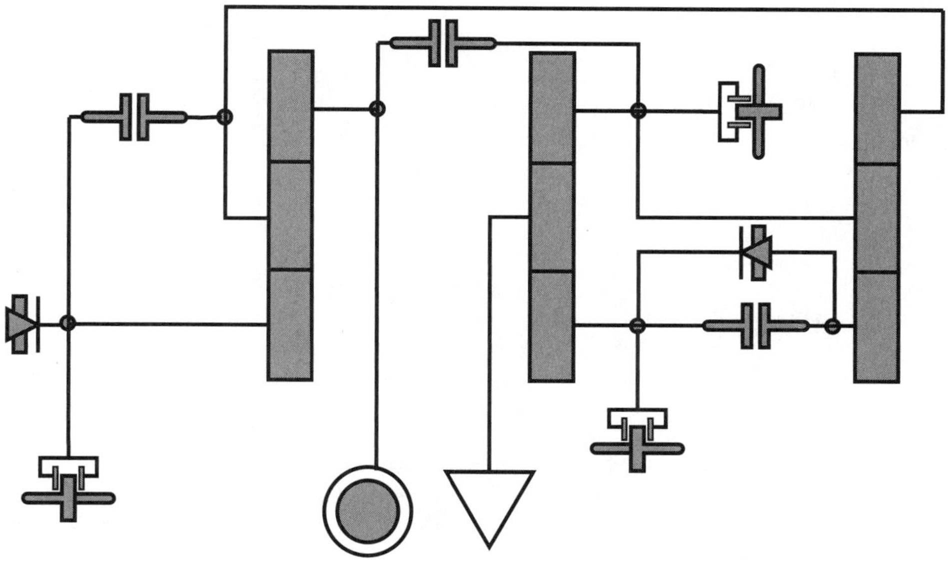

Advances in Design and Control

SIAM's Advances in Design and Control series consists of texts and monographs dealing with all areas of design and control and their applications. Topics of interest include shape optimization, multidisciplinary design, trajectory optimization, feedback, and optimal control. The series focuses on the mathematical and computational aspects of engineering design and control that are usable in a wide variety of scientific and engineering disciplines.

Series Volumes

Biegler, Lorenz T., Campbell, Stephen L., and Mehrmann, Volker, eds., *Control and Optimization with Differential-Algebraic Constraints*

Delfour, M. C. and Zolésio, J.-P., *Shapes and Geometries: Metrics, Analysis, Differential Calculus, and Optimization, Second Edition*

Hovakimyan, Naira and Cao, Chengyu, *$\mathcal{L}_1$ Adaptive Control Theory: Guaranteed Robustness with Fast Adaptation*

Speyer, Jason L. and Jacobson, David H., *Primer on Optimal Control Theory*

Betts, John T., *Practical Methods for Optimal Control and Estimation Using Nonlinear Programming, Second Edition*

Shima, Tal and Rasmussen, Steven, eds., *UAV Cooperative Decision and Control: Challenges and Practical Approaches*

Speyer, Jason L. and Chung, Walter H., *Stochastic Processes, Estimation, and Control*

Krstic, Miroslav and Smyshlyaev, Andrey, *Boundary Control of PDEs: A Course on Backstepping Designs*

Ito, Kazufumi and Kunisch, Karl, *Lagrange Multiplier Approach to Variational Problems and Applications*

Xue, Dingyü, Chen, YangQuan, and Atherton, Derek P., *Linear Feedback Control: Analysis and Design with MATLAB*

Hanson, Floyd B., *Applied Stochastic Processes and Control for Jump-Diffusions: Modeling, Analysis, and Computation*

Michiels, Wim and Niculescu, Silviu-Iulian, *Stability and Stabilization of Time-Delay Systems: An Eigenvalue-Based Approach*

Ioannou, Petros and Fidan, Barış, *Adaptive Control Tutorial*

Bhaya, Amit and Kaszkurewicz, Eugenius, *Control Perspectives on Numerical Algorithms and Matrix Problems*

Robinett III, Rush D., Wilson, David G., Eisler, G. Richard, and Hurtado, John E., *Applied Dynamic Programming for Optimization of Dynamical Systems*

Huang, J., *Nonlinear Output Regulation: Theory and Applications*

Haslinger, J. and Mäkinen, R. A. E., *Introduction to Shape Optimization: Theory, Approximation, and Computation*

Antoulas, Athanasios C., *Approximation of Large-Scale Dynamical Systems*

Gunzburger, Max D., *Perspectives in Flow Control and Optimization*

Delfour, M. C. and Zolésio, J.-P., *Shapes and Geometries: Analysis, Differential Calculus, and Optimization*

Betts, John T., *Practical Methods for Optimal Control Using Nonlinear Programming*

El Ghaoui, Laurent and Niculescu, Silviu-Iulian, eds., *Advances in Linear Matrix Inequality Methods in Control*

Helton, J. William and James, Matthew R., *Extending H^∞ Control to Nonlinear Systems: Control of Nonlinear Systems to Achieve Performance Objectives*

Control and Optimization with Differential-Algebraic Constraints

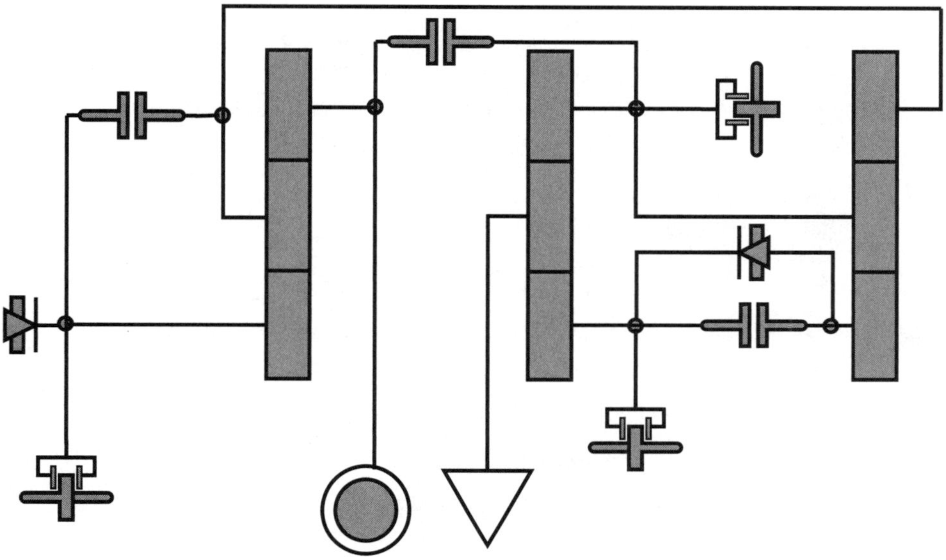

Edited by
Lorenz T. Biegler
Carnegie Mellon University
Pittsburgh, Pennsylvania

Stephen L. Campbell
North Carolina State University
Raleigh, North Carolina

Volker Mehrmann
Technische Universität Berlin
Berlin, Germany

siam.
Society for Industrial and Applied Mathematics
Philadelphia

10 9 8 7 6 5 4 3 2 1

Figure 2.1 courtesy of Dr. Ming Hou.
Figure 2.2 reprinted with permission of Springer Science+Business Media.
Figures 6.1-2 © 2010 IEEE. Reprinted with permission from IEEE Proceedings.
Figure 6.3 reprinted with permission from the International Federation of Automatic Control.
Figure 8.1 reprinted with permission from Nauka Publishing.
Figure 9.1 © 2006 John Wiley & Sons, Ltd. Reprinted with permission.
Figures 13.2-5 reprinted with permission from Cambridge University Press.
Figure 14.5 © 2012 IEEE. Reprinted with permission from IEEE Control Systems Magazine.

Library of Congress Cataloging-in-Publication Data
Control and optimization with differential-algebraic constraints / edited by Lorenz T. Biegler, Stephen L. Campbell, Volker Mehrmann.
p. cm. – (Advances in design and control)
Includes bibliographical references and index.
ISBN 978-1-611972-24-5
1. Differential-algebraic equations. 2. Control theory. 3. Mathematical optimization. I. Biegler, Lorenz T. II. Campbell, S. L. (Stephen La Vern) III. Mehrmann, V. L. (Volker Ludwig), 1955-
QA372.5.C66 2012
512'.56–dc23

2012016258

List of Contributors

Uri M. Ascher
Department of Computer Science
University of British Columbia
Vancouver, British Columbia, Canada

Michael Baldea
Department of Chemical Engineering
The University of Texas at Austin
Austin, Texas

Thomas Berger
Institut für Mathematik
Technische Universität Ilmenau
Ilmenau, Germany

John T. Betts
Partner, Applied Mathematical Analysis, LLC

Lorenz T. Biegler
Chemical Engineering Department
Carnegie Mellon University
Pittsburgh, Pennsylvania

Jason J. Boisvert
Department of Computer Science
University of Saskatchewan
Saskatoon, Saskatchewan, Canada

Francesco Borrelli
Department of Mechanical Engineering
University of California at Berkeley
Berkeley, California

Stephen L. Campbell
Department of Mathematics
North Carolina State University
Raleigh, North Carolina

Weifeng Chen
College of Information Technology
Zhejiang University of Technology
Hangzhou, People's Republic of China

Prodromos Daoutidis
Department of Chemical Engineering and Materials Science
University of Minnesota
Minneapolis, Minnesota

Matthew W. Donaldson
Department of Computer Science
University of Saskatchewan
Saskatoon, Saskatchewan, Canada

Matthias Gerdts
Department of Aerospace Engineering
Universität der Bundeswehr, Neubiberg
München, Germany

Achim Ilchmann
Institut für Mathematik
Technische Universität Ilmenau
Ilmenau, Germany

Pedher Johansson
Department of Computing Science and HPC2N
Umeå University
Umeå, Sweden

Stefan Johansson
Department of Computing Science and HPC2N
Umeå University
Umeå, Sweden

Bo Kågström
Department of Computing Science and HPC2N
Umeå University
Umeå, Sweden

Peter Kunkel
Mathematisches Institut
Universität Leipzig
Leipzig, Germany

Galina Kurina
Department of Mathematics
Voronezh State University
Voronezh, Russia

Vu Hoang Linh
Department of Mathematics, Mechanics, and Informatics
Vietnam National University
Thanh Xuan, Hanoi, Vietnam

Yudong Ma
Department of Mechanical Engineering
University of California at Berkeley
Berkeley, California

Roswitha März
Institut für Numerische Mathematik, retired
Humboldt Universität zu Berlin
Berlin, Germany

Volker Mehrmann
Institut für Mathematik
Technische Universität Berlin
Berlin, Germany

Srinivas Palanki
Department of Chemical and Biomolecular Engineering
University of South Alabama
Mobile, Alabama

Timo Reis
Fachbereich Mathematik
Universität Hamburg
Hamburg, Germany

Stefan Richter
Automatic Control Laboratory
ETH Zürich
Zürich, Switzerland

Sebastian Sager
Interdisciplinary Center for Scientific Computing
Ruprecht-Karls Universitat
Heidelberg, Germany

Zhijiang Shao
Department of Control Science and Engineering
Zhejiang University
Hangzhou, People's Republic of China

Raymond J. Spiteri
Department of Computer Science
University of Saskatchewan
Saskatoon, Saskatchewan, Canada

Karmethia C. Thompson
Department of Mathematics
North Carolina State University
Raleigh, North Carolina

Kees van den Doel
Department of Computer Science
University of British Columbia
Vancouver, British Columbia, Canada

Andreas Varga
Institute of Robotics and Mechatronics
German Aerospace Center Oberpfaffenhofen
Wessling, Germany

Kexin Wang
Department of Control Science and Engineering
Zhejiang University
Hangzhou, People's Republic of China

Contents

Preface

Differential-algebraic equations (DAEs) are mixed systems of differential and algebraic equations. They form one of the most elegant and simple ways to model a physical system because they allow for the creation of separate models for subcomponents that can then be pasted together via a network. Based on this feature, the analytical and numerical work on DAEs was originally driven by the simulation community, where the desire was to simulate the behavior of a complex system which could be a combination of electrical, mechanical, chemical, or other physical components.

Once a system can be modeled and simulated, this DAE model provides an ideal tool to control or optimize the performance of the underlying physical system. Tremendous advances in the theory and numerical methods have led to applications of model-based control and optimization in almost all areas of science and technology, including many industrial research departments. In this way, model-based optimization and control of DAE modeled processes has become an important driver of innovation.

In spite of the large importance of the interconnected areas of DAE-based simulation, optimization, and control, there had never been a workshop dedicated to this topic.

To meet this need, a week-long workshop, *Control and Optimization with Differential-Algebraic Constraints*, was held at the Banff International Research Station (BIRS) in Banff, Canada, from Oct. 24 to Oct. 29, 2010. Participants came from academia and industry. Their backgrounds included mathematics, scientific computing, chemical engineering, aerospace engineering, mechanical engineering, and computer science. They represented seven countries and three continents.

That highly successful workshop and its accompanying energy and enthusiasm were the genesis for this volume, whose editors organized the original workshop. All of the chapters in this book were written by attendees, and in most cases an earlier version of the paper was presented at the workshop. The first chapter is written by the workshop organizers and is both a survey of the area and an introduction to the papers that follow.

The editors would like to thank the staff at Banff for their help in planning and running the workshop. We would also like to thank SIAM for making this volume a reality and providing needed assistance and guidance. A special thanks from one of us goes to John Rogosich of Techsetters, who was able to get all of the contributions smoothly working together within the special style being used in spite of the contributions altering hidden commands inside AMSTeX and the style file having undocumented restrictions.

Larry Biegler, Steve Campbell, and Volker Mehrmann

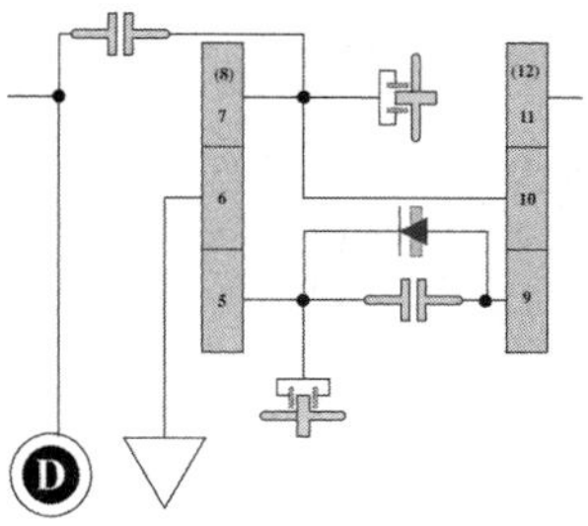

Chapter 1
DAEs, Control, and Optimization

Lorenz Biegler, Stephen Campbell, and Volker Mehrmann

This chapter introduces the topic of differential-algebraic equations (DAEs) and provides a brief overview of DAE representations and properties as well as the control and optimization of DAE systems. All of these issues are explored in more detail in the subsequent chapters, which are also described briefly here.

1.1 Introduction

The advent of powerful computers has led to fundamental changes in the way that scientists and engineers work. Numerical simulation has become the third pillar of scientific discovery, along with theory and experiment. Furthermore, today, the design of a system is typically envisioned, formulated, simulated, and then modified on a computer. This leads not only to more convenient work processes and faster design cycles, but it is increasingly necessary as more complex systems are being considered. Desires for increased system performance, exotic new materials, or reduced cost make consideration of the nonlinear behavior of complex higher-dimensional systems essential.

Very often, the dynamic behavior of a system is most naturally modeled as a system of differential and algebraic equations that describes different physical components that are coupled via a network. In the most general form [5, 13, 14, 15, 16, 36, 37, 49] such a system of DAEs can be expressed as

$$F(\dot{z}, z, u, p, t) = 0, \tag{1.1}$$

possibly combined with an output equation $y = G(z, u, p, t)$. Here z is an n-dimensional "state" vector that is used to describe the system and $\dot{z} = dz/dt$. The variable u represents any control or source terms, y represents an output that describes, for example, measured variables, t is a real variable (often time), and p comprises various design parameters and constants. Typically F is ℓ-dimensional, and cases with $\ell > n$ and $\ell < n$ both occur in applications, but here we will mostly just discuss the case that $\ell = n$. In many applications z is even infinite-dimensional, and then the finite-dimensional case arises from a discretization of z.

We have designated certain variables as source or control variables. However, in the DAE formulation (1.1) of control problems it is easy for a designer to experiment with different choices for the control variables. Usually the term DAE refers to $F_{\dot{z}} = \partial F/\partial \dot{z}$ as being singular, as opposed to the case when $F_{\dot{z}}$ is invertible except at certain isolated (turning) points. In the control context the DAE (1.1) is often called a *descriptor system*. Other names are *implicit, degenerate, semistate*, and *constrained*. Both initial value and boundary value problems are important in practice.

To consider examples, for a standard electronic circuit, (1.1) would represent Kirchhoff's laws and the equations determining the device characteristics. For mechanical systems, (1.1) could arise from the variational equations. In chemical processes, DAEs arise from conservation laws, constitutive equations, and process restrictions.

Initially in practice, the DAE (1.1) almost never arises directly in the form of an ordinary differential equation (ODE),

$$\dot{x} = f(x,u,p,t), \quad x(0) = x_0, \tag{1.2}$$

for which a multitude of simulation, control, and optimization techniques are available. For a DAE it is not necessarily the case that all of these techniques are immediately available. We shall often omit the p in what follows to simplify the formulas.

Classically, this difficulty was overcome by applying some elimination process and reformulating the DAE as an ODE. For some particular classes of problems, such as linear circuits, it was possible to do this algorithmically. This approach has also been utilized under various guises, such as *index reduction*, or *generalized coordinate partitioning*, for the simulation of certain classes of mechanical systems.

It was, however, realized very early on that there are tremendous advantages to being able to solve the original DAE rather than rewriting it as an ODE. A few of the advantages of a DAE formulation are the following:

- it may be difficult or impossible to reformulate the problem as an ODE when nonlinearities are present;
- the algebraic equations typically describe conservation laws or explicit equality constraints, and they should be kept invariant;
- it is easier to vary design parameters in an implicit model;
- the implicit model does not require the modeling simplifications often necessary to get an ODE;
- the variables keep their original physical interpretation;
- the system structure can be exploited by problem-specific solvers, i.e., numerical integrators;
- less specialized mathematical expertise is required on the part of the designer.

An important property of DAEs is that they consist not only of parts that integrate source functions or inhomogeneities that act in a smoothing manner like an ODE does, but they also have parts that differentiate source functions or inhomogeneities. The latter property leads to serious problems for many solvers or control and optimization techniques, in particular when these parts are not explicitly given in the model. The necessary regularity requirements are typically described by a so-called index (there are many such index concepts; here we use the differentiation index [13]) which is discussed in the next section.

DAE applications arise in almost all areas of science and engineering and range from low-dimensional to very high-dimensional nonlinear DAEs that arise from discretization or approximation of partial differential equations (PDEs). More generally, they can consist of mixed systems of PDEs, ODEs, functional and delay equations, and algebraic constraints. These are sometimes called partial differential-algebraic equations (PDAEs) or delay-differential-algebraic equations (DDAEs).

In almost all applications there is an important interaction between DAE modeling, their analytical properties, and the numerical approximation and optimization algorithms. Some of these interactions will be discussed in the later chapters.

1.2 DAEs

Modeling, simulation, and optimization with DAEs have become essential tools in many areas in science and engineering. Here, we give only a cursory literature review and omit many contributions. Additional references may be found in [5, 13, 15, 16, 22, 27, 29, 32, 33, 37]. Before discussing the connections to control and optimization some terminology is needed.

Intuitively, a DAE is *solvable* on an interval $\mathcal{I}$ if there is a smooth manifold $\mathcal{M}$ of solutions which are uniquely determined by their value at any t in $\mathcal{I}$. More precise definitions can be found in [13, 15, 17, 20, 22]. An initial condition (t_0, z_0) is *consistent* if $(t_0, z_0) \in \mathcal{M}$. The DAE (1.1) is *fully implicit* if the constraints are not explicit as in (1.6b). If (1.1) is fully implicit in $\dot{z}$, then an initial condition is usually taken to be a vector $(t_0, z_0, \dot{z}_0)$. Explicit characterizations of the manifold of consistent initial conditions have been given for linear [22] and nonlinear DAEs [21]. On the other hand, inconsistent initial conditions, when physically meaningful, can give rise to distributional solutions, which were initially studied from a state-space viewpoint [25]. Distributional solutions are used, for example, to explain the behavior of backward differentiation formulas in the initial boundary layer. More recently, researchers have turned to the computation of impulsive solutions [53], solutions with discontinuous right-hand sides, and multiple solutions [37]. In this volume, our interest is in piecewise smooth solutions.

The basic theory for the linear time-invariant case,

$$E\dot{z} + Az = f(t), \tag{1.3}$$

is given by the Kronecker canonical form of the *matrix pencil* $\lambda E + A$. If the pencil is regular, i.e., $\lambda E + A$ is nonsingular for all but a finite number of values λ, then the DAE (1.3) is solvable for all sufficiently differentiable right-hand sides and all consistent initial values. A pencil may fail to be regular either because the determinant of $\lambda E + A$ does not exist (E, A are not square) or because the determinant is identically zero. If the pencil is not regular, then the DAE may be solvable only for specific right-hand sides or the solution may not be uniquely determined by a consistent initial equation or both.

In the regular case there exist constant coordinate changes that transform the system (1.3) into the form

$$\begin{bmatrix} I & 0 \\ 0 & N \end{bmatrix} \begin{bmatrix} \dot{z}_1 \\ \dot{z}_2 \end{bmatrix} + \begin{bmatrix} C & 0 \\ 0 & I \end{bmatrix} \begin{bmatrix} z_1 \\ z_2 \end{bmatrix} = \begin{bmatrix} f_1 \\ f_2 \end{bmatrix}. \tag{1.4}$$

Here N is nilpotent of index k, i.e., $N^k = 0$ and $N^{k-1} \neq 0$. The top equation $\dot{z}_1 + Cz_1 = f_1$ is an ODE, and $z_1(0)$ can be specified to be any vector of appropriate size. The second equation $N\dot{z}_2 + z_2 = f_2$ has only one solution $z_2 = \sum_{i=0}^{k-1}(-N)^i f_2^{(i)}$, and no initial condition can be chosen since $x_2(0)$ is completely determined by f. Note that the z_2 part of the solution requires $f_2(t)$ to be differentiated $k-1$ times.

While the linear variable coefficient and nonlinear cases are more complicated, a number of key features of DAEs are already apparent in the linear time-invariant case. There are restrictions on consistent initial conditions, solutions live on a submanifold, and solutions can depend on derivatives of the forcing function f. In the linear variable coefficient and nonlinear cases, the solutions of a DAE can also depend on derivatives of coefficients. Thus, the question is often not whether to differentiate but rather where to do so. If we differentiate the DAE (1.1) k times with respect to t, we obtain the *derivative array* equations

$$\begin{aligned} F(\dot{z}, z, t) &= 0 \\ \frac{d}{dt} F = F_1(z^{(2)}, \dot{z}, z, t) &= 0 \\ &\vdots \\ \frac{d^k}{dt^k} F = F_k(z^{(k+1)}, z^{(k)}, \ldots, z, t) &= 0. \end{aligned} \tag{1.5}$$

The smallest k such that (1.5) uniquely determines $\dot{z}$ in terms of (z,t), which are consistent initial conditions for the original DAE, is the *(differentiation) index of the DAE* (1.1). It implicitly defines a relationship $\dot{z} = Q(z,t)$. For fully implicit problems, $\dot{z}$ is unique in a relatively open set containing $\dot{z}_0$. An implicit ODE has index zero. Intuitively, the index is a measure of how many differentiations would be required to convert the DAE into an ODE. The actual description of what takes place is technically much more difficult, since constraints may not be explicit and one need not have constant ranks [17, 20, 21]. Also, in numerical algorithms one wants to avoid differentiating computed quantities. DAEs which are not index one or not semi-explicit always have additional hidden constraints. Moreover, other index definitions include the strangeness index [37], the tractability index [29], and the perturbation index [32]. Some of these indices are closely related [37, 42]. However, [20] shows that different index concepts can differ greatly for unstructured nonlinear DAEs.

An explicit ODE whose solutions include all the solutions of the DAE (1.1) is called a *completion* of (1.1). Completions [48] have traditionally been used in numerical simulation methods but more recently have been used in observer design [9].

Motivated by applications, especially from mechanics, various special types of structured DAEs have been defined. A special case of (1.1) is the *semi-explicit* DAE

$$\dot{z}_1 = f(z_1, z_2, t), \tag{1.6a}$$

$$0 = g(z_1, z_2, t), \tag{1.6b}$$

which is of index one if and only if $\partial g/\partial z_2$ is nonsingular. A semi-explicit index-one system is usually the easiest case to work with. Fortunately, many applications satisfy (or can be reformulated to satisfy) this assumption. Moreover, note that z_2 need not be as smooth as z_1 in (1.6). Similarly, not all elements of z need be differentiable in (1.1). This property is discussed in Chapter 3 along with an introduction to regularity properties and their relationship to linearizations.

A well-studied class of higher-index systems are the *Hessenberg* DAEs:

$$\dot{z}_1 = f(z_1, z_2, z_3, t), \tag{1.7a}$$
$$\dot{z}_2 = g(z_1, z_2, t), \tag{1.7b}$$
$$0 = h(z_2, t). \tag{1.7c}$$

The system (1.7c) and (1.7b) is an index-two Hessenberg system if $(\partial h/\partial z_2) \times (\partial g/\partial z_1)$ is nonsingular. The system (1.7c)–(1.7a) is an index-three Hessenberg system if $(\partial h/\partial z_2) \times (\partial g/\partial z_1) \times (\partial f/\partial z_3)$ is nonsingular.

In general, the direct numerical simulation of high-index DAEs requires more than z and $\dot{z}$ as initial value information, and some methods must carry higher derivatives along throughout the integration. Because of this, both the nonlinear initialization and time stepping equations are underdetermined. Global convergence issues arise in the initialization phase of solvers for high-index DAEs, where the derivative data needed to start the integration is almost certainly not available. This is an active research topic even in the index-one case. The problems are more acute in the high-index case because the nonlinear equations that must be solved are underdetermined and the resulting nonuniqueness of solutions must be resolved by using other information.

Finally, a wide variety of numerical solution methods have been developed for DAEs including [13, 32, 33, 56, 57]. None of the variants of standard methods, such as the various implicit Runge–Kutta, extrapolation, or multistep methods, works even on general index-two problems. For high-index problems all of these methods require special structure, and they also usually require the index to be no more than two. An exception is RADAU5, which works on index-three Hessenberg systems [33]. All the methods that apply to more general DAE systems, including those described in [6, 23, 37, 40, 41], are based on some type of reformulation using a variant of the derivative array equations (1.5).

To summarize these introductory remarks, DAEs are a powerful framework for the modeling of complex systems with a network-based structure of components. However, this simple modeling concept may lead to substantial difficulties in the analytical and numerical treatment of these equations. These difficulties increase further in the development of appropriate control and optimization techniques for DAEs. The chapters of this volume illustrate these difficulties and describe steps to overcome some of these difficulties.

To introduce this volume and describe the subsequent chapters, we divide the remainder of this chapter into five parts. The next section deals with general control questions and formulations, while Section 1.4 focuses on linear DAE systems where both theory and solution strategies are more widely developed for sophisticated control applications. Section 1.5 discusses optimal control properties and solution strategies for nonlinear DAEs, while Section 1.6 briefly describes software contributions for DAEs, particularly within the chapters in this volume. Finally, Section 1.7 describes the chapters devoted to real-world DAE applications.

1.3 DAEs and Control

DAEs and control problems intertwine in a number of ways. In control problems the starting point is often a system of the form

$$\dot{x} = f(x,u,t), \tag{1.8a}$$
$$y = g(x,u,t). \tag{1.8b}$$

Here (1.8a) is the process or plant, u is the control, x is the state, and y is the output. Important control theoretic questions that need to be studied include the question of controllability, observability, or stabilization by feedback.

The most obvious way that a DAE enters the picture occurs when the process model takes the form (1.1) instead of (1.8a) resulting in a system

$$0 = F(\dot{x},x,u,t), \tag{1.9a}$$
$$y = g(x,u,t). \tag{1.9b}$$

There is a vast literature on this case especially when (1.9a) is linear time-invariant, as in (1.10); see, e.g., [27, 37].

However, DAEs also occur in another way. Sometimes we want to specify y in (1.8b). For example, (1.8a) could describe the flight of an aerospace vehicle, where (1.8b) specifies a particular reentry path. Equation (1.8b) is then a constraint and (1.8) becomes a DAE. This is a prescribed path control problem [12], and the presence and structure of these constraints has a large influence on the theory and the numerical solution. Of course, y could also be prescribed in (1.9).

In problems when the constraints arise due to physical considerations, the DAEs may have a special structure such as Hessenberg. This is exploited in the simulation of mechanical systems where the constraint results in a force. However, in a prescribed path control problem, or many other problems where the constraints are desired operational, and not physical constraints, the DAEs that arise are often no longer Hessenberg. Instead the control has to try to meet the constraints without a specific structure to the DAEs, and the index can be higher than the usual two or three. This can pose a major issue for some numerical integrators. This issue can also be seen in nonlinear control with the important ideas of relative degree and zero dynamics, as discussed in Chapters 13 and 16. These two concepts are closely related to the index of the prescribed path control problem and the dimension of the solution manifold when $y = 0$ and $u = 0$, respectively.

The next two chapters contain general discussions about DAEs and how this general theory relates to some control issues. Often, a reformulation of the models or even a regularization is necessary to avoid failure of the computational DAE method and to make it suitable for its incorporation as an optimal control problem that can be solved with current software. Chapter 2 discusses difficulties with DAE models particularly for optimal control. A general framework is considered for nonlinear DAEs, with analysis and regularization strategies discussed to make the system equivalent to semi-explicit, index one. This strategy is further specialized to linear descriptor systems. A review of the theory is presented for both cases, and several problem instances are presented to illustrate the reformulation strategies.

Linearization also plays an important role in control theory in several aspects, from determining stability of equilibria or controlled trajectories to creating linear feedbacks for nonlinear systems. The obvious linearization of a DAE need not be the correct approach to consider for a general DAE [18], since the linearization may not lie in the solution manifold. However, the situation is much better when considering DAEs in Hessenberg and similar forms. Linearization also plays a role in optimal control. Chapter 3 presents a DAE formulation that allows for less smoothness in the solutions, discusses linearizations in the context of this formulation, and relates them to regularity regions. These linearization concepts are also extended to necessary conditions for optimal control problems.

1.4 Analysis and Control of Linear DAE Systems

The next four chapters focus on the analysis and control of linear descriptor systems, given by

$$E\dot{x} = Ax + Bu, \tag{1.10a}$$

$$y = Cx + Du, \tag{1.10b}$$

including the linear time-varying case when the coefficient matrices depend on t. Chapter 4 surveys the spectral analysis of linear time-varying DAEs. For linear time-varying and nonlinear problems it is no longer sufficient to just consider eigenvalues. Instead, basic spectral notions such as Lyapunov and Sacker–Sell spectra are introduced. Results are presented that extend these notions from ODEs to DAEs, but only under additional conditions. This has consequences concerning the boundedness of solutions of inhomogeneous equations. Also, linear subspaces of leading directions are characterized, which are associated with spectral intervals and which generalize eigenvectors and invariant subspaces as they are used in the linear time-invariant setting.

System structure and robustness with respect to perturbations are important parts of linear descriptor system analysis. System characteristics of (1.10) (e.g., poles, zeros, controllability, and observability) involve computing the canonical structure information of an associated system pencil. In Chapter 5 descriptor system models are analyzed via their associated system pencils, represented through closure hierarchy, or stratification, graphs. The underlying theory is reviewed, and it is illustrated how the resulting stratification algorithms can be used to analyze descriptor system models. These also provide a deeper understanding of the control system dynamics and its characteristics under perturbations.

Most modern systems require some type of fault detection and isolation. Chapter 6 considers $\mathcal{H}_{2/\infty}$-optimal fault detection and isolation problems. A detailed computational framework for the calculation of faults in linear descriptor systems is discussed in both discrete and continuous time. Underlying numerical algorithms that compute rational null-space bases, inner-outer factorizations, or proper coprime factorizations are based on descriptor system representations and rely on orthogonal matrix pencil reductions. Integrated computational approaches developed in this chapter fully exploit these structural aspects at each solution step and produce fault detectors of least order.

Chapter 7 explores controller synthesis strategies for linear descriptor systems (1.10). Structural properties, including normal forms, zero dynamics, and high-gain stabilizability, are analyzed to gain insight into the system classes and also to solve the output regulation problem by funnel control, where a class of reference signals is tracked within a prespecified funnel. Such controllers are simpler to design, as the output error is prespecified, and neither an internal model nor an identification mechanism is required. The results are illustrated by position and velocity control of a mechanical system with springs, masses, and dampers. Many of the concepts mentioned in this chapter, such as zero dynamics, are important when considering nonlinear systems.

Chapter 8 considers two linear-quadratic optimal control problems with DAEs. The first has discontinuous coefficients and the state equations are switched systems, while the second has a cost functional that depends on state variable values at intermediate points. Optimal feedback controls are determined from discontinuous solutions of a special implicit operator Riccati equation, and procedures are developed to obtain asymptotic solutions for singularly perturbed problems with switched state equations and discontinuous coefficients.

1.5 Optimal Control with DAEs

Once a system can be controlled, then inevitably the desire is to do so in as efficient a manner as possible. One may want to maximize a function of the outputs, such as downrange glide, or payload in aerospace systems, or perhaps minimize time, energy consumed, or off-spec products in manufacturing processes. In all cases an optimal control problem can result. Consideration of DAEs arises in a number of ways when solving optimal control problems.

Of course, the original process model can be a DAE, but there are several other ways that DAEs are fundamental. Suppose that we have the cost and process given by the system

$$J(u) = \int_0^T L(x,u,t)\,dt + \phi(x(T),T), \tag{1.11a}$$

$$\dot{x} = f(x,u,t). \tag{1.11b}$$

Here (1.11b) is the model of the process which is assumed to be an ODE. J in (1.11a) is the cost. The ϕ term in (1.11a) is the cost of the terminal state value. The goal is to choose the control u to minimize J. In addition to (1.11) there will be additional initial and terminal conditions such as a starting value for $x(t)$ or the requirement that $x(T)$ lie on a particular surface. However, in practice there frequently are also additional equality and inequality operational constraints that must hold during the time period in question. Thus, the actual problem is to minimize J subject to the following constraints:

$$\min\ J(u) = \int_0^T L(x,u,t)\,dt + \phi(x(T),T), \tag{1.12a}$$

$$\dot{x} = f(x,u,t), \tag{1.12b}$$

$$0 = g(x,u,t), \tag{1.12c}$$

$$0 \leq h(x,u,t). \tag{1.12d}$$

Equations (1.12b)–(1.12d) form a DAE if (1.12c) is present or if an inequality constraint in (1.12d) is active.

However, DAEs are present in other ways even if (1.12c) and (1.12d) are absent. To illustrate this, consider the simple case where we have just (1.12a)–(1.12b) and u appears in L and u is not constrained. Then, ignoring all the various boundary conditions, the necessary conditions for a minimum (or maximum) are given by a boundary value problem in the state x, Lagrange multiplier λ, and control u,

$$\dot{x} = f(x,u,t), \tag{1.13a}$$
$$-\dot{\lambda} = f_x(x,u,t)^T\lambda + L_x(x,u,t), \tag{1.13b}$$
$$0 = f_u(x,u,t)^T\lambda + L_u(x,u,t), \tag{1.13c}$$

which is a DAE in x,λ,u with boundary conditions for x at 0 and λ at T. This DAE will be higher index if L_{uu} is a singular matrix. The boundary value problems which are obtained as necessary optimality conditions for an optimal control problem with a DAE constraint have a similar structure [38, 39]. If inequality constraints (1.12d) are present in (1.12) and if they involve u, then (1.13c) is replaced by the more general condition that u minimizes the Hamiltonian $H = L(x,u,t)+\lambda^T f(x,u,t)$.

Most optimal control problems do not admit an analytic solution and must be solved numerically. There are several ways that this numerical solution can proceed. The most obvious approach is to solve the necessary conditions (1.13) as a boundary value problem. This variational or *indirect approach* has seen a number of studies. But it usually becomes impractical when inequality constraints are present, as then multipliers, or adjoint variables, $\lambda(t)$ may be discontinuous or might not even exist on some subintervals of t.

If the dynamics and state constraints form a DAE that can be integrated for a given value of u, then it is often possible to evaluate the cost function with a simulation. Using some type of finite-dimensional parameterization of u, one then gets the cost in terms of a finite number of parameters and one can then determine an approximate optimal solution with a general purpose nonlinear programming (NLP) package such as SNOPT [28]. We shall refer to this as *control parameterization*. This approach comes in a number of variations. It is popular if the DAE is integrable, since one is just coupling two available codes, the NLP code and the DAE simulator. On the other hand, control parameterization may have difficulties with inequality constraints, in order to determine when they become active and when they become inactive over t. Difficulties also arise in resolving details like touch points and singular arcs, even if they are not numerically significant.

A third option is to transfer both the state and control profiles and the resulting large NLP problem to an optimization solver. This is called *direct transcription*. Direct transcription is widely used in industry. It has the advantage that the necessary conditions need not be formulated and the adjoint λ need not be calculated directly. Thus, if one can write the problem down, one can try to solve it, and useful answers are often obtained for very difficult problems. Moreover, the investigation of how to solve ever wider classes of problems by direct transcription is an active research topic. On the other hand, the resulting NLP problem from direct transcription can be quite large, and it is important that nonuniform grids be carefully generated. It is also essential to exploit highly tuned sparse linear algebra solvers for the NLP formulation and solver. So while the codes can be quite useful it takes some effort to design efficient and robust direct transcription codes.

Several chapters of this volume deal with formulation and solution of optimal control problems. Chapter 9 considers the optimal control of an important class of hybrid systems, which are becoming increasingly widespread in industrial applications. Hybrid problems involve a mixture of continuous processes, discrete-time processes, and processes where the states or controls can take on only a finite number of values. This chapter presents necessary conditions for optimality of hybrid systems with index-one DAEs and integer restrictions. These results generalize previous work for the case of ODEs and allow integer controls to be obtained from reformulations or relaxations of purely continuous control functions. They can also be used for efficient numerical calculation of optimal solutions. The results are illustrated with a mixed-integer nonlinear optimal control benchmark problem.

The study of delay-differential equations (DDEs) as well as delay-differential-algebraic equations (DDAEs) has provided challenging analytic and numerical problems for over thirty years. In Chapter 10 a direct transcription optimal control code ($\mathbb{SOCX}$) is extended to included delays in state variables. These delayed profiles are interpolated using Hermite polynomials and incorporated within the direct transcription solver. Here, some of the computational issues are discussed and the control of a reaction-diffusion system using both constant and time-varying delays is presented.

An important question with the direct transcription philosophy is the choice of discretization. This is also a topic of active consideration. In the approach of Chapter 10 the control and state are assumed continuous. While this approach can solve control constrained problems, there will tend to be some chatter near control discontinuities unless phases are used. Phases are helpful, but they require some user expertise. In Chapter 11 an alternative discretization is considered. Discontinuity at a grid point is allowed, and the grid points are allowed to move. For certain types of problems one sees better performance at the price of additional user expertise and some restrictions on the problem complexity.

Finally, an example or two can be enough for illustration purposes, and sometimes the solution of a specific example is the question of interest. But once a software package reaches a certain maturity it is important that it be evaluated on large test sets and compared to other approaches, when a comparison is possible. In Chapter 12 another fundamental application of optimization software is discussed—the parameter estimation problem. More discussion on Chapter 12 is also given in the next section.

1.6 Software for DAEs

The development of software to deal with DAE systems has been progressing for several years, although the area is still far from mature. The first DAE software packages were developed for simulation. Some such as DASSL [50] or RADAU5 [33] were general purpose integrators, which assumed that the problem had the appropriate structure and index. COLDAE [7] is a related boundary value solver for DAEs. Other codes were designed for specific classes of applications such as mechanics or electrical engineering. More recently, a number of more general integrators have been developed [6, 23, 40, 41], but some of them are more in the form of prototype codes and have not been extensively tested on large classes of problems.

However, one of the driving forces behind DAEs has been progress on object-oriented modeling where the model of the process is assembled, perhaps graphically, from models and subproblems and then the software runs the simulation if possible. Moreover, for a piece of software to reach its maximum user base it is important that good interfaces are developed, that the software has been extensively tested, and that good user advice is available. Such interfaces are popular in engineering, with Simulink as a major example. Although many of these packages do not accept DAE models, nor do they generate DAE models, there are some exceptions, such as DYMOLA [2] and related codes which use the Modelica modeling language. One open source example is Scicos, developed at INRIA [19], which has implicit modeling blocks. Scicos can also accept and generate hybrid DAE models.

Finally, there are a number of other aspects of analyzing DAE models or solving control problems that are best done in software. These include optimal control, sensitivity analysis, and parameter estimation. A number of sensitivity codes are available including DDASAC [24], DASPK [44], and DASPK/Adjoint [45]. These tasks are also incorporated in a number of commercial dynamic process modeling packages including ASPEN Custom Modeler [1], gPROMS [3], and JACOBIAN [4]. Generally, the DAS** family of software works on fully implicit index-one systems but can sometimes work on some classes of index-two problems. As other examples of DAE optimization software we mention COOPT [58] based on the DAS** family, the multiple shooting approach of Bock and colleagues [10, 11], and the software packages DAEPACK by Barton and colleagues [4].

A number of the above chapters deal with software descriptions. Chapter 5 describes the StratiGraph software package and illustrates how it can be used to analyze large matrix pencils under perturbation. The software provides a deeper understanding of the control system dynamics and its characteristic behavior under perturbations and for sensitivity analysis.

In Chapter 6, Descriptor Systems and Fault Detection toolboxes in MATLAB® are described. In addition to providing a detailed description of the calculation approach, references are made to additional MATLAB toolboxes and an example system is provided.

The optimal control software $\mathbb{SOCX}$ (Sparse Optimal Control software eXtended) is discussed in Chapter 10. An interface for $\mathbb{SOCX}$ is described in Chapter 12, and results on a set of test problems from parameter estimation are presented.

1.7 DAE Applications

Throughout the development of DAE theory and software, applications have been a driving force to present challenges and suggest what types of assumptions are appropriate to make. Examples of applications are mentioned throughout this volume, mostly to illustrate newly developed results and algorithms. For instance, Chapter 2 gives a mechanical example, while Chapter 6 discusses failure detection and Chapter 12 addresses parameter identification. Chapters 9 and 11 also present and discuss examples from chemical engineering, while Chapter 10 has an example from mathematical biology.

In contrast, the final four chapters are devoted to specific real-world applications. Chapter 13 describes the synthesis of a chemical process control strategy, especially for integrated design configurations that feature material recycling and energy recovery. The decomposition leads to a reduced order modeling strategy with the help of singular perturbation tools. This leads to a low-dimensional DAE representation of the fast core process dynamics, followed by postulating a hierarchical control framework for slower dynamics. This approach allows the design of well-conditioned controllers with desirable closed-loop properties, as illustrated with a detailed chemical process case study.

Chapter 14 addresses the control and optimization for building temperature regulation, using nonlinear model predictive control. This application contains a number of challenges for DAE modeling, control, and optimization, as the solvers for online control and optimization need to be especially efficient and reliable. Here the nonlinear model predictive control (MPC) problem is solved with a specialized algorithm that operates in a parallel, distributed manner. This approach is presented with a model validated on an actual building example.

In Chapter 15 a biomedical imaging application, electromyography (EMG) is considered as an inverse problem. Here the goal is to reliably determine the level of electric activity of individual muscles in a human limb using surface EMG (sEMG) voltage measurements. Applications include prosthetic control and muscle function assessment. The resulting model consists of a PDE and requires boundary value solutions, and it takes both linear and nonlinear forms. For the former the inverse problem is solved with a preconditioned conjugate gradient (CG) least squares method. On the other hand, the nonlinear inverse problem is addressed through shape optimization and level set methods.

Finally, Chapter 16 deals with the optimization of batch and semibatch processes that are widespread in the specialty chemicals and pharmaceutical industry. This process is favored when production volumes are low, product isolation is required, materials are hard to handle, or the process is not well characterized. Described by DAE models, the objective is defined by the states at final time. A real-time optimization scheme must accommodate changes in initial conditions, process disturbances, and modeling uncertainties. Consequently, a feedback control-theoretic approach is used to solve the endpoint optimization problem, drawing from tools from nonlinear systems theory and differential geometry, and illustrated with the optimization of a semibatch reactor.

These applications are not only of interest in their own right but are illustrative of the wide range of applications where DAEs and control arise. In fact DAEs are ubiquitous throughout the sciences and engineering.

Acknowledgments

Stephen Campbell's research was supported in part by NSF grant DMS-0907832. Volker Mehrmann's research was supported by the European Research Council through ERC Advanced grant MODSIMCONMP.

Bibliography

[1] *Aspen Custom Modeler User's Guide*, Technical report, Aspen Technology, http://www.aspentech.com, 2002.

[2] *Dymola User's Guide*, Technical report, Process Systems Enterprises Limited, http://www.dymola.com, 2011

[3] *gPROMS User's Guide*, Technical report, Process Systems Enterprises Limited, http://www.psenterprise.com, 2002.

[4] *JACOBIAN dynamic modeling and optimization software*, Technical report, Numerica Technology LLC, http://www.numericatech.com, 2005.

[5] J. D. Aplevich, *Implicit Linear Systems*, Springer-Verlag, Berlin, 1991.

[6] C. Arevalo, S. L. Campbell, and M. Selva, *Unitary partitioning in general constraint preserving DAE integrators*, Math. Comput. Modelling, 40 (2004), pp. 1273–1284.

[7] U. M. Ascher and R. J. Spiteri, *Collocation software for boundary value differential-algebraic equations*, SIAM J. Sci. Comput., 15 (1994), pp. 938–952.

[8] J. T. BETTS, *Path constrained trajectory optimization using sparse sequential quadratic programming*, J. Guidance, Control Dyn., 16 (1993), pp. 59–68.

[9] K. BOBINYEC, S. L. CAMPBELL, AND P. KUNKEL, *Maximally reduced observers for linear time varying DAEs*, in Proceedings of the IEEE Multi-Conference on Systems and Control (2011 MSC), Denver, 2011, pp. 1373–1378.

[10] H. G. BOCK, M. M. DIEHL, D. B. LEINEWEBER AND J. P. SCHLÖDER, *A direct multiple shooting method for real-time optimization of nonlinear DAE processes*, Nonlinear Model Predictive Control (Ascona, 1998), Progr. Systems Control Theory, 26, Birkhäuser, Basel, 2000, pp. 245–267.

[11] H. G. BOCK, R. W. LONGMAN, AND M. C. STEINBACH, *Time-optimal extension and retraction of robots: Numerical analysis of the switching structure*, J. Optim. Theory Appl., 84 (1995), pp. 589–616.

[12] K. E. BRENAN, *Differential-algebraic issues in the direct transcription of path constrained optimal control problems*, Ann. Numer. Math., 1 (1994), pp. 247–263.

[13] K. E. BRENAN, S. L. CAMPBELL, AND L. R. PETZOLD, *Numerical Solution of Initial-Value Problems in Differential-Algebraic Equations*, SIAM, Philadelphia, 1995.

[14] G. D. BYRNE AND W. E. SCHIESSER, EDS., *Recent Developments in Numerical Methods and Software for ODEs/DAEs/PDEs*, World Scientific, London, 1991.

[15] S. L. CAMPBELL, *Singular Systems of Differential Equations*, Pitman, New York, 1980.

[16] S. L. CAMPBELL, *Singular Systems of Differential Equations* II, Pitman, New York, 1982.

[17] S. L. CAMPBELL, *A general form for solvable linear time varying singular systems of differential equations*, SIAM J. Math. Anal., 18 (1987), pp. 1101–1115.

[18] S. L. CAMPBELL, *Linearization of DAEs along trajectories*, Z. Angew. Math. Phys. (ZAMP), 46 (1995), pp. 70–84.

[19] S. L. CAMPBELL, J.-P. CHANCELIER, AND R. NIKOUKHAH, *Modeling and Simulation in Scilab/Scicos with ScicosLab* 4.4, Springer, Berlin, 2009.

[20] S. L. CAMPBELL AND C. W. GEAR, *The index of general nonlinear DAEs*, Numer. Math., 72 (1995), pp. 173–196.

[21] S. L. CAMPBELL AND E. GRIEPENTROG, *Solvability of general differential algebraic equations*, SIAM J. Sci. Comput., 16 (1995), pp. 257–270.

[22] S. L. CAMPBELL AND C. D. MEYER, *Generalized Inverses of Linear Transformations*, SIAM, Philadelphia, 2009.

[23] S. L. CAMPBELL, E. MOORE, AND Y. ZHONG, *Constraint preserving integrators for unstructured higher index DAEs*, Z. Angew. Math. Mech. (ZAMM), suppl. 1 (1996), pp. 83–86.

[24] M. Caracotsios and W. E. Stewart, *Sensitivity analysis of initial value problems with mixed ODEs and algebraic equations*, Comp. Chem. Eng., 4 (1985), pp. 350–365.

[25] J. D. Cobb, *On the solutions of linear differential equations with singular coefficients*, J. Differential Equations, 46 (1982), pp. 310–323.

[26] J. E. Cuthrell and L. T. Biegler, *On the optimization of differential-algebraic process systems*, AIChE J., 33 (1987), pp. 1257–1270.

[27] L. Dai, *Singular Control Problems*, Springer-Verlag, New York, 1989.

[28] P. E. Gill, W. Murray, and M. A. Saunders, *SNOPT: An SQP algorithm for large-scale constrained optimization*, SIAM Rev., 47 (2005), pp. 99–131.

[29] E. Griepentrog and R. März, *Differential-Algebraic Equations and Their Numerical Treatment*, Teubner-Texte Math. 88, Teubner, Leipzig, 1986.

[30] D. M. Gritsis, C. C. Pantelides, and R. W. H. Sargent, *Optimal control of systems described by index two differential-algebraic equations*, SIAM J. Sci. Comput., 16 (1995), pp. 1349–1366.

[31] M. Günther and U. Feldmann, *The DAE-index in electric circuit simulation*, Math. Comput. Simulation, 39 (1995), pp. 573–582.

[32] E. Hairer, C. Lubich, and M. Roche, *The Numerical Solution of Differential-algebraic Systems by Runge-Kutta Methods*, Lecture Notes in Math. 1409, Springer, Berlin, 1989.

[33] E. Hairer and G. Wanner, *Solving Ordinary Differential Equations* II: *Stiff and Differential-Algebraic Problems*, Springer-Verlag, Berlin, 1991.

[34] M. Hou, Th. Schmidt, R. Schüpphaus, and P. C. Müller, *Normal form and Luenberger observer for linear mechanical descriptor systems*, J. Dynamic Sys., Meas., Con., 115 (1993), pp. 611–620.

[35] H. Krishnan and N. H. McClamroch, *Computation of state realizations for control systems described by a class of differential-algebraic equations*, Int. J. Control, 55 (1992), pp. 1424–1441.

[36] A. Kumar and P. Daoutidis, *Control of nonlinear differential-algebraic process systems*, in Proceedings of the American Control Conference, 1994, pp. 330–334.

[37] P. Kunkel and V. Mehrmann, *Differential-Algebraic Equations, Analysis and Numerical Solution*, European Mathematical Society, Zürich, 2006.

[38] P. Kunkel and V. Mehrmann, *Optimal control for unstructured nonlinear differential-algebraic equations of arbitrary index*. Math. Control Signals Systems, 20 (2008), pp. 227–269.

[39] P. Kunkel and V. Mehrmann, *Formal adjoints of linear DAE operators and their role in optimal control*, Electron. J. Linear Algebra, 22 (2011), pp. 672–693.

[40] P. KUNKEL, V. MEHRMANN, W. RATH, AND J. WEICKERT, *A new software package for linear differential-algebraic equations*, SIAM J. Sci. Comput., 18 (1997), pp. 115–138.

[41] P. KUNKEL, V. MEHRMANN, AND I. SEUFER, *GENDA: A Software Package for the Numerical Solution of General Nonlinear Differential-Algebraic Equations*, Preprint series of the Institute of Mathematics, Preprint 730, Technische Universität Berlin, Berlin, 2002.

[42] R. LAMOUR, *A Projector Based Representation of the Strangeness Index Concept*, Preprint 07-03, Institut für Mathematik, Humboldt-Universität zu Berlin, Berlin, 2007.

[43] A. LEFKOPOULOS AND M. A. STADTHERR, *Index analysis of unsteady-state chemical process systems- I: An algorithm for problem formulation*, J. Comp. Chem. Eng., 17 (1993), pp. 399–413.

[44] S. LI AND L. PETZOLD, *Software and algorithms for sensitivity analysis of large-scale differential-algebraic systems*, J. Comput. Appl. Math., 125 (2000), pp. 131–145.

[45] S. LI AND L. PETZOLD, *Description of DASPKADJOINT: An Adjoint Sensitivity Solver for Differential-Algebraic Equations*, Technical report, University of California, Department of Computer Science, Santa Barbara, CA, 2001.

[46] T. MALY AND L. R. PETZOLD, *Numerical methods and software for sensitivity analysis of differential-algebraic systems*, Appl. Numer. Math., 20 (1996), pp. 57–79.

[47] N. H. MCCLAMROCH, *Feedback stabilization of control systems described by a class of nonlinear differential-algebraic equations*, Systems Control Lett., 15 (1990), pp. 53–60.

[48] I. OKAY, S. L. CAMPBELL, AND P. KUNKEL, *Completions of implicitly defined linear time varying vector fields*, Linear Algebra Appl., 431 (2009), pp. 1422–1438.

[49] C. C. PANTELIDES, D. GRITSIS, K. R. MORISON, AND R. W. H. SARGENT, *The mathematical modeling of transient systems using differential-algebraic equations*, Comp. Chem. Eng., 12 (1988), pp. 449–454.

[50] L. R. PETZOLD, *A Description of DASSL: A Differential/Algebraic System Solvers*, Technical Report SAND82-8637, Sandia National Laboratory, Albuquerque, NM, 1982.

[51] M. DO R. DE PINHO AND R. B. VINTER, *Necessary conditions for optimal control problems involving differential algebraic equations*, J. Math. Anal. Appl., 212 (1997), pp. 493–516.

[52] R. PYTLAK, *Numerical Methods for Optimal Control Problems with State Constraints*, Springer, Berlin, 1999.

[53] P. J. Rabier and W. C. Rheinboldt, *Classical and generalized solutions of time-dependent linear differential algebraic equations*, Linear Algebra Appl., 245 (1996), pp. 1241–1256.

[54] P. J. Rabier and W. C. Rheinboldt, *Nonholonomic Motion of Rigid Mechanical Systems from a DAE Viewpoint*, SIAM, Philadelphia, 2000.

[55] J. G. Renfro, A. M. Morshedi, and O. A. Asbjornsen, *Simultaneous optimization and solution of systems described by differential/algebraic equations*, Comp. Chem. Eng., 11 (1987), pp. 503–517.

[56] W. C. Rheinboldt, *MANPAK: a set of algorithms for computations on implicitly defined manifolds*, Comput. Math. Appl., 32 (1996), pp. 15–28.

[57] W. C. Rheinboldt, *Solving algebraically explicit DAEs with the MANPAK-manifold-algorithms*, Comput. Math. Appl., 33 (1997), pp. 31–43.

[58] R. Serban and L.R. Petzold, *COOPT—A software package for optimal control of large-scale differential-algebraic equation systems*, Math. Comput. Simulation, 56 (2001), pp. 187–203.

[59] J. E. Tolsma and P. I. Barton, *Hidden discontinuities and parametric sensitivity calculations*, SIAM J. Sci. Comput., 23 (2002), pp. 1861–1874.

[60] R. von Schwerin, *Multibody System Simulation: Numerical Methods, Algorithms, and Software*, Lect. Notes Comput. Sci. Eng., Springer, Berlin, 1999.

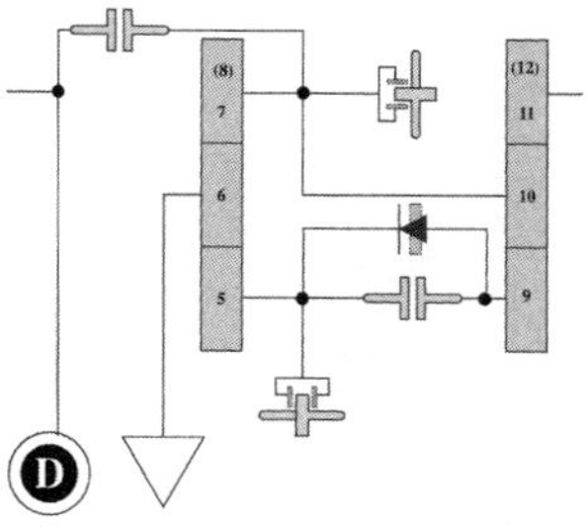

Chapter 2
Regularization of Linear and Nonlinear Descriptor Systems

Stephen L. Campbell, Peter Kunkel, and Volker Mehrmann

DAEs present today the state of the art in dynamical systems arising from automated modularized modeling in almost all areas of science and engineering. While the modeling becomes more and more convenient, the resulting models are typically not easy to treat with current numerical simulation, control, and optimization methods. In many cases a reformulation of the models or even a regularization is necessary to avoid failure of the computational methods. In this chapter we will discuss general DAE control problems and how they can be systematically reformulated and regularized so that the resulting system can be used in control and optimization procedures without much further difficulty.

2.1 Introduction

Modern modeling and simulation packages such as Modelica (https://www.modelica.org) or MATLAB/Simulink (http://www.mathworks.com) offer excellent tools for the automated generation of dynamic system models. Modeling is done in a modularized way, based on a network of subsystems which again consists of simple standardized subcomponents. Such approaches have been the industrial standard in circuit simulation for decades and have also made their way into industrially used packages in other areas such as multiphysics problems from different physical domains including mechanical, mechatronic, fluidic, thermic, hydraulic, pneumatic, elastic, plastic, or electric components [1, 7, 9, 22, 27, 28, 29, 30, 31, 32]. These automatically generated models are today used for the simulation, control, and optimization of complex technological systems, in particular in the design phase.

This concept of modeling dynamical systems leads to systems with differential-algebraic character; it is a DAE when the components are semidiscretized (i.e., discretized with respect to all variables but the time variable). Such systems contain differential equations that model the dynamical behavior and algebraic equations that model constraints, interface and boundary conditions, conservation laws, as well as balance equations.

When the system arises from automated modeling, then typically all the mathematical difficulties arising in the numerical simulation, control, and optimization are pushed onto the numerical methods. A reformulation, remodeling, or regularization may be required to deal with these difficulties.

A classical and nowadays again popular approach to overcome these difficulties in numerical simulation methods is to resolve all algebraic constraints and interface conditions (often with computer algebra methods [8]) and to turn the system into an explicit differential equation. For this system then one applies standard integration or control techniques. As simple as it looks, this approach, however, has several drawbacks. First of all, it is known that computer-algebra-based methods cannot be applied to large-scale problems and that they often produce formulas with bad numerical properties. Second, this procedure leads to variables that lose their interpretation as physical quantities and furthermore, since the algebraic constraints are no longer explicitly available, often the numerical methods deviate from the constraints and interface conditions and produce physically meaningless results.

A very successful alternative approach [4, 17] uses the system and some of its derivatives to produce an equivalent (so-called strangeness-free) system with the same solution set, but where all explicit and implicit constraints and therefore the complete constraint manifold are available. This approach keeps the physical meaning of all the variables and allows easy initialization. Furthermore, since the constraint manifolds are available, it is easier to guarantee that the numerical solution stays on these manifolds. Related approaches have been developed using dummy variables [24, 25]. The terms minimally extended system or index reduction are also sometimes used. In the context of control and optimization methods, however, further difficulties arise requiring the reformulation, remodeling, and regularization of the given problem. It is the purpose of this chapter to present these ideas and to discuss their advantages and disadvantages.

In the general nonlinear case the resulting DAE mathematical model can be written in the form

$$F(t,x,\dot{x},u,y)=0,\ x(\underline{t})=\underline{x}. \tag{2.1}$$

Denoting by $C^0(\mathbf{I},\mathbf{R}^m)$ the set of continuous functions from a compact time interval $\mathbf{I}=[\underline{\mathbf{t}},\bar{\mathbf{t}}]\subseteq\mathbf{R}$ to $\mathbf{R}^m$, here x represents the state, u the input, and y the output of the system. Although more general function spaces can be considered, we assume that $F\in C^0(\mathbf{I}\times\mathbf{D}_x\times\mathbf{D}_{\dot{x}}\times\mathbf{D}_u\times\mathbf{D}_y,\mathbf{R}^\ell)$ is sufficiently smooth, and that $\mathbf{D}_x,\mathbf{D}_{\dot{x}}\subseteq\mathbf{R}^n$, $\mathbf{D}_u\subseteq\mathbf{R}^m$, $\mathbf{D}_y\subseteq\mathbf{R}^p$ are open sets.

In many practical applications the output is given explicitly; i.e., the system has the form

$$F(t,x,\dot{x},u)=0,\ x(\underline{t})=\underline{x}, \tag{2.2a}$$

$$y=G(t,x,u). \tag{2.2b}$$

Then we assume that $F\in C^0(\mathbf{I}\times\mathbf{D}_x\times\mathbf{D}_{\dot{x}}\times\mathbf{D}_u,\mathbf{R}^\ell)$ and $G\in C^0(\mathbf{I}\times\mathbf{D}_x\times\mathbf{D}_u,\mathbf{R}^p)$.

To avoid confusion, in the following we use L for the number of equations, where $L=\ell$ if as in (2.2) the output equation is included in the system of equations or if the output equation is kept separately, and $L=\ell+p$ if the output equation is given as a separate equation but is included in the total system.

Throughout this chapter we assume that all functions are sufficiently smooth, i.e., sufficiently often continuously differentiable, whenever this is needed.

In the following we make use of a *behavior approach* [26], which introduces a *descriptor vector* z that includes the state variables, the input variables, and also the output variables in (2.2). If the output equation is given explicitly as in (2.2b), then it often suffices to build the descriptor vector just from the state and input variables in (2.2a) and add the output equation to the system later again.

The behavior approach does not distinguish the meaning of variables and is ideally suited for the analysis and also for some of the numerical methods. In some cases, however, it is necessary (and we will do so) to analyze the representation in terms of the original variables x, u, y.

Whether an output equation is included or not, the resulting behavior system has the form

$$\mathcal{F}(t, z, \dot{z}) = 0, \tag{2.3}$$

with $\mathcal{F} \in C^0(\mathbf{I} \times \mathbf{D}_z \times \mathbf{D}_{\dot{z}}, \mathbf{R}^L)$ sufficiently smooth, together with a set of equations for the initial conditions

$$Hz(\underline{t}) = \underline{z}, \tag{2.4}$$

with a projection matrix H, that are compiled from the given initial conditions. Here $z \in C^0(\mathbf{I}, \mathbf{R}^N)$, with $N = n + m + p$ if the output variables are included or $N = n + m$ if not.

We use the following solution concept in the behavior context. A function $z : \mathbf{I} \to \mathbf{R}^N$ is called a *solution* of (2.3) if $z \in C^0(\mathbf{I}, \mathbf{R}^N)$ is differentiable as required by the structure of (2.3) and if z satisfies the equation pointwise. It is called a *solution of the initial value problem* if it furthermore satisfies the initial conditions (2.4), which are then called *consistent*. We say that the system is *regular* if it has a unique solution for every sufficiently smooth input function u and every initial value that is consistent for the system with this u.

Example 2.1. A simple example of such a descriptor system is the model of a two-dimensional, three-link mobile manipulator from [10] (see Figure 2.1); see also [3]. The

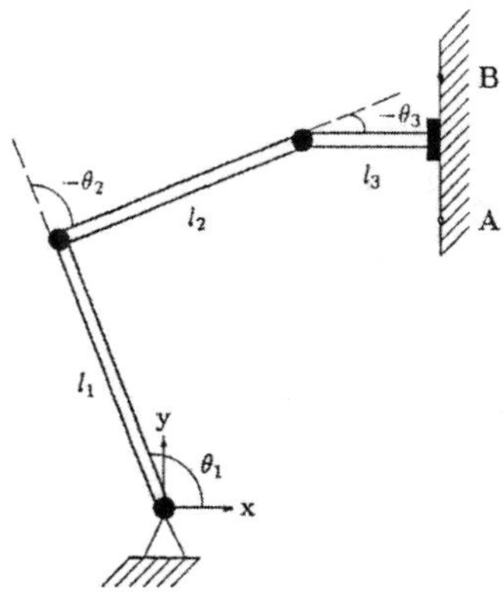

Figure 2.1. Three-link mobile manipulator model [11].

Lagrangian equations of motion take the form

$$\tilde{M}(\Theta)\ddot{\Theta}+\tilde{C}(\Theta,\dot{\Theta})+\tilde{G}(\Theta)=\tilde{u}+\Psi^T\lambda,$$
$$\psi(\Theta)=0,$$

where $\Theta=\begin{bmatrix}\Theta_1 & \Theta_2 & \Theta_3\end{bmatrix}^T$ is the vector of joint displacements, $\tilde{u}$ is the three-component vector of control torques applied at the joints, $\tilde{M}$ is a 3×3 mass matrix, $\tilde{C}$ is the vector of centrifugal and Coriolis forces, and $\tilde{G}$ is the gravity vector. The constraint function ψ is given by

$$\psi(\Theta)=\begin{bmatrix} l_1\cos(\Theta_1)+l_2\cos(\Theta_1+\Theta_2)+l_3\cos(\Theta_1+\Theta_2+\Theta_3)l_3-l \\ \Theta_1+\Theta_2+\Theta_3 \end{bmatrix}.$$

Here $\Psi=\frac{\partial\psi}{\partial\Theta}$, the Lagrange multiplier λ has two components and $\Psi^T\lambda$ is the generalized constraint force. Typically a tracking output $y=\tilde{C}(\Theta)$ is used for the system as in [10]. ■

Example 2.2. Another simple example from the mathematical modeling of chemical reactions [25] is a first order isomerization reaction which is externally cooled.

With c_0 the given feed reactant concentration, T_0 the initial temperature, $c(t)$ and $T(t)$ the concentration and temperature at time t, and R the reaction rate per unit volume, the model takes the form

$$\begin{bmatrix} 1 & 0 & 0 \\ 0 & 1 & 0 \\ 0 & 0 & 0 \end{bmatrix}\begin{bmatrix} \dot{c} \\ \dot{T} \\ \dot{R} \end{bmatrix}=\begin{bmatrix} k_1(c_0-c)-R \\ k_1(T_0-T)+k_2R-k_3(T-T_C) \\ R-k_3\exp\left(-\frac{k_4}{T}\right)c \end{bmatrix}. \tag{2.5}$$

Here the cooling temperature T_C may be used as input and k_1,k_2,k_3,k_4 are constants. ■

We will also discuss in detail the case of linear systems, which either are the original model equations or are obtained after linearization along trajectories; see [5] for details. Then the control system takes the form

$$E(t)\dot{x}=A(t)x+B(t)u+f(t),\ x(\underline{t})=\underline{x}, \tag{2.6a}$$
$$y=C(t)x+D(t)u+g(t), \tag{2.6b}$$

with $E,A\in C^0(\mathbf{I}\times\mathbf{R}^{\ell\times n})$, $B\in C^0(\mathbf{I},\mathbf{R}^{\ell\times m})$, $C\in C^0(\mathbf{I},\mathbf{R}^{p\times n})$, $D\in C^0(\mathbf{I},\mathbf{R}^{p\times m})$, $f\in C^0(\mathbf{I},\mathbf{R}^{\ell})$, and $g\in C^0(\mathbf{I},\mathbf{R}^{p})$. For better readability we will omit in the following the argument t of the involved coefficient functions.

In the linear case the behavior system takes the form

$$\mathcal{E}\dot{z}=\mathcal{A}z+\phi, \tag{2.7}$$

where $\mathcal{E},\mathcal{A}\in C^0(\mathbf{I},\mathbf{R}^{L\times N})$, $\phi\in C^0(\mathbf{I},\mathbf{R}^{L})$, and $L=\ell+p$ if the output equation is included or $L=\ell$ otherwise. Again $z\in C^0(\mathbf{I},\mathbf{R}^{N})$, with $N=n+m+p$ if the output variables are included or $N=n+m$ if not. In either case we have an initial condition of the form (2.4).

Example 2.3. Considering Example 2.1, rewriting the joint displacements in Cartesian coordinates for positions p, and linearizing around a nonstationary solution yields a linear model of the form

$$M(t)\delta\ddot{\tilde{p}} + D(t)\delta\dot{\tilde{p}} + K(t)\delta\tilde{p} = S\delta\tilde{u} + G(t)^T\delta\lambda,$$
$$G(t)\delta p = 0.$$

Transforming to first order and letting $x = \begin{bmatrix} \delta\tilde{p} & \delta\dot{\tilde{p}} & \delta\lambda \end{bmatrix}^T$ and $u = \delta\tilde{u}$, one obtains a linear descriptor system of the form $E\dot{x} = Ax + Bu, y = Cx$, with

$$E = \begin{bmatrix} I_3 & 0 & 0 \\ 0 & M & 0 \\ 0 & 0 & 0 \end{bmatrix}, \quad A = \begin{bmatrix} 0 & I_3 & 0 \\ -K & -D & G^T \\ G & 0 & 0 \end{bmatrix}, \quad B = \begin{bmatrix} 0 \\ S \\ 0 \end{bmatrix}. \quad \blacksquare$$

Example 2.4. A simple example that comes directly as a constant coefficient linear system is the simple RLC electrical circuit from [6] in Figure 2.2. Here R, L, and C are the resistance, inductance, and capacitance, respectively. The corresponding voltage drops are denoted by $v_R(t)$, $v_L(t)$, and $v_C(t)$, respectively, and $I(t)$ denotes the current. Typically one would consider the voltage source $v_s(t)$ as the control input.

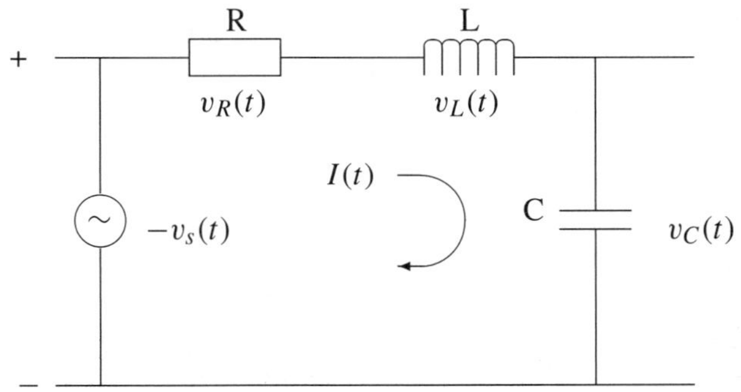

Figure 2.2. A simple RLC circuit [6].

Applying Kirchhoff's laws we obtain the following circuit equation:

$$\begin{bmatrix} L & 0 & 0 & 0 \\ 0 & 0 & 1 & 0 \\ 0 & 0 & 0 & 0 \\ 0 & 0 & 0 & 0 \end{bmatrix} \begin{bmatrix} \dot{I} \\ \dot{v}_L \\ \dot{v}_C \\ \dot{v}_R \end{bmatrix} = \begin{bmatrix} 0 & 1 & 0 & 0 \\ 1/C & 0 & 0 & 0 \\ -R & 0 & 0 & 1 \\ 0 & 1 & 1 & 1 \end{bmatrix} \begin{bmatrix} I \\ v_L \\ v_C \\ v_R \end{bmatrix} + \begin{bmatrix} 0 \\ 0 \\ 0 \\ -1 \end{bmatrix} v_s.$$

If we measure the voltage at the capacitor as output, we have

$$y = \begin{bmatrix} 0 & 0 & 1 & 0 \end{bmatrix} \begin{bmatrix} I \\ v_L \\ v_C \\ v_R \end{bmatrix}. \quad \blacksquare$$

2.2 Regularization of Nonlinear Systems

As we have already noted, the general formulation of the system as it arises in automated modeling is typically not adequate for numerical simulation, control, or optimization. For this reason, the first step in any treatment of descriptor systems is a reformulation of the system that turns the system into a so-called *strangeness-free formulation* of the system [17].

For this, consider the general nonlinear system (2.3) and construct a derivative array, as introduced in [4], of the form

$$\mathcal{F}_\mu(t,z,\dot{z},\dots,z^{(\mu+1)})=0, \tag{2.8}$$

which stacks the original equation and all its derivatives up to level μ in one large system. Here, partial derivatives of $\mathcal{F}_\mu$ with respect to selected variables ζ from $(t,z,\dot{z},\dots,z^{(\mu+1)})$ are denoted by $\mathcal{F}_{\mu;\zeta}$. The solution set of the nonlinear algebraic equation associated with the derivative array $\mathcal{F}_\mu$ for some integer μ (considering the variables as algebraic variables) is denoted by $\mathbf{L}_\mu=\{z_\mu\in\mathbf{I}\times\mathbf{R}^N\times\cdots\times\mathbf{R}^N \mid \mathcal{F}_\mu(z_\mu)=0\}$.

As is typical for the treatment of nonlinear problems, we must assume some kind of regularity of the equation, for example to exclude bifurcation phenomena. We therefore assume the following hypothesis for the rest of the paper; see [17].

Hypothesis 1. *Consider the general system of nonlinear DAEs* (2.3). *There exist integers* μ, r, a, d, *and* v *such that* $\mathbf{L}_\mu$ *is not empty and such that for every* $z_\mu^0=(t_0,z_0,\dot{z}_0,\dots,z_0^{(\mu+1)})\in\mathbf{L}_\mu$ *there exists a (sufficiently small) neighborhood in which the following properties hold:*

1. *The set* $\mathbf{L}_\mu\subseteq\mathbf{R}^{(\mu+2)(N)+1}$ *forms a manifold of dimension* $(\mu+2)N+1-r$.

2. *We have* $\operatorname{rank}\mathcal{F}_{\mu;z,\dot{z},\dots,z^{(\mu+1)}}=r$ *on* $\mathbf{L}_\mu$.

3. *We have* $\operatorname{corank}\mathcal{F}_{\mu;z,\dot{z},\dots,z^{(\mu+1)}}-\operatorname{corank}\mathcal{F}_{\mu-1;z,\dot{z},\dots,z^{(\mu)}}=v$ *on* $\mathbf{L}_\mu$, *where the corank is the dimension of the corange and the convention is used that* corank *of* $\mathcal{F}_{-1;z}$ *is* 0.

4. *We have* $\operatorname{rank}\mathcal{F}_{\mu;\dot{z},\dots,z^{(\mu+1)}}=r-a$ *on* $\mathbf{L}_\mu$ *such that there exist smooth full rank matrix functions* Z_2 *and* T_2 *of size* $(\mu+1)l\times a$ *and* $N\times(N-a)$, *respectively, satisfying* $Z_2^T\mathcal{F}_{\mu;\dot{z},\dots,z^{(\mu+1)}}=0$, $\operatorname{rank}Z_2^T\mathcal{F}_{\mu;z}=a$, *and* $Z_2^T\mathcal{F}_{\mu;z}T_2=0$ *on* $\mathbf{L}_\mu$.

5. *We have* $\operatorname{rank}\mathcal{F}_{\dot{z}}T_2=d=L-a-v$ *on* $\mathbf{L}_\mu$ *such that there exists a smooth full rank matrix function* Z_1 *of size* $N\times d$ *satisfying* $\operatorname{rank}Z_1^T\mathcal{F}_{\dot{z}}T_2=d$.

Note that not every system will satisfy Hypothesis 1 but many of those from applications do. In order to reduce the computational costs, μ should be chosen as small as possible. If μ is not known in advance, one can start with $\mu=0$ and successively enlarge μ until Hypothesis 1 holds. This can be checked computationally at given points.

The quantity v measures the number of equations in the original system that give rise to trivial equations $0=0$; i.e., it counts the number of redundancies in the system. Of course, trivial equations can be simply removed without altering the solution set and we assume from now on that this has been done.

The smallest possible μ for which Hypothesis 1 holds is called the *strangeness index* of (2.3); see [17]. It generalizes the concept of differentiation index [2] to arbitrary over- and underdetermined systems.

It has been shown in [15] that Hypothesis 1 implies locally (via the implicit function theorem) the existence of a *reduced system* given by

$$\hat{\mathcal{F}}_1(t,z,\dot{z}) = 0, \tag{2.9a}$$

$$\hat{\mathcal{F}}_2(t,z) = 0, \tag{2.9b}$$

with $\hat{\mathcal{F}}_1 = Z_1^T \mathcal{F}$ describing the dynamics of the system, while $\hat{\mathcal{F}}_2(t,z) = 0$ contains all the algebraic constraints and can be used to parameterize the solution manifold. Consider the following example from [16].

Example 2.5. Example 2.1 (written in Cartesian coordinates and first order form) is a special case of a control problem for a multibody system of the form

$$\begin{aligned} \dot{p} &= v, \\ M(p)\dot{v} &= f(p,q,u) + g_p(p)^T \lambda, \\ g(p) &= 0, \\ y &= C(p), \end{aligned}$$

with position variable p, velocity v, mass matrix $M(p)$, forcing function f, control u, constraint $g(p)$, Lagrange multiplier λ, and output y. Here for given u, the coupling between the constraint and the dynamical equation leads to a strangeness index 2 (differentiation index 3). Instead of using the whole derivative array to compute the reduced strangeness-free form, using the structure of the system it suffices here to differentiate only the constraint $g(p) = 0$ twice and to add these equations to the system to obtain a *reduced* derivative array [16, 17].

Assuming that the Jacobian matrix $g_p(p)$ has full row rank and that $M(p)$ is symmetric and positive definite one can (locally) determine a permutation matrix $\Pi = [\Pi_1\ \Pi_2]$ such that for $g_p(p)$ we have $g_p(p)[\Pi_1\ \Pi_2] = [G_1\ G_2]$, with G_2 being square and nonsingular. Partitioning

$$[\Pi_1\ \Pi_2]^T p = \begin{bmatrix} p_1 \\ p_2 \end{bmatrix}, \quad [\Pi_1\ \Pi_2]^T v = \begin{bmatrix} v_1 \\ v_2 \end{bmatrix} \tag{2.10}$$

conformally and replacing every occurrence of $\dot{p}_2$ by the auxiliary variable w_1 and every occurrence of $\dot{v}_2$ by the auxiliary variable w_2 gives the system

$$\begin{aligned} \dot{p}_1 &= v_1, \\ w_1 &= v_2, \\ M\Pi \begin{bmatrix} \dot{v}_1 \\ w_2 \end{bmatrix} &= f(p,v,u) + g_p(p)^T \lambda, \\ 0 &= g(p), \\ 0 &= g_p(p)v, \\ 0 &= g_{pp}(p)(v,v) + g_p(p)\Pi \begin{bmatrix} \dot{v}_1 \\ w_2 \end{bmatrix}, \\ y &= C(p), \end{aligned}$$

from which then we can (by inverting M) remove the auxiliary variables w_1, w_2 and obtain the system

$$\dot{p}_1 = v_1, \tag{2.11a}$$

$$\dot{v}_1 = \Pi_1^T M(p)^{-1}(f(p,v,u) + g_p(p)^T \lambda), \tag{2.11b}$$

$$0 = g(p), \tag{2.11c}$$

$$0 = g_p(p)v, \tag{2.11d}$$

$$0 = g_{pp}(v,v) + g_p(p)\Pi_1^T M(p)^{-1}(f(p,v,u) + g_p(p)^T \lambda), \tag{2.11e}$$

$$y = C(p), \tag{2.11f}$$

together with (2.10). This system is regular and strangeness-free for given u near the initial value, since the Jacobian of (2.11c)–(2.11e) with respect to p_2, v_2, λ has full row rank [16]. ■

Based on the construction of reduced strangeness-free systems we are now able to characterize consistency and solvability of the system and also perform, if necessary, a regularization.

2.2.1 Regularization of Initial Conditions

It is clear that for an initial condition to be consistent, equation (2.9b) has to be compatible with the initial condition (2.4); i.e., the combined system

$$\hat{\mathcal{F}}_2(\underline{t}, z(\underline{t})) = 0, \; Hz(\underline{t}) = \underline{z} \tag{2.12}$$

must be uniquely solvable to obtain a unique solution for a given input u. If this is not the case, then a modification of the initial conditions is necessary, so that (2.12) has a unique solution. This typically can be done in many different ways, e.g., by changing the initial value $\underline{x}$ and with this $\underline{z}$, so that (2.12) becomes solvable. A good choice should certainly also be justified from the underlying physical problem.

Example 2.6. Consider Example 2.5 with initial conditions $p(0) = \underline{p}$ and $v(0) = \underline{v}$. Then due to the two extra constraints that are obtained from the derivative array the following conditions have to be satisfied for the initial value to be consistent:

$$\begin{aligned} g(\underline{p}) &= 0, \\ g_p(\underline{p})\underline{v} &= 0, \\ g_{pp}(\underline{v},\underline{v}) + g_p(\underline{p})M(\underline{p})^{-1}(f(\underline{p},\underline{v},u(0)) + g_p(\underline{p})^T\lambda(0)) &= 0. \end{aligned}$$

Note that due to the assumptions, the third condition simply fixes the initial value $\lambda(0)$ of the Lagrange parameter. ■

2.2.2 Reinterpretation of Variables

In some applications it turns out that the original choice of input and state variables was not appropriate. In the general behavior approach [26] such an interpretation of variables should be made after the analysis.

In order to (locally) get an interpretation of the variables, we use a suitable splitting of the unknown z as $z = (z_1, z_2, z_3) \in \mathbf{R}^d \times \mathbf{R}^{N-d-a} \times \mathbf{R}^a$ in such a way that part 4 of Hypothesis 1 guarantees that (2.9b) can be solved for z_3 according to $z_3 = \mathcal{R}(t, z_1, z_2)$. Eliminating z_3 and $\dot{z}_3$ in (2.9a) with the help of this relation and its derivative then leads to

$$\hat{\mathcal{F}}_1(t, z_1, z_2, \mathcal{R}(t, z_1, z_2), \dot{z}_1, \dot{z}_2, \mathcal{R}_t(t, z_1, z_2) + \mathcal{R}_{z_1}(t, z_1, z_2)\dot{z}_1 + \mathcal{R}_{z_2}(t, z_1, z_2)\dot{z}_2) = 0. \tag{2.13}$$

By part 5 of Hypothesis 1, we may assume without loss of generality that this system can (locally) be solved for $\dot{z}_1$ leading to a system

$$\dot{z}_1 = \mathcal{L}(t, z_1, z_2, \dot{z}_2), \tag{2.14a}$$

$$z_3 = \mathcal{R}(t, z_1, z_2). \tag{2.14b}$$

When this system is interpreted as a DAE, $z_2 \in C^1(\mathbf{I}, \mathbf{R}^{N-d-a})$ can be chosen arbitrarily (at least when staying in the domain of definition of $\mathcal{R}$ and $\mathcal{L}$). This means that the variables in z_2 can be considered as controls, while the resulting system has locally a unique solution for z_1 and z_3, provided a consistent initial condition as in (2.12) is given.

The variables in z_3 represent algebraic variables, and if output variables were included, then they are part of z_3. In the general behavior framework we are now free to choose output variables, but clearly this should be done according to the underlying physical model, in particular if the output variables are meant to be variables that are measured.

In summary, by appropriately renaming the variables, it follows that after this step we have obtained a regularization of the system in the form of a strangeness-free system

$$\hat{\mathcal{F}}_1(t, \hat{x}, \dot{\hat{x}}, \hat{u}, \dot{\hat{u}}, \hat{y}) = 0, \tag{2.15a}$$

$$\hat{\mathcal{F}}_2(t, \hat{x}, \hat{u}, \hat{y}) = 0, \tag{2.15b}$$

together with initial conditions that are modified so that they satisfy (2.12).

If the output equation was kept as a separate equation, then (by appropriately renaming the variables and functions) we obtain a system

$$\hat{\mathcal{F}}_1(t, \hat{x}, \dot{\hat{x}}, \hat{u}, \dot{\hat{u}}) = 0, \tag{2.16a}$$

$$\hat{\mathcal{F}}_2(t, \hat{x}, \hat{u}) = 0, \tag{2.16b}$$

$$\hat{y} = \hat{G}(t, \hat{x}, \hat{u}). \tag{2.16c}$$

Note that in both formulations derivatives of the chosen input variables may arise. In applications it is often desirable to avoid derivatives of chosen inputs if possible. Sometimes this dependency can be avoided by an alternative choice of control. However, if there is an explicit dependency on $\dot{\hat{u}}$, then one can introduce $\tilde{u} = \dot{\hat{u}}$ and add this differential equation for $\hat{u}$ as a new dynamical equation to the system. In this case appropriate initial conditions for $\hat{u}$ have to be added as well. Again we can rename the variables appropriately and can assume in the following that the state equations do not contain explicit derivatives of the input equations; i.e., $\dot{\hat{u}}$ does not occur.

If the variables of the original system are given as states x and inputs u, and if we keep these variables, then we obtain a system

$$\hat{\mathcal{F}}_1(t,x,\dot{x},u) = 0, \tag{2.17a}$$

$$\hat{\mathcal{F}}_2(t,x,u) = 0, \tag{2.17b}$$

$$\hat{y} = \hat{G}(t,x,u), \tag{2.17c}$$

with $L = d + a$ state equations and unknowns of size n for x and m for u.

Then several possibilities can arise. If $d + a \neq n$, then for given u we cannot expect a unique solution; i.e., the system is not regular. In a general behavior setting this does not matter, since the variables are not distinguished and one can make a decision which variables one wants to consider as inputs and states at this point; see [12, 13, 26]. If, however, the application clearly defines which variables are input or state variables, then a reinterpretation of these variables is necessary. If $d + a > n$, then we just introduce a new vector $\tilde{x}$ by attaching $d + a - n$ of the input variables in u to the vector x and considering a new input vector $\tilde{u}$ which contains the remaining variables. If $d + a < n$, then we attach $n - d + a$ variables from x to u to obtain $\hat{u}$ and $\hat{x}$.

This renaming of variables yields a system

$$\tilde{\mathcal{F}}_1(t,\tilde{x},\dot{\tilde{x}},\tilde{u}) = 0, \tag{2.18a}$$

$$\tilde{\mathcal{F}}_2(t,\tilde{x},\tilde{u}) = 0, \tag{2.18b}$$

$$\tilde{y} = \tilde{G}(t,\tilde{x},\tilde{u}). \tag{2.18c}$$

There is a lot of freedom in the choice of these new variables. From a theoretical point of view any choice is just fine, but from a practical point of view there may be preferred choices. Also, different choices may lead to different scaling and stability properties of the resulting system and it may be necessary to take this into consideration.

2.2.3 Feedback Regularization

In general, the system given by (2.18) is not strangeness-free as a free system, i.e., when $\tilde{u} = 0$ is chosen as an input, although as we have seen, there is always a possibility of reinterpreting the variables so that is the case. If, however, the original variables x and u were kept, $d + a = n$, and the system is not strangeness-free as a free system, then a classical approach [3, 20] is to choose a feedback such that the resulting reduced problem is regular and strangeness-free.

In the nonlinear case, a state feedback has the form

$$\tilde{u} = K(t,\tilde{x}) + w, \tag{2.19}$$

leading to a closed-loop reduced problem

$$\tilde{\mathcal{F}}_1(t,\tilde{x},\dot{\tilde{x}},K(t,\tilde{x})+w) = 0, \tag{2.20a}$$

$$\tilde{\mathcal{F}}_2(t,\tilde{x},K(t,\tilde{x})+w) = 0, \tag{2.20b}$$

$$\tilde{y} = \tilde{G}_1(t,\tilde{x},K(t,\tilde{x})+w). \tag{2.20c}$$

This system is regular and strangeness-free if

$$\begin{bmatrix} \tilde{\mathcal{F}}_{1;\dot{\tilde{x}}} \\ \tilde{\mathcal{F}}_{2;\tilde{x}} + \tilde{\mathcal{F}}_{2;\tilde{u}} K_{\tilde{x}} \end{bmatrix}$$

is nonsingular. Since the reduced system is defined only locally, it is sufficient to satisfy this condition only locally. Thus, one can use linear feedbacks and it has been shown in [20] that then there (locally) exists a state feedback $\tilde{u} = K(t,\tilde{x}) + w$ satisfying $\tilde{u}(\underline{t}) = K(\underline{t},\underline{x})$ and $\dot{u}(\underline{t}) = K_t(\underline{t},\underline{x}) + K_x(\underline{t},\underline{x})\dot{x}(\underline{t}) + \dot{w}$ such that the closed-loop reduced problem is regular and strangeness-free.

It has also been discussed in [17, 20] under which conditions such a regularization can be achieved via output control. Determining such an output feedback, however, usually requires changes of variables.

It should be noted, however, that it depends on the application whether this feedback is necessary. For example, in optimal control problems it suffices if the system is strangeness-free in the behavior sense; see [19].

Example 2.7. Consider the control problem

$$F(t,x,u,\dot{x}) = \begin{bmatrix} \dot{x}_2 \\ \log x_2 + \sin u \end{bmatrix} = 0,$$

with $n = 2$ and $\ell = 1$. The corresponding behavior system reads as

$$\mathcal{F}(t,z,\dot{z}) = \begin{bmatrix} \dot{z}_2 \\ \log z_2 + \sin z_3 \end{bmatrix} = 0.$$

This system satisfies Hypothesis 1 with $\mu = 0$, $v = 0$, $a = 1$, and $d = 1$ and is already in strangeness-free form. The free system with $u = 0$, however, is not strangeness-free. Moreover, x_1 is not even present.

To get a regular and strangeness-free closed-loop reduced problem, we may choose $u = x_1 + w$ and obtain the system $\dot{x}_2 = 0$, $\log x_2 + \sin(x_1 + w) = 0$, which is strangeness-free as a behavior as well as a free system with $w = 0$. For more details see [17]. ∎

2.2.4 General Procedure

Considering the system with separate output equation (2.2), we can summarize the procedure as follows:

$$\boxed{\begin{aligned} F(t,x,\dot{x},u) &= 0,\ x(\underline{t}) = \underline{x}, \\ y &= G(t,x,u) \end{aligned}}$$

$\mu \neq 0$ ↓ index reduction in behavior

$$\boxed{\begin{aligned} \hat{\mathcal{F}}_1(t,z,\dot{z}) &= 0,\ Hz(\underline{t}) = \underline{z}, \\ \hat{\mathcal{F}}_2(t,z) &= 0, \\ y &= \hat{G}(t,z) \end{aligned}}$$

$\hat{\mathcal{F}}_2(\underline{t},z(\underline{t})) = 0,\ Hz(\underline{t}) = \underline{z}$ not solvable ↓ force consistency

$a+d \neq N$ ↓ reinterpret variables

$\dot{u}$ occurs ↓ transform to remove $\dot{u}$

$$\boxed{\begin{aligned} \tilde{\mathcal{F}}_1(t,\tilde{x},\dot{\tilde{x}},\tilde{u}) &= 0, \\ \tilde{\mathcal{F}}_2(t,\tilde{x},\tilde{u}) &= 0, \\ \tilde{y} &= \tilde{G}(t,\tilde{x},\tilde{u}) \end{aligned}}$$

not strangeness-free for $u = 0$ ↓ feedback $\tilde{u} = K(t,\tilde{x}) + w$

$$\boxed{\begin{aligned} \tilde{\mathcal{F}}_1(t,\tilde{x},\dot{\tilde{x}},K(t,\tilde{x})+w) &= 0, \\ \tilde{\mathcal{F}}_2(t,\tilde{x},K(t,\tilde{x})+w) &= 0, \\ \tilde{y} &= \tilde{G}_1(t,\tilde{x},K(t,\tilde{x})+w). \end{aligned}}$$

In the case where the equations have some structure, such as with multibody systems, the procedure can be simplified substantially by making use of the structure available in the equations.

If we want to keep the original variables and no reinterpretation of variables is necessary, then the procedure looks the same except that the "reinterpret variables" and "transform" lines are removed along with the third block. In all cases, the final system is strangeness-free as a free system and as a behavior system and by the implicit function theorem it can locally be expressed as

$$\dot{\tilde{x}}_1 = \mathcal{L}(t,\tilde{x}_1,w), \tag{2.21a}$$

$$\tilde{x}_2 = \mathcal{R}(t,\tilde{x}_1,w), \tag{2.21b}$$

$$\tilde{y} = \tilde{G}(t,\tilde{x}_1,\tilde{x}_2). \tag{2.21c}$$

This system is in the right form to carry out the tasks in numerical simulation [17], stability analysis [18, 23], or optimal control [19].

2.3 Regularization of Linear Descriptor Systems

In the linear case we can be more specific and also consider a more general situation, since we do not have to make the assumption that the solution set is nonempty. In this case the derivative array of (2.7) is given by

$$\mathcal{M}_\mu \dot{z}_\mu = \mathcal{N}_\mu z_\mu + \phi_\mu, \tag{2.22}$$

where

$$(\mathcal{M}_\mu)_{i,j} = \binom{i}{j}\mathcal{E}^{(i-j)} - \binom{i}{j+1}\mathcal{A}^{(i-j-1)}, \; i,j = 0,\ldots,\mu,$$

$$(\mathcal{N}_\mu)_{i,j} = \begin{cases} \mathcal{A}^{(i)} & \text{for } i = 0,\ldots,\mu, \; j = 0, \\ 0 & \text{otherwise,} \end{cases}$$

$$(z_\mu)_j = z^{(j)}, \; j = 0,\ldots,\mu,$$

$$(\phi_\mu)_i = \phi^{(i)}, \; i = 0,\ldots,\mu.$$

In the linear case Hypothesis 1 can be verified under some constant-rank assumptions; see [14, 17]. It then has the following form.

Hypothesis 2. *There exist integers μ, d, a, and v such that the pair $(\mathcal{M}_\mu, \mathcal{N}_\mu)$ associated with* (2.22) *has the following properties:*

1. *For all $t \in \mathbf{I}$ we have* $\operatorname{rank} \mathcal{M}_\mu(t) = (\mu+1)L - a - v$. *This implies the existence of a smooth matrix function Z of size $(\mu+1)L \times (a+v)$ and pointwise maximal rank satisfying $Z^T \mathcal{M}_\mu = 0$.*

2. *For all $t \in \mathbf{I}$ we have* $\operatorname{rank} Z^T \mathcal{N}_\mu [I_N \, 0 \cdots 0]^T = a$. *This implies that without loss of generality Z can be partitioned as $Z = [\, Z_2 \; Z_3 \,]$, with Z_2 of size $(\mu+1)L \times a$ and Z_3 of size $(\mu+1)L \times v$, such that $\hat{A}_2 = Z_2^T \mathcal{N}_\mu [I_N \, 0 \cdots 0]^T$ has full row rank a and $Z_3^T \mathcal{N}_\mu [I_n \, 0 \cdots 0]^T = 0$. Furthermore, there exists a smooth matrix function T_2 of pointwise maximal rank satisfying $\hat{A}_2 T_2 = 0$.*

3. *For all $t \in \mathbf{I}$ we have* $\operatorname{rank} \mathcal{E}(t) T_2(t) = d$. *This implies the existence of a smooth matrix function Z_1 of size $L \times d$ and pointwise maximal rank satisfying* $\operatorname{rank} \hat{E}_1 = d$ *with $\hat{E}_1 = Z_1^T \mathcal{E}$.*

The following theorem adapted to the setting in this chapter is from [20].

Theorem 2.8. *If Hypothesis 2 holds for the inflated pair $(\mathcal{M}_\mu, \mathcal{N}_\mu)$ associated with $(\mathcal{E}, \mathcal{A})$, then system* (2.7) *has the same solution set as the system*

$$\begin{bmatrix} \hat{E}_1 \\ 0 \\ 0 \end{bmatrix} \dot{z} = \begin{bmatrix} \hat{A}_1 \\ \hat{A}_2 \\ 0 \end{bmatrix} z + \begin{bmatrix} \hat{f}_1 \\ \hat{f}_2 \\ \hat{f}_3 \end{bmatrix}, \tag{2.23}$$

where $\hat{E}_1 = Z_1^T \mathcal{E}$, $\hat{A}_1 = Z_1^T \mathcal{A}$, $\hat{f}_1 = Z_1^T \phi$, $\hat{f}_i = Z_i^T \phi_\mu$ for $i = 2,3$.

The coefficients of the differential-algebraic system (2.23) can be computed either analytically or numerically at every time point t by using three nullspace computations, as long as this is feasible in the available computing environment. It is a *reformulation* of (2.7) (using the original model and its derivatives) without changing the solution set. Note that no transformation or splitting of the vector z into parts has been made, which means that the variables still have their original physical meaning.

If the original system was in the form (2.6a) and if we ignore (for the moment) the fact that the vector z in (2.7) is composed of parts that may have quite different orders of differentiability, then the constructed submatrices $\hat{A}_1$ and $\hat{A}_2$ have been obtained from the block matrix

$$\begin{bmatrix} A & B \\ \dot{A} & \dot{B} \\ \vdots & \vdots \\ A^{(\mu)} & B^{(\mu)} \end{bmatrix}$$

by transformations from the left. This has two immediate consequences [20]. First, this means that derivatives of the input function u are nowhere needed, just derivatives of the coefficient matrices; i.e., although formally the derivatives of u occur in the derivative array, they are not used for the form (2.23), and hence we do not need any additional smoothness requirements for the input function u.

Second, it follows from the construction of $\hat{A}_1$ and $\hat{A}_2$ that the partitioning into the part stemming from the original states x and the original controls u is preserved. Including again the output equation (2.6b), we then have obtained a system of the form

$$E_1 \dot{x} = A_1 x + B_1 u + \hat{f}_1, \tag{2.24a}$$

$$0 = A_2 + B_2 u + \hat{f}_2, \tag{2.24b}$$

$$0 = \hat{f}_3, \tag{2.24c}$$

$$y = Cx + Du + g, \tag{2.24d}$$

where

$$E_1 = \hat{E}_1 \begin{bmatrix} I_n \\ 0 \end{bmatrix}, \quad A_i = \hat{A}_i \begin{bmatrix} I_n \\ 0 \end{bmatrix}, \quad B_i = \hat{A}_i \begin{bmatrix} 0 \\ I_m \end{bmatrix}, \quad i = 1,2.$$

Here E_1, A_1 have size $d \times n$, while E_2, A_2 are of size $a \times n$.

2.3.1 Regularization of Equations and Initial Values

In the following we continue with the general form (2.23). In contrast to the nonlinear case where we have assumed solvability, here this can be checked by considering the function $\hat{f}_3$. If $\hat{f}_3$ is identically 0 in $\mathbf{I}$, then we have solvability and we can just leave off the third equation in (2.23). If $\hat{f}_3$ is nonzero, then the model is not consistent. This actually happens frequently in applications due to modeling errors or when overdetermined systems are considered which are consistent in exact computation but due to modeling errors, measurement errors, or round-off errors one has nonzero $\hat{f}_3$.

If $\hat{f}_3$ is nonzero, then we can regularize the system by setting $\hat{f}_3 \equiv 0$, and by removing the third equation of (2.23) which is then a redundant equation. We then continue with the modified model of $\tilde{L} = d + a$ equations

$$\begin{bmatrix} \hat{E}_1 \\ 0 \end{bmatrix} \dot{z} = \begin{bmatrix} \hat{A}_1 \\ \hat{A}_2 \end{bmatrix} z + \begin{bmatrix} \hat{f}_1 \\ \hat{f}_2 \end{bmatrix}, \tag{2.25}$$

together with initial conditions (2.4).

Consistency of the initial values can again be easily checked, since these have to satisfy the equation

$$\hat{A}_2(\underline{t}) z(\underline{t}) + \hat{f}_2(\underline{t}) = 0, \quad H z(\underline{t}) = \underline{z}. \tag{2.26}$$

If the initial values are not consistent, then there are again several possibilities to regularize the system. One can, e.g., compute a consistent least squares approximation $\hat{\underline{z}}$ of $\underline{z}$ or one can choose a consistent initial value $\underline{\hat{z}}$ by making sure that important user requirements on the initial value are satisfied exactly; see [21] for a detailed discussion and also software implementations of this regularization step.

2.3.2 Reinterpretation of Variables

After having made the system and the initial conditions consistent, we can analyze the uniqueness of the resulting solution of the strangeness-free behavior system. With $\tilde{L} = d + a$ equations and N unknowns, it is clear that to obtain uniqueness, it is necessary that $N = \tilde{L} = d + a$. If $N > d + a$, then we have free variables in the system which can be considered as controls or inputs.

If $N = \tilde{L}$, then as in the nonlinear case, an appropriate transformation of the variables $\hat{x} = Tx$ can be performed so that we obtain a set of $a + d$ (in the whole interval $\mathbf{I}$) linearly independent columns of the matrix function $\tilde{E} := \begin{bmatrix} \hat{E}_1 T \\ \hat{A}_2 T \end{bmatrix}$ and so that in $\hat{A}_2 T$ the columns numbered $d + 1, \ldots, d + a$ are linearly independent (in the whole interval $\mathbf{I}$). Denoting the variables associated with these columns in $\hat{A}_2 T$ as x_2, the variables associated with the first d columns of $\tilde{E}$ as x_1, and the remaining $N - d - a$ variables as $u = x_3$, we obtain a new system

$$\begin{bmatrix} \tilde{E}_{11} & \tilde{E}_{12} & \tilde{E}_{13} \\ 0 & 0 & 0 \end{bmatrix} \begin{bmatrix} \dot{x}_1 \\ \dot{x}_2 \\ \dot{x}_3 \end{bmatrix} = \begin{bmatrix} \tilde{A}_{11} & \tilde{A}_{12} & \tilde{A}_{13} \\ \tilde{A}_{21} & \tilde{A}_{22} & \tilde{A}_{23} \end{bmatrix} \begin{bmatrix} x_1 \\ x_2 \\ x_3 \end{bmatrix} + \begin{bmatrix} \tilde{f}_1 \\ \tilde{f}_2 \end{bmatrix}, \tag{2.27}$$

where $\tilde{A}_{22}$ is invertible. This choice of variables makes sure that the free system obtained by setting $u = x_3 = 0$ itself is strangeness-free.

Typically, there is a lot of freedom in the choice of the variables x_1, x_2, x_3 and this is a place where the physical background of the system has to come into play. If, for example, in the original system the variables had a clear definition as input and state variables and if the number of state variables was $d + a$, then it seems a reasonable choice to pick this partitioning. Another motivation for the choice of variables may come from the fact that we may not want derivatives of input functions to appear in the model. In this case (if possible) the choice of the state variables should be made in such a way that the block $\tilde{E}_{13}$ is zero, because then no derivative of $u = x_3$ occurs.

Example 2.9. In the electrical circuit of Example 2.4 we have used the voltage source as input, but from a mathematical point of view we may as well choose the variable v_L as input. ∎

In the case that the reduced system has the form (2.24), we now have $d + a$ equations in (2.24a) and (2.24b) and still n variables in x and m variables in u. In order for this system to be regular, i.e., uniquely solvable for all sufficiently smooth inputs u, all sufficiently smooth inhomogeneities $\hat{f}_1, \hat{f}_2$, and all consistent initial conditions $\underline{x}$, we would need that $d + a = n$.

If $d + a < n$, then for given $u, \hat{f}_1, \hat{f}_2$ we cannot expect a unique solution; i.e., the system is not regular. In a general behavior setting this does not matter, since the variables are not distinguished and one can make a decision which variables one wants to consider as inputs and states [12, 13, 26]. If, however, the application clearly defines which variables are input or state variables, then a reinterpretation of these variables is necessary. If $d + a >$

n, then we just introduce a new vector $\hat{x}$ by attaching $d+a-n$ of the input variables in u to the vector x and considering a new input vector $\hat{u}$ which contains the remaining variables and if $d+a<n$ we attach $n-d+a$ variables from x to u to obtain $\hat{u}$ and $\hat{x}$. There is freedom in the choice of these variables. As already discussed in the nonlinear case, different choices are possible and the physical problem as well as scaling and stability properties should be considered.

When performing this reinterpretation of variables, we must also change the output equation by moving appropriate columns from D to C or vice versa.

In both the general and the special case after an appropriate renaming of coefficients we obtain a new system

$$\tilde{E}_1\dot{\tilde{x}} = \tilde{A}_1\tilde{x} + \tilde{B}_1\tilde{u} + \tilde{f}_1, \tag{2.28a}$$

$$y = \tilde{C}\tilde{x} + \tilde{D}\tilde{u} + \tilde{g}, \tag{2.28b}$$

where now the matrix functions $\left[\begin{smallmatrix}\tilde{E}_1\\0\end{smallmatrix}\right]$ and $\left[\begin{smallmatrix}\tilde{A}_1\\\tilde{A}_2\end{smallmatrix}\right]$ are square of size $\tilde{n}=d+a$ and $\left[\begin{smallmatrix}\tilde{B}_1\\\tilde{B}_2\end{smallmatrix}\right]$ is of size $\tilde{n}\times\tilde{m}$ with $\tilde{m}=n+m-\tilde{n}$.

2.3.3 Removing Feedthrough Terms

In the linear case a feedthrough can be easily removed by increasing the size of the state vector. Suppose that $\tilde{D}$ has constant rank p_1 in the interval under consideration; then we can find (see, e.g., [17, Thm 3.9]) orthogonal matrix functions P, Q with the same smoothness as $\tilde{D}$ such that $P^T\tilde{D}Q = \left[\begin{smallmatrix}\tilde{D}_1 & 0\\0 & 0\end{smallmatrix}\right]$, with $\tilde{D}_1$ being $p_1\times p_1$ and pointwise invertible. Setting (partitioned accordingly)

$$P^T y = \begin{bmatrix}\tilde{y}_1\\ \tilde{y}_2\end{bmatrix},\ P^T\tilde{C} = \begin{bmatrix}\tilde{C}_1\\ \tilde{C}_2\end{bmatrix},\ P^T g = \begin{bmatrix}\tilde{g}_1\\ \tilde{g}_2\end{bmatrix},\ \tilde{u} = Qu = \begin{bmatrix}\tilde{u}_1\\ \tilde{u}_2\end{bmatrix},$$

with $\tilde{C}_1$ of size $p_1\times d+a$ and $\tilde{u}_2$ of size $m-p_1$, then we obtain a new system without feedthrough term of the form

$$\bar{E}\dot{\bar{x}} = \bar{A}\bar{x} + \bar{B}\bar{u} + \bar{f}, \tag{2.29a}$$

$$\bar{y} = \bar{C}\bar{x} + \bar{g}, \tag{2.29b}$$

with data

$$\bar{x} = \begin{bmatrix}\tilde{x}\\ \tilde{y}_1\end{bmatrix},\ \bar{y}=\tilde{y}_2,\ \bar{u}=\tilde{u}_2,\ \bar{f} = \begin{bmatrix}\tilde{f}_1\\ \tilde{f}_2\\ \tilde{g}_1\end{bmatrix},\ \bar{g}=\tilde{g}_2,$$

$$\bar{E} = \begin{bmatrix}\tilde{E}_1 & 0\\ 0 & 0\\ 0 & 0\end{bmatrix},\ \bar{A} = \begin{bmatrix}\tilde{A}_1 & 0\\ \tilde{A}_2 & 0\\ \tilde{C}_1 & -I_{p_1}\end{bmatrix},\ \bar{B} = \begin{bmatrix}\tilde{B}_1\\ \tilde{B}_2\\ \tilde{D}_1\end{bmatrix},\ \bar{C}=\tilde{C}_2.$$

2.3.4 Feedback Regularization

As in the nonlinear case, if no reinterpretation of variables was made, i.e., the system is still in terms of the original x and u, and if the output equation was kept separately, we may have that the resulting system is not strangeness-free as a free system with $\bar{u} = 0$. It has been shown in [20] that then there exists a linear feedback $\bar{u} = K\bar{x} + w$, with $K \in C^0(\mathbf{I}, \mathbf{R}^{\bar{m}\times\bar{n}})$, such that in the closed-loop system

$$\bar{E}\dot{\bar{x}} = (\bar{A} + \bar{B}K)\bar{x} + \bar{B}w + \bar{f}, \quad \bar{x}(\underline{t}) = \underline{\bar{x}}, \tag{2.30}$$

the matrix function $(\bar{A}_2 + \bar{B}_2 K)\bar{T}_2'$ is pointwise nonsingular, where $\bar{T}_2'$ is a matrix-valued function that spans the kernel of $\bar{E}_1$. This implies that the DAE in (2.30) is regular and strangeness-free as a free system with $w = 0$. The output equation then changes to $\bar{y} = \bar{C}\bar{x} + \bar{g}$.

2.3.5 General Procedure

We can summarize the whole procedure in the following table.

$$\boxed{\begin{aligned} E\dot{x} &= Ax + Bu + f,\ x(\underline{t}) = \underline{x}, \\ y &= Cx + Du + g \end{aligned}}$$

$$\mu \neq 0 \ \downarrow \ \text{index reduction in behavior}$$

$$\boxed{\begin{aligned} E_1\dot{x} &= A_1x + B_1u + \hat{f}_1, \\ 0 &= A_2x + B_2u + \hat{f}_2, \\ 0 &= \hat{f}_3, \\ y &= Cx + Du + g \end{aligned}}$$

$$\hat{f}_3 \neq 0 \ \downarrow \ \text{force } \hat{f}_3 = 0$$

$$0 \neq A_2(\underline{t})\underline{x} + B_2(\underline{t})u(\underline{t}) + \hat{f}_2(\underline{t}) \ \downarrow \ \text{force consistency}$$

$$a + d \neq n \ \downarrow \ \text{reinterpret variables}$$

$$\boxed{\begin{aligned} \tilde{E}_1\dot{\tilde{x}} &= \tilde{A}_1\tilde{x} + \tilde{B}_1\tilde{u} + \tilde{f}_1, \\ 0 &= \tilde{A}_2\tilde{x} + \tilde{B}_2\tilde{u} + \tilde{f}_2, \\ \tilde{y} &= \tilde{C}\tilde{x} + \tilde{D}\tilde{u} + \tilde{g} \end{aligned}}$$

$$\tilde{D} \neq 0 \ \downarrow \ \text{remove feedthrough}$$

$$\boxed{\begin{aligned} \bar{E}_1\dot{\bar{x}} &= \bar{A}_1\bar{x} + \bar{B}_1\bar{u} + \bar{f}_1, \\ 0 &= \bar{A}_2\bar{x} + \bar{B}_2\bar{u} + \bar{f}_2, \\ \bar{y} &= \bar{C}\bar{x} + \bar{g} \end{aligned}}$$

$$\text{not strangeness-free for } u = 0 \ \downarrow \ \text{feedback } \bar{u} = K\bar{x} + w$$

$$\boxed{\begin{aligned} \bar{E}_1\dot{\bar{x}} &= (\bar{A}_1 + \bar{B}_1K)\bar{x} + \bar{B}_1w + \bar{f}_1, \\ 0 &= (\bar{A}_2 + \bar{B}_2K)\bar{x} + \bar{B}_2w + \bar{f}_2, \\ \bar{y} &= \bar{C}\bar{x} + \bar{g}. \end{aligned}}$$

If no reinterpretation of variables is necessary, and if we want a strangeness-free system as a behavior and as a free system, then the preceding procedure is followed except the reinterpretation of variables and the third block is omitted. Otherwise the procedure is identical.

2.4 Conclusions

We have described detailed regularization procedures for general linear and nonlinear descriptor systems and shown that under some general hypotheses every system can be reformulated as a system that is strangeness-free as a behavior system as well as a free system. Regularization procedures for inconsistent systems and initial conditions have been described as well as procedures for the construction of regularizing state feedbacks and for the removal of feedthrough terms. In all cases it is possible to achieve a reformulated systems which allows the application of standard simulation, control, and optimization techniques.

Acknowledgments

Stephen L. Campbell's research was supported in part by NSF grant DMS-0907832. Peter Kunkel's research was supported by Research-in-Pairs Program at Mathematisches Forschungsinstitut Oberwolfach and by Deutsche Forschungsgemeinschaft through grant KU964/7-1. Volker Mehrmann's research was supported by Research-in-Pairs Program at Mathematisches Forschungsinstitut Oberwolfach and the European Research Council through ERC Advanced grant MODSIMCONMP.

Bibliography

[1] J. BALS, G. HOFER, A. PFEIFFER, AND C. SCHALLERT, *Virtual iron bird – A multidisciplinary modelling and simulation platform for new aircraft system architectures*, in Deutscher Luft- und Raumfahrtkongress, Friedrichshafen, Germany, 2005.

[2] K. E. BRENAN, S. L. CAMPBELL, AND L. R. PETZOLD, *Numerical Solution of Initial-Value Problems in Differential-Algebraic Equations*, SIAM, Philadelphia, 1996.

[3] A. BUNSE-GERSTNER, R. BYERS, V. MEHRMANN, AND N. K. NICHOLS, *Feedback design for regularizing descriptor systems*, Linear Algebra Appl., 299 (1999), pp. 119–151.

[4] S. L. CAMPBELL, *A general form for solvable linear time varying singular systems of differential equations*, SIAM J. Math. Anal., 18 (1987), pp. 1101–1115.

[5] S. L. CAMPBELL, *Linearization of DAE's along trajectories*, Z. Angew. Math. Phys., 46 (1995), pp. 70–84.

[6] L. DAI, *Singular Control Systems*, Springer-Verlag, Berlin, 1989.

[7] E. EICH-SOELLNER AND C. FÜHRER, *Numerical Methods in Multibody Systems*, Teubner Verlag, Stuttgart, Germany, 1998.

[8] P. FRITZON, *Principles of Object-Oriented Modeling and Simulation with Modelica 2.1*, Wiley-IEEE Press, New York, 2003.

[9] M. HILLER AND K. HIRSCH, *Multibody system dynamics and mechatronics*, Z. Angew. Math. Mech., 86 (2006), pp. 87–109.

[10] M. HOU AND P. C. MÜLLER, *LQ and Tracking Control of Descriptor Systems with Application to Constrained Manipulator*, Technical report, Sicherheitstechnische Regelungs- und Meßtechnik, Universität Wuppertal, Wuppertal 1, Germany, 1994.

[11] M. HOU, *Descriptor Systems: Observers and Fault Diagnosis*, Fortschr. Ber. VDI, Reihe 8, Nr. 482, VDI-Verlag, Düsseldorf, 1995.

[12] A. ILCHMANN AND V. MEHRMANN, *A behavioral approach to time-varying linear systems. Part 1: General theory*, SIAM J. Control Optim., 44 (2005), pp. 1725–1747.

[13] A. ILCHMANN AND V. MEHRMANN, *A behavioral approach to time-varying linear systems. Part 2: Descriptor systems*, SIAM J. Contol Optim., 44 (2005), pp. 1748–1765.

[14] P. KUNKEL AND V. MEHRMANN, *A new class of discretization methods for the solution of linear differential-algebraic equations with variable coefficients*, SIAM J. Numer. Anal., 33 (1996), pp. 1941–1961.

[15] P. KUNKEL AND V. MEHRMANN, *Analysis of over- and underdetermined nonlinear differential-algebraic systems with application to nonlinear control problems*, Math. Control Signals Systems, 14 (2001), pp. 233–256.

[16] P. KUNKEL AND V. MEHRMANN, *Index reduction for differential-algebraic equations by minimal extension*, Z. Angew. Math. Mech., 84 (2004), pp. 579–597.

[17] P. KUNKEL AND V. MEHRMANN, *Differential-Algebraic Equations. Analysis and Numerical Solution*, EMS Publishing House, Zürich, Switzerland, 2006.

[18] P. KUNKEL AND V. MEHRMANN, *Stability properties of differential-algebraic equations and spin-stabilized discretizations*, Electron. Trans. Numer. Anal., 26 (2007), pp. 385–420.

[19] P. KUNKEL AND V. MEHRMANN, *Optimal control for unstructured nonlinear differential-algebraic equations of arbitrary index*, Math. Control Signals Systems, 20 (2008), pp. 227–269.

[20] P. KUNKEL, V. MEHRMANN, AND W. RATH, *Analysis and numerical solution of control problems in descriptor form*, Math. Control Signals Systems, 14 (2001), pp. 29–61.

[21] P. KUNKEL, V. MEHRMANN, W. RATH, AND J. WEICKERT, *A new software package for linear differential-algebraic equations*, SIAM J. Sci. Comput., 18 (1997), pp. 115–138.

[22] M. LANDWEHR, U. LEFARTH, AND E. WASSMUTH, *Parameter identification and optimization of nonlinear dynamic systems, exemplified by mechatronic systems*, in Computational systems analysis, Elsevier, Amsterdam, The Netherlands, 1992, pp. 257–262.

[23] V. H. LINH AND V. MEHRMANN, *Lyapunov, Bohl and Sacker-Sell spectral intervals for differential-algebraic equations*, J. Dynam. Differential Equations, 21 (2009), pp. 153–194.

[24] S. E. MATTSSON AND G. SÖDERLIND, *Index reduction in differential-algebraic equations using dummy derivatives*, SIAM J. Sci. Statist. Comput., 14 (1993), pp. 677–692.

[25] C. C. PANTELIDES, *The consistent initialization of differential-algebraic systems*, SIAM J. Sci. Statist. Comput., 9 (1988), pp. 213–231.

[26] J. W. POLDERMAN AND J. C. WILLEMS, *Introduction to Mathematical Systems Theory: A Behavioural Approach*, Springer-Verlag, New York, 1998.

[27] R. E. ROBERSON AND R. SCHWERTASSEK, *Dynamics of Multibody Systems*, Springer-Verlag, Heidelberg, 1988.

[28] W. SCHIEHLEN, *Multibody Systems Handbook*, Springer-Verlag, Heidelberg, 1990.

[29] W. SCHIEHLEN, *Advanced Multibody System Dynamics*, Kluwer Academic Publishers, Stuttgart, Germany, 1993.

[30] K. SCHLACHER AND A. KUGI, *Automatic control of mechatronic systems*, Int. J. Math. Comput. Sci., 11 (2001), pp. 131–164.

[31] K. SCHLACHER, A. KUGI, AND R. SCHEIDL, *Tensor analysis based symbolic computation for mechatronic systems*, Math. Comput. Simulation, 46 (1998), pp. 517–525.

[32] W. TRAUTENBERG, *Simpack* 8.9 *Manual*, INTEC GmbH, Wessling, Germany.

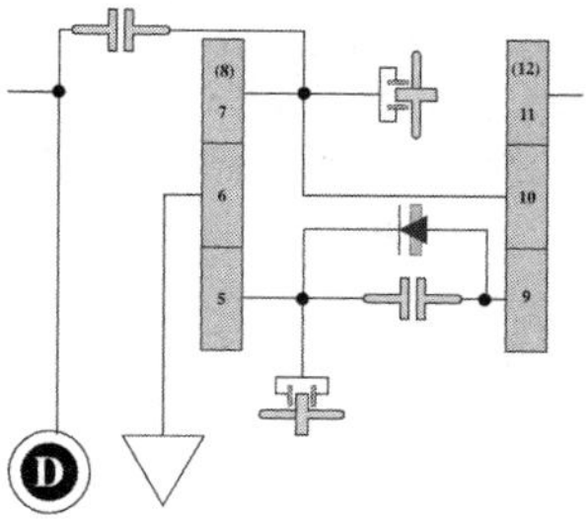

Chapter 3

Notes on Linearization of DAEs and on Optimization with Differential-Algebraic Constraints

Roswitha März

We regard different linearization phenomena which newly emerge from the DAE context and relate them to regularity regions. Moreover, we consider the surjectivity of linearizations of underdetermined DAEs, which in turn yields necessary extremal conditions for optimization problems. We investigate also the linearization of the resulting optimality DAE and provide constitutive properties by clear conditions in terms of the original data.

3.1 Introduction

Linearization is used throughout analysis, optimization, and control—and in particular, if DAEs are involved. Linearized DAEs play their role as approximations of nonlinear models (e.g., [7, Section 1.5]). In many cases, papers either apply to linear DAEs or base their analysis on linear DAEs. Frequently even time-invariant linear DAEs are supposed. Linearized DAEs also play their role in extremal conditions (e.g., [1]) as approximations of nonlinear equations in Newton–Kantorovich, Gauss–Newton, and sequential quadratic programming (SQP) methods (e.g., [21]). Linearized DAEs and their adjoints, respectively, are utilized in the sensitivity analysis (e.g., [6]), in the stability analysis (e.g., [16]), and when considering observability (e.g., [22]).

Even though linearization represents a generally important mathematical tool, only few papers (e.g., [4, 15]) address general relations between a given nonlinear DAE and its linearizations along trajectories.

In general, one expects a linearized DAE to provide local information on the nonlinear DAE the linearized originates from. However, this can turn out to be an error. Linearizations of DAEs feature quite different phenomena. With this chapter we intend to direct attention to linearization phenomena of square DAEs and to the role of regularity regions. Furthermore, we apply linearization to obtain an extremal condition for an optimization problem with DAE constraints together with the so-called optimality DAE which constitutes the background of indirect optimization. The associated criteria as well as the optimality DAE are given in terms of the original data.

In our analysis we pursue a different view on DAEs. While usually the obvious and hidden constraints come to the fore, we primarily consider the definition domain of the given DAE to decompose in several (maximal) regularity regions with inherent characteristics, such as the index, which are bordered by critical points; see [13, 19]. Solutions may remain within such a region, but they also may cross the borders or stay there. We show the close correspondence between regularity regions and linearizations along reference functions standing in this regularity region.

We solely address optimization aspects in the context of linearizations. For a more general discussion we refer the reader to [1, 11]. Nevertheless, the given below clear algebraic criteria in terms of the original data are to enable an advanced utilization of the modeling latitude for obtaining an optimality DAE that is locally regular with index one. This way one secures indirect optimization methods.

The chapter is organized as follows. In Section 3.2 we describe DAEs with properly involved derivatives and their linearizations. In Section 3.3, underdetermined and regular index-one DAEs are introduced and a solvability assertion is provided, which plays its role later on when formulating extremal conditions. Section 3.4 describes a collection of linearization phenomena in several examples. Regularity regions are introduced in Section 3.5. In Section 3.6 we consider an optimization problem comprising a DAE constraint. We provide a necessary extremal condition and algebraic conditions ensuring that the optimality DAE is locally regular with index one and its linearization has an inherent Hamiltonian flow. These results are specified for controlled square DAEs in Section 3.7.

3.2 Preliminaries

In this chapter we deal with DAEs of the form

$$f((D(t)x(t))', x(t), t) = 0, \tag{3.1}$$

where $f(y,x,t) \in \mathbf{R}^k$, $D(t) \in L(\mathbf{R}^m, \mathbf{R}^n)$, $y \in \mathbf{R}^n$, $x \in \mathcal{D}_f \subseteq \mathbf{R}^m$, $t \in \mathcal{I}_f \subseteq \mathbf{R}$. $\mathcal{D}_f$ is open in $\mathbf{R}^m$ and $\mathcal{I}_f$ is an interval. The function f is supposed to be continuous on the definition domain $\mathbf{R}^n \times \mathcal{D}_f \times \mathcal{I}_f$, together with the partial derivatives f_y, f_x. The leading Jacobian $f_y(y,x,t)$ is everywhere singular. The matrix function D is continuously differentiable on $\mathcal{I}_f$ and has constant rank r.

A solution of (3.1) is a function given on a certain interval $\mathcal{I} \subseteq \mathcal{I}_f$, with values in $\mathcal{D}_f$, which belongs to the function space

$$\mathcal{C}^1_D(\mathcal{I}, \mathbf{R}^m) := \{x \in \mathcal{C}(\mathcal{I}, \mathbf{R}^m) : Dx \in \mathcal{C}^1(\mathcal{I}, \mathbf{R}^n)\}$$

and satisfies the DAE pointwise on $\mathcal{I}$.

Obviously, all solution values at time t have to belong to the so-called obvious constraint set

$$\mathcal{M}_0(t) = \{x \in \mathcal{D}_f : \exists y \in \mathbf{R}^n : f(y,x,t) = 0\}. \tag{3.2}$$

The extra matrix function D figuring out the derivative term entails an enhanced DAE model (e.g., [13, 18, 19]). In contrast to standard form DAEs

$$\mathfrak{f}(x'(t), x(t), t) = 0, \tag{3.3}$$

equation (3.1) precisely indicates in which way derivatives of the unknown function are actually involved. Instead of the commonly used $\mathcal{C}^1$-solutions one can tolerate $\mathcal{C}^1_D$-solutions.

For standard form DAEs (3.3) there is often a singular incidence or projector matrix $D \in L(\mathbf{R}^m)$ such that the identity $\mathfrak{f}(x^1,x,t) \equiv \mathfrak{f}(Dx^1,x,t)$ is valid, and hence the standard DAE (3.3) can be interpreted as

$$\mathfrak{f}((D(t)x(t))',x(t),t) = 0.$$

In particular, any semi-explicit system of m_1 and m_2 equations

$$x_1'(t)+b_1(x_1(t),x_2(t),t) = 0, \tag{3.4}$$

$$b_2(x_1(t),x_2(t),t) = 0 \tag{3.5}$$

can be naturally rewritten as

$$\begin{bmatrix} I_{m_1} \\ 0 \end{bmatrix} \left(\begin{bmatrix} I_{m_1} & 0 \end{bmatrix} x\right)'(t) + \begin{bmatrix} b_1(x_1(t),x_2(t),t) \\ b_2(x_1(t),x_2(t),t) \end{bmatrix} = 0. \tag{3.6}$$

In case of a semi-explicit DAE, one has good reason to ask for solutions having a continuously differentiable first component x_1, but accept a continuous second component x_2, as it is prescribed by the associated function space

$$\{x \in \mathcal{C}(\mathcal{I},\mathbf{R}^m) : x_1 \in \mathcal{C}^1(\mathcal{I},\mathbf{R}^{m_1})\}.$$

Definition 3.1. *The DAE* (3.1) *has a properly involved derivative if there is a projector-valued function* $K \in \mathcal{C}^1(\mathcal{J}_f, L(\mathbf{R}^n))$ *such that*

$$\ker f_y(y,x,t) = \ker K(t), \quad \operatorname{im} D(t) = \operatorname{im} K(t), \quad (y,x,t) \in \mathbf{R}^n \times \mathcal{D}_f \times \mathcal{I}_f. \tag{3.7}$$

The DAE is in full-rank proper form if $n = r$ *and*

$$\ker f_y(y,x,t) = \{0\}, \quad \operatorname{im} D(t) = \mathbf{R}^n, \quad (y,x,t) \in \mathbf{R}^n \times \mathcal{D}_f \times \mathcal{I}_f. \tag{3.8}$$

In case of a properly involved derivative, the partial Jacobian $f_y(y,x,t)$ has constant rank and its nullspace is independent of y and x. In contrast, a so-called *quasi-proper* involved derivative permits rank changes in $f_y(y,x,t)$; see [13, 20]. The full-rank form is associated with the projector $K(t) = I_n$, as, for instance, in (3.6) with $n = m_1$. Full-rank formulations are known to be advantageous in view of numerical integration. Also, having in mind indirect optimization, for ensuring Hamiltonian structure, one is well advised to choose this form (see [2]).

In earlier papers, when dealing with quasi-linear DAEs, instead of the phrase *properly involved derivative* the wording *properly stated leading term* is introduced. We apply both notations also for fully nonlinear DAEs.

For each arbitrary reference function $x_* \in \mathcal{C}^1_D(\mathcal{I},\mathbf{R}^m)$, with graph in $\mathcal{D}_f \times \mathcal{I}_f$, the coefficient functions

$$A_*(t) := f_y((D(t)x_*(t))',x_*(t),t), \quad B_*(t) := f_x((D(t)x_*(t))',x_*(t),t), \quad t \in \mathcal{I},$$

of the linear DAE

$$A_*(t)(D(t)x(t))' + B_*(t)x(t) = q(t), \quad t \in \mathcal{I}, \tag{3.9}$$

are continuous and the DAE (3.9) inherits from the nonlinear DAE the special form of the leading term. Which further relations between the nonlinear DAE (3.1) and the linear DAEs (3.9) can we expect?

Definition 3.2. *The linear DAE* (3.9) *is called the linearization of the nonlinear DAE* (3.1) *along x_* or along the trajectory of x_*.*

It should be emphasized that the reference function $x_* \in \mathcal{C}^1_D(\mathcal{I},\mathbf{R}^m)$ is not necessarily a solution of (3.1).

If the DAE (3.1) is autonomous and the reference function is a constant one, then the coefficients A_*, D, B_* are time-invariant and the resulting linear constant coefficient DAE (3.9) can be treated via the Kronecker canonical form of the matrix pencil $\lambda A_* D + B_*$. However, in general, the original DAE (3.1) is explicitly time-dependent and the reference function varies in time so that first and foremost time-varying linear DAEs (3.9) result.

We stress that here the linearization along x_* of the nonlinear DAE (3.1) represents a linear DAE (3.9) with unassigned general right-hand side q. In contrast, at times one restricts the interest to linearized DAEs with specific right-hand side

$$A_*(t)(D(t)x(t))' + B_*(t)x(t) = -f((D(t)x_*(t))', x_*(t), t), \quad t \in \mathcal{I}. \tag{3.10}$$

If x_* solves a square DAE (3.1), the resulting homogeneous linear DAE (3.10) is expected to afford information concerning the local flow stability. In the context of Gauss–Newton and SQP procedures, x_* represents merely an approximate solution or initial guess. Then, the specified linearized DAE (3.10) serves to improve the approximation x_* by $x_* + x$.

While, trivially, the linearization of an explicit ODE is again an explicit ODE, the situation in case of DAEs is complex. The basic structure may change unlike the ODE case. If one intends to benefit from linearization, in case of DAEs, not only are solutions of the specified DAEs (3.10) to be considered but also structural properties of the general linear DAE (3.9) such as surjectivity and regularity characteristics.

3.3 Linearizations of Index-One DAEs

Definition 3.3. *Let the DAE* (3.1) *have a properly involved derivative and let $\mathfrak{G} \subseteq \mathbf{R}^n \times \mathcal{D}_f \times \mathcal{I}_f$ be an open connected set. If the rank condition*

$$\operatorname{rank}\{f_y(y,x,t)D(t) + f_x(y,x,t)(I_m - D(t)^+ D(t))\} = k \tag{3.11}$$

is satisfied for all $(y,x,t) \in \mathfrak{G}$, then the DAE (3.1) *is said to be underdetermined on $\mathfrak{G}$ with tractability index one. If, additionally, $k = m$, then the square DAE* (3.1) *is said to be regular on $\mathfrak{G}$ with tractability index one.*

The semi-explicit DAE (3.6) is regular with index one, exactly if the partial Jacobian $b_{2,x_2}(x_1,x_2,t)$ is square and nonsingular. The DAE (3.6) is underdetermined with index one if $b_{2,x_2}(x_1,x_2,t)$ is rectangular with full row rank. For general square equations (3.1) with properly involved derivative, for $k = m$, condition (3.11) is valid if and only if the local matrix pencil $\lambda f_y(y,x,t)D(t) + f_x(y,x,t)$ is regular with Kronecker index one (e.g., [12, page 49]).

Proposition 3.4. *Let the DAE* (3.1) *be given with properly stated leading term:*

(1) *If the DAE* (3.1) *has index one on the open connected set* $\mathfrak{G} \subseteq \mathbf{R}^n \times \mathfrak{D}_f \times \mathfrak{I}_f$*, then each linearization* (3.9) *along a function* $x_* \in \mathcal{C}_D^1(\mathfrak{I}, \mathbf{R}^m)$ *with graph in* $\mathfrak{G}$ *is also an index-one DAE.*

(2) *If* $\mathfrak{I} \subseteq \mathfrak{I}_f$*,* $x_* \in \mathcal{C}_D^1(\mathfrak{I}, \mathbf{R}^m)$ *has values in* $\mathfrak{D}_f$ *and the linearization* (3.9) *is an index-one DAE, such that*

$$\operatorname{rank}\{A_*(t)D(t) + B_*(t)(I - D(t)^+ D(t))\} = k,\ t \in \mathfrak{I}, \tag{3.12}$$

then there is an open connected set $\mathfrak{G}$ *enclosing the graph of* x_* *and the DAE* (3.1) *has index one on* $\mathfrak{G}$*, where* $D(t)^+$ *denotes the Moore–Penrose inverse of* $D(t)$*.*

Proof. The first part is a direct consequence of the definition and the second one follows from continuity arguments. □

Next we add to the DAE the initial condition

$$D(t_0)x(t_0) = z \tag{3.13}$$

and consider the solvability of initial value problems (IVPs).

Proposition 3.5. *Let the DAE* (3.1) *have a properly stated leading term,* $t_0 \in \mathfrak{I} \subseteq \mathfrak{I}_f$*, and let* $x_* \in \mathcal{C}_D^1(\mathfrak{I}, \mathbf{R}^m)$ *have values in* $\mathfrak{D}_f$*. Let the linearization* (3.9) *satisfy condition* (3.12)*. Then*

(1) *for each arbitrary* $q \in \mathcal{C}(\mathfrak{I}, \mathbf{R}^k)$*,* $z \in \operatorname{im} D(t_0)$*, the linear IVP* (3.9), (3.13) *possesses at least one solution in* $\mathcal{C}_D^1(\mathfrak{I}, \mathbf{R}^m)$*;*

(2) *for square DAEs, with* $k = m$*, the IVP* (3.9), (3.13) *is uniquely solvable.*

Proof. Here we drop the argument t of the functions. Denote $Q_0 := I_m - D^+D$, $G_{*1} := A_*D + B_*Q_0$ and introduce the reflexive generalized inverse D^- such that $D^-D = I_m - Q_0$, $DD^- = K$. Since G_{*1} has full row rank k, G_{*1}^+ is continuous and $G_{*1}G_{*1}^+ = I_k$. This allows us to rewrite the DAE (3.9) as

$$\underbrace{A_*DD^-}_{G_{*1}D^-}(Dx)' + \underbrace{B_*Q_0}_{G_{*1}Q_0}x + G_{*1}G_{*1}^+(B_*D^-Dx - q) = 0,$$

therefore

$$G_{*1}[D^-(Dx)' + Q_0x + G_{*1}^+B_*D^-Dx - G_{*1}^+q] = 0,$$

and thus

$$D^-(Dx)' + Q_0x + G_{*1}^+B_*D^-Dx - G_{*1}^+q = w, \tag{3.14}$$

with arbitrary functions w such that $G_{*1}w = 0$. Multiplying by D and Q_0, respectively, we split equation (3.14) into

$$(Dx)' - K'Dx + DG_{*1}^+B_*D^-Dx - DG_{*1}^+q = Dw, \tag{3.15}$$

$$Q_0x + Q_0G_{*1}^+B_*D^-Dx - Q_0G_{*1}^+q = Q_0w. \tag{3.16}$$

Set $w = 0$. For each given $z \in \operatorname{im} D(t_0)$ and continuous q, the resulting IVP (3.15), (3.13) is uniquely solvable and provides a continuously differentiable Dx. Having Dx we determine $Q_0 x$ by (3.16). This way we generate a solution $x = D^- Dx + Q_0 x$ of the IVP (3.9), (3.13), which proves the first assertion. If $m = k$, then G_{*1} is also injective, and thus the unique solvability is evident and so is the second assertion. □

The linear map $\mathfrak{L} : \mathcal{C}^1_D \to \mathcal{C} \times \operatorname{im} D(t_0)$ being determined by

$$\mathfrak{L}x := (A_*(Dx)' + B_* x,\ D(t_0)x(t_0)), \quad x \in \mathcal{C}^1_D, \tag{3.17}$$

is surjective in Proposition 3.5(1). This property of a rectangular DAE (3.1) plays its role if (3.1) serves as a constraint in an optimization problem (cf. Section 3.6). At this place it is worth mentioning that condition (3.12) allows a linear splitting transformation of the variable x into a state and a control part such that the resulting controlled DAE is regular with index one locally around the reference function (cf. [5] and the examples in Section 3.6).

Proposition 3.5(2) is related to square systems and means then the bijectivity of the map $\mathfrak{L}$. In turn, the bijectivity of the map $\mathfrak{L}$ is most important when applying indirect optimization methods via the square optimality DAE (cf. Section 3.6) and also when considering the control of DAEs (cf. Section 3.7).

Furthermore, also famous linearization results of Perron and Lyapunov concerning asymptotical stability apply to regular index-one DAEs (3.1), under certain further conditions and in slightly modified form: if the linearization along a solution $x_* \in \mathcal{C}^1_D[0,\infty)$ is strictly contractive, then the reference solution x_* is asymptotically stable (see [13]; cf. also [16]).

Altogether, in case of index-one problems, linearization works as expected: The linearized DAE inherits relevant properties of the nonlinear DAE, and conversely, the linearized DAE provides local information on the nonlinear DAE. Unfortunately, the situation becomes different in a more general setting, as we see in the next section.

3.4 Linearization: Case Studies

In this part we demonstrate by case studies that linearizations may reflect properties of the original nonlinear DAE as expected coming from regular ODEs, but they may also show astonishingly different phenomena unlike regular ODEs. This happens for reference functions being solutions of the given nonlinear DAE and also for arbitrary reference functions. In particular, Example 3.7/Case 3 shows a behavior fully consistent with the (regular) ODE theory. In contrast, a linearized DAE may

- suggest by mistake stability, observability, etc., see Example 3.6,
- show both lower index and higher index than the original DAE seems to have, see Examples 3.6 and 3.7/Case 4, and fail to remain regular at all, see Example 3.7/Case 2,
- show an index which varies on subintervals, causing serious flow singularities, see Example 3.8,
- show a singular flow which emerges from a singular inherent ODE, see Example 3.7/Case 1.

Those kinds of negative phenomena cannot be observed in adequate regular ODEs, but they newly emerge from DAE structures. The regularity regions (see [13, 18, 19]) which we introduce in the next section are to comprehend what is going on. It seems that regularity regions actually constitute a helpful tool for investigating the linearization phenomena. In the following examples, we simply quote the regularity regions and their structural characteristics. Later on in Section 3.5 we explain how those regions are determined.

Example 3.6. The autonomous DAE [16]

$$\begin{aligned} x_1'(t) - x_2(t) &= 0, \\ x_1(t) - x_2(t)^3 &= 0, \\ x_3'(t) - \alpha x_3(t) &= 0, \\ -x_2(t) + x_3(t) + x_4(t) &= 0 \end{aligned}$$

has the form (3.1) with a properly stated leading term, $m = k = 4, r = n = 2$, $\mathcal{D}_f = \mathbf{R}^4$, $\mathcal{I}_f = [0, \infty)$, a real parameter $\alpha < 0$, and

$$D(t) = \begin{bmatrix} 1 & 0 & 0 & 0 \\ 0 & 0 & 1 & 0 \end{bmatrix}, \quad f(y,x,t) = \begin{bmatrix} 1 & 0 \\ 0 & 0 \\ 0 & 1 \\ 0 & 0 \end{bmatrix} y + \begin{bmatrix} -x_2 \\ x_1 - x_2^3 \\ -\alpha x_3 \\ -x_2 + x_3 + x_4 \end{bmatrix}.$$

The obvious constraint set $\mathcal{M}_0 = \{x \in \mathbf{R}^4 : x_1 = x_2^3, x_2 = x_3 + x_4\}$ is time-invariant. Besides the obvious zero solution, the solutions of this DAE are given by the formulae

$$x_1(t) = \left(c^2 + \frac{2}{3}t\right)^{\frac{3}{2}}, \; x_2(t) = \left(c^2 + \frac{2}{3}t\right)^{\frac{1}{2}}, \; x_3(t) = \mathrm{e}^{\alpha t} d, \; x_4(t) = -\mathrm{e}^{\alpha t} d + \left(c^2 + \frac{2}{3}t\right)^{\frac{1}{2}},$$

whereby $c, d \in \mathbf{R}$ denote the arbitrary integration constants. Obviously, all these solutions grow unboundedly if $t \to \infty$. We take the stationary solution $x_* = 0$ as reference function. The homogeneous version of the linearized DAE along x_* is

$$\underbrace{\begin{bmatrix} 1 & 0 \\ 0 & 0 \\ 0 & 1 \\ 0 & 0 \end{bmatrix}}_{A_*} \left(\begin{bmatrix} 1 & 0 & 0 & 0 \\ 0 & 0 & 1 & 0 \end{bmatrix} x \right)'(t) + \underbrace{\begin{bmatrix} 0 & -1 & 0 & 0 \\ 1 & 0 & 0 & 0 \\ 0 & 0 & -\alpha & 0 \\ 0 & -1 & 1 & 1 \end{bmatrix}}_{B_*} x(t) = 0.$$

This linear constant coefficient DAE is regular with Kronecker index two, and its general solution is given as

$$x_1(t) = x_2(t) = 0, \quad x_4(t) = -x_3(t) = -\mathrm{e}^{\alpha t} x_3(0),$$

which shows the asymptotical stability of the linearized DAE and which induces to conjecture stability of the reference solution of the original DAE.

Since all nonzero solutions of the nonlinear DAE grow unboundedly if $t \to \infty$, the zero solution x_* of the nonlinear DAE is far from being stable, as one—by mistake—might conclude from the stability of the linearization.

The definition domain $\mathcal{D}_f \times \mathcal{I}_f = \mathbf{R}^4 \times [0,\infty)$ of the original DAE decomposes into the two maximal regularity regions

$$\mathcal{G}_- = \{(x,t) \in \mathcal{D}_f \times \mathcal{I}_f : x_2 < 0\} \quad \text{and} \quad \mathcal{G}_+ = \{(x,t) \in \mathbf{R}^4 \times \mathbf{R} : x_2 > 0\}.$$

The border between these regions, the set $x_2 = 0$, consists of somehow critical points. Actually the DAE flow bifurcates at these critical points. Namely, for each $d \in \mathbf{R}$, there are two solutions

$$\bar{x}(t) = \begin{bmatrix} 0 \\ 0 \\ e^{\alpha t} d \\ -e^{\alpha t} d \end{bmatrix} \text{ and } \bar{\bar{x}}(t) = \begin{bmatrix} (\frac{2}{3}t)^{\frac{3}{2}} \\ (\frac{2}{3}t)^{\frac{1}{2}} \\ e^{\alpha t} d \\ -e^{\alpha t} d + (\frac{2}{3}t)^{\frac{1}{2}} \end{bmatrix} \text{ such that } \bar{x}(0) = \bar{\bar{x}}(0) = \begin{bmatrix} 0 \\ 0 \\ d \\ -d \end{bmatrix}.$$

In particular, for $d = 0$, besides the zero solution x_*, there is also a nontrivial solution that satisfies the initial condition $x(0) = 0$.

Observe that our reference function resides on that critical border.

The nonlinear DAE is regular with tractability index one only on the two regularity regions but not on its entire definition domain. If a reference function remains within one of the regularity regions, then the linearization along this function is also regular with index one. ■

Example 3.7. The semi-explicit DAE [13, 19]

$$\begin{aligned} x_1'(t) - x_3(t) &= 0, \\ x_2(t)(1 - x_2(t)) - \frac{1}{4} + t^2 &= 0, \\ x_1(t)x_2(t) + x_3(t)(1 - x_2(t)) - t &= 0, \end{aligned}$$

with $k = m = 3$, $r = n = 1$, $D(t) = [1 0 0]$, and

$$f(y,x,t) = \begin{bmatrix} 1 \\ 0 \\ 0 \end{bmatrix} y + \begin{bmatrix} -x_3 \\ x_2(1 - x_2) - \frac{1}{4} + t^2 \\ x_1 x_2 + x_3(1 - x_2) - t \end{bmatrix}, \quad y \in \mathbf{R},\ x \in \mathbf{R}^3,\ t \in \mathbf{R},$$

has full-rank proper form. As we will see below, its linearizations

$$\begin{bmatrix} 1 \\ 0 \\ 0 \end{bmatrix} \left(\begin{bmatrix} 1 & 0 & 0 \end{bmatrix} x(t) \right)' + \begin{bmatrix} 0 & 0 & -1 \\ 0 & 1 - 2x_{*2}(t) & 0 \\ x_{*2}(t) & x_{*1}(t) - x_{*3}(t) & 1 - x_{*2}(t) \end{bmatrix} x(t) = q(t) \tag{3.18}$$

behave manifestly different for different reference functions x_*.

Case 1: Letting $x_{*2}(t) = \frac{1}{2} + t$ such that the second equation of the original DAE is satisfied, the resulting linear DAE (3.18) reads in detail as

$$\begin{aligned} x_1'(t) - x_3(t) &= q_1(t), \\ 2t\, x_2(t) &= q_2(t), \\ \left(\frac{1}{2} + t\right) x_1(t) + (x_{*1}(t) - x_{*3}(t))x_2(t) + \left(\frac{1}{2} - t\right) x_3(t) &= q_3(t). \end{aligned}$$

On all intervals that include neither $t = 0$ nor $t = \frac{1}{2}$, this linear DAE is regular with tractability index one, and its inherent ODE reads as

$$x_1'(t) = -\frac{1+2t}{1-2t}x_1(t) + q_1(t) + \frac{2}{1-2t}q_3(t) - \frac{1}{(1-2t)t}q_2(t)(x_{*1}(t) - x_{*3}(t)).$$

Take a closer look at the homogeneous linear DAE, with identically vanishing right-hand sides q. Now a singular homogeneous inherent ODE results, with a singularity at $t = \frac{1}{2}$. The solutions are given by $x_1(t) = (1-2t)e^t c$, $c \in \mathbf{R}$ being arbitrary, and $x_1(\frac{1}{2}) = 0$ is valid for all solutions, which indicates a singular flow at $t = \frac{1}{2}$.

The solutions of the inhomogeneous linearization may grow unboundedly if t approaches the critical point $t = 0$ or $t = \frac{1}{2}$.

Case 2: Letting $x_{*2}(t) = \frac{1}{2}$, the linear DAE (3.18), that is,

$$\begin{aligned} x_1'(t) - x_3(t) &= q_1(t), \\ 0 &= q_2(t), \\ \frac{1}{2}x_1(t) + (x_{*1}(t) - x_{*3}(t))x_2(t) + \frac{1}{2}x_3(t) &= q_3(t), \end{aligned}$$

fails to be regular at all. The second equation is a consistency condition for q, while the solution component x_2 can be fixed arbitrarily.

Case 3: Letting $x_{*2}(t) = 0$, the linearization (3.18) has the form

$$\begin{aligned} x_1'(t) - x_3(t) &= q_1(t), \\ x_2(t) &= q_2(t), \\ (x_{*1}(t) - x_{*3}(t))x_2(t) + x_3(t) &= q_3(t), \end{aligned}$$

which is a regular DAE with index one.

Case 4: Letting $x_{*2}(t) = 1$, the linearization

$$\begin{aligned} x_1'(t) - x_3(t) &= q_1(t), \\ -x_2(t) &= q_2(t), \\ x_1(t) + (x_{*1}(t) - x_{*3}(t))x_2(t) &= q_3(t) \end{aligned}$$

is a regular DAE with index two.

What is going on here? The given nonlinear DAE seems to have index one, but this is not correct. More precisely, the domain $\mathcal{D}_f \times \mathcal{I}_f = \mathbf{R}^3 \times \mathbf{R}$ splits into the three maximal regularity regions

$$\mathcal{G}_1 = \left\{(x,t) \in \mathbf{R}^3 \times \mathbf{R} : x_2 < \frac{1}{2}\right\},$$
$$\mathcal{G}_2 = \left\{(x,t) \in \mathbf{R}^3 \times \mathbf{R} : \frac{1}{2} < x_2 < 1\right\},$$
$$\mathcal{G}_3 = \left\{(x,t) \in \mathbf{R}^3 \times \mathbf{R} : 1 < x_2\right\}.$$

The DAE is regular with tractability index one on each region $\mathcal{G}_\ell$, $\ell = 1,2,3$. Through each point $(\bar{x},\bar{t}) \in \mathcal{G}_\ell$ that satisfies the obvious constraint (the second and third equations of the DAE) passes a locally unique solution. Linearizations along reference functions which remain in one of the regularity regions are also regular with index one; see Case 3. The border points indicate a critical flow behavior for the original nonlinear DAE as well as a certain degeneration of the linearization; see Cases 1, 2, and 4 and also the figures in [13, 19]. ∎

Example 3.8. Consider the semi-explicit DAE

$$\begin{aligned} x_1'(t) - x_2(t) + x_3(t) &= 0, \\ x_2'(t) + x_1(t) &= 0, \\ x_1(t)^3 + \alpha(x_1(t))x_3(t) - (\sin t)^3 &= 0, \end{aligned}$$

where

$$\alpha(s) := \begin{cases} s^3 \text{ if } s \geq 0, \\ 0 \text{ if } s < 0, \end{cases}$$

which yield the linearizations

$$\underbrace{\begin{bmatrix} 1 & 0 \\ 0 & 1 \\ 0 & 0 \end{bmatrix}}_{A_*}\left(\underbrace{\begin{bmatrix} 1 & 0 & 0 \\ 0 & 1 & 0 \end{bmatrix}}_{D} x\right)'(t) + \underbrace{\begin{bmatrix} 0 & -1 & 1 \\ 1 & 0 & 0 \\ \alpha'(x_{*1}(t))x_{*3}(t) + 3x_{*1}^2(t) & 0 & \alpha(x_{*1}(t)) \end{bmatrix}}_{B_*} x(t) = q(t).$$

We choose the periodic solution of the original DAE as reference function $x_*(t) = (\sin t, \cos t, 0)^T$. Then the linearization reads in detail as

$$\begin{aligned} x_1'(t) - x_2(t) + x_3(t) &= q_1(t), \\ x_2'(t) + x_1(t) &= q_2(t), \\ 3(\sin t)^2 x_1(t) + \alpha(\sin t)x_3(t) &= q_3(t). \end{aligned}$$

The original DAE has the two maximal stability regions

$$\mathcal{G}_+ = \{(x,t) \in \mathcal{D}_f \times \mathcal{I}_f : x_1 > 0\} \quad \text{and} \quad \mathcal{G}_- = \{(x,t) \in \mathcal{D}_f \times \mathcal{I}_f : x_1 < 0\}.$$

The DAE is regular with index one on $\mathcal{G}_+$ but regular with index two on $\mathcal{G}_-$. The periodic solution x_* switches between the index-one region and the index-two region. Accordingly, the linearized along this solution DAE has index one on the intervals $(2j\pi,(2j+1)\pi)$ but index two on intervals $((2j+1)\pi,(2j+2)\pi)$. Observe that also the dynamical degree of freedom changes between two and one such that the flow becomes severely discontinuous. ■

3.5 Linearizations and Regularity Regions

Regarding our examples we do not expect a general DAE (3.1) to show uniform structure on its entire domain $\mathcal{D}_f \times \mathcal{I}_f$. As suggested by the case studies, it is rather natural that the domain $\mathcal{D}_f \times \mathcal{I}_f$ decomposes into several maximal regularity regions bordered by critical points. It may well happen that the structural characteristics of the DAE are different on different regularity regions. However, in each regularity region there must be uniform structural characteristics. A solution of the DAE can enter a regularity region with new characteristic values only after passing a critical point. Now it is time to explain how a regularity region is actually determined.

Turn for a moment to the matrix pencil $\lambda G + B$ given by the $m \times m$ matrices G and B. We construct a sequence of matrices by special smart projectors starting with $G_0 := G$, $B_0 := B$, $P_0 := I_m - Q_0$, and Q_0 being a projector matrix onto $N_0 = \ker G_0$. One can simply set $Q_0 = I_m - G^+ G$. Then, for $i \geq 1$, we put

$$G_i := G_{i-1} + B_{i-1} Q_{i-1}, \tag{3.19}$$

$$\begin{aligned} &\text{choose a nontrivial subspace} \quad N_i \subseteq \ker G_i,\ \widehat{N_i} := N_0 + \cdots + N_{i-1}, \\ &\text{choose a projector } Q_i \ \text{ such that } \ker Q_i \supseteq X_i,\ \operatorname{im} Q_i = N_i, \\ &\text{with a complement } X_i \subseteq N_0 + \cdots + N_{i-1},\ X_i \oplus \widehat{N_i} = N_0 + \cdots + N_{i-1}, \\ &P_i := I_m - Q_i, \end{aligned} \tag{3.20}$$

$$B_i := B_{i-1} P_{i-1}. \tag{3.21}$$

Theorem 3.9 (See [10, 12]). *Denote $r_i := \operatorname{rank} G_i$. The following assertions are equivalent:*

(1) *The matrix pencil $\lambda G + B$ is regular with Kronecker index μ.*

(2) *Each matrix sequence* (3.19)–(3.21), *built in such a way that $N_i = \ker G_i$ is valid at each level, shows $r_0 \leq \cdots \leq r_{\mu-1} < r_\mu = m$.*

(3) *Each matrix sequence* (3.19)–(3.21) *shows $r_0 \leq \cdots \leq r_{\kappa-1} < r_\kappa = m$ with a certain number $\kappa \geq \mu$.*

Owing to the maximal choice $N_i = \ker G_i$, the numbers $r_0, \ldots, r_\mu$ in Theorem 3.9(2) characterize the detailed structure of the Weierstraß–Kronecker canonical form of the regular matrix pencil. In contrast, these numbers lose the structural meaning if only subspaces $N_i \subset \ker G_i$ are used instead of the nullspaces itself. From this point of view the choice $N_i = \ker G_i$ is most beneficial.

Return to the DAE (3.1). We introduce the basic matrix functions

$$\begin{aligned} A(x^1, x, t) &:= f_y(D(t)x^1 + D_t(t)x, x, t), \\ B(x^1, x, t) &:= f_x(D(t)x^1 + D_t(t)x, x, t), \quad x^1 \in \mathbf{R}^m, x \in \mathcal{D}_f, t \in \mathcal{I}_f, \end{aligned}$$

and form pointwise a sequence of continuous matrix functions analogously to (3.19)–(3.21), but drop the arguments of the matrix functions. We start by

$$G_0 := AD,\ B_0 := B,$$
$$\text{choose a projector function } Q_0 \text{ onto } N_0 := \ker D,\ P_0 := I_m - Q_0,\ \Pi_0 := P_0.$$

Then, for $i \geq 1$, as long as the expressions exist, we put

$$G_i := G_{i-1} + B_{i-1}Q_{i-1}, \tag{3.22}$$

$$\text{choose a nontrivial } \mathcal{C}\text{-subspace } N_i \subseteq \ker G_i,\ \widehat{N_i} := N_i \cap (N_0 + \cdots + N_{i-1}),$$
$$\text{choose a projector function } Q_i \text{ such that } \operatorname{im} Q_i = N_i,\ \ker Q_i \supseteq X_i, \tag{3.23}$$
$$\text{with a complement } X_i \subseteq N_0 + \cdots + N_{i-1},\quad X_i \oplus \widehat{N_i} = N_0 + \cdots + N_{i-1},$$
$$P_i := I - Q_i,\quad \Pi_i := \Pi_{i-1}P_i,$$
$$B_i := B_{i-1}P_{i-1} - G_iD^-(D\Pi_iD^-)'D\Pi_{i-1}. \tag{3.24}$$

For several examples of those matrix function sequences we refer the reader to [20].

The expression $(D\Pi_iD^-)'$ in formula (3.24) means the total derivative in jet variables; see [13, 19]. Formally, $D\Pi_iD^-$ may depend on $t,x,x^1,\dots,x^i$, and then $(D\Pi_iD^-)'$ depends on $t,x,x^1,\dots,x^i$ and x^{i+1}. The matrix function sequence (3.22)–(3.24) clearly generalizes the matrix sequence (3.19)–(3.21). The nonlinearity and time dependence is now encoded in the new extra term in (3.24).

To have at each level a continuous matrix function G_i, we suppose the projector functions $Q_0,\dots,Q_{i-1}$ to be *admissible* also in the sense that Π_{i-1} is supposed to be continuous and $D\Pi_{i-1}D^-$ to be continuously differentiable.

By definition, a $\mathcal{C}$-subspace in $\mathbf{R}^m$ is such that the orthoprojector function onto this subspace is continuous. Any $\mathcal{C}$-subspace has constant dimension, and hence the construction with $N_i = \ker G_i$ requires constant-rank matrix functions G_i. Mind at this point that, in the linear time-invariant case, the rank values r_i determine the structure of the Weierstraß–Kronecker canonical form if at each level $N_i = \ker G_i$. Here, in the context of general nonlinear DAEs we can benefit from the associated constant-rank conditions to describe regularity regions and to detect critical points.

Definition 3.10. *Let $\mathcal{G} \subseteq \mathcal{D}_f \times \mathcal{I}_f$ be open connected; further let $m = k$.*

(1) *The DAE* (3.1) *with proper leading term is said to be regular on $\mathcal{G}$ if there is a number $\mu \in \mathbf{N}$ such that on $\mathcal{G}$ a matrix function sequence* (3.22)–(3.24) *can be formed up to level μ, with $N_i = \ker G_i$, $r_i := \operatorname{rank} G_i$, $i = 0,\dots,\mu$, and $r_0 \leq \cdots \leq r_{\mu-1} < r_\mu = m$. $\mathcal{G}$ is then named a regularity region.*

The number μ is named tractability index, and the constant rank values $r_0,\dots,r_\mu$ are said to be characteristic values of the DAE on $\mathcal{G}$.

(2) *A point $(\bar{x},\bar{t}) \in \mathcal{D}_f \times \mathcal{I}_f$ is a regular point if there is a neighborhood being a regularity region, and a critical point otherwise.*

Roughly speaking, critical points are those points where the constant-rank conditions supporting the matrix functions, with N_i being the nullspace of G_i, at each level fail to be valid. Critical points can cause serious flow singularities as in Examples 3.6, 3.7, and

3.8. A critical point is *harmless* if it is at the same time a *quasi-regular* point such that, in a smooth setting, the flow is noncritical. Quasi regularity is associated with the sequence construction with $N_i \subseteq \ker G_i$ (see [13, 20]).

The following properties justify our regularity notion and allow a deeper comprehension at the same time. We underline once more that neither the existence of solutions nor any knowledge concerning the constraints is presupposed to determine regularity regions.

Properties of regularity regions:

* If $\mathcal{G}$ is a regularity region of the DAE (3.1), with characteristics $r_0 \leq \cdots \leq r_{\mu-1} < r_\mu = m$, then each open connected subset $\tilde{\mathcal{G}} \subset \mathcal{G}$ is a regularity region, too, and it has the same characteristics.

* A regularity region consists of regular points with uniform characteristics.

* The union of intersecting regularity regions is again a regularity region.

* Regularity regions, regular and critical points are unchanged if one turns from the original DAE (3.1) to its perturbed version

$$f((D(t)x(t))', x(t), t) = q(t). \tag{3.25}$$

* Regularity, in particular the characteristics $r_0 \leq \cdots \leq r_{\mu-1} < r_\mu = m$, is invariant with respect to coordinate changes, to refactorizations of the leading term, as well as to the special choice of the admissible projector functions Q_i; see [13].

Theorem 3.11 (Linearization theorem). *Let the DAE* (3.1) *have a proper leading term. Let $\mathcal{G} \subseteq \mathcal{D}_f \times \mathcal{I}_f$ be open connected. Then the following three assertions are equivalent:*

(1) *$\mathcal{G}$ is a regularity region of the DAE* (3.1).

(2) *Each linearization* (3.9) *of the DAE* (3.1) *along a sufficiently smooth function x_* with graph in $\mathcal{G}$ is a regular linear DAE.*

(3) *All linearizations* (3.9) *of the DAE* (3.1) *along sufficiently smooth functions x_* with graph in $\mathcal{G}$ are regular with uniform characteristics.*

Proof. The implications (1) $\to$ (3) $\to$ (2) are due to the construction. The implication (2) $\to$ (1) is proved in [13] by means of the so-called widely orthogonal projector functions. □

3.6 Optimality Condition

Consider the cost

$$J(x) = \int_{t_0}^{t_f} h(x(t),t)dt + g(D(t_f)x(t_f)) \tag{3.26}$$

to be minimized on functions $x \in \mathcal{C}^1_D(\mathcal{I}, \mathbf{R}^m)$, $\mathcal{I} = [t_0, t_f]$, subject to the constraints

$$f((D(t)x(t))', x(t), t) = 0,\ t \in \mathcal{I}, \quad D(t_0)x(t_0) = z_0 \in \mathbf{R}^n. \tag{3.27}$$

Let the DAE (3.27) have a full-rank proper leading term ($r = n$; see (3.8)) and let it satisfy the basic assumptions in Section 3.2. The DAE (3.27) comprises $k \le m$ equations, usually $k < m$. Additionally, let the real functions $h(x,t)$ and $g(\eta)$ depend continuously differentiable on their arguments. Later on, in Theorem 3.14 we suppose also continuous partial derivatives h_{xx} and f_{xx}.

Theorem 3.12 (Necessary optimality condition). *Let $x_* \in \mathcal{C}^1_D(\mathcal{I},\mathbf{R}^m)$ be a local solution of the optimization problem* (3.26), (3.27). *Let, for each arbitrary $z \in \operatorname{im} D(t_0)$, $q \in \mathcal{C}(\mathcal{I},\mathbf{R}^k)$, the linear IVP*

$$A_*(t)(D(t)x(t))' + B_*(t)x(t) = q(t),\ t \in \mathcal{I}, \quad D(t_0)x(t_0) = z \tag{3.28}$$

possess a solution in $\mathcal{C}^1_D(\mathcal{I},\mathbf{R}^m)$. Then the terminal value problem

$$-D(t)^T(A_*^T(t)\lambda(t))' + B_*(t)^T\lambda(t) = h_x(x_*(t),t)^T, \quad t \in \mathcal{I}, \tag{3.29}$$

$$D(t_f)^T A_*(t_f)^T\lambda(t_f) = (g_\eta(D(t_f)x_*(t_f))D(t_f))^T \tag{3.30}$$

has a solution $\lambda_* \in \mathcal{C}^1_{A_*^T}(\mathcal{I},\mathbf{R}^k)$.

Proof. For the case of quasi-linear DAEs with $f(y,x,t) = A(x,t)y + b(x,t)$ the assertion is proved in [1, pages 121–139] by applying the famous Lyusternik theorem [14], providing a representation of functionals on $\mathcal{C}^1_D(\mathcal{I},\mathbf{R}^m)$ and then a representation of the Lagrange multiplier. The same arguments apply also in the slightly more general case discussed now. □

The required solvability of the linear IVP (3.28) is guaranteed by Proposition 3.5(1). Actually, this means the surjectivity of the linear operator $\mathfrak{L}$ defined in (3.17). Thereby, the full row-rank condition (3.12) plays its role. The operator $\mathfrak{L}$ is surjective, exactly if condition (3.12) is valid. In turn, the surjectivity (closed-range property) of the linear operator $\mathfrak{L}$ plays an essential part in the Lyusternik theorem.

Condition (3.12) is necessary for the existence of a solution of the terminal value problem (3.29), (3.30), as it is demonstrated in [1, pages 50–52].

If this full-rank condition is not given in a problem, then it might be a good idea to reformulate or reduce the problem so that the reduced DAE meets the condition. Another way consists in exploiting given structural peculiarities with the aim to obtain surjectivity of the operator $\mathfrak{L}$ in specially adjusted function spaces, for instance, in case of controlled Hessenberg size-2 DAEs; cf. [3, 8, 9]. Note that different function spaces may lead to different representations of the Lagrange multiplier and hence yield another terminal value problem than (3.29), (3.30).

Corollary 3.13. *Let the DAE in* (3.27) *be underdetermined index-one tractable such that condition* (3.11) *is valid. Then, if $x_* \in \mathcal{C}^1_D(\mathcal{I},\mathbf{R}^m)$ is a local solution of the optimization problem* (3.26), (3.27), *the terminal value problem* (3.29), (3.30) *is solvable on $\mathcal{C}^1_{A_*^T}(\mathcal{I},\mathbf{R}^k)$.*

Proof. This is a direct consequence of Theorem 3.12. □

Indirect optimization methods rely on the boundary value problem (BVP) for the composed so-called *optimality DAE*

$$f((D(t)x(t))',x(t),t) = 0, \tag{3.31}$$

$$\begin{aligned}-D(t)^T(f_y((D(t)x(t))',x(t),t)^T\lambda(t))'\\ + f_x((D(t)x(t))',x(t),t)^T\lambda(t) = h_x(x(t),t)^T,\end{aligned} \tag{3.32}$$

completed by the boundary conditions from (3.27) and (3.30). Owing to Theorem 3.12 this BVP is solvable. By introducing the new function $y = (Dx)'$ and collecting the components λ, x, y in $\tilde{x}$, the DAE (3.31), (3.32) can be put into the more prevalent form

$$\tilde{f}((\tilde{d}(\tilde{x}(t),t))',\tilde{x}(t),t) = 0,$$

with properly involved derivative and nonlinear derivative term. Those equations are investigated in [13]. Here we restrict our interest to the easier quasi-linear case

$$f(y,x,t) = A(t)y + b(x,t), \tag{3.33}$$

which naturally comprises the semi-explicit systems (3.4), (3.5). For (3.33), the optimality DAE simplifies to

$$A(t)(D(t)x(t))' + b(x(t),t) = 0, \tag{3.34}$$

$$-D(t)^T(A^T(t)\lambda(t))' + b_x(x(t),t)^T\lambda(t) = h_x(x(t),t)^T. \tag{3.35}$$

The optimality DAE combines $k+m$ equations for the same number of unknown functions. In view of a reliable practical treatment, when applying an indirect optimization method, it would be a great advantage to know whether the DAE is regular with index one. For this aim we consider the linearization of the DAE (3.34), (3.35) along (λ_*, x_*), namely

$$\begin{bmatrix} A(t) & 0 \\ 0 & D(t)^T \end{bmatrix} \left(\begin{bmatrix} 0 & D(t) \\ -A(t)^T & 0 \end{bmatrix} \begin{bmatrix} \lambda(t) \\ x(t) \end{bmatrix} \right)' + \begin{bmatrix} 0 & B_*(t) \\ B_*(t)^T & -H_*(t) \end{bmatrix} \begin{bmatrix} \lambda(t) \\ x(t) \end{bmatrix} = 0, \tag{3.36}$$

with the continuous matrix functions

$$H_*(t) := h_{xx}(x_*(t),t) - (b_x^T(x,t)\lambda_*(t))_x(x_*(t),t), \quad B_*(t) := b_x(x_*(t),t). \tag{3.37}$$

Theorem 3.14 (Properties of the optimality DAE). *Let the DAE in* (3.27) *have the special form given by* (3.33). *Let the functions b and h have the additional second continuous partial derivatives b_{xx}, h_{xx}. Let $x_* \in \mathcal{C}_D^1(\mathcal{I}, \mathbf{R}^m)$ be a local solution of the optimization problem* (3.26), (3.27). *Put $Q_0(t) = I_m - D(t)^+D(t)$, $W_0(t) = I_k - A(t)A(t)^+$, $t \in \mathcal{I}$. Let the rank condition* (3.12) *be satisfied, that is,*

$$\operatorname{rank}[A(t)D(t) + B_*(t)Q_0(t)] = k, \quad t \in \mathcal{I}, \tag{3.38}$$

and let λ_* be the solution of the terminal value problem (3.29), (3.30).

(1) *Then the optimality DAE* (3.34), (3.35) *is regular with index one in a neighborhood of the graph of (λ_*, x_*), exactly if*

$$\begin{aligned}&(A(t)D(t) + W_0(t)B_*(t)Q_0(t))z = 0,\\ &\qquad H_*(t)Q_0(t)z \in \ker(A(t)D(t) + W_0(t)B_*(t)Q_0(t))^{\perp}\\ &\qquad\qquad \textit{imply} \quad z = 0 \quad \forall t \in \mathcal{I}.\end{aligned} \tag{3.39}$$

(2) *If condition* (3.39) *is given, then the linearized DAE* (3.36) *is self-adjoint and its inherent regular ODE has Hamiltonian structure such that*

$$\Theta' = \begin{bmatrix} 0 & -I_n \\ I_n & 0 \end{bmatrix} \mathcal{E}\,\Theta, \quad \Theta := \begin{bmatrix} Dx \\ -A^T\lambda \end{bmatrix}, \tag{3.40}$$

with a symmetric continuous matrix function $\mathcal{E}$ *of size* $2n \times 2n$.

(3) *If* $Q_0(t)H_*(t)Q_0(t)$ *is semidefinite for all* $t \in \mathcal{I}$, *then condition* (3.39) *simplifies to the full-rank condition*

$$\operatorname{rank} \begin{bmatrix} A(t)D(t) + W_0(t)B_*(t)Q_0(t) \\ Q_0(t)H_*(t)Q_0(t) \end{bmatrix} = m, \quad t \in \mathcal{I}. \tag{3.41}$$

Proof. The assertions (1) and (3) are verified in [20, pages 19–20]. (2) Owing to [2, Theorem 4.3], the linearized DAE (3.36) is self-adjoint. As a self-adjoint index-one DAE it has Hamiltonian structure due to [2, Theorem 4.5]. □

Notice that the Hamiltonian structure can get lost if the leading term is properly stated but not full-rank proper (cf. Definition 3.1), as is demonstrated in [2, Example 4.7].

Example 3.15. [1, pages 144–146] Minimize the cost

$$J(x) = \frac{1}{2}\int_0^{t_f} (x_3(t)^2 + (x_4(t) - R^2)^2)\,dt$$

subject to the constraint

$$\begin{aligned} x_1'(t) + x_2(t) &= 0, \\ x_2'(t) - x_1(t) - x_3(t) &= 0, \\ -x_1(t)^2 - x_2(t)^2 + x_4(t) &= 0, \quad x_1(0) = r,\ x_2(0) = 0, \end{aligned}$$

with constants $r > 0$, $R > 0$. If $x_3(t)$ vanishes identically, the remaining IVP has a unique solution. Then the point $(x_1(t), x_2(t))$ orbits the origin with radius r and $x_4(t) = r$. By optimizing in view of the cost, the point $(x_1(t), x_2(t))$ becomes driven to the circle of radius R, with low cost of $x_3(t)$. The resulting optimality DAE is everywhere regular with index one; see [20, Example 6.4]. ■

Our next example underlines that the only property that matters for the extremal condition is the surjectivity of the operator $\mathfrak{L}$. However, *to obtain a regular index-one optimality DAE* (3.34), (3.35) *the cost must be somehow consistent with the DAE* describing the constraint.

Example 3.16. Minimize the cost

$$J(x) = \frac{1}{2}\int_0^{2\pi} ((x_1(t) - \sin t)^2 + (x_2(t) - \cos t)^2 + \gamma x_3(t)^2 + x_4(t)^2)\,dt$$

subject to the constraint

$$\begin{aligned} x_1'(t) - x_2(t) + x_3(t) &= 0, \\ x_2'(t) + x_1(t) &= 0, \\ x_1(t)^3 + \alpha(x_1(t))x_3(t) - (\sin t)^3 - x_4(t) &= 0, \quad x_1(0) = 0,\ x_2(0) = 1, \end{aligned}$$

with the real function α given in Example 3.8 and a constant $\gamma \geq 0$. Considering x_4 as a control and letting $x_4 = 0$, the DAE from Example 3.8 reappears.

The unique optimal solution is

$$x_{*1}(t) = \sin t,\ x_{*2}(t) = \cos t,\ x_{*3}(t) = 0,\ x_{*4}(t) = 0.$$

We have $m = 4$, $k = 3$, $n = 2$, and

$$f(y,x,t) = \begin{bmatrix} 1 & 0 \\ 0 & 1 \\ 0 & 0 \end{bmatrix} y + \begin{bmatrix} -x_2 + x_3 \\ x_1 \\ x_1^3 + \alpha(x_1)x_3 - (\sin t)^3 - x_4 \end{bmatrix}, \quad D = \begin{bmatrix} 1 & 0 & 0 & 0 \\ 0 & 1 & 0 & 0 \end{bmatrix}$$

such that the matrix function

$$f_y D + f_x(I - D^+ D) = \begin{bmatrix} 1 & 0 & 1 & 0 \\ 0 & 1 & 0 & 0 \\ 0 & 0 & \alpha(x_1) & -1 \end{bmatrix}$$

results, which has full row rank independently of the behavior of $\alpha(x_1)$. The optimality DAE reads as

$$\begin{aligned} x_1'(t) - x_2(t) + x_3(t) &= 0, \\ x_2'(t) + x_1(t) &= 0, \\ x_1(t)^3 + \alpha(x_1(t))x_3(t) - (\sin t)^3 - x_4(t) &= 0, \\ -\lambda_1'(t) + \lambda_2(t) + (3x_1(t)^2 + \alpha'(x_1(t))x_3(t))\lambda_3(t) &= x_1(t) - \sin t, \\ -\lambda_2'(t) - \lambda_1(t) &= x_2(t) - \cos t, \\ \lambda_1(t) + \alpha(x_1(t))\lambda_3(t) &= \gamma x_3(t), \\ -\lambda_3(t) &= x_4(t). \end{aligned}$$

It holds that

$$AD + W_0 B_* Q_0 = \begin{bmatrix} 1 & 0 & 0 & 0 \\ 0 & 1 & 0 & 0 \\ 0 & 0 & \alpha(x_1) & -1 \end{bmatrix},$$

$$H_*(t) = \begin{bmatrix} 1 - 6x_{*1}(t)\lambda_{*3}(t) & 0 & -\alpha'(x_{*1}(t))\lambda_{*3}(t) & 0 \\ 0 & 1 & 0 & 0 \\ -\alpha'(x_{*1}(t))\lambda_{*3}(t) & 0 & \gamma & 0 \\ 0 & 0 & 0 & 1 \end{bmatrix}, \quad Q_0 H_*(t) Q_0 = \begin{bmatrix} 0 & 0 & 0 & 0 \\ 0 & 0 & 0 & 0 \\ 0 & 0 & \gamma & 0 \\ 0 & 0 & 0 & 1 \end{bmatrix}.$$

Condition (3.41) requires $\gamma + (\alpha(x_1))^2 \neq 0$. Therefore, the optimality DAE is regular with index one in case of $\gamma > 0$.

If $\gamma = 0$, then only the set

$$\mathcal{G}_+ = \{(z,t) \in \mathbf{R}^7 \times (t_0,t_f) : z_1 > 0\}$$

is a regularity region with characteristic $\mu = 1$. There are two further regions with $\mu = 3$. Unfortunately, the optimal solution does not remain in the index-one region such that one has to expect numerical difficulties caused by the discontinuous flow. ∎

Altogether, when intending to apply an indirect optimization method, it seems to be a good idea to make use of the modeling flexibility to reach an *optimality DAE which is regular with index one* or, at least, to reach the situation that the expected solution stands in a regularity region with characteristic $\mu = 1$.

3.7 Specification for Controlled DAEs

In the present section we specify results of the previous section for the important case of constraints described by controlled DAEs. Here the DAE and the cost depend on a pair of functions, the *state* $x \in \mathcal{C}^1_D(\mathcal{I},\mathbf{R}^m)$ and the *control* $u \in \mathcal{C}(\mathcal{I},\mathbf{R}^l)$. Now the DAE comprises m equations so that, for each fixed control, a square m-dimensional DAE results. Consider the cost

$$J(x,u) = \int_{t_0}^{t_f} h(x(t),u(t),t)dt + g(D(t_f)x(t_f)) \tag{3.42}$$

to be minimized on pairs $(x,u) \in \mathcal{C}^1_D(\mathcal{I},\mathbf{R}^m) \times \mathcal{C}(\mathcal{I},\mathbf{R}^l)$, subject to the constraints

$$f((D(t)x(t))',x(t),u(t),t) = 0,\ t \in \mathcal{I}, \quad D(t_0)x(t_0) = z_0 \in \mathbf{R}^n, \tag{3.43}$$

with $f(y,x,u,t) \in \mathbf{R}^m$, $y \in \mathbf{R}^n$, $x \in \mathbf{R}^m$, $u \in \mathbf{R}^l$, $t \in \mathcal{I} = [t_0,t_f]$, $D(t) \in L(\mathbf{R}^m,\mathbf{R}^n)$, and $\operatorname{rank} D(t) = r = n$. Assume analogous smoothness as in Section 3.6. Moreover, as in Section 3.6 the DAE in (3.43) is supposed to have a full-rank proper leading term (see (3.8)).

Denote

$$\begin{aligned} A_*(t) &= f_y((Dx_*)'(t),x_*(t),u_*(t),t), \\ B_*(t) &= f_x((Dx_*)'(t),x_*(t),u_*(t),t), \\ C_*(t) &= f_u((Dx_*)'(t),x_*(t),u_*(t),t), \quad t \in \mathcal{I}, \end{aligned}$$

such that now the linearization along (x_*,u_*) reads as

$$A_*(t)(D(t)x(t))' + B_*(t)x(t) + C_*(t)u(t) = q(t), \quad t \in \mathcal{I}. \tag{3.44}$$

Our present problem constitutes a special case with partitioned variables of the general optimization problem considered in the previous section. The following assertion is a straightforward consequence of Theorem 3.12.

Theorem 3.17 (Necessary optimality condition). *Let the optimization problem* (3.42), (3.43) *have the local solution* $(x_*,u_*) \in \mathcal{C}^1_D(\mathcal{I},\mathbf{R}^m) \times \mathcal{C}(\mathcal{I},\mathbf{R}^l)$*. Then, if the full-rank condition*

$$\operatorname{rank}\,[A_*(t)D(t) + B_*(t)(I - D(t)^+D(t)),\ C_*(t)] = m, \quad t \in \mathcal{I}, \tag{3.45}$$

is valid, the terminal value problem

$$-D(t)^T(A_*^T(t)\lambda(t))' + B_*(t)^T\lambda(t) = h_x(x_*(t),u_*(t),t)^T, \tag{3.46}$$

$$C_*(t)^T\lambda(t) = h_u(x_*(t),u_*(t),t)^T, \quad t \in \mathcal{I}, \tag{3.47}$$

$$D(t_f)^T A_*(t_f)^T\lambda(t_f) = D(t_f)^T(g_\eta(D(t_f)x_*(t_f)))^T \tag{3.48}$$

has a solution $\lambda_* \in \mathcal{C}^1_{A_*}(\mathcal{I},\mathbf{R}^m)$.

If the controlled DAE is regular with index ≤ 1, then $A_*D + B_*(I - D^+D)$ is necessarily nonsingular such that condition (3.45) follows. In this connection it does not matter what C_* looks like. However, in all other cases, condition (3.45) entails structural requirements concerning C_*. On the other hand, no regularity and index conditions for the given controlled DAE are required. For instance, in Example 3.15, one might consider x_3 or x_4 to serve as the control. In the first case, the resulting controlled DAE is regular with index one; in the second case it is regular with index two on the two regularity regions given by $x_2 > 0$ and $x_2 < 0$. Both versions result in the same regular index-one optimality DAE.

At this place we stress once more that our criteria are clearly represented *algebraic conditions in terms of the original optimization problem.* In contrast, in [11] an analogous optimization problem with DAE constraint

$$\mathfrak{f}(x'(t),x(t),u(t),t) = 0, \quad t \in \mathcal{I},$$

is treated by transforming this equation first into the so-called reduced form

$$x_1'(t) - \mathcal{L}(x_1(t),x_2(t),u(t),t), \quad x_2(t) = \mathcal{R}(x_1(t),u(t),t), \tag{3.49}$$

and not until then formulating an extremal condition and the optimality DAE in terms of (3.49). This prehandling is based on demanding assumptions (e.g., [11, Hypothesis 1]) and it needs considerable effort. In turn, the reduced system (3.49) represents a special case of a semi-explicit controlled regular index-one DAE such that condition (3.45) is given. The optimality DAE for the optimization problem with constraint DAE (3.49) is then the corresponding special case of the DAE (3.51)–(3.53) below.

As a consequence of Theorem 3.17, the BVP composed from the IVP (3.43) and the terminal value problem (3.46)–(3.48) is solvable. Indirect optimization relies on this BVP. Then, for practical reasons, the question arises whether the associated *optimality DAE* is regular with index one. We give an answer for the clearly arranged quasi-linear case

$$f(y,x,u,t) = A(t)y + b(x,u,t), \tag{3.50}$$

so that the optimality DAE simplifies to

$$A(t)(D(t)x(t))' + b(x(t),u(t),t) = 0, \tag{3.51}$$

$$-D(t)^T(A(t)^T\lambda(t))' + b_x(x(t),u(t),t)^T\lambda(t) = h_x(x(t),u(t),t)^T, \tag{3.52}$$

$$b_u(x(t),u(t),t)^T\lambda(t) = h_u(x(t),u(t),t)^T. \tag{3.53}$$

The optimality DAE (3.51)–(3.53) has the linearization (the argument t is dropped)

$$\begin{bmatrix} A & 0 \\ 0 & D^T \\ 0 & 0 \end{bmatrix}\left(\begin{bmatrix} 0 & D & 0 \\ -A^T & 0 & 0 \end{bmatrix}\begin{bmatrix} \lambda \\ x \\ u \end{bmatrix}\right)' + \begin{bmatrix} 0 & B_* & C_* \\ B_*^T & -\mathcal{W}_* & -\mathcal{S}_* \\ C_*^T & -\mathcal{S}_*^T & -\mathcal{R}_* \end{bmatrix}\begin{bmatrix} \lambda \\ x \\ u \end{bmatrix} = 0, \tag{3.54}$$

with continuous matrix functions

$$\begin{aligned}
\mathcal{W}_*(t) &:= h_{xx}(x_*(t),t) - (b_x^T(x,t)\lambda_*(t))_x(x_*(t),t),\\
\mathcal{S}_*(t) &:= h_{xu}(x_*(t),t) - (b_x^T(x,t)\lambda_*(t))_u(x_*(t),t),\\
\mathcal{R}_*(t) &:= h_{uu}(x_*(t),t) - (b_u^T(x,t)\lambda_*(t))_u(x_*(t),t),\\
B_*(t) &:= b_x(x_*(t),u_*(t),t), \quad C_*(t) = b_u(x_*(t),u_*(t)t), \quad t \in \mathcal{I}.
\end{aligned}$$

Theorem 3.18 (Properties of the optimality DAE). *Let the DAE in* (3.43) *have the special form given by* (3.50)*. Let the functions b and h have the additional second continuous partial derivatives needed. Let* $(x_*,u_*) \in \mathcal{C}_D^1(\mathcal{I},\mathbf{R}^m) \times \mathcal{C}(\mathcal{I},\mathbf{R}^l)$ *be a local solution of the optimization problem* (3.42), (3.43)*. Denote* $Q_0(t) = I_m - D(t)^+D(t)$, $W_0(t) = I_m - A(t)A(t)^+$, $t \in \mathcal{I}$*. Let the full-rank condition* (3.45) *be satisfied and let* λ_* *denote the solution of the terminal value problem* (3.46)–(3.48)*:*

(1) *Then the optimality DAE* (3.51)–(3.53) *is regular with index one in a neighborhood of the graph of* (λ_*,x_*,u_*)*, exactly if*

$$[A(t)D(t) + W_0(t)B_*(t)Q_0(t),\ W_0(t)C_*(t)]z = 0,$$

$$\begin{bmatrix} \mathcal{W}_*(t)Q_0(t) & \mathcal{S}_*(t) \\ \mathcal{S}_*(t)^T Q_0(t) & \mathcal{R}_*(t) \end{bmatrix} z \in \ker[A(t)D(t) + W_0(t)B_*(t)Q_0(t),\ W_0(t)C_*(t)]^\perp$$

$$\textit{imply} \quad z = 0 \quad \forall t \in \mathcal{I}. \tag{3.55}$$

(2) *If condition* (3.55) *is given, then the linearized DAE* (3.54) *is self-adjoint and its inherent regular ODE has Hamiltonian structure such that*

$$\Theta' = \begin{bmatrix} 0 & -I_n \\ I_n & 0 \end{bmatrix} \mathcal{E}\,\Theta, \quad \Theta := \begin{bmatrix} Dx \\ -A^T\lambda \end{bmatrix}, \tag{3.56}$$

with a symmetric continuous matrix function $\mathcal{E}$ *of size* $2n \times 2n$.

(3) *If the matrix*

$$\begin{bmatrix} Q_0(t)W_*(t)Q_0(t) & Q_0(t)S_*(t) \\ \mathcal{S}_*(t)^T Q_0(t) & \mathcal{R}_*(t) \end{bmatrix}$$

is semidefinite for all $t \in \mathcal{I}$*, then condition* (3.55) *simplifies to the full-rank condition*

$$\operatorname{rank} \begin{bmatrix} A(t)D(t) + W_0(t)B_*(t)Q_0(t) & W_0(t)C_*(t) \\ Q_0(t)\mathcal{W}_*(t)Q_0(t) & Q_0(t)\mathcal{S}_*(t) \\ \mathcal{S}_*(t)^T Q_0(t) & \mathcal{R}_*(t) \end{bmatrix} = m + l, \quad t \in \mathcal{I}. \tag{3.57}$$

Proof. These assertions are special cases of the corresponding assertions in Theorem 3.14. □

We refer the reader to [1, 2, 5, 17] for further index relations and for the consistency with well-known facts in the context of linear-quadratic optimal control problems.

Bibliography

[1] A. BACKES, *Extremalbedingungen für Optimierungs-Probleme mit Algebro-Differentialgleichungen*, Logos Verlag, Berlin, 2006. Dissertation, Humboldt-University Berlin, October 2005/January 2006.

[2] K. BALLA, G. A. KURINA, AND R. MÄRZ, *Index criteria for differential algebraic equations arising from linear-quadratic optimal control problems*, J. Dyn. Control Syst., 12 (2006), pp. 289–311.

[3] R. CALLIES, *Some aspects of optimal control of nonlinear differential-algebraic equations*, in Differential-Algebraic Equations, S. L. Campbell, R. März, L. R.Petzold, and P. Rentrop, eds., Mathematisches Forschungsinstitut Oberwolfach, Report 18/2006, 2006, pp. 19–21.

[4] S. L. CAMPBELL, *Linearization of DAES along trajectories*, Z. Angew. Math. Phys., 46 (1995), pp. 70–84.

[5] S. L. CAMPBELL AND R: MÄRZ, *Direct transcription solution of high index optimal control problems and regular Euler Lagrange equations*, J. Comput. Appl. Math., 202 (2007), pp. 186–202.

[6] Y. CAO, S. LI, AND L. R. PETZOLD, *Adjoint sensitivity for differential-algebraic equations: Algorithms and software*, J. Comput. Appl. Math., 149 (2002), pp. 171–192.

[7] E. EICH-SOELLNER AND C. FÜHRER, *Numerical Methods in Multibody Dynamics*, B.G. Teubner, Stuttgart, Germany, 1998.

[8] M. GERDTS, *Local minimum principle for optimal control problems subject to index-two differential-algebraic equations*, J. Optim. Theory Appl., 130 (2006), pp. 443–462.

[9] M. GERDTS, *Representation of Lagrange multipliers for optimal control problems subject to index-two differential-algebraic equations*, J. Optim. Theory Appl., 130 (2006), pp. 231–251.

[10] E. GRIEPENTROG AND R. MÄRZ, *Basic properties of some differential-algebraic equations*, Z. Anal. Anwendungen, 8 (1989), pp. 25–41.

[11] P. KUNKEL AND V. MEHRMANN, *Optimal control for unstructured nonlinear differential-algebraic equations of arbitrary index*, Math. Control Signals Systems, 20 (2008), pp. 227–269.

[12] R. LAMOUR, R. MÄRZ, AND C. TISCHENDORF, *Projector Based Treatment of Linear Constant Coefficient DAEs*, Technical Report 2011-15, Humboldt University Berlin, Institute of Mathematics; available online from www.math.hu-berlin.de/publ/pre/2011/M-11-15.html, 2011.

[13] R. LAMOUR, R. MÄRZ, AND C. TISCHENDORF, *Projector Based DAE Analysis*, manuscript, 2011.

[14] L. A. LYUSTERNIK, *Ob uslovnykh ehkhstremumakh funktsionalov*, Mat. Sbornik, 41 (1934), pp. 390–401 (in Russian).

[15] R. MÄRZ, *On linear differential-algebraic equations and linearizations*, Appl. Numer. Math., 18 (1995), pp. 267–292.

[16] R. MÄRZ, *Criteria for the trivial solution of differential algebraic equations with small nonlinearities to be asymptotically stable*, J. Math. Anal. Appl., 225 (1998), pp. 587–607.

[17] R. MÄRZ, *Differential algebraic equations in optimal control problems*, in Proceedings of the International Conference "Control Problems and Applications (technology, industry, economics)," Minsk, 2005, pp. 22–31.

[18] R. MÄRZ, *Projector based DAE analysis*, in Differential-Algebraic Equations, S. L. Campbell, R. März, L. R. Petzold, and P. Rentrop, eds., Mathematisches Forschungsinstitut Oberwolfach, Report 18/2006, 2006, pp. 49–52.

[19] R. MÄRZ, *Regularity regions of differential algebraic equations*, in Numerical Analysis and Applied Mathematics, Vol. 2, T. E. Simos, G. Psihoyios, and Ch. Tsitouras, eds., AIP Conference Proceedings, Melville, NY, 2009, pp. 1029–1032.

[20] R. MÄRZ, *Notes on Linearization of Differential-Algebraic Equations and on Optimization with Differential-Algebraic Constraints*, Technical Report 2011-16, Humboldt University Berlin, Institute of Mathematics; available online from www.math.hu-berlin.de/publ/pre/2011/M-11-16.html, 2011.

[21] T. PETRY, *Realisierung des Newton-Kantorovich-Verfahrens für nichtlineare Algebro-Differentialgleichungen mittels Abramov-Transfer*, Logos Verlag, Berlin, 1998. Dissertation, Humboldt-University Berlin, June 1998.

[22] W. J. TERRELL, *Observability of nonlinear differential algebraic systems*, Circuits Systems Signal Process., 14 (1997), pp. 271–285.

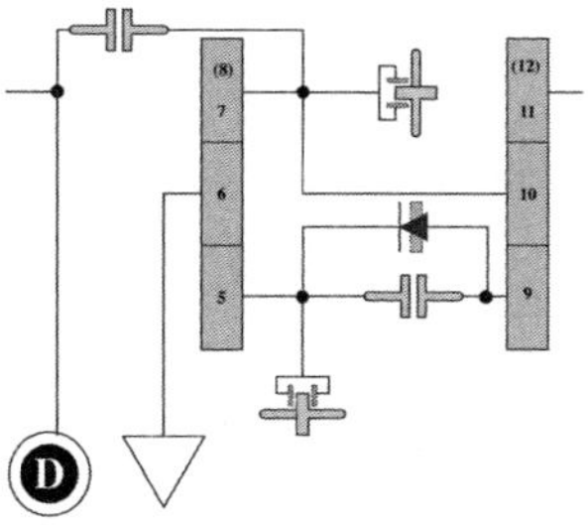

Chapter 4

Spectra and Leading Directions for Linear DAEs

Vu Hoang Linh and Volker Mehrmann

The stabilization of systems or processes that have become unstable is an important task in control and optimization. While for linear time-invariant descriptor systems instability can be checked by computing eigenvalues and eigenvectors, already for linear variable coefficient descriptor systems different spectral concepts are needed which have only recently been introduced. To characterize the asymptotic behavior and the growth rate of solutions, basic spectral notions such as Lyapunov and Bohl exponents and Sacker–Sell spectra are discussed. For DAEs in strangeness-free form, the results extend those for ODEs, but only under additional conditions. This has consequences concerning the boundedness of solutions of inhomogeneous equations. Also, linear subspaces of leading directions are characterized, which are associated with spectral intervals and which generalize eigenvectors and invariant subspaces as they are used in the linear time-invariant setting. The computation of these spectra then again leads to hard optimization problems for DAEs.

4.1 Introduction

DAEs are a very convenient modeling concept in many different application areas; see [5, 25, 28, 30, 43, 44] and the references therein. But many numerical difficulties arise due to the fact that the solution depends on derivatives of the data and that the dynamics is constrained to a manifold, which often is given only implicitly. In order to identify when such a system becomes unstable it is necessary to extend the spectral concepts such as Lyapunov, Bohl, and Sacker–Sell spectra that are well known for ODEs. The computation of these spectra leads to extremely difficult and expensive optimization problems for differential equations. The difficulties are even larger when it comes to the stabilization of DAE systems by control and optimization methods.

In this chapter we survey the spectral theory for linear DAEs with variable coefficients of the form

$$E(t)\dot{x} = A(t)x + f(t), \tag{4.1}$$

on the half-line $\mathbf{I} = [0,\infty)$, together with an initial condition

$$x(0) = x_0. \tag{4.2}$$

We assume that $E, A \in C(\mathbf{I}, \mathbf{R}^{n\times n})$, and $f \in C(\mathbf{I}, \mathbf{R}^n)$ are sufficiently smooth, using the notation $C(\mathbf{I}, \mathbf{R}^{n\times n})$ to denote the space of continuous functions from $\mathbf{I}$ to $\mathbf{R}^{n\times n}$. In the following we leave off the explicit dependence of the coefficients on the time t.

Linear systems of the form (4.1) arise directly in many applications and, via linearization [8], under some further regularity assumptions, they describe the local behavior in the neighborhood of a solution for general implicit nonlinear system of DAEs

$$F(t, x, \dot{x}) = 0, \quad t \in \mathbf{I}. \tag{4.3}$$

For linear systems with constant coefficients, the asymptotic behavior and the directions of growth or decay as well as oscillatory behavior can be characterized via the eigenvalues and eigenvectors of the matrix pencil $\lambda E - A$ and many good numerical methods are available; see [26, 46]. For systems with variable coefficients, however, different concepts are necessary such as the Lyapunov [41], Bohl [1, 4, 13], and Sacker–Sell spectra [45], which were designed for ODEs to analyze the qualitative behavior of solutions of differential equations as time tends to infinity. For a long time the numerical computation of these spectra in the variable coefficient ODE case was considered unfeasible, since this requires the solution of difficult optimization problems in unbounded domains, but recently considerable progress has been made; see [15, 19, 21] and the references therein.

The spectral theory and numerical methods for the computation of spectra in the case of DAEs, however, are still in their infancy. In [11, 12] results on Lyapunov exponents and Lyapunov regularity were studied, in [35] the concept of exponential dichotomy was used in the numerical solution of DAE BVPs, and in [23, 24] robustness results in the context of exponential stability and Bohl exponents were studied. All these papers use the tractability index approach of [27, 42] and consider linear systems of DAEs of tractability index at most one, only. Recently, in [31, 36, 37, 39, 40], the classical spectral theory and the numerical methods (such as *QR* and *SVD* methods) for the computation of Lyapunov, Bohl, and Sacker–Sell spectra were developed for general DAEs in so-called strangeness-free formulation. We systematically survey these results and emphasize the difficulties that arise for DAEs with some simple illustrative examples.

After a brief review of the DAE theory based on the strangeness-index approach [30], we present the basic properties of Lyapunov characteristic exponents in Section 4.3. Then, in Section 4.4, we introduce the notions for characterizing the uniform growth rate: Bohl exponents, exponential dichotomy, and Sacker–Sell spectrum. The connection between the spectra and the existence of bounded solutions for both homogeneous and inhomogeneous equations is also described. The analysis of leading directions and leading subspaces is given in Section 4.5.

4.2 A Review of DAE Theory

In this section we briefly recall some concepts from the theory of DAEs; see, e.g., [5, 30, 43]. We follow [30] in notation and style.

A function $x : \mathbf{I} \to \mathbf{R^n}$ is called a *solution* of (4.1) if $x \in C^1(\mathbf{I}, \mathbf{R^n})$ and x satisfies (4.1) pointwise. It is called a *solution of the IVP* (4.1)–(4.2) if x is a solution of (4.1) and satisfies (4.2). An initial condition (4.2) is called *consistent* if the corresponding IVP has at least one solution.

In this chaper, we restrict ourselves to *regular* DAEs; i.e., we require that (4.1) (or (4.3) locally) have a unique solution for sufficiently smooth E, A, f (F) and appropriately chosen (consistent) initial conditions.

4.2.1 Reduction to Strangeness-Free DAEs

The concept of strangeness index is based on *derivative arrays* associated with (4.1) as first introduced in [7]. Consider the inflated system

$$M_\ell \dot{z}_\ell = N_\ell z_\ell + g_\ell, \tag{4.4}$$

where

$$\begin{aligned}
(M_\ell)_{i,j} &= \tbinom{i}{j} E^{(i-j)} - \tbinom{i}{j+1} A^{(i-j-1)}, \quad i,j = 0,\ldots,\ell,\\
(N_\ell)_{i,j} &= \begin{cases} A^{(i)} & \text{for } i = 0,\ldots,\ell,\ j = 0,\\ 0 & \text{otherwise,} \end{cases}\\
(z_\ell)_j &= x^{(j)}, \quad j = 0,\ldots,\ell,\\
(g_\ell)_i &= f^{(i)}, \quad i = 0,\ldots,\ell,
\end{aligned}$$

using the convention that $\binom{i}{j} = 0$ for $i < 0$, $j < 0$, or $j > i$. To guarantee existence and uniqueness of solutions, the following hypothesis is used; see [29, 30].

Hypothesis 1. *There exist integers μ, a, and d such that the inflated pair (M_μ, N_μ) in* (4.4) *associated with the given pair of matrix functions (E, A) has the following properties:*

(1) *For all $t \in \mathbf{I}$ we have* $\operatorname{rank} M_\mu(t) = (\mu+1)n - a$ *such that there exists a smooth matrix function Z_2 of size $(\mu+1)n \times a$ and pointwise maximal rank satisfying $Z_2^T M_\mu = 0$.*

(2) *For all $t \in \mathbf{I}$ we have* $\operatorname{rank} \hat{A}_2(t) = a$, *where $\hat{A}_2 = Z_2^T N_\mu [I_n\, 0 \cdots 0]^T$ such that there exists a smooth matrix function T_2 of size $n \times d$, $d = n - a$, and pointwise maximal rank satisfying $\hat{A}_2 T_2 = 0$.*

(3) *For all $t \in \mathbf{I}$ we have* $\operatorname{rank} E(t)T_2(t) = d$ *such that there exists a smooth matrix function Z_1 of size $n \times d$ and pointwise maximal rank satisfying* $\operatorname{rank} \hat{E}_1 T_2 = d$ *with $\hat{E}_1 = Z_1^T E$.*

The smallest possible μ for which Hypothesis 1 holds is called the *strangeness index* of (4.1). Systems with vanishing strangeness index are called *strangeness-free*. It has been shown in [29, 30] that the strangeness index is closely related to the *differentiation index*, see [5], but allowing over- and underdetermined systems, and that under some constant-rank conditions, every uniquely solvable (regular) linear DAE of the form (4.1) with sufficiently smooth E, A satisfies Hypothesis 1 and that there exists a (pointwise) numerically computable *reduced system*

$$\begin{aligned}
&\text{(a)} \quad E_1 \dot{x} = A_1 x + f_1,\\
&\text{(b)} \quad 0 = A_2 x + f_2,
\end{aligned} \tag{4.5}$$

with $E_1 = Z_1^T E$, $A_1 = Z_1^T A \in C(\mathbf{I}, \mathbf{R}^{d \times n})$, $A_2 = Z_2^T N_\mu [\, I_n \; 0 \; \cdots \; 0]^T \in C(\mathbf{I}, \mathbf{R}^{a \times n})$, $f_1 = Z_1^T f \in C(\mathbf{I}, \mathbf{R}^d)$, and $f_2 = Z_2^T g_\mu \in C(\mathbf{I}, \mathbf{R}^a)$. This implies that the matrix function

$$\bar{E} := \begin{bmatrix} E_1 \\ A_2 \end{bmatrix} \tag{4.6}$$

is invertible for all $t \in \mathbf{I}$ and therefore also that E_1 and A_2 are of full row rank.

In the case of a uniquely solvable (regular) linear DAE, system (4.5) is just a reformulation of system (4.1) of differentiation index less than or equal to one that displays all the algebraic constraints explicitly in (4.5 (b)). However, unlike other publications on this topic we do not have to assume that the system is of differentiation index at most one, but that the system can be reformulated into this form. In the numerical computation of Lyapunov and Sacker–Sell spectra as in [36, 39, 40] this can be done locally at every time step.

From (4.5 (b)) it follows that an initial vector $x_0 \in \mathbf{R}^n$ is *consistent* if and only if $A_2(0)x_0 + f_2(0) = 0$. It has also been shown in [29, 30] that for this system implicit Runge–Kutta and BDF methods behave as for ODEs and that this approach can also be extended to over- and underdetermined systems as well as locally to general nonlinear systems.

4.2.2 Essential Underlying Implicit ODEs

In the following, we assume that the homogeneous linear DAE in consideration

$$E\dot{x} = Ax, \quad t \in \mathbf{I}, \tag{4.7}$$

is regular and strangeness-free, i.e., of differentiation index less than or equal to one, and the coefficients are as in (4.5). But it is often convenient to transform (4.7) into another form which is easier to handle but its solutions have the same asymptotic behavior as those of (4.7). Suppose that $P \in C(\mathbf{I}, \mathbf{R}^{n \times n})$ and $Q \in C^1(\mathbf{I}, \mathbf{R}^{n \times n})$ are nonsingular matrix functions such that Q and Q^{-1} are bounded. Then the transformed DAE system

$$\tilde{E}(t)\dot{\tilde{x}} = \tilde{A}(t)\tilde{x}, \tag{4.8}$$

with $\tilde{E} = PEQ$, $\tilde{A} = PAQ - PE\dot{Q}$, and $x = Q\tilde{x}$, is called the *globally kinematically equivalent* to (4.7) and the transformation is called a *global kinematic equivalence transformation.* If $P \in C^1(\mathbf{I}, \mathbf{R}^{n \times n})$ and, furthermore, also P and P^{-1} are bounded, then we call this a *strong global kinematic equivalence transformation.*

The following key lemma is a modification of [36, Lemma 7].

Lemma 4.1. *Consider a regular strangeness-free DAE system of the form* (4.7) *with continuous coefficients E, A. Let $U \in C^1(\mathbf{I}, \mathbf{R}^{n \times d})$ be an arbitrary orthonormal basis of the solution space of* (4.7). *Then there exists a matrix function $V \in C(\mathbf{I}, \mathbf{R}^{n \times d})$ with pointwise orthonormal columns such that by the change of variables $x = Uz$ and multiplication of both sides of* (4.7) *from the left by V^T, one obtains the system*

$$\mathcal{E}\dot{z} = \mathcal{A}z, \tag{4.9}$$

where $\mathcal{E} := V^T E U$, $\mathcal{A} := V^T A U - V^T E \dot{U}$, and $\mathcal{E}$ is upper triangular.

Proof. The proof is given in [37], [39] and is similar to that of [40]. □

System (4.9) is called *essential underlying implicit ODE system* (EUODE) of (4.7). It can be made explicit by multiplying by $\mathcal{E}^{-1}$ from the left; see also [2] for constructing EUODEs of so-called *properly stated DAEs*.

Note that for a fixed U, the matrix function V that leads to the EUODE is not unique. In fact, any V for which $V^T EU$ is invertible yields an implicit EUODE. However, obviously $\mathcal{E}^{-1}\mathcal{A}$ is unique; i.e., with a given basis, the explicit EUODE provided by Lemma 4.1 is unique.

We also often use the special case of regular semi-implicit strangeness-free DAEs with coefficients of the form

$$E := \begin{bmatrix} E_{11} & 0 \\ 0 & 0 \end{bmatrix}, \quad A := \begin{bmatrix} A_{11} & A_{12} \\ A_{21} & A_{22} \end{bmatrix}, \tag{4.10}$$

with E_{11} pointwise nonsingular. This DAE is strangeness-free, i.e., has differentiation index less than or equal to one, if and only if A_{22} is pointwise invertible, and by inserting $x_2 = -A_{22}^{-1}A_{21}x_1$ into the first equation, we obtain an implicit ODE

$$E_{11}\dot{x}_1 = \tilde{A}_{11}x_1, \tag{4.11}$$

where $\tilde{A}_{11} = A_{11} - A_{12}A_{22}^{-1}A_{21}$. It is easy to show that if $A_{22}^{-1}A_{21}$ is bounded, then (4.11) and (4.9) are globally kinematically equivalent. Furthermore, if E_1 is sufficiently smooth, then (4.7) can always be transformed into the form (4.10) by an appropriate kinematical equivalence transformation [36].

4.3 Lyapunov Spectral Theory for DAEs

In this section we review results on the qualitative behavior of solutions of DAEs from [36, 37, 40]. For a nonvanishing function $f : [0,\infty) \longrightarrow \mathbf{R}^n$, the quantities $\chi^u(f) = \limsup_{t\to\infty} \frac{1}{t}\ln\|f(t)\|$, $\chi^\ell(f) = \liminf_{t\to\infty}\frac{1}{t}\ln\|f(t)\|$ are called *upper and lower Lyapunov exponents of* f, respectively. It is well known, see e.g., [1], how the Lyapunov exponents characterize the growth of a function. Let $f : [0,\infty) \longrightarrow \mathbf{R}$ be a nonvanishing function. Then $\chi^u(f) = \alpha \neq \pm\infty$ if and only if for any $\varepsilon > 0$ the following two conditions hold simultaneously:

$$\lim_{t\to\infty} \frac{|f(t)|}{\exp(\alpha+\varepsilon)t} = 0, \quad \limsup_{t\to\infty} \frac{|f(t)|}{\exp(\alpha-\varepsilon)t} = \infty.$$

Analogously, $\chi^\ell(f) = \beta \neq \pm\infty$ if and only if for any $\varepsilon > 0$ the following two conditions hold simultaneously:

$$\liminf_{t\to\infty} \frac{|f(t)|}{\exp(\beta+\varepsilon)t} = 0, \quad \lim_{t\to\infty} \frac{|f(t)|}{\exp(\beta-\varepsilon)t} = \infty.$$

For a constant $c \neq 0$ and nonvanishing functions $f, f_1, \dots, f_n$, the Lyapunov exponents satisfy

$$\chi^u(cf) = \chi^u(f), \quad \chi^\ell(cf) = \chi^\ell(f),$$

and if $f_1, \dots, f_n$ are nonvanishing scalar functions, then

$$\chi^u\left(\sum_{i=1}^n f_i\right) \le \max_{i=1,\dots,n} \chi^u(f_i), \quad \chi^\ell\left(\sum_{i=1}^n f_i\right) \le \max_{i=1,\dots,n} \chi^\ell(f_i),$$

where equality holds if the maximal upper/lower Lyapunov exponent is attained by only one function.

4.3.1 Lyapunov Exponents and Spectral Intervals

To study the qualitative behavior of DAE solutions we determine the Lyapunov exponents of the columns of fundamental solution matrices.

A matrix function $X \in C^1(\mathbf{I}, \mathbf{R}^{n\times k})$, with $d \le k \le n$, is called *a fundamental solution matrix of the strangeness-free DAE* (4.7) if each of its columns is a solution to (4.7) and $\operatorname{rank} X(t) = d$ for all $t \ge 0$. A fundamental solution matrix is said to be *minimal* if $k = d$.

One may construct a minimal fundamental matrix solution by solving IVPs for (4.7) with d linearly independent, consistent initial vectors. Given an EUODE (4.9), any minimal fundamental solution matrix X of (4.7) has the form $X = UZ$, where Z is the corresponding fundamental solution matrix of (4.9).

Definition 4.2 (See [36]). *For a given minimal fundamental solution matrix X of a regular strangeness-free DAE system of the form* (4.7), *and for* $1 \le i \le d$, *we introduce*

$$\lambda_i^u = \limsup_{t\to\infty} \frac{1}{t} \ln \|X(t)e_i\|, \quad \lambda_i^\ell = \liminf_{t\to\infty} \frac{1}{t} \ln \|X(t)e_i\|,$$

where e_i denotes the ith unit vector and $\|\cdot\|$ denotes the Euclidean norm. The columns of a minimal fundamental solution matrix form a normal basis *if $\Sigma_{i=1}^d \lambda_i^u$ is minimal. The λ_i^u, $i = 1, 2, \ldots, d$, belonging to a normal basis are called* (upper) Lyapunov exponents. *If $\Sigma_{i=1}^d \lambda_i^\ell$ is minimal as well, then the intervals $[\lambda_i^\ell, \lambda_i^u]$, $i = 1, 2, \ldots, d$, are called* Lyapunov spectral intervals. *The union of the Lyapunov spectral intervals is called the* Lyapunov spectrum *of* (4.7) *and denoted by* Σ_L.

Example 4.3. If (4.7) is time-invariant, i.e., E, A are constant matrices, then the Lyapunov spectrum of (4.7) is the set of the real parts of generalized eigenvalues, i.e., $\Sigma_L = \{\operatorname{Re}\lambda, \det(\lambda E - A) = 0\}$. ∎

From the properties of the Lyapunov exponents, it is easy to establish the following observations:

- If the largest upper Lyapunov exponent is negative, then (4.7) is asymptotically stable.
- If the largest upper Lyapunov exponent is positive, then (4.7) is unstable.
- If at least one upper Lyapunov exponent is negative, then there exists a bounded solution of (4.7), which tends to 0 exponentially.
- If the smallest lower Lyapunov exponent is positive, then all solutions of (4.7) are unbounded and exponentially increasing.
- If the largest upper Lyapunov exponent is zero, then nothing can be concluded about the stability, since the system (4.7) may have unbounded solutions.

Similarly as in the case of ODEs, a normal basis for (4.7) exists and it can be constructed from any (minimal) fundamental matrix solution.

Proposition 4.4 (See [37, 39]). *For any given minimal fundamental matrix X of* (4.7), *for which the Lyapunov exponents of the columns are ordered decreasingly, there exists a constant, nonsingular, and upper triangular matrix $C \in \mathbf{R}^{d\times d}$ such that the columns of XC form a normal basis for* (4.7).

Example 4.5. For the DAE

$$\begin{aligned} \dot{x}_1 &= x_1, \\ \dot{x}_2 &= -x_2, \\ 0 &= x_1 + x_2 - x_3, \end{aligned}$$

we have fundamental solution matrices

$$X_1(t) = \begin{bmatrix} e^t & 0 \\ 0 & e^{-t} \\ e^t & e^{-t} \end{bmatrix}, \quad X_2(t) = \begin{bmatrix} e^t & e^t \\ 0 & e^{-t} \\ e^t & e^{-t}+e^t \end{bmatrix}.$$

Here X_1 is normal and has $\Sigma_L = \{\pm 1\}$ and X_2 is not normal, but $X_1 = X_2 C$ with $C = \left[\begin{smallmatrix} 1 & -1 \\ 0 & 1 \end{smallmatrix}\right]$. ∎

Global kinematic equivalence transformations preserve the Lyapunov exponents as well as the normality of a basis formed by the columns of a fundamental solution matrix.

Theorem 4.6 (See [37, 40]). *Let X be a normal basis for* (4.7). *Then the Lyapunov spectrum of the DAE* (4.7) *and that of the ODE* (4.9) *are the same. If $\mathcal{E}, \mathcal{A}$ are as in* (4.9) *and if $\mathcal{E}^{-1}\mathcal{A}$ is bounded, then all the Lyapunov exponents of* (4.7) *are finite. Furthermore, the spectrum of* (4.9) *does not depend on the choice of the basis U and the matrix function V.*

As a consequence of this theorem and the Lyapunov inequality [41, Theorem 2.5.1], we have the following corollary.

Corollary 4.7. *Let $\{\lambda_i^u\}_{i=1}^d$ be the upper Lyapunov exponents of* (4.7). *Then*

$$\sum_{i=1}^{d} \lambda_i^u \geq \limsup_{t\to\infty} \frac{1}{t} \int_0^t \operatorname{tr} \mathcal{E}^{-1}\mathcal{A}(s)\, ds. \tag{4.12}$$

Here $\operatorname{tr} \mathcal{E}^{-1}\mathcal{A}(s)$ *denotes the trace of the matrix function.*

As in [12], we say that the DAE system (4.7) is *Lyapunov-regular* if its EUODE (4.9) is Lyapunov-regular, i.e., if

$$\sum_{i}^{d} \lambda_i^u = \liminf_{t\to\infty} \frac{1}{t} \int_0^t \operatorname{tr} \mathcal{E}^{-1}\mathcal{A}(s)\, ds.$$

Remark 4.8. *The Lyapunov regularity of a regular strangeness-free DAE system* (4.7) *is well defined, since it does not depend on the construction of* (4.9)*; i.e., the choice of the pair* U,V. *If* (4.7) *is Lyapunov-regular, then for any nontrivial solution* x, *the limit* $\lim_{t\to\infty}\frac{1}{t}\ln\|x(t)\|$ *exists. Hence, we have* $\lambda_i^l=\lambda_i^u$; *i. e., the Lyapunov spectrum of* (4.7) *is a point spectrum. Note that unlike in* [12], *where certain inherent ODEs of the same size as the original DAE are used, here the spectral analysis is based on the essential underlying ODEs, which have reduced size and can be constructed numerically.*

In the following we consider the adjoint equation of (4.7), given by

$$-E^T\dot{y}=(A+\dot{E})^T y, \tag{4.13}$$

see, e.g., [9, 32, 33], and also a slightly different formulation in [2].

The following result gives the relation between the EUODEs of (4.7) and (4.13).

Proposition 4.9. *Let the orthonormal columns of the matrix* U *form a basis of the solution subspace of* (4.7). *Then there exists* $V\in C^1(\mathbf{I},\mathbf{R}^{n\times d})$ *such that the columns of* V *form an orthonormal basis for the solution subspace of* (4.13). *Furthermore, via the change of variables* $y=Vw$ *and multiplication of both sides of* (4.13) *by* U^T, *the EUODE for the adjoint system* (4.13) *is given by*

$$-\mathcal{E}^T\dot{w}=(\mathcal{A}+\dot{\mathcal{E}})^T w, \tag{4.14}$$

which is exactly the adjoint of (4.9). *If* U *is such that the matrix* $\mathcal{E}$ *is upper triangular with positive diagonal elements, then the corresponding* V *is unique.*

Proof. To prove uniqueness, suppose that there exist matrix functions V and $\hat{V}$ with orthonormal columns such that $\mathcal{E}\dot{z}=\mathcal{A}z$ and $\hat{\mathcal{E}}\dot{z}=\hat{\mathcal{A}}z$, respectively, where both $\mathcal{E}$ and $\hat{\mathcal{E}}$ are upper triangular with positive diagonal elements. Since the columns of V and $\hat{V}$ are bases of the same subspace, there exists $S\in C(\mathbf{I},\mathbf{R}^{n\times d})$ such that $V=\hat{V}S$. We have $S^TS=V^T(\hat{V}\hat{V}^TV)=V^TV=I_d$, and thus S is orthogonal. On the other hand, by the construction of the EUODE, we have $S^T\hat{\mathcal{E}}=\mathcal{E}$ and $\hat{\mathcal{E}}$ is invertible, which implies that S^T is upper triangular. Hence, S is a diagonal matrix with diagonal elements $+1$ or -1. But since $\mathcal{E}$ and $\hat{\mathcal{E}}$ have positive diagonals, then $S=I$.

For the existence we give a constructive proof. From the proof of Lemma 4.1, V is determined via $EU=V\mathcal{E}$. Due to the special form of E, we can first determine an auxiliary pair $\tilde{V},\tilde{\mathcal{E}}$ such that $EU=\left[\begin{smallmatrix}E_1U\\0\end{smallmatrix}\right]=\tilde{V}\tilde{\mathcal{E}}$, which implies that $\tilde{V}=\left[\begin{smallmatrix}\tilde{V}_1\\0\end{smallmatrix}\right]$. Here $\tilde{V}_1$ and $\tilde{\mathcal{E}}$ are determined by, e.g., a smooth QR decomposition of E_1U. Unfortunately, in general such a $\tilde{V}$ is not a basis of the solution subspace of (4.13) yet. But we observe that we can replace the zero block by any $\tilde{V}_2$ and the relation $\tilde{V}^TEU=\mathcal{E}$ still holds. Hence, we look for an appropriate $\tilde{V}_2$ block so that $\tilde{V}=\left[\begin{smallmatrix}\tilde{V}_1\\\tilde{V}_2\end{smallmatrix}\right]$ satisfies the algebraic constraint of (4.13). It remains to orthonormalize the columns of $\tilde{V}$. The adjoint DAE (4.13) has the form

$$\begin{bmatrix}E_{11}^T & 0\\ E_{12}^T & 0\end{bmatrix}\begin{bmatrix}\dot{y}_1\\ \dot{y}_2\end{bmatrix}=\begin{bmatrix}-(A_{11}+\dot{E}_{11})^T & -A_{21}^T\\ -(A_{12}+\dot{E}_{12})^T & -A_{22}^T\end{bmatrix}\begin{bmatrix}y_1\\ y_2\end{bmatrix},$$

and, since the adjoint of a regular strangeness-free DAE of the form (4.7) is again regular and strangeness-free, see [9, 32, 33], we can reorder the equations so that the left upper

$d \times d$ block of the coefficient matrix on the left-hand side is nonsingular and then eliminate the left lower block giving

$$\begin{bmatrix} \tilde{E}_{11}^T & 0 \\ 0 & 0 \end{bmatrix} \begin{bmatrix} \dot{y}_1 \\ \dot{y}_2 \end{bmatrix} = \begin{bmatrix} -\tilde{A}_{11}^T & -\tilde{A}_{21}^T \\ -\tilde{A}_{12}^T & -\tilde{A}_{22}^T \end{bmatrix} \begin{bmatrix} y_1 \\ y_2 \end{bmatrix}.$$

In this system, the algebraic constraint is given explicitly by $0 = \tilde{A}_{12}^T y_1 + \tilde{A}_{22}^T y_2$, where $\tilde{A}_{22}^T$ is nonsingular. Hence, let $\tilde{V}_2 = -\tilde{A}_{22}^{-T} \tilde{A}_{12}^T \tilde{V}_1$. Finally, applying a Gram–Schmidt orthogonalization to $\tilde{V}$, we obtain a basis of (4.13), denoted by V, which also fulfils $V^T EU = \mathcal{E}$, where $\mathcal{E}$ is upper triangular. One easily verifies that the so-obtained EUODE of the adjoint equation (4.13) is exactly the adjoint of EUODE (4.9) of the original DAE (4.7). □

With these preparations we obtain a generalization of [36, Theorem 19].

Theorem 4.10. *Suppose that the matrix function $\mathcal{E} = V^T EU$ and its inverse are bounded on* **I**, *where the columns of U, V form bases of the solution spaces in Proposition* 4.9. *System* (4.7) *is Lyapunov regular if and only if* (4.13) *is Lyapunov regular, and in this case we have the* Perron identity

$$\lambda_i + \mu_i = 0, \quad i = 1, 2, \ldots, d,$$

where λ_i are the Lyapunov exponents of (4.7) *in decreasing order and μ_i are the Lyapunov exponents of the adjoint system* (4.13) *in increasing order.*

Proof. Due to Proposition 4.9, it suffices to consider two implicit EUODEs which are adjoint of each other. The assertion then follows from the *Lagrange identity* $W^T(t)\mathcal{E}(t)Z(t) = W^T(0)\mathcal{E}(0)Z(0)$, where Z and W are fundamental solutions of EUODE (4.9) and its adjoint (4.14), respectively. □

Note that if either $\mathcal{E}$ or its inverse is not bounded, then the Perron identity may not hold.

Example 4.11. Consider the ODE

$$\begin{aligned} \dot{x}_1 &= x_1, \\ \dot{x}_2 &= (\sin(\ln t) + \cos(\ln t))x_2 \end{aligned}$$

together with its adjoint system and the fundamental solution matrices

$$X_1 = \begin{bmatrix} e^t & 0 \\ 0 & e^{t \sin \ln t} \end{bmatrix}, \quad X_2 = \begin{bmatrix} e^t & e^t \\ 0 & e^{t \sin \ln t} \end{bmatrix};$$

then X_1 is normal and $\Sigma_L = \{1, [-1, 1]\}$. X_1^{-T} is a normal basis of the adjoint system whose Lyapunov spectrum is $\{-1, [-1, 1]\}$. The columns of X_2 form a normal basis as well, but the columns of

$$X_2^{-T} = \begin{bmatrix} e^{-t} & 0 \\ e^{-t \sin \ln t} & e^{-t \sin \ln t} \end{bmatrix}$$

do not form a normal basis of the adjoint system. ■

4.3.2 Stability of Lyapunov Exponents

It is, in general, very difficult to compute Lyapunov exponents via numerical methods, since they may be very sensitive under small changes in the coefficient matrices. In order to study this sensitivity for DAEs, we consider a DAE in the form (4.7) and the specially perturbed system

$$[E+G]\dot{x} = [A+H]x, \quad t \in \mathbf{I}, \tag{4.15}$$

where

$$G = \begin{bmatrix} G_1 \\ 0 \end{bmatrix}, \quad H(t) = \begin{bmatrix} H_1(t) \\ H_2(t) \end{bmatrix},$$

and where G_1 and H_1, H_2 are assumed to have the same order of smoothness as E_1 and A_1, A_2, respectively.

Perturbations of this special structure are called *admissible perturbations*, generalizing the concept for constant coefficient DAEs studied in [6]. The strangeness-free DAE (4.7) is said to be *robustly strangeness-free* if it stays strangeness-free under all sufficiently small admissible perturbations. It is easy to see that the DAE (4.7) is robustly strangeness-free under admissible perturbations if and only if the matrix function $\bar{E}$ in (4.6) is boundedly invertible.

The upper Lyapunov exponents $\lambda_1^u \geq \cdots \geq \lambda_d^u$ of (4.7) are said to be *stable* if for any $\epsilon > 0$, there exists $\delta > 0$ such that the conditions $\sup_t \|G(t)\| < \delta$, $\sup_t \|H(t)\| < \delta$, and $\sup_t \left\|\dot{H}_2(t)\right\| < \delta$ on the perturbations imply that the perturbed DAE system (4.15) is strangeness-free and

$$|\lambda_i^u - \gamma_i^u| < \epsilon \quad \forall i = 1, 2, \ldots, d,$$

where the γ_i^u are the ordered upper Lyapunov exponents of an admissibly perturbed system (4.15).

It is clear that the stability of upper Lyapunov exponents is invariant under strong global kinematic equivalence transformations. Compared with the ODE case, however, the boundedness of $\dot{H}_2$ is needed.

Example 4.12 (See [37, 39]). Consider the system $\dot{x}_1 = x_1, \quad 0 = x_2$, which is easily seen to be robustly strangeness-free and Lyapunov regular with Lyapunov exponent $\lambda = 1$. For the perturbed DAE

$$(1+\varepsilon^2 \sin(2nt))\dot{x}_1 - \varepsilon\cos(nt)\dot{x}_2 = x_1, \quad 0 = -2\varepsilon\sin(nt)x_1 + x_2, \tag{4.16}$$

where ε is a small parameter and n is a given integer, from the second equation of (4.16), we obtain $x_2 = 2\varepsilon \sin nt\, x_1$. Differentiating this expression for x_2 and inserting the result into the first equation, after some elementary calculations, we obtain $\dot{x}_1 = (1 + n\varepsilon^2 + n\varepsilon^2\cos(2nt))x_1$. Explicit integration yields $x_1 = e^{(1+n\varepsilon^2)t+\varepsilon^2\sin(2nt)/2}$, from which the only Lyapunov exponent $\hat{\lambda} = 1 + n\varepsilon^2$ is calculated. Clearly, though ε is small (hence the perturbations in the coefficient matrices are small), the difference between two Lyapunov exponents may be made arbitrarily large by choosing large n. ∎

A minimal fundamental solution matrix X for (4.7) is called *integrally separated* if for $i = 1, 2, \ldots, d-1$ there exist constants $c_1 > 0$ and $c_2 > 0$ such that

$$\frac{\|X(t)e_i\|}{\|X(s)e_i\|} \cdot \frac{\|X(s)e_{i+1}\|}{\|X(t)e_{i+1}\|} \geq c_2 e^{c_1(t-s)}$$

for all t, s with $t \geq s \geq 0$.

The integral separation property is invariant under strong global kinematic equivalence transformations. Furthermore, if a fundamental solution X of (4.7) is integrally separated, then so is the corresponding fundamental solution Z of (4.9) and vice versa.

By using a global kinematic equivalence transformation, (4.7) can be transformed to a special structured form, where the block A_{21} becomes zero; see [36, Remark 13]. The advantage of this form is that the associated EUODE then reads as $E_{11}\dot{x}_1 = A_{11}x_1$. Therefore, for the perturbation analysis, we may assume that (4.7) is already given with $A_{21} = 0$.

Theorem 4.13 (See [36]). *Consider* (4.7) *with* $A_{21} = 0$. *Suppose that the matrix* $\bar{E}$ *in* (4.6) *is boundedly invertible and that* $E_{11}^{-1}A_{11}$, $A_{12}A_{22}^{-1}$, *and the derivative of* A_{22} *are bounded on* $[0,\infty)$. *Then, the upper Lyapunov exponents of* (4.7) *are distinct and stable if and only if the system has the integral separation property.*

Remark 4.14. *Example* 4.12 *and Theorem* 4.13 *demonstrate that, unlike in the perturbation analysis of time-invariant DAEs* [6], *that of time-varying DAEs requires more restrictive conditions. However, for some classes of structured problems, see* [10] *and* [36, *Section* 3.2], *part of these conditions can be relaxed.*

4.4 Bohl Exponents and Sacker–Sell Spectrum

Since in general Lyapunov spectra are unstable, other spectral concepts such as Bohl exponents [4, 13] and Sacker–Sell spectra [45] were introduced. The extension of these concepts to DAEs was first presented in [36].

Definition 4.15. *Let x be a nontrivial solution of* (4.7). *The* (upper) Bohl exponent $\kappa_B^u(x)$ *of this solution is the greatest lower bound of all those values ρ for which there exist constants* $N_\rho > 0$ *such that*

$$\|x(t)\| \leq N_\rho e^{\rho(t-s)} \|x(s)\|$$

for any $t \geq s \geq 0$. If such numbers ρ do not exist, then one sets $\kappa_B^u(x) = +\infty$. Similarly, the lower Bohl exponent $\kappa_B^\ell(x)$ *is the least upper bound of all those values ρ' for which there exist constants $N'_\rho > 0$ such that*

$$\|x(t)\| \geq N'_\rho e^{\rho'(t-s)} \|x(s)\|, \quad 0 \leq s \leq t.$$

Lyapunov exponents and Bohl exponents are related via

$$\kappa_B^\ell(x) \leq \lambda^\ell(x) \leq \lambda^u(x) \leq \kappa_B^u(x);$$

i.e., the Bohl spectral intervals include the Lyapunov intervals. If the largest upper Bohl exponent of all solutions of (4.7) is negative, then the system is (uniformly) exponentially stable. Formulas characterizing Bohl exponents for ODEs, see e.g. [13], immediately extend to DAEs; i.e.,

$$\kappa_B^u(x) = \limsup_{s,t-s\to\infty} \frac{\ln\|x(t)\| - \ln\|x(s)\|}{t-s}, \quad \kappa_B^\ell(x) = \liminf_{s,t-s\to\infty} \frac{\ln\|x(t)\| - \ln\|x(s)\|}{t-s}.$$

Moreover, unlike the Lyapunov exponents, the Bohl exponents are stable for admissible perturbations without the integral separation assumption; see [10, 36]. The third spectral concept of Sacker–Sell spectra is defined via exponential dichotomy.

Definition 4.16. *The DAE* (4.7) *is said to have* exponential dichotomy *if for any minimal fundamental solution X there exist a projection $\Pi \in \mathbf{R}^{d\times d}$ and positive constants K and α such that*

$$\begin{aligned} \left\|X(t)\Pi X^+(s)\right\| &\le Ke^{-\alpha(t-s)}, \quad t\ge s,\\ \left\|X(t)(I_d-\Pi)X^+(s)\right\| &\le Ke^{\alpha(t-s)}, \quad s>t, \end{aligned} \tag{4.17}$$

where X^+ denotes the generalized Moore–Penrose inverse of X.

Since for a fundamental solution matrix X of (4.7) and an orthonormal basis of the solution subspace represented by a matrix U we have that $X = UZ$, where Z is the fundamental solution matrix of (4.9) and hence invertible, it follows that $X^+ = Z^{-1}U^T$, and hence $\left\|X(t)\Pi X^+(s)\right\| = \left\|Z(t)\Pi Z^{-1}(s)\right\|$, $\left\|X(t)(I_d-\Pi)X^+(s)\right\| = \left\|Z(t)(I_d-\Pi)Z^{-1}(s)\right\|$. From this it follows that the DAE (4.7) has exponential dichotomy if and only if its corresponding EUODE (4.9) has exponential dichotomy.

The projector Π can be chosen to be orthogonal [15, 19]. It projects to a subspace of the complete solution subspace in which all the solutions are uniformly exponentially decreasing, while the solutions belonging to the complementary subspace are uniformly exponentially increasing.

In order to extend the concept of exponential dichotomy spectrum to DAEs, we need *shifted DAE systems*

$$E(t)\dot{x} = [A(t)-\lambda E(t)]x, \quad t\in \mathbf{I}, \tag{4.18}$$

where $\lambda\in\mathbf{R}$, which immediately leads to a shifted EUODE $\mathcal{E}\dot{z} = (\mathcal{A}-\lambda\mathcal{E})z$.

Definition 4.17. *The* Sacker–Sell (or exponential dichotomy) spectrum *of the DAE system* (4.7) *is defined by*

$$\Sigma_S := \{\lambda\in\mathbf{R},\ \textit{the shifted DAE (4.18) does not have an exponential dichotomy}\}.$$

The complement of Σ_S is called the resolvent set *for the DAE system* (4.7)*, denoted by $\rho(E,A)$.*

Theorem 4.18 (See [36]). *The Sacker–Sell spectrum of* (4.7) *is exactly the Sacker–Sell spectrum of its EUODE* (4.9)*. Furthermore, the Sacker–Sell spectrum of* (4.7) *consists of at most d closed intervals.*

Note that the rightmost endpoint of the Sacker–Sell spectral interval is the largest Bohl exponent of the system.

Example 4.19. Consider the following DAE:

$$\begin{aligned} \dot{x}_1 &= \lambda x_1,\\ \dot{x}_2 &= (\sin(\ln t)+\cos(\ln t))x_2, \qquad \lambda\in\mathbf{R}, t\ge t_0>0,\\ 0 &= x_1+x_2-x_3. \end{aligned}$$

A simple computation gives $\Sigma_L = \{\lambda\} \cup [-1,1]$ and $\Sigma_S = \{\lambda\} \cup [-\sqrt{2}, \sqrt{2}]$, i.e., $\Sigma_L \subset \Sigma_S$. If $\lambda \in [-\sqrt{2}, \sqrt{2}]$, then the Sacker–Sell spectrum reduces to one single interval. For $\lambda > \sqrt{2}$ or $\lambda < -\sqrt{2}$, the endpoints of the Sacker–Sell spectral intervals are the lower/upper Bohl exponents of the columns of the fundamental solution matrix

$$X(t) = \begin{bmatrix} e^{\lambda t} & 0 \\ 0 & e^{t \sin \ln t} \\ e^{\lambda t} & e^{t \sin \ln t} \end{bmatrix}. \quad \blacksquare$$

It is shown in [36, Section 3.4] that under some boundedness conditions, the Sacker–Sell spectrum of the DAE (4.7) is stable with respect to admissible perturbations. Theorem 50 in [36] also states that if X is an integrally separated fundamental matrix of (4.7), then the Sacker–Sell spectrum of the system is exactly given by the d (not necessarily disjoint) Bohl intervals associated with the columns of X.

In the remainder of the chapter, we assume that Σ_S consists of $p \leq d$ pairwise disjoint spectral intervals; i.e., $\Sigma_S = \bigcup_{i=1}^{p} [a_i, b_i]$, and $b_i < a_{i+1}$ for all $1 \leq i \leq p$.

4.4.1 Boundedness of Solutions of Inhomogeneous Equations

The Sacker–Sell spectrum can be used to study the boundedness of solutions of the inhomogeneous equation (4.1). For simplicity of presentation, we assume that the coefficient matrices E and A are given in the semi-implicit form (4.10). Decomposing f into two parts f_1 and f_2, solving for x_2, and inserting gives the underlying inhomogeneous ODE

$$E_{11}\dot{x}_1 = \tilde{A}_{11}x_1 + f_1 - A_{12}A_{22}^{-1}f_2, \tag{4.19}$$

where $\tilde{A}_{11}$ is defined as in (4.11).

Theorem 4.20. *Suppose that E_{11} and A_{22} are boundedly invertible and that A_{11}, A_{12}, A_{21} are bounded. Then, for all continuous bounded functions f, the following statements hold:*

(i) *If $\Sigma_S \cap [0,\infty) = \emptyset$, i.e., the largest Bohl exponent is negative, then all the solutions of (4.1) are bounded.*

(ii) *If $0 \notin \Sigma_S$, i.e., system (4.7) has an exponential dichotomy, then there exists at least one bounded solution for (4.1). If in addition $\Sigma_S \cap (-\infty, 0) \neq \emptyset$ i.e., the projection Π in (4.17) is not trivially zero, then there exist infinitely many bounded solutions of (4.1) by adding an arbitrary bounded nontrivial solution of the corresponding homogeneous equation.*

Proof. Since $A_{22}^{-1}A_{21}$ is bounded, the Sacker–Sell spectrum of (4.10) and that of (4.11) are the same. Furthermore, the boundedness of x_1 implies that of x_2. Thus, it suffices to consider the underlying inhomogeneous ODE (4.19). Under the given assumptions, (4.19) reduces to

$$\dot{x}_1 = E_{11}^{-1}\tilde{A}_{11}x_1 + E_{11}^{-1}\left(f_1 - A_{12}A_{22}^{-1}f_2\right),$$

where $E_{11}^{-1}\tilde{A}_{11}$ and $E_{11}^{-1}(f_1 - A_{12}A_{22}^{-1}f_2)$ are bounded. By invoking [13, Theorem 5.2, p. 129], it follows that all the solutions x_1 of (4.19) are bounded if the largest Bohl exponent is negative. Since the algebraic component x_2 is bounded as well, this implies the boundedness of all the solutions of (4.1).

Similarly, if (4.11) has an exponential dichotomy, then due to [13, Theorem 3.2, p. 168], (4.19) has at least a bounded solution, which can be explicitly defined by the Green function of (4.11). As before, the boundedness of x_1 implies that of x_2 and the whole solution x.

If $\Sigma_S \cap (-\infty, 0) \neq \emptyset$, then there exists at least one negative upper Lyapunov exponent, which implies the existence of bounded nontrivial solutions of the corresponding homogeneous equation. Furthermore, these bounded solutions create a linear subspace [15]. Finally, due to the solution theory of linear ODEs, adding an arbitrary (bounded) nontrivial solution of the corresponding homogeneous equation to a (bounded) solution of the inhomogeneous equation gives another (bounded) solution. $\square$

4.5 Leading Directions Associated with Spectral Intervals

As we have noted before, initial vectors of (4.7) must be chosen in a consistent way and they form a d-dimensional subspace in $\mathbf{R}^n$. Furthermore, the solutions of (4.7) also form a d-dimensional subspace of functions in $C^1(\mathbf{I}, \mathbf{R}^n)$. We denote these spaces by $\mathbf{S_0}$ and $\mathbf{S(t)}$, respectively, and for $x_0 \in \mathbf{S_0}$ we denote by $x(t; x_0)$ the (unique) solution of (4.7) that satisfies $x(0; x_0) = x_0$.

Assume that the upper Lyapunov exponents $\{\lambda_i^u\})_{i=1}^d$ are given in decreasing order. For $j = 1, \ldots, d$, define the set W_j of all consistent initial conditions w such that the upper Lyapunov exponent of the solution $x(t; w)$ of (4.7) satisfies $\chi^u(x(\cdot; w)) \leq \lambda_j^u$; i.e.,

$$W_j = \left\{ w \in \mathbf{S_0} : \chi^{\mathbf{u}}(\mathbf{x}(\cdot; \mathbf{w})) \leq \lambda_{\mathbf{j}}^{\mathbf{u}} \right\}, \quad j = 1, \ldots, d.$$

Let the columns of $U(\cdot)$ form a smoothly varying basis of the solution subspace $\mathbf{S}(\cdot)$ of (4.7) and consider an associated EUODE (4.9). Then instead of W_j, the corresponding set of all initial conditions for (4.9) that lead to Lyapunov exponents not greater than λ_j^u can be considered. In this way it is obvious that all results for ODEs in [15] apply to EUODEs of the form (4.9) and, as a consequence of Theorem 4.6, we obtain analogous statements for (4.7).

Proposition 4.21. *Let d_j be the largest number of linearly independent solutions x of* (4.7) *such that* $\limsup_{t\to\infty} \frac{1}{t} \ln \|x(t)\| = \lambda_j^u$. *Then W_j is a d_j-dimensional linear subspace of* $\mathbf{S_0}$. *Furthermore, the spaces W_j, $j = 1, 2, \ldots$, form a* filtration *of* $\mathbf{S_0}$*; i.e., if p is the number of distinct upper Lyapunov exponents of the system, then we have*

$$\mathbf{S}_0 = W_1 \supset W_2 \supset \cdots \supset W_p \supset W_{p+1} = \{0\}.$$

It follows that $\limsup_{t\to\infty} \frac{1}{t} \ln \|x(t; w)\| = \lambda_j^u$ if and only if $w \in W_j \backslash W_{j+1}$. Moreover, if we have d distinct upper Lyapunov exponents, then the dimension of W_j is $d - j + 1$.

If Y_j is defined as the orthogonal complement of W_{j+1} in W_j, i.e.,

$$W_j = W_{j+1} \oplus Y_j, \quad Y_j \perp W_{j+1},$$

then $\mathbf{S}_0 = Y_1 \oplus Y_2 \oplus \cdots \oplus Y_p$, and

$$\limsup_{t\to\infty} \frac{1}{t} \ln \|x(t; w)\| = \lambda_j^u \text{ if and only if } w \in Y_j.$$

Thus, if we have $p = d$ distinct Lyapunov exponents, then $\dim(Y_j) = 1$ for all $j = 1,\dots,d$. Generalizing results of [15, 16] to DAEs, in [37] a numerical method for computing the spaces Y_j via smooth singular value decompositions of fundamental solutions was suggested.

Considering the resolvent set $\rho(E,A)$, for $\mu \in \rho(E,A)$, we can define the *stable* set associated with (4.7) as

$$\mathcal{S}_\mu = \left\{ w \in \mathbf{S}_0 : \lim_{t\to\infty} e^{-\mu t} \|x(t;w)\| = 0 \right\}.$$

Then, for $\mu_1, \mu_2 \in \rho(E,A)$, $\mu_1 < \mu_2$, we have $\mathcal{S}_{\mu_1} \subseteq \mathcal{S}_{\mu_2}$.

In the following we study the EUODE (4.9) associated with (4.7), and for simplicity, we assume that Z is the *principal matrix solution*, i.e., $Z(0) = I_d$. This can always be achieved by an appropriate kinematic equivalence transformation.

Following the construction for ODEs in [15, 45], we can characterize the stable and unstable subspaces

$$\mathcal{S}_\mu^d = \left\{ v \in \mathbf{R}^d : \lim_{t\to\infty} e^{-\mu t} \|Z(t)v\| = 0 \right\}, \quad \mathcal{U}_\mu^d = \left\{ v \in \mathbf{R}^d : \lim_{t\to\infty} e^{\mu t} \left\| Z(t)^{-T} v \right\| = 0 \right\}$$

associated with (4.9).

Recalling that p is the number of disjoint spectral intervals, we choose a set of values $\mu_0 < \mu_1 < \cdots < \mu_p$ such that $\mu_j \in \rho(E,A)$ and $\Sigma_S \cap (\mu_{j-1},\mu_j) = [a_j,b_j]$ for $j = 1,\dots,p$; i.e.,

$$\mu_0 < a_1 \le b_1 < \mu_1 < \cdots < \mu_{j-1} < a_j \le b_j < \mu_j < \cdots < \mu_{p-1} < a_p \le b_p < \mu_p.$$

Let U be an orthonormal basis of the solution subspace for (4.7) and introduce the sets

$$\mathcal{N}_j = U(0)\mathcal{N}_j^d = \left\{ w \in \mathbf{S}_0 : w = U(0)v,\ v \in \mathcal{N}_j^d \right\}, \quad j = 1,\dots,p. \tag{4.20}$$

Proposition 4.22 (See [37]). *Consider the EUODE* (4.9) *associated with* (4.7) *and the sets* $\mathcal{N}_j$ *defined in* (4.20), $j = 1,\dots,p$. *If* $w \in \mathcal{N}_j \setminus \{0\}$ *and*

$$\limsup_{t\to\infty} \frac{1}{t} \ln \|x(t;w)\| = \chi^u, \quad \liminf_{t\to\infty} \frac{1}{t} \ln \|x(t;w)\| = \chi^\ell,$$

then $\chi^\ell, \chi^u \in [a_j,b_j]$.

This result means that $\mathcal{N}_j$ is the subspace of initial conditions associated with solutions of (4.7) whose upper and lower Lyapunov exponents are inside $[a_j,b_j]$.

Theorem 4.23. *Consider the EUODE* (4.9) *associated with* (4.7) *and the sets* $\mathcal{N}_j$ *defined in* (4.20). *Then* $w \in \mathcal{N}_j \setminus \{0\}$ *if and only if*

$$\frac{1}{K_{j-1}} e^{a_j(t-s)} \le \frac{\|x(t;w)\|}{\|x(s;w)\|} \le K_j e^{b_j(t-s)} \quad \forall t \ge s \ge 0 \tag{4.21}$$

and for some positive constants K_{j-1}, K_j.

Proof. Due to the construction of the EUODE (4.9), see Lemma 4.1, we have $x(t;w) = U(t)Z(t)v$, where $v = U(0)^T w$, and thus $\|x(t;w)\| = \|Z(t)v\|$. Theorem 3.9 and Remark 3.10 of [15] state that $v \in \mathcal{N}_j^d$ if and only if

$$\frac{1}{K_{j-1}} e^{a_j(t-s)} \le \frac{\|Z(t)v\|}{\|Z(s)v\|} \le K_j e^{b_j(t-s)} \quad \forall t \ge s \ge 0$$

and for some positive constants K_{j-1}, K_j. Hence, (4.21) follows immediately. For more details see [37]. □

The sets N_j and the Bohl exponents are closely related.

Corollary 4.24. *Consider the EUODE* (4.9) *associated with* (4.7) *and the sets* $\mathcal{N}_j$ *defined in* (4.20). *Then for all* $j = 1,\dots,p$, *one has* $w \in \mathcal{N}_j \setminus \{0\}$ *if and only if* $a_j \le \kappa^\ell(x(\cdot;w)) \le \kappa^u(x(\cdot;w)) \le b_j$, *where* κ^ℓ, κ^u *are the Bohl exponents.*

Proof. The proof follows from Theorem 4.23 and Definition 4.15. □

The stable and unstable sets are then characterized as follows.

Proposition 4.25. *Consider the EUODE* (4.9) *associated with* (4.7). *For all* $j = 1,\dots,p$, *we have the following:*

(i) $\mathcal{S}_{\mu_j} = U(0)\mathcal{S}_{\mu_j}^d$.

(ii) *Let the unstable sets for* (4.7) *be defined by* $\mathcal{U}_{\mu_j} = U(0)\mathcal{U}_{\mu_j}^d$. *Then* $\mathcal{S}_{\mu_j} \oplus \mathcal{U}_{\mu_j} = \mathbf{S}_0$ *and* $\mathcal{N}_j = \mathcal{S}_{\mu_j} \cap \mathcal{U}_{\mu_{j-1}}$.

(iii) $\mathbf{S}_0 = \mathcal{N}_1 \oplus \mathcal{N}_2 \oplus \cdots \oplus \mathcal{N}_p$.

Proof. (i) First we prove that

$$U(0)\mathcal{S}_{\mu_j}^d \subseteq \mathcal{S}_{\mu_j}.$$

To this end, take an arbitrary $w \in U(0)\mathcal{S}_{\mu_j}^d$. Then the corresponding initial value for (4.9) defined by $v = U(0)^T w$ clearly belongs to $\mathcal{S}_{\mu_j}^d$ and $w = U(0)v$ holds. By considering the one-to-one relation between the solutions of (4.7) and those of its associated EUODE (4.9) and invoking the equality $\|x(t;w)\| = \|Z(t)v\|$, $v \in \mathcal{S}_{\mu_j}^d$ implies $w \in \mathcal{S}_{\mu_j}$. Conversely, take an arbitrary $w \in \mathcal{S}_{\mu_j}$. Then there exists a unique $v \in \mathbf{R}^d$ which satisfies $w = U(0)v$. Using again that $\|x(t;w)\| = \|Z(t)v\|$, from the definition of $\mathcal{S}_{\mu_j}$ and that of $\mathcal{S}_{\mu_j}^d$, the claim $v \in \mathcal{S}_{\mu_j}^d$ follows.

(ii) As a consequence of Theorem 3.4 in [15], we have $\mathcal{S}_{\mu_j}^d \oplus \mathcal{U}_{\mu_j}^d = \mathbf{R}^d$. Since $U(0)$ consists of orthonormal columns, we have

$$U(0)\mathcal{S}_{\mu_j}^d \oplus U(0)\mathcal{U}_{\mu_j}^d = \operatorname{range} U(0) = \mathbf{S}_0,$$

from which the first equality immediately follows. The second equality is obvious from (i) and the definition of $\mathcal{N}_j$.

(iii) The equality follows from the relation between $\mathcal{N}_j$ and $\mathcal{N}_j^d$ and the result for ODEs [15, 37]. □

Example 4.26. Consider the DAE in Example 4.19 with $\lambda < 1$. The distinct upper Lyapunov exponents are λ and 1 and $d = p = 2$. The subspaces of consistent initial vectors associated with the upper Lyapunov exponents are $\mathbf{S}_0 = W_1 = \{[a,b,a+b]^T,\ a,b \in \mathbf{R}\}$, $W_2 = \{[a,0,a]^T,\ a \in \mathbf{R}\}$, and $Y_1 = \{[0,b,b]^T,\ b \in \mathbf{R}\}$, $Y_2 = W_2$. The case $\lambda > 1$ is similar by exchanging the roles of Y_1 and Y_2. For $\lambda = 1$, we have $p = 1$, $\mathbf{S}_0 = W_1$ as above, and $W_2 = \{[0,0,0]^T\}$.

Now taking the concrete value $\lambda = -3$, then the endpoints of the spectral intervals are $a_1 = b_1 = -3$, $a_2 = -\sqrt{2}$, $b_2 = \sqrt{2}$. Choosing any triplet $\mu_0 < -3$, $\mu_1 \in (-3, -\sqrt{2})$, and $\mu_2 > \sqrt{2}$, then $\mathcal{S}_{\mu_0} = \{[0,0,0]^T\}$, $\mathcal{S}_{\mu_1} = \{[a,0,a]^T,\ a \in \mathbf{R}\}$, $\mathcal{S}_{\mu_2} = \{[a,b,a+b]^T,\ a,b \in \mathbf{R}\}$, $\mathcal{N}_1 = \{[a,0,a]^T,\ a \in \mathbf{R}\}$, and $\mathcal{N}_2 = \{[0,b,b]^T,\ b \in \mathbf{R}\}$. ∎

4.6 Discussion

We have summarized the spectral theory for linear time-varying DAEs which was recently given in [36, 37, 39, 40]. It is shown that most spectral notions and results can be extended from ODEs to general regular DAEs that have a well-defined strangeness index, but with some extra conditions posed on the coefficients. Based on a suitable construction of EUODEs and the presented analysis, numerical methods for computing spectral intervals can be constructed as in [37, 40]. As future work, an analysis for quasi-linear and/or nonlinear DAEs, together with efficient methods for approximating spectral intervals of linearized DAEs, is of interest. Another challenging problem is an extension of the spectral theory from finite-dimensional systems to infinite-dimensional ones, i.e., for partial differential-algebraic equations (PDAEs).

Acknowledgments

We thank the anonymous referee for useful suggestions that led to improvements of the chapter. Vu Hoang Linh's work was supported by *Alexander von Humboldt Foundation* and partially by VNU's Project QG 10-01. Volker Mehrmann's research was supported by the European Research Council through ERC Advanced grant MODSIMCONMP.

Bibliography

[1] L. YA. ADRIANOVA, *Introduction to Linear Systems of Differential Equations*, Trans. Math. Monogr. 146, AMS, Providence, RI, 1995.

[2] K. BALLA AND V. H. LINH, *Adjoint pairs of differential-algebraic equations and Hamiltonian systems*, Appl. Numer. Math., 53 (2005), pp. 131–148.

[3] P. BENNER AND R. BYERS, *An arithmetic for matrix pencils: Theory and new algorithms*, Numer. Math., 103 (2006), pp. 539–573.

[4] P. BOHL, *Über Differentialungleichungen*, J. f. d. Reine und Angew. Math., 144 (1913), pp. 284–313.

[5] K. E. BRENAN, S. L. CAMPBELL, AND L. R. PETZOLD, *Numerical Solution of Initial-Value Problems in Differential-Algebraic Equations*, SIAM, Philadelphia, 1996.

[6] R. Byers and N. K. Nichols, *On the stability radius of a generalized state-space system*, Linear Algebra Appl., 188/189 (1993), pp. 113–134.

[7] S. L. Campbell, *A general form for solvable linear time varying singular systems of differential equations*, SIAM J. Math. Anal., 18 (1987), pp. 1101–1115.

[8] S. L. Campbell, *Linearization of DAE's along trajectories*, Z. Angew. Math. Phys., 46 (1995), pp. 70–84.

[9] S. L. Campbell, N. K. Nichols, and W. J. Terrell, *Duality, observability, and controllability for linear time-varying descriptor systems*, Circuits Systems Signal Process., 10 (1991), pp. 455–470.

[10] C. J. Chyan, N. H. Du, and V. H. Linh, *On data-dependence of exponential stability and the stability radii for linear time-varying differential-algebraic systems*, J. Differential Equations, 245 (2008), pp. 2078–2102.

[11] N. D. Cong and H. Nam, *Lyapunov's inequality for linear differential algebraic equation*, Acta Math. Vietnam, 28 (2003), pp. 73–88.

[12] N. D. Cong and H. Nam, *Lyapunov regularity of linear differential algebraic equations of index 1*, Acta Math. Vietnam, 29 (2004), pp. 1–21.

[13] J. L. Daleckii and M. G. Krein, *Stability of Solutions of Differential Equations in Banach Spaces*, American Mathematical Society, Providence, RI, 1974.

[14] L. Dieci and T. Eirola, *On smooth decompositions of matrices*, SIAM J. Matrix Anal. Appl., 20 (1999), pp. 800–819.

[15] L. Dieci and C. Elia, *The singular value decomposition to approximate spectra of dynamical systems. Theoretical aspects*, J. Differential Equations, 230 (2006), pp. 502–531.

[16] L. Dieci and C. Elia, *SVD algorithms to approximate spectra of dynamical systems*, Math. Comput. Simulation, 79 (2008), pp. 1235–1254.

[17] L. Dieci, R. D. Russell, and E. S. Van Vleck, *On the computation of Lyapunov exponents for continuous dynamical systems*, SIAM J. Numer. Anal., 34 (1997), pp. 402–423.

[18] L. Dieci and E. S. Van Vleck, *Computation of a few Lyapunov exponents for continuous and discrete dynamical systems*, Appl. Numer. Math., 17 (1995), pp. 275–291.

[19] L. Dieci and E. S. Van Vleck, *Lyapunov spectral intervals: Theory and computation*, SIAM J. Numer. Anal., 40 (2002), pp. 516–542.

[20] L. Dieci and E. S. Van Vleck, *On the error in computing Lyapunov exponents by QR methods*, Numer. Math., 101 (2005), pp. 619–642.

[21] L. Dieci and E. S. Van Vleck, *Lyapunov and Sacker-Sell spectral intervals*, J. Dynam. Differential Equations, 19 (2007), pp. 265–293.

[22] L. DIECI AND E. S. VAN VLECK, *On the error in QR integration*, SIAM J. Numer. Anal., 46 (2008), pp. 1166–1189.

[23] N. H. DU AND V. H. LINH, *Robust stability of implicit linear systems containing a small parameter in the leading term*, IMA J. Math. Cont. Inf., 23 (2006), pp. 67–84.

[24] N. H. DU AND V. H. LINH, *Stability radii for linear time-varying differential-algebraic equations with respect to dynamic perturbations*, J. Differential Equations, 230 (2006), pp. 579–599.

[25] E. EICH-SOELLNER AND C. FÜHRER, *Numerical Methods in Multibody Systems*, Teubner Verlag, Stuttgart, Germany, 1998.

[26] G. H. GOLUB AND C. F. VAN LOAN, *Matrix Computations*, 3rd ed., The Johns Hopkins University Press, Baltimore, MD, 1996.

[27] E. GRIEPENTROG AND R. MÄRZ, *Differential-Algebraic Equations and Their Numerical Treatment*, Teubner Verlag, Leipzig, Germany, 1986.

[28] E. HAIRER AND G. WANNER, *Solving Ordinary Differential Equations* II: *Stiff and Differential-Algebraic Problems*, 2nd ed., Springer-Verlag, Berlin, 1996.

[29] P. KUNKEL AND V. MEHRMANN, *A new class of discretization methods for the solution of linear differential-algebraic equations with variable coefficients*, SIAM J. Numer. Anal., 33 (1996), pp. 1941–1961.

[30] P. KUNKEL AND V. MEHRMANN, *Differential-Algebraic Equations. Analysis and Numerical Solution*, EMS Publishing House, Zürich, Switzerland, 2006.

[31] P. KUNKEL AND V. MEHRMANN, *Stability properties of differential-algebraic equations and spin-stabilized discretization*, Electron. Trans. Numer. Anal., 26 (2007), pp. 383–420.

[32] P. KUNKEL AND V. MEHRMANN, *Optimal control for unstructured nonlinear differential-algebraic equations of arbitrary index*, Math. Control Signals Systems, 20 (2008), pp. 227–269.

[33] P. KUNKEL AND V. MEHRMANN, *Formal adjoints of linear DAE operators and their role in optimal control*, Electron. J. Linear Algebra, 22 (2011), pp. 672–693.

[34] P. KUNKEL, V. MEHRMANN, W. RATH, AND J. WEICKERT, *A new software package for linear differential-algebraic equations*, SIAM J. Sci. Comput., 18 (1997), pp. 115–138.

[35] M. LENTINI AND R. MÄRZ, *Conditioning and dichotomy in differential algebraic equations*, SIAM J. Numer. Anal., 27 (1990), pp. 1519–1526.

[36] V. H. LINH AND V. MEHRMANN, *Lyapunov, Bohl and Sacker-Sell spectral intervals for differential-algebraic equations*, J. Dynam. Differential Equations, 21 (2009), pp. 153–194.

[37] V. H. Linh and V. Mehrmann, *Approximation of spectral intervals and leading directions for differential-algebraic equation via smooth singular value decompositions*, SIAM J. Numer. Anal., 49 (2011), pp. 1810–1835.

[38] V. H. Linh and V. Mehrmann, *Efficient Integration of Matrix-Valued Non-stiff DAEs by Half-Explicit Methods*, Preprint 2011/14, Institut für Mathematik, TU Berlin, Berlin, 2011.

[39] V. H. Linh and V. Mehrmann, *Spectral analysis for linear differential-algebraic systems*, in Proceedings of the 8th AIMS International Conference on Dynamical Systems, Differential Equations and Applications, Dresden, 2010, AIMS Proceedings Series, 2011 (in print).

[40] V. H. Linh, V. Mehrmann, and E. Van Vleck, *QR methods and error analysis for computing Lyapunov and Sacker-Sell spectral intervals for linear differential-algebraic equations*, Adv. Comput. Math., 35 (2011), pp. 281–322.

[41] A. M. Lyapunov, *The general problem of the stability of motion*, Internat. J. Control, 55 (1992), pp. 521–790.

[42] R. März, *The index of linear differential algebraic equations with properly stated leading terms*, Res. in Math., 42 (2002), pp. 308–338.

[43] P. J. Rabier and W. C. Rheinboldt, *Theoretical and Numerical Analysis of Differential-Algebraic Equations*, Handb. Numer. Anal. VIII, Elsevier, Amsterdam, The Netherlands, 2002.

[44] R. Riaza, *Differential-Algebraic Systems: Analytical Aspects and Circuit Applications*, World Scientific, Hackensack, NJ, 2008.

[45] R. J. Sacker and G. R. Sell, *A spectral theory for linear differential systems*, J. Differential Equations, 27 (1978), pp. 320–358.

[46] D. S. Watkins, *The Matrix Eigenvalue Problem: GR and Krylov Subspace Methods*, SIAM, Philadelphia, 2007.

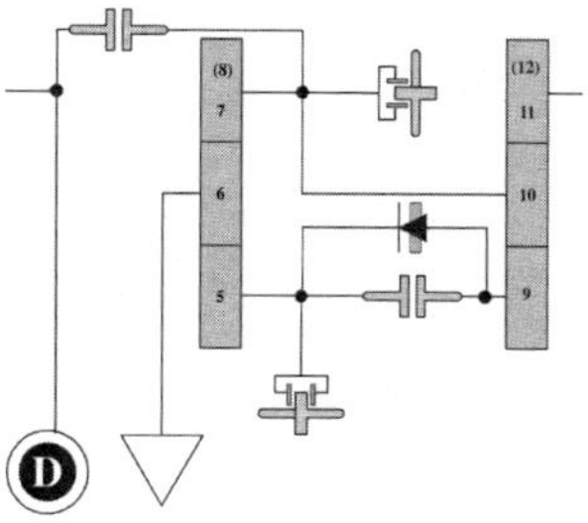

Chapter 5
StratiGraph Tool: Matrix Stratifications in Control Applications

Bo Kågström, Stefan Johansson, and Pedher Johansson

In this chapter, the software tool StratiGraph for computing and visualizing closure hierarchy graphs associated with different orbit and bundle stratifications is presented. In addition, we review the underlying theory and illustrate how StratiGraph can be used to analyze descriptor system models via their associated system pencils. The stratification theory provides information for a deeper understanding of how the dynamics of a control system and its system characteristics behave under perturbations.

5.1 Introduction

Dynamical systems described by linear time-invariant DAEs can often be expressed as *descriptors* (or *generalized state-space*) models of the following form [5, 46]:

$$\begin{aligned} E\dot{x} &= Ax + Bu, \\ y &= Cx + Du, \end{aligned} \tag{5.1}$$

where E, $A \in \mathbf{C}^{n\times n}$, $B \in \mathbf{C}^{n\times m}$, $C \in \mathbf{C}^{p\times n}$, and $D \in \mathbf{C}^{p\times m}$. Determining the system characteristics (like poles, zeros, controllability, and observability) of a model (5.1) involves computing the canonical structure information of an associated system pencil. In the general case, the *system pencil* has the form

$$\mathbf{S}(\lambda) = \begin{bmatrix} A & B \\ C & D \end{bmatrix} - \lambda \begin{bmatrix} E & 0 \\ 0 & 0 \end{bmatrix}, \tag{5.2}$$

where $\mathbf{S}(\lambda)$ is of size $(n+p)\times(n+m)$. Such information, including the dynamics of the DAE system $E\dot{x} = Ax + Bu$, is revealed by various canonical forms of the complete system pencil or parts of $\mathbf{S}(\lambda)$. All these features are ill-posed problems in the sense that small perturbations of the matrices (E, A, B, C, D) defining the model can drastically change the system characteristics. The stratification theory for orbits and bundles of matrices, matrix

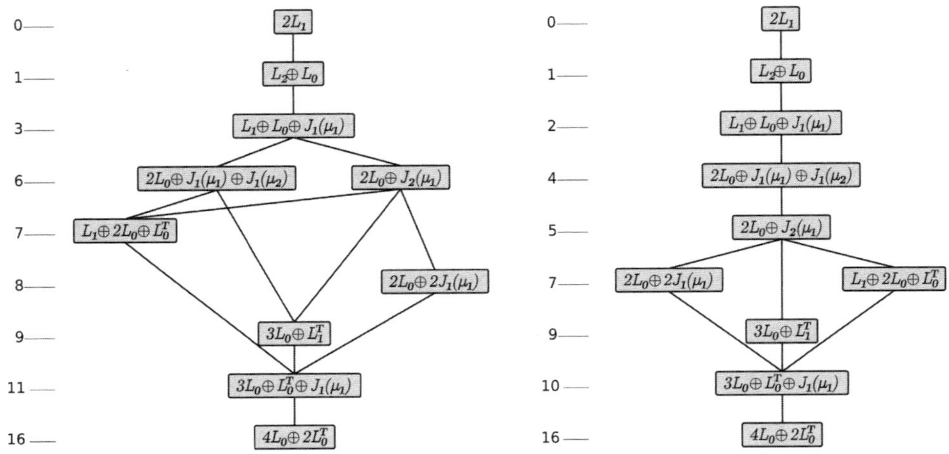

Figure 5.1. Orbit (left graph) and bundle (right graph) stratifications of 2×4 matrix pencils computed by StratiGraph.

pencils, and various system pencils provides information for a deeper understanding of how such system transitions take place under perturbations of the involved matrices.

We start by considering a general matrix pencil $G - \lambda H$ of size 2×4 in order to introduce and illustrate some concepts and results. In Figure 5.1, closure hierarchy (or stratification) graphs of the orbits and bundles of general 2×4 matrix pencils under equivalence transformations are shown. Each node represents an equivalence orbit (or bundle) with its unique canonical structure, and each edge between two nodes corresponds to a closure relation. The topmost node (of full row rank) corresponds to the *most generic* canonical structure, while the bottom node (pencil with only zero elements) corresponds to the *most degenerate* canonical structure. Since the pencil is rectangular (2×4), it is singular by definition and the structure elements are revealed by the Kronecker canonical form of $G - \lambda H$. In the graphs of Figure 5.1, L_k, L_k^T, and $J_k(\mu_i)$ are right singular, left singular, and Jordan blocks, respectively. A more complete definition of different concepts and canonical attributes appears later in the chapter.

Traversing an orbit (or bundle) closure hierarchy gives information about how structure transitions can take place under perturbations. First, traversing the graph upwards, e.g., moving from one orbit to a more generic orbit, can always be done with arbitrary small perturbations of matrices in the descriptor model. Think about setting some zero elements to very small numbers which will change the canonical structure of the perturbed pencil (at least in finite precision arithmetic). Indeed, adding random noise to all elements of the zero pencil (typically) gives a generic pencil. On the other side, moving from a given orbit to a more degenerate model will in general require much larger perturbations of the matrices involved. For example, in control applications it is important to understand how a system model changes from a controllable one to different uncontrollable system models or how perturbations may change the dynamics of the underlying DAE system.

So what is the difference between the orbit and bundle closure hierarchy graphs in Figure 5.1? For the orbit case, all eigenvalues are kept fixed—only their elementary divisors (Jordan block sizes) may change—and an eigenvalue may even disappear or a new may appear. For the bundle case, specified eigenvalues may as well coalesce or a multiple eigenvalue can split apart in different eigenvalues (not in this example).

Before we go into any further details and explanations, we outline the content of the rest of this chapter. In Section 5.2, the concept of stratification and some relevant background theory used in this chapter is introduced. Section 5.3 presents the software tool StratiGraph for computing and visualizing closure hierarchy graphs associated with different orbit and bundle stratifications. For a given orbit, a stratification graph gives information of nearby canonical forms and associated system models. In Section 5.4, the stratification theory is applied and we illustrate how StratiGraph can be used to analyze sample DAE systems and in control applications. Finally, Section 5.5 gives an overview of possible new features and problem setups to StratiGraph. Our related ongoing and planned work, including distance information to more degenerate systems, is reviewed and illustrated. For a controllable system, a more degenerate system can be the closest uncontrollable one.

5.2 Stratification Theory—Some Background

The *theory of stratification* reveals the qualitative information that the closure hierarchy of orbits and bundles provide (e.g., see [14, 15, 19, 28] and the references therein). The closure hierarchy is determined by the closure and cover relations among orbits or bundles, where a cover relation guarantees that two orbits or bundles are nearest neighbors in the closure hierarchy. The orbit, for example, of a matrix pencil $G - \lambda H$ consists of all pencils with the same eigenvalues and the same canonical form as $G - \lambda H$; see Section 5.2.3. A bundle is the union of all orbits with the same canonical form but with unspecified eigenvalues [1]. Figure 5.1 illustrates an example of orbit and bundle closure hierarchy graphs, where a cover relation is represented by an edge between two nodes. In a stratification, an orbit can never be covered by a less or equally generic orbit. This implies that structures within the closure hierarchy can be ordered by their dimension (or their codimension).

We continue in Section 5.2.1 to define integer partitions, which are used as a tool for representing the structural information presented in Section 5.2.2. In Section 5.2.3 the geometry of the matrix pencil space is considered, and we end by reviewing the stratification rules for matrix pencils in Section 5.2.4.

5.2.1 Integer Partitions

Edelman, Elmroth, and Kågström [15] show how canonical structural information (defined in the next section) can be represented as integer partitions such that the closure relations of the various orbits and bundles are revealed by applying a simple set of rules. Below, we define integer partitions and introduce the combinatorial rules which are used in the rest of this chapter.

An *integer partition* $\kappa = (\kappa_1, \kappa_2, \kappa_3, \ldots)$ such that $\kappa_1 \geq \kappa_2 \geq \cdots \geq 0$ is said to dominate another partition ν, i.e., $\kappa > \nu$ if $\kappa_1 + \kappa_2 + \cdots + \kappa_i \geq \nu_1 + \nu_2 + \cdots + \nu_i$ for $i = 1, 2, \ldots$, where $\nu \neq \kappa$. Different partitions of an integer can in this way form a dominance ordering. If $\kappa > \nu$ and there is no partition μ such that $\kappa > \mu > \nu$, then κ is said to *cover* ν. Furthermore, the *conjugate partition* of κ, $\nu = \text{conj}(\kappa)$, is defined such that ν_j are the number of integers in $\kappa \geq j$.

The integer partitions can also be represented as piles of coins in a table; i.e., an integer partition $\kappa = (\kappa_1, \ldots, \kappa_n)$ can be seen as n piles of coins where pile i has κ_i coins.

The covering relation between two integer partitions can then easily be determined: If an integer partition ν can be obtained from κ by moving one coin in κ *one* column rightward or *one* row downward and ν remains monotonic decreasing, then κ *covers* ν. This defines a *minimum rightward coin move* [15]; see Figure 5.2. The minimal *leftward* coin move is defined analogously.

Figure 5.2. Minimum rightward and leftward coin moves illustrate that $\kappa = (3,2,2,1)$ covers $\nu = (3,2,1,1,1)$ and $\kappa = (3,2,2,1)$ is covered by $\tau = (3,3,1,1)$ [19].

5.2.2 Structure Information and Kronecker Canonical Form

The system characteristics of a descriptor system (5.1) are described by the (canonical) *structural elements* of an associated system pencil. The structural elements we consider in this chapter are (1) the *right minimal indices* $(\epsilon_1,\dots,\epsilon_{r_0})$, with $\epsilon_1 \geq \epsilon_2 \geq \cdots \geq \epsilon_{r_1} > \epsilon_{r_1+1} = \cdots = \epsilon_{r_0} = 0$; (2) the *left minimal indices* $(\eta_1,\dots,\eta_{l_0})$, with $\eta_1 \geq \eta_2 \geq \cdots \geq \eta_{l_1} > \eta_{l_1+1} = \cdots = \eta_{l_0} = 0$; (3) the *finite elementary divisors* of the form

$$(\lambda-\mu_1)^{h_1^{(1)}},\dots,(\lambda-\mu_1)^{h_{g_1}^{(1)}},\dots,(\lambda-\mu_q)^{h_1^{(q)}},\dots,(\lambda-\mu_q)^{h_{g_q}^{(q)}},$$

with $h_1^{(i)} \geq \cdots \geq h_{g_i}^{(i)} \geq 1$ for each of the q distinct finite eigenvalues μ_i, $i = 1,\dots,q$; (4) the *infinite elementary divisors* on the form $1/\lambda^{s_1}, 1/\lambda^{s_2},\dots,1/\lambda^{s_{g_\infty}}$, with $s_1 \geq \cdots \geq s_{g_\infty} \geq 1$.

For matrix pencils this structural information is revealed by the *Kronecker canonical form* (KCF) [20]. Any $m_{\text{p}} \times n_{\text{p}}$ matrix pencil $G - \lambda H$ can be transformed into KCF in terms of an equivalence transformation such that

$$\begin{aligned} &U(G-\lambda H)V^{-1} \\ &= \operatorname{diag}(L_{\epsilon_1},\dots,L_{\epsilon_{r_0}},J(\mu_1),\dots,J(\mu_q),N_{s_1},\dots,N_{s_{g_\infty}},L_{\eta_1}^T,\dots,L_{\eta_{l_0}}^T), \end{aligned} \tag{5.3}$$

where $J(\mu_i) = \operatorname{diag}(J_{h_1}(\mu_i),\dots,J_{h_{g_i}}(\mu_i))$, $i = 1,\dots,q$.

The blocks $J_{h_k}(\mu_i)$ are $h_k \times h_k$ *Jordan blocks* associated with each distinct finite eigenvalue μ_i, where each block corresponds to a *finite elementary divisor* of degree h_k, namely $(\lambda-\mu_i)^{h_k}$. The blocks N_{s_k} are $s_k \times s_k$ Jordan blocks for matrix pencils associated with the eigenvalue at infinity, where each block corresponds to an *infinite elementary divisor* of degree s_k, namely $1/\lambda^{s_k}$. Moreover, g_i is the geometric multiplicity of the finite eigenvalues μ_i and g_∞ is the geometric multiplicity of the infinite eigenvalue. These two types of blocks constitute the *regular part* of a matrix pencil and are defined by

$$J_{h_k}(\mu_i) \equiv \begin{bmatrix} \mu_i-\lambda & 1 & & \\ & \ddots & \ddots & \\ & & \ddots & 1 \\ & & & \mu_i-\lambda \end{bmatrix} \quad \text{and} \quad N_{s_k} \equiv \begin{bmatrix} 1 & -\lambda & & \\ & \ddots & \ddots & \\ & & \ddots & -\lambda \\ & & & 1 \end{bmatrix}.$$

If $m_p \neq n_p$ or $\det(G - \lambda H) \equiv 0$ for all $\lambda \in \overline{\mathbf{C}} \equiv \mathbf{C} \cup \{\infty\}$, then the matrix pencil also includes a *singular part* and we say that the matrix pencil is singular. The singular part of the KCF consists of the r_0 *right singular blocks* L_{ϵ_k} of size $\epsilon_k \times (\epsilon_k + 1)$, corresponding to the *right minimal indices* ϵ_k, and the l_0 *left singular blocks* $L_{\eta_k}^T$ of size $(\eta_k + 1) \times r_k$, corresponding to the *left minimal indices* η_k. These blocks are defined by

$$L_{\epsilon_k} \equiv \begin{bmatrix} -\lambda & 1 & & \\ & \ddots & \ddots & \\ & & -\lambda & 1 \end{bmatrix} \quad \text{and} \quad L_{\eta_k}^T \equiv \begin{bmatrix} -\lambda & & \\ 1 & \ddots & \\ & \ddots & -\lambda \\ & & 1 \end{bmatrix}.$$

An L_0 and an L_0^T block are of size 0×1 and 1×0, respectively, and each of them contributes to a column or row of zeros, respectively, in the matrix representation of the KCF.

In the following, we also use a more compact notation by writing the KCF as a direct sum of blocks:

$$U(G - \lambda H)V^{-1} \equiv \mathbf{L} \oplus \mathbf{L}^T \oplus \mathbf{J}(\mu_1) \oplus \cdots \oplus \mathbf{J}(\mu_q) \oplus \mathbf{N},$$

where

$$\mathbf{L} = \bigoplus_{j=1}^{r_0} L_{\epsilon_j}, \quad \mathbf{L}^T = \bigoplus_{j=1}^{l_0} L_{\eta_j}^T, \quad \mathbf{J}(\mu_i) = \bigoplus_{j=1}^{g_i} J_{h_j}(\mu_i), \quad \text{and} \quad \mathbf{N} = \bigoplus_{j=1}^{g_\infty} N_{s_j}.$$

The most robust way of computing the canonical structure information is to use *staircase-type algorithms*, which apply unitary transformations to determine Weyr-type characteristics [9, 10, 32, 42, 44]. For Jordan blocks Weyr and Segre characteristics are closely related. In summary, the canonical structural information can be expressed as the *structure integer partitions*:

- $\mathcal{R} = (r_0, r_1, \ldots, r_{\epsilon_1})$, where $r_i = \#L_k$ blocks with $k \geq i$.

- $\mathcal{L} = (l_0, l_1, \ldots, l_{\eta_1})$, where $l_i = \#L_k^T$ blocks with $k \geq i$.

- $\mathcal{J}_{\mu_i} = (j_1, j_2, \ldots)$, where $j_t = \#J_k(\mu_i)$ blocks with $k \geq t$. $\mathcal{J}_{\mu_i}$ is known as the *Weyr characteristics* of the finite eigenvalue μ_i. $\mathcal{J}_{\mu_i}$ is the conjugate partition of the *Segre characteristics* $h = (h_1, \ldots, h_{g_i})$ defined by h_k, the exponents of the finite elementary divisors.

- $\mathcal{N} = (n_1, n_2, \ldots)$, where $n_t = \#N_k$, with $k \geq t$. $\mathcal{N}$ is known as the *Weyr characteristics* of the infinite eigenvalue, where $\mathcal{N}$ is the conjugate partition of the *Segre characteristics* $s = (s_1, \ldots, s_{g_\infty})$ defined by s_k, the exponents of the infinite elementary divisors.

Above we have used the notation $G - \lambda H$, with $\lambda \in \overline{\mathbf{C}}$, for a general matrix pencil. However, from a computational point of view it is more appropriate to consider $\beta G - \alpha H$ for all $(\alpha, \beta) \in \mathbf{C}^2$, where the pair (α, β) is a *generalized eigenvalue* of $G - \lambda H$. If $\beta \neq 0$, then the pair represents the finite eigenvalue α/β, and if $\alpha \neq 0$ and $\beta = 0$, then (α,β) represents an infinite eigenvalue. This notation also applies to general DAE systems $E\dot{x} = Ax + f$, when we want to compute the eigenvalues of $A - \lambda E$. In this chapter, we mainly consider regular descriptor systems where A and E are square. We also remark that if E is nonsingular, the descriptor system can be transformed into a standard state-space form. However, this type of transformation should be done only if E is a well-conditioned matrix. Otherwise, we should keep the DAE formulation and treat it as a descriptor system.

We are also using the concept of *normal-rank* (e.g., see [44]), which can be defined as $r = n_{\text{p}} - r_0 = m_{\text{p}} - l_0$, where r_0 and l_0 are the number of right and left singular blocks, respectively. Therefore, a square regular $n_{\text{p}} \times n_{\text{p}}$ matrix pencil has full normal-rank (it has no singular blocks).

5.2.3 Matrix Pencil Space

An $n \times n$ matrix A can be seen as a point in an n^2-dimensional (matrix) space, one dimension for each parameter of A. Consequently, the union of all $n \times n$ matrices constitutes the entire matrix space, and an orbit of a matrix is a manifold in the space. Similarly, an $m_{\text{p}} \times n_{\text{p}}$ matrix pencil belongs to a $2m_{\text{p}}n_{\text{p}}$-dimensional space, and an $(n+m) \times (n+p)$ system pencil (5.2) belongs to an $(n+m)(n+p)+n^2$-dimensional space. In the system pencil case, the dimension count is done with the assumption that the zero matrices in the λ-part of $\mathbf{S}(\lambda)$ are fixed, while E (typically singular) is not.

In the following, we use a general matrix pencil to illustrate the concepts of orbit, bundle, and their (co)dimensions. The *orbit* of an $m_{\text{p}} \times n_{\text{p}}$ matrix pencils is the manifold of equivalent matrix pencils:

$$\mathcal{O}(G - \lambda H) = \{U(G - \lambda H)V^{-1} \; : \; \det(U) \cdot \det(V) \neq 0\}.$$

A *bundle* defines the union of all orbits with the same canonical form but with the eigenvalues unspecified, $\bigcup_{\mu_i} \mathcal{O}(G - \lambda H)$ [1]. We denote the bundle of $G - \lambda H$ by $\mathcal{B}(G - \lambda H)$.

The dimension of $\mathcal{O}(G - \lambda H)$ is equal to the dimension of the tangent space to $\mathcal{O}(G - \lambda H)$ at $G - \lambda H$, which can be expressed by the pencils of the form

$$T_G - \lambda T_H = X(G - \lambda H) - (G - \lambda H)Y,$$

where X is an $m_{\text{p}} \times m_{\text{p}}$ matrix and Y is an $n_{\text{p}} \times n_{\text{p}}$ matrix. In practice, it is more convenient to work with the dimension of the *normal space*, which is the orthogonal complement of the tangent space. The dimension of the normal space is called the *codimension* and is uniquely determined by the Kronecker structure [7, 14]:

$$\operatorname{cod}(G - \lambda H) = c_{\text{Right}} + c_{\text{Left}} + c_{\text{Sing}} + c_{\text{Jor}} + c_{\text{Jor,Sing}}, \tag{5.4}$$

where

$$c_{\text{Right}} = \sum_{\epsilon_i > \epsilon_j} (\epsilon_i - \epsilon_j - 1), \qquad c_{\text{Left}} = \sum_{\eta_i > \eta_j} (\eta_i - \eta_j - 1),$$

$$c_{\text{Sing}} = \sum_{\epsilon_i, \eta_j} (\epsilon_i + \eta_j + 2), \qquad c_{\text{Jor}} = \sum_{i=1}^{q} \sum_{j=1}^{g_i} (2j-1)h_j^{(i)} + \sum_{j=1}^{g_\infty} (2j-1)s_j,$$

and

$$c_{\text{Jor,Sing}} = (r_0 + l_0)\left(\sum_{i=1}^{q}\sum_{j=1}^{g_i} h_j^{(i)} + \sum_{j=1}^{g_\infty} s_j\right).$$

Notably, the codimension of an orbit can also be determined, without knowing the KCF, by computing the dimension of the kernel of Z, where Z is a matrix Kronecker product representation of the tangent space [14, 15]. A reliable way is to apply the singular value decomposition to Z.

Under perturbations the dimension of the tangent space of a matrix pencil orbit may change. Perturbations where the dimension decreases (the codimension increases), which corresponds to less generic (more degenerate) cases, are of special interest in applications especially when the impact can be disastrous. For the most generic rectangular pencil the tangent space of the orbit spans the complete space and hence the codimension is zero. For the most degenerate case these dimensions are reversed for the tangent and normal spaces, respectively.

Since a bundle does not have the eigenvalues specified (or fixed), corresponding to one more degree of freedom for each eigenvalue, the tangent space of a bundle spans one extra dimension for each distinct eigenvalue. In conclusion, the codimension of a bundle is equal to the codimension of the corresponding orbit minus the number of distinct fixed eigenvalues [14, 15].

5.2.4 Cover Rules for Matrix Pencils

The closure decision problem for orbits of general matrix pencils was solved by Pokrzywa [38]; see also later reformulations in [2, 6]. The necessary conditions for an orbit or a bundle of two matrix pencils to be closest neighbors (i.e, has a cover relation) were derived in [2, 6, 38], which was later completed with the sufficient conditions in [15]. These results were also expressed in [15] as *stratification rules*, i.e., combinatorial rules acting on the structure integer partitions defined earlier. The stratification theory has further been developed with the stratification rules for matrix pairs (A,B) [19, 28] and most recently for full normal-rank polynomial matrices $P(\lambda) := P_d\lambda^d + \cdots + P_1\lambda + P_0$ of degree d [29].

Here, we state the theorem of the stratification rules for finding the closest more degenerate (less generic) matrix pencil(s) in the orbit and bundle closure hierarchies, respectively.

Theorem 5.1 (See [15]). *Given the structure integer partitions $\mathcal{L}$, $\mathcal{R}$, and $\mathcal{J}_{\mu_i}$ of $G - \lambda H$, where $\mu_i \in \overline{\mathbf{C}}$, one of the following* if-and-only-if *rules finds $\widetilde{G} - \lambda\widetilde{H}$ such that we have the following:*

A. *$\mathcal{O}(G - \lambda H)$ covers $\mathcal{O}(\widetilde{G} - \lambda\widetilde{H})$:*

(1) *Minimum* rightward *coin move in $\mathcal{R}$ (or $\mathcal{L}$).*

(2) *If the rightmost column in $\mathcal{R}$ (or $\mathcal{L}$) is one single coin, move that coin to a new rightmost column of some $\mathcal{J}_{\mu_i}$ (which may be empty initially).*

(3) *Minimum* leftward *coin move in any $\mathcal{J}_{\mu_i}$.*

(4) *Let k denote the total number of coins in all of the longest (= lowest) rows from all of the $\mathcal{J}_{\mu_i}$. Remove these k coins, add one more coin to the set, and distribute $k+1$ coins to r_p, $p = 0,\ldots,t$, and l_q, $q = 0,\ldots,k-t-1$, such that at least all nonzero columns of $\mathcal{R}$ and $\mathcal{L}$ are given coins.*

B. *$\mathcal{B}(G-\lambda H)$ covers $\mathcal{B}(\widetilde{G}-\lambda\widetilde{H})$:*

(1) *Same as rule 1 on the left.*

(2) *Same as rule 2 on the left, except it is allowed only to start a new set corresponding to a new eigenvalue (i.e., no appending to nonempty sets).*

(3) *Same as rule 3 on the left.*

(4) *Same as rule 4 on the left, but apply only if there exists only one set of coins corresponding to one eigenvalue, or if all sets corresponding to each eigenvalue have at least two rows of coins.*

(5) *Let any pair of eigenvalues coalesce; i.e., take the union of their sets of coins.*

Rules 1 and 2 are not allowed to make coin moves that affect r_0 (or l_0).

For theorems and explicit formulations of the stratification rules for finding covering orbits we refer the reader to [17, 27]. In the examples presented in Section 5.4, we make explicit references to the rules in Theorem 5.1 (e.g., A.(3) when referring to the orbit rule number 3) and we also briefly explain differences with rules for matrix pairs and full normal-rank polynomial matrices in the context of each example. The StratiGraph tool is based on the theory and stratification rules discussed above for different types of problem setups.

5.3 StratiGraph

StratiGraph[1] is a tool for the computation and visualization of closure hierarchy graphs of orbits and bundles of matrices, matrix pencils, and various system pencils (e.g., see [25, 26] and further references therein). Over the years, StratiGraph has grown to a flexible and extendable software tool with a broad range of functionality. In this section, we give a short introduction to the current functionality of StratiGraph version 3.0 and the possibility for extensions. For illustration, the orbit stratification of 2×4 matrix pencils in Figure 5.1 is used.

5.3.1 StratiGraph User Interface

Figure 5.3 presents the StratiGraph user interface including an example of a closure hierarchy graph. In the left margin of the graph, the different codimensions of the equivalence orbits are listed. The orbit of the most generic structure has codimension 0, and the most degenerate structure has codimension 16 ($= 2mn$). We remark that different orbits can have the same codimension (here two with codimension 6) and may (but not necessarily so) include different orbits in their closure hierarchies. For example, $\mathcal{O}(2L_0 \oplus 2J_1(\mu_1))$ is included in the closure of $\mathcal{O}(2L_0 \oplus J_2(\mu_1))$ but not in the closure of $\mathcal{O}(2L_0 \oplus J_1(\mu_1) \oplus J_1(\mu_2))$.

By right-clicking an edge, information of the structural change between two connected nodes is shown. In the figure, we see an example where a 2×2 Jordan block splits into two 1×1 Jordan blocks corresponding to the same eigenvalue μ_1 when going downwards in the hierarchy (to less generic structures).

[1] StratiGraph is an acronym for *Strati*fication *Graph*.

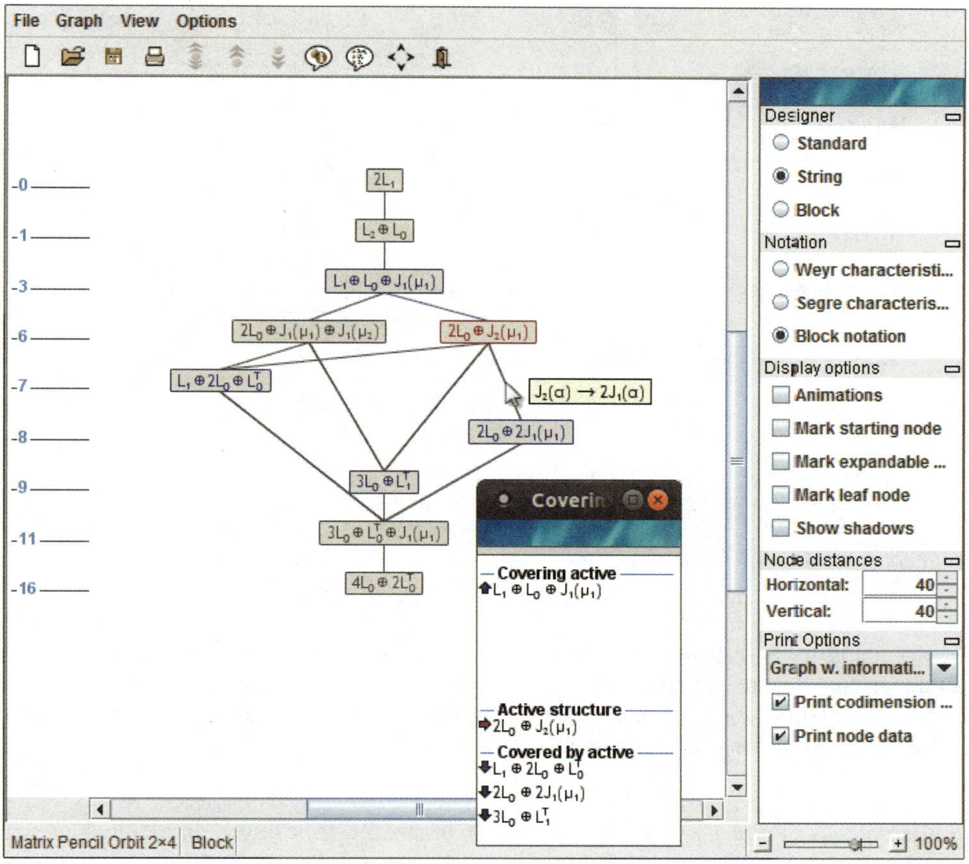

Figure 5.3. Sample user interface of a StratiGraph window showing the complete closure hierarchy of orbits of 2×4 matrix pencils.

In addition to the main window, two dialog windows can be opened that show information on orbit (or bundle) structure relations and a complete list of structures in the stratification, ordered with respect to increasing codimension. In Figure 5.3 one of these windows is shown, labeled “Covering structures.” The information includes the active structure (defined by clicking on a selected node) and the orbits that are covered by and are covering the active orbit, respectively. This feature is especially useful when there are many nodes and edges in the stratification graph.

In the preference side-bar under “Notation” of Figure 5.3, we see that block notation is chosen; i.e., the canonical structures of the orbits are listed as direct sums of blocks in the associated KCF. The notation can be changed by using the representation of the canonical blocks as indices representing Weyr and Segre characteristics, respectively.

The indices representing Segre characteristics are the numbers and the sizes of the different blocks in the KCF, where $\mathcal{R}$ denotes right singular blocks, $\mathcal{L}$ denotes left singular blocks, and $\mathcal{J}(\mu_1)$ denotes Jordan blocks corresponding to eigenvalue μ_1. For the Weyr characteristics, the minimal indices are represented as an ordered integer partition (see definitions in Section 5.2.1). For example, $\mathcal{J}(\mu_1):\ 2,\ 2,\ 1$ is read from left to right as 2

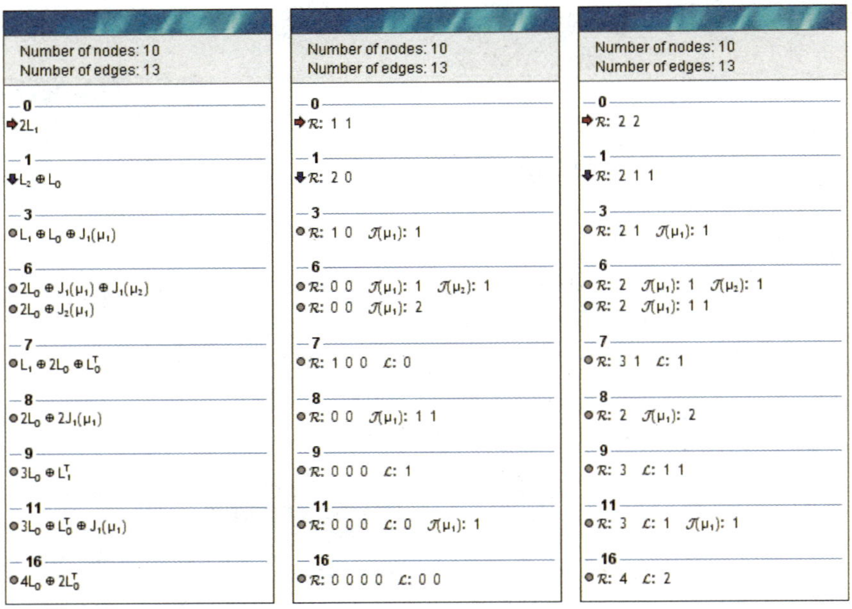

Figure 5.4. A dialog window showing the canonical information of all expanded nodes in the closure hierarchy of Figure 5.3 in three different notations. In the middle, the representation using Segre characteristics is shown and to the right the corresponding information using Weyr characteristics.

Jordan blocks of size 1×1 or larger, 2 blocks of size 2×2 or larger, and 1 block of size 3×3 or larger; i.e., $\mathcal{J}(\mu_1)$ corresponds to $J_2(\mu_1) \oplus J_3(\mu_1)$. Similarly, $\mathcal{R}$: 2, 2, 1 for right singular blocks corresponds to $L_1 \oplus L_2$.

In Figure 5.4, the complete list of canonical information of the closure hierarchy in Figure 5.3 is presented in the different notations provided in StratiGraph. These are obtained from the second dialog window.

5.3.2 Working with StratiGraph

The normal way to use StratiGraph is via the built-in wizard for specifying the type of input problem to be considered (see Figure 5.5). This includes a choice between matrices, matrix pencils, or matrix pairs, a choice between the stratification for orbits or bundles, and finally the starting canonical structure, which will be presented and appear as the starting node in the stratification graph. In the figure, we also see an option for full normal-rank polynomial matrices that is available as a prototype plug-in (see Section 5.3.3).

Graph Expansion

Given a starting structure and applying the stratification rules reviewed in Section 5.2, the graph can be expanded both upwards or downwards from each node, making the graph

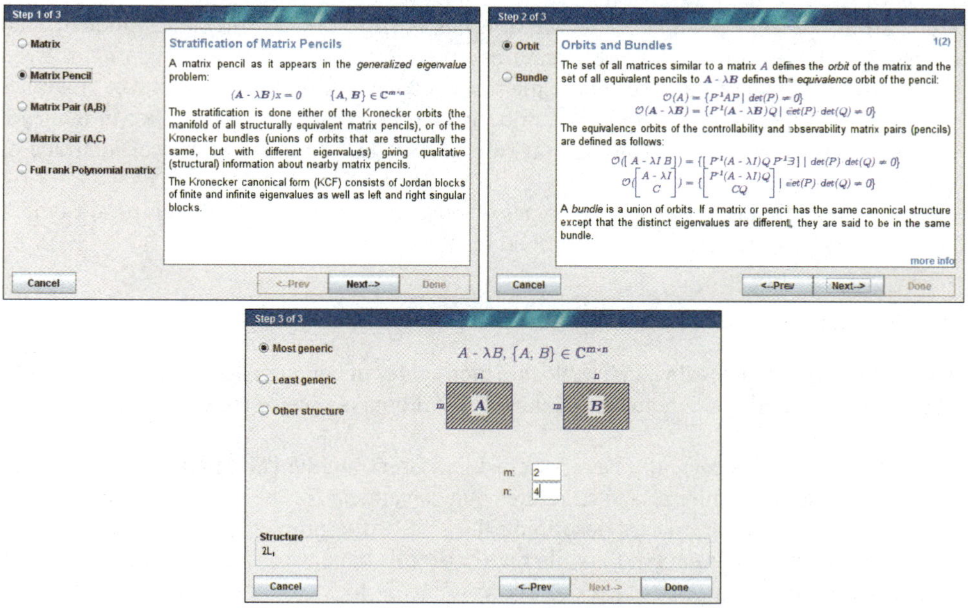

Figure 5.5. Wizard for specifying a starting canonical structure. First, the problem setup is specified, then the choice between orbit and bundle stratifications, and finally the size of the structure. The choices include the most or least generic structure of a specified size or an arbitrary structure for which the blocks are specified in a new window. In the example, the most generic orbit of size 2×4 is specified.

larger and larger. A symbol on the node indicates whether the node can be further expanded or not. The complete stratification can also be rendered. However, the number of nodes grows exponentially with the problem size. Already a modest problem size can result in thousands of nodes and edges. Therefore, the user will be warned if the operation will be very time and memory consuming.

When the graph is expanded, StratiGraph tries to place the nodes to avoid crossing of edges and edges crossing nodes they are not connected to. However, an optimal node placement regarding the least number of crossings is an NP-complete problem, and depending on the problem size we may not be able to afford computing the complete stratification ahead. The algorithm to place nodes is a compromise between computations, optimal node placement, and trying to let already placed nodes stay at a given position. Still the algorithm does not always give a satisfactory result, or we may wish to arrange nodes in a different way. Therefore, StratiGraph provides the functionality to manually move nodes around on each codimension level in the closure hierarchy graph.

5.3.3 Extendable Software

StratiGraph is developed in Java and consists of almost 450 classes and interfaces. Its purpose to compute and visualize stratification graphs means that we must be able to enter a starting canonical structure, show structural and other information, expand the graph and finally layout, visualize, save, and print it.

From a design point of view, the aim has been to create a flexible, modular, and extendable architecture. Most of the features are controlled in a plug-in like system. Even the built-in problem setups like square matrices and pencils are handled just like any other plug-in would be handled. Modules can be loaded at run-time, and the plug-in manager notifies different components in the software of changes, be it the graphical interface or new sets of choices.

StratiGraph has basically three types of plug-ins, namely problem setups, designers, and extensions, which are briefly described below.

Problem Setups

The ability to compute and visualize closure hierarchies of canonical structures, represented both as a graph and in mathematical notation, is, of course, one of the basic functionalities of StratiGraph.

Problem setups basically have four tasks: (i) defining the valid parameters for different kinds of canonical blocks and structures that can appear for the given problem type; (ii) defining valid input parameters describing the size of the problem; e.g., square matrices have only one input parameter, n, while matrix pencils have two, m and n (see Figure 5.5); (iii) given a problem size or a list of valid canonical blocks, define a canonical structure object that will become the start node for further expansion of the stratification graph; (iv) implement a set of stratification rules that specifies structural changes between connected nodes in the expansion upwards and downwards.

Presently, StratiGraph has three built-in problem setups, namely for matrices, matrix pencils, and matrix pairs. Available is also a prototype plug-in for full normal-rank polynomial matrices.

Designers

Designers control the visualization of nodes and edges. Given a set of graphical properties, like color sets and fonts, and a canonical structure, they define how it will be visualized. The procedure is the same regardless of whether it is for printing or for the screen.

StratiGraph has a number of built-in designers, for example, three designers for nodes: a round ball shaped representation (used in Figure 5.10), a node showing canonical information using different representations, and a node that shows the canonical structure represented as blocks in a matrix. The last one is most instructive for structures with few eigenvalues.

Extensions

A program extension plug-in is a fully integrated piece of software that adds functionality to StratiGraph. They can be loaded and unloaded at run-time and interact fully with the internal flow of events while loaded. Even parts of the GUI can be altered.

One extension built in with StratiGraph is used to control the internal commands, like specifying a new starting structure, etc. A prototype of an extension that makes it possible to interact with MATLAB has also been developed. The aim is to be able to compute the canonical structure of an input setup of matrices in MATLAB and then use StratiGraph to investigate nearby structures. Interaction with MATLAB also makes it possible to compute quantitative information between pairs of canonical structures as further discussed in Section 5.5.

5.4 StratiGraph in Applications

In this section, we apply the theory of stratification and illustrate how StratiGraph can be used to analyze qualitative information of some control applications.

The first two examples are problem setups that relate to the solution of linear time-invariant DAE systems $E\dot{x}(t) = Ax(t) + f(t)$. Then follows a study of an electrical circuit descriptor model from [5]. Finally, we briefly review how controllability characteristics of two mechanical system models can be analyzed using stratification theory for matrix pair and full normal-rank polynomial matrix representations [19, 29].

5.4.1 High Nilpotency Index System

Consider a DAE system $E\dot{x}(t) = Ax(t) + f(t)$ with the associated matrix pencil $A - \lambda E = N_6$. The pencil $A - \lambda E$ is regular and has only infinite eigenvalues (its KCF has one block, namely N_6). Indeed, the *(nilpotency) index* of the DAE system is $n_{\text{ind}} = 6$ and the DAE system has only an algebraic part (A-part) which requires that $f(t)$ be sufficiently differentiable (at least $n_{\text{ind}} - 1$ times). Similar differentiability of $f(t)$ is also a requirement for admissible initial conditions of a general DAE system with regular $A - \lambda E$.

Since a square regular matrix pencil is a first order ($d = 1$) square polynomial matrix of full normal-rank, we can compute the stratification using the derived rules in [29]. These rules coincide with the rules in Theorem 5.1, with the restrictions that no singular blocks can exist and there can exist at most n Jordan blocks for each eigenvalue (including infinity).

The orbit closure hierarchy graph computed by StratiGraph is shown in Figure 5.6. By considering the orbit stratification of $A - \lambda E \equiv N_6$ we allow only perturbations that preserve the regularity and keep all eigenvalues at infinity. Starting at $\mathcal{O}(N_6)$ of codimension 6 ($= n$), the Segre (and Weyr) characteristics of the computed canonical structures in the stratification correspond to the dominance (and reversed dominance) ordering of the integer 6. So $\mathcal{O}(6N_1)$ of codimension 36 ($= n^2$) corresponds to the most degenerate canonical structure (E is the zero matrix). For example, to obtain the Weyr characteristics, only minimal leftward coin moves (rule A.(3) in Theorem 5.1) are applied to the sequence $(1,1,1,1,1,1) = \text{conj}(6)$. In addition, the DAE systems corresponding to the pencils in the orbit stratification have their index equal to the size of the largest N_k block in its KCF. Notably, by transversing the closure hierarchy graph downwards we pass (more degenerate) orbits of increasing codimensions while the associated DAE systems have decreasing index. Since all degenerate orbits in the graph belong to the closure of $\mathcal{O}(N_6)$, these cases can be made more generic with arbitrary small perturbations. This is in general not the case for moving downwards in the closure hierarchy.

5.4.2 DAE System with Singular Neighbors

Next we consider a DAE system with both a differential part (D-part) and an A-part. The DAE system is represented by a regular 8×8 pencil $A - \lambda E$ with a known KCF $2J_1(\mu) \oplus 2N_3$. For this study we make no restrictions on the perturbations, which, for

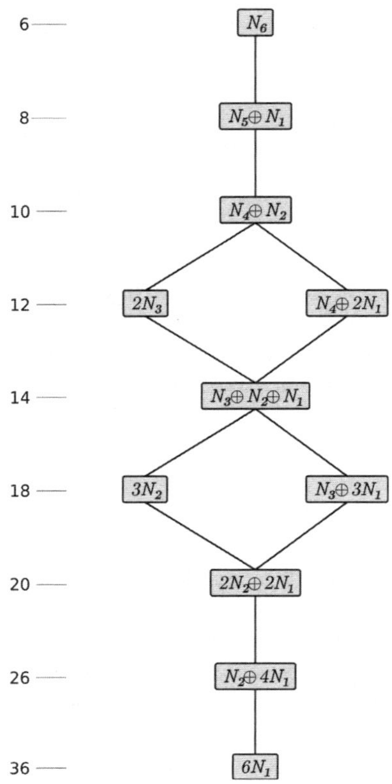

Figure 5.6. The orbit closure hierarchy graph of N_6, a 6×6 matrix pencil of full normal-rank.

example, means that the stratification may include singular pencils. Already for this small-sized system the complete stratification graph contains 1247 nodes and 4015 edges, which is not easily analyzed. In many cases, it is enough to be able to obtain information about neighboring orbits, and here we demonstrate how the stratification rules are used to compute the orbits in the cover of $\mathcal{O}(A - \lambda E)$.

Figure 5.7 shows the orbit of $A - \lambda E$ with its nearest neighbors (above and below) in the closure hierarchy. The active node is $\mathcal{O}(2J_1(\mu) \oplus 2N_3)$ with codimension 16. The orbits downwards in the graph (covered by $\mathcal{O}(A - \lambda E)$) are computed using the A-rules of Theorem 5.1. The orbits upwards in the graph are obtained from the corresponding set of rules for covering orbits (see [15, 27]). For edges going downwards, the corresponding stratification rule and the changes in the KCF are displayed. Actually, only two of the rules are used for $A - \lambda E$. First, rule A.(4) applied to $2J_1(\mu) \oplus 2N_3$ gives four different results: $J_1(\mu) \oplus N_3 \rightarrow L_k \oplus L_{3-k}^T$, where $k = \{0, 1, 2, 3\}$. Then rule A.(3) applied to $2J_1(\mu) \oplus 2N_3$ tells us to do a minimum leftward coin move on $\mathcal{N} = (2, 2, 2)$ giving $\mathcal{N} = (3, 2, 1)$, i.e., $2N_3 \rightarrow N_3 \oplus N_2 \oplus N_1$. Rules A.(1) and A.(2) cannot be applied since the pencil does not have any singular blocks.

We remark that the subgraph in Figure 5.7 has been generated with a not yet official version of StratiGraph. Current version 3.0 considers all eigenvalues in $\overline{\mathbf{C}}$ similarly;

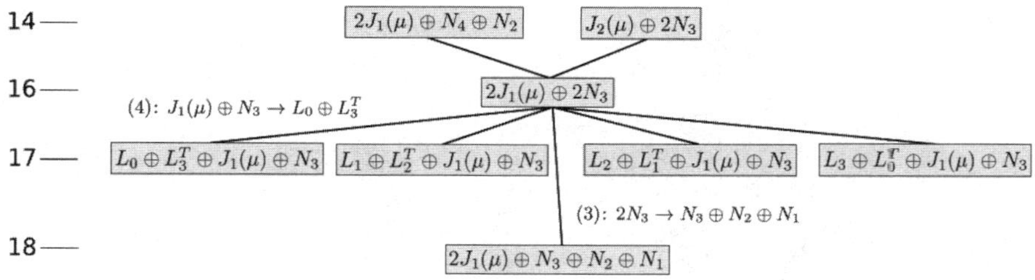

Figure 5.7. Subgraph of the orbit closure hierarchy of the 8×8 matrix pencil $A - \lambda E = 2J_1(\mu) \oplus 2N_3$ with codimension 16. The A-rules in Theorem 5.1 applied for producing orbits covered by $A - \lambda E$ are marked at two of the edges. Rule A.(4) produces the four singular cases with codimension 17. Rule A.(3) produces the regular case with codimension 18.

i.e., finite eigenvalues ($J_1(\mu)$ Jordan blocks) and infinite eigenvalues (N_k blocks) are not automatically treated separately.

In Figure 5.7, the four nodes at codimension level 17 are singular pencils ($\det(A - \lambda E) \equiv 0$ for all λ) and the corresponding $E\dot{x}(t) = Ax(t) + f(t)$ are *singular systems*. Such systems may or may not have a solution, and can even have infinitely many solutions (e.g., see [8, 20] for details).

5.4.3 Electrical Circuit Descriptor Model

In the following example, we consider the electrical circuit in Figure 5.8, where the control input u is the voltage source u_e. The resistor, inductor, and capacitors are denoted by R, L, C_1, and C_2, respectively. Let the state vector be

$$x = \begin{bmatrix} u_{C_1} & u_{C_2} & I_1 & I_2 \end{bmatrix}^T,$$

where u_{C_1} and u_{C_2} are the voltages over C_1 and C_2, respectively, with the corresponding currents I_1 and I_2. Then according to Kirchoff's second law we obtain a descriptor model of the circuit:

$$E\dot{x} = Ax + Bu, \tag{5.5}$$

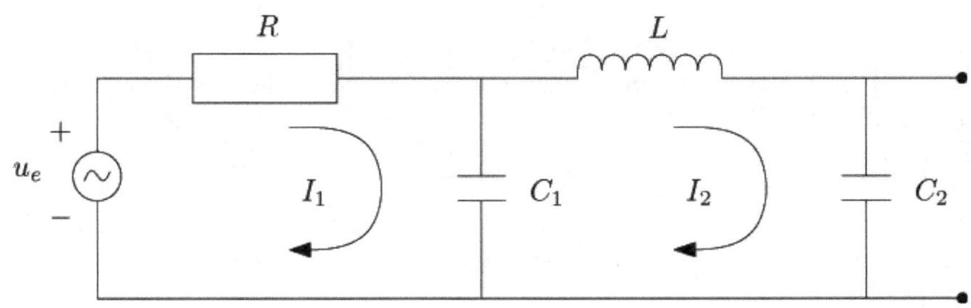

Figure 5.8. Electrical circuit from [5].

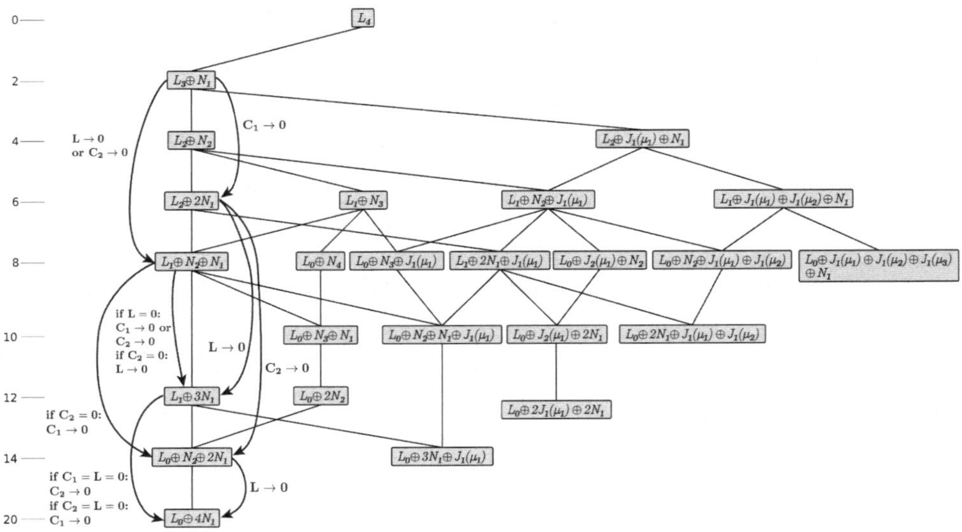

Figure 5.9. Orbit closure hierarchy graph of a 4×5 full normal-rank polynomial matrix. Transitions obtained by changing the parameters in (5.6) are shown in the graph with bold arrowed edges (added manually).

with

$$E = \begin{bmatrix} C_1 & 0 & 0 & 0 \\ 0 & C_2 & 0 & 0 \\ 0 & 0 & -L & 0 \\ 0 & 0 & 0 & 0 \end{bmatrix}, \quad A = \begin{bmatrix} 0 & 0 & 0 & 1 \\ 0 & 0 & 1 & 0 \\ -1 & 1 & 0 & 0 \\ 1 & 0 & 0 & R \end{bmatrix}, \quad B = \begin{bmatrix} 0 \\ 0 \\ 0 \\ -1 \end{bmatrix}. \tag{5.6}$$

The $n \times (n+m)$ system pencil associated with this descriptor model is 4×5 and has the form

$$\mathbf{S}(\lambda) = \begin{bmatrix} A & B \end{bmatrix} - \lambda \begin{bmatrix} E & 0 \end{bmatrix}. \tag{5.7}$$

If R, L, C_1, and C_2 are nonzero, the pencil $A - \lambda E$ is regular. Moreover, the descriptor circuit model (5.5) is *completely controllable* since $\operatorname{rank}\begin{bmatrix} \lambda E - A, & B \end{bmatrix} = n$ for all $\lambda \in \mathbf{C}$ and $\operatorname{rank}\begin{bmatrix} E, & B \end{bmatrix} = n$ [47].

The orbit closure hierarchy graph of a general 4×5 matrix pencil does not take account of any special structure of the system pencil. Instead, we study the system pencil $\mathbf{S}(\lambda)$ in the form of a first order polynomial matrix $P(\lambda) = H\lambda - G$, with $G = \begin{bmatrix} A & B \end{bmatrix}$ and $H = \begin{bmatrix} E & 0 \end{bmatrix}$. $P(\lambda)$ has full normal-rank ($= 4$), and we can make use of the results in [29] to compute the orbit closure hierarchy graph of a first order full normal-rank polynomial matrix, which is shown in Figure 5.9. The restrictions of the stratification rules for the full normal-rank polynomial matrices compared to the rules for general matrix pencils are the following: No left singular blocks (L_k^T) can exist, the number of right singular blocks (L_k) must be m, there can exist at most n Jordan blocks for each eigenvalue (including infinite eigenvalues).

Consider the circuit with the parameters $R = 330$, $L = 1.1$, and $C_1 = C_2 = 10^{-4}$. Then $\mathbf{S}(\lambda)$ (5.7) has the KCF $L_3 \oplus N_1$, it is completely controllable, and the pair (E, A)

is of index one with KCF $3J_1(\alpha) \oplus N_1$. In the graph of Figure 5.9, this circuit belongs to $\mathcal{O}(L_3 \oplus N_1)$ with codimension 2. Note that we could omit the most generic $\mathcal{O}(L_4)$ in the graph since the matrix E in (5.6) is always singular. However, in the current version of StratiGraph no such constraints can be specified or imposed automatically.

What happens when we let the parameters R, L, C_1, and C_2 approach zero? By setting one parameter after another to zero we get canonical structures that belong to less generic orbits further down in the closure hierarchy graph. For example, if we let $L \to 0$ or $C_2 \to 0$, the pencil (5.7) will approach $\mathcal{O}(L_1 \oplus N_2 \oplus N_1)$ of codimension 8. All transitions are shown in Figure 5.9. Note that the value of R does not play any role for the computed eigenstructure; i.e., R can be arbitrary. We leave it to the reader to interpret the results from the application's point of view.

A natural question one may ask is, Why do we not get systems that belong to all orbits in the closure hierarchy graph by varying the parameters? The reason is that the matrices in the circuit model (5.6) are structured with the entries 0 and ± 1 fixed.

5.4.4 Other Examples—Two Mechanical System Models

In this section, we briefly review how the stratification of two mechanical systems can be handled. The first is a linearized model of a uniform platform with two degrees of freedom and the second is a half-car suspension model with four degrees of freedom. These models are analyzed via the state-space model (where $\det(E) \neq 0$ in (5.1)) and via a polynomial matrix representation with a nonsingular highest degree coefficient matrix. Only partial results are presented below. For the complete stories we refer the reader to two recent papers [19, 29].

Uniform Platform Model with Two Degrees of Freedom

First we discuss the stratification of a linearized model of a uniform platform supported in both ends by springs. The linearization of the equations of motion can be written either on the form of a state-space model $\dot{x} = Ax + Bu$, where $A \in \mathbf{C}^{4\times 4}$ and $B \in \mathbf{C}^{4\times 1}$, or as a second order differential equation of the form $M\ddot{x} + C\dot{x} + Kx = Su$, where M, C, $K \in \mathbf{C}^{2\times 2}$, $S \in \mathbf{C}^{2\times 1}$, and $\det(M) \neq 0$. In the most general form, when analyzing the controllability of the platform the stratification can be performed by considering $\begin{bmatrix} A - \lambda I & B \end{bmatrix}$ as a general 4×5 matrix pencil. However, this leads to canonical structures which cannot exist for the platform appearing in the closure hierarchy. In [19], the stratification of the state-space model is studied via the closure hierarchy of the associated 4×5 matrix pair (A, B). In [29], the corresponding study is instead performed on the right linearization

$$\left[\begin{array}{c|c} A - \lambda I_4 & B \end{array}\right] = \left[\begin{array}{cc|c} \lambda I_2 & M^{-1}K & M^{-1}S \\ -I_2 & \lambda I_2 + M^{-1}C & 0 \end{array}\right]$$

of the associated 2×3 full normal-rank polynomial matrix to $M\ddot{x} + C\dot{x} + Kx = Su$.

The bundle closure hierarchy graph of a general 4×5 matrix pencil is shown in Figure 5.10. The light gray area marks the closure hierarchy of a 4×5 matrix pair and the dark gray area marks the closure hierarchy of the right linearization of a 2×3 full normal-rank polynomial matrix. As we see, the latter hierarchies form subgraphs of the complete graph and of each other. This follows since the corresponding system pencils have a special predetermined structure which imposes restrictions on possible canonical structures.

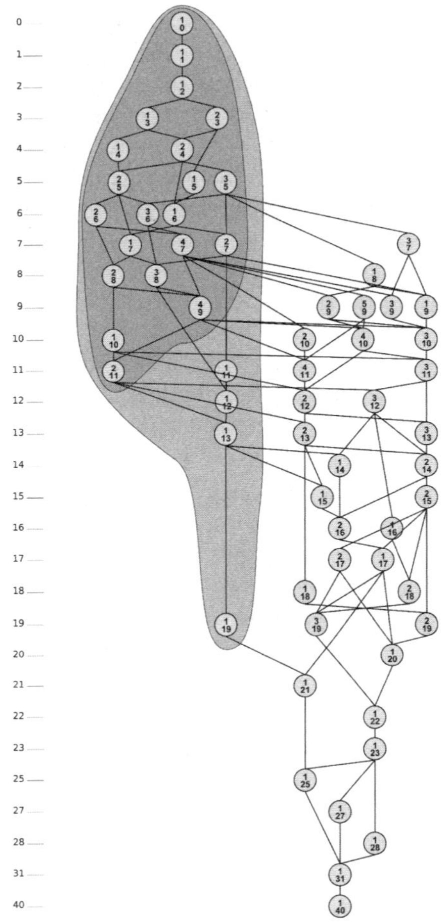

Figure 5.10. Bundle closure hierarchy graph of a general 4×5 matrix pencil. The light gray area marks the closure hierarchy of a 4×5 matrix pair (A, B), and the dark gray area marks the closure hierarchy of the right linearization of a 2×3 full normal-rank polynomial matrix of order two.

The nodes in Figure 5.10 are shown using the round-ball representation labeled with an edge number (at the top) and the codimension of the bundle below. The associated canonical structure information of each node can be obtained in a dialog window (see examples in Figure 5.4).

Half-Car Suspension Model with Four Degrees of Freedom

A half-car passive suspension model is studied in [29]. The suspension model can be expressed as a fourth order differential equation

$$P_4 x^{(4)} + P_3 x^{(3)} + P_2 x^{(2)} + P_1 x^{(1)} + P_0 x = Q_2 u^{(2)} + Q_1 u^{(1)} + Q_0 u,$$

where $P_k \in \mathbf{C}^{3\times 3}$, $Q_k \in \mathbf{C}^{3\times 3}$, and $\det(P_4) \neq 0$. The resulting right linearization of the associated 3×6 polynomial matrix is

$$\begin{bmatrix} A - \lambda I_{12} \mid B \end{bmatrix} = \left[\begin{array}{cccc|c} \lambda I_3 & & & P_4^{-1}P_0 & P_4^{-1}Q_0 \\ -I_3 & \lambda I_3 & & P_4^{-1}P_1 & P_4^{-1}Q_1 \\ & -I_3 & \lambda I_3 & P_4^{-1}P_2 & P_4^{-1}Q_2 \\ & & -I_3 & \lambda I_3 + P_4^{-1}P_3 & 0 \end{array} \right],$$

where $A \in \mathbf{C}^{12\times 12}$ and $B \in \mathbf{C}^{12\times 3}$. The complete stratification of this linearization is rather large. So a possibility when analyzing, for example, the controllability of the suspension model is to compute the subgraph representing all the controllable orbits (or bundles) together with the closest uncontrollable ones. This subgraph has a moderate size of 35 nodes and 61 edges. For further details we refer the reader to [29], where also the stratification graph is presented.

5.5 Future Work and Some Open Problems

In this section, we discuss possible new features and setups to StratiGraph and related ongoing and planned work as well as some open problems.

One feature is a plug-in for importing and exporting matrix and system pencils between MATLAB and StratiGraph. MATLAB can also be used as a computing engine to StratiGraph, e.g., to compute quantitative information like bounds on the distance between orbits and bundles in the stratification graph.

Existing problem setups applicable to descriptor systems (5.1) with a singular E are the matrix pencil setup and in some cases the full normal-rank polynomial matrix setup (currently available only as a prototype). Examples of future setups are general polynomial matrices, descriptor (or singular) system pencils (5.2), and subpencils corresponding to particular systems.

Version 3.0 of StratiGraph does not separate the finite and infinite eigenvalues when computing and visualizing closure hierarchy graphs. This separation is especially important when we consider general descriptor systems, and it will be available in a coming release. A prototype has been implemented in the full normal-rank polynomial matrix setup, as illustrated in Figure 5.6. In the following, we discuss some of these topics.

5.5.1 Matrix Canonical Structure Toolbox

For the computational routines and the interaction between MATLAB and StratiGraph the *Matrix Canonical Structure* (MCS) toolbox has been developed [24]. The prototype toolbox includes a framework with new data-type objects for representing canonical structures and several routines for handling the interface to StratiGraph.

The current version of MCS toolbox includes routines for computing the canonical structure information using staircase algorithms. These are prototype MATLAB implementations based on the existing GUPTRI routines [9, 10, 30] for matrix pencils and the controllability and observability staircase forms [43]. Hence, given a linearized model we can compute its Kronecker structure and then let StratiGraph determine and visualize nearby structures in the closure hierarchy. The toolbox also includes prototype implementations of imposed GUPTRI forms, which impose a given canonical structure on an input (matrix, matrix pencil, or matrix pair) with respect to specified deflation tolerances (*tol* and *gap*).

Other MCS functionality includes routines for computing the distance (measured in the Frobenius norm) between two sets of inputs (matrices, matrix pencils, or matrix pairs), and routines for computing bounds of the distance to a less (or more) generic orbit/bundle [18].

5.5.2 Bounds and Other Types of Quantitative Information

The lower bounds implemented in MCS toolbox are of Eckart–Young type and derived from the matrix representation of the tangent space of the orbit of a matrix or a system pencil [14, 18]. The upper bounds are based on staircase regularizing perturbations [18], which make use of the routines for computing an imposed GUPTRI form. It is well known that these bounds are sometimes too conservative. Future work includes improving the algorithms for computing the bounds for matrices and matrix pencils, especially the upper bound. And it is still open how to compute upper and lower bounds for polynomial matrices and some other particular systems. Another future extension of StratiGraph could be the possibility to compute pseudospectra information (e.g., see [40]) along with distance bounds in the closure hierarchy graph.

Important quantitative information in control applications includes the distances to uncontrollability and unobservability [16, 37]. In [19], a linearized nominal longitudinal model of a Boeing 747 [45] is studied with regard to its controllability characteristics; i.e., the stratification is done on the controllability pair (A, B). The model has 5 states and 5 inputs, and the KCF of (A, B) with the chosen parameters is $2L_2 \oplus L_1 \oplus 2L_0$; i.e., the system is controllable. Using the MATLAB routines of Mengi [35], the computed distance to uncontrollability is within the interval $\mathcal{I} = (3.0323 \cdot 10^{-2},\ 3.0332 \cdot 10^{-2})$. The results in [19] show that there exist three orbits in the closure hierarchy for which $\mathcal{I}$ is included in the intervals defined by the computed lower and upper bounds. These are the orbits with KCF $L_2 \oplus 2L_1 \oplus 2L_0 \oplus J_1(\mu)$, $2L_2 \oplus 3L_0 \oplus J_1(\mu)$, and $L_3 \oplus L_1 \oplus 3L_0 \oplus J_1(\mu)$, respectively. In Figure 5.11, a selection of orbits covered by $\mathcal{O}(A, B) \equiv \mathcal{O}(2L_2 \oplus L_1 \oplus 2L_0)$ in the orbit closure hierarchy of 5×10 matrix pairs is shown. We can also see bound information in the figure delivered by MCS toolbox. For example, the interval $\mathcal{I}$ is within the bounds of $L_2 \oplus 2L_1 \oplus 2L_0 \oplus J_1(\mu_1)$ and $2L_2 \oplus 3L_0 \oplus J_1(\mu_1)$. We remark that these bounds are not as tight as the bounds to the more degenerate uncontrollable system $L_4 \oplus 4L_0 \oplus J_1(\mu_1)$ or the controllable system $L_3 \oplus L_2 \oplus 3L_0$, which has one less input than (A, B) for controlling the states.

5.5.3 Future Problem Setups in StratiGraph

Finally, we introduce examples of new problem setups that we aim to include in future versions of StratiGraph. The theory and definitions that are needed to create a problem setup, say for the system pencil $\mathbf{S}(\lambda)$ in (5.2), are the following: (1) The orbit of $\mathbf{S}(\lambda)$ must be determined and well defined (all and only those $\mathbf{S}(\lambda)$ with the same structural elements belong to the same orbit). (2) The codimension of the orbit must be determined and computable from the structural information of $\mathbf{S}(\lambda)$. (3) The closure relation between all possible orbits must be determined. (4) Finally, the cover relation between all pairs of orbits must be derived and expressed as stratification rules on associated integer partitions.

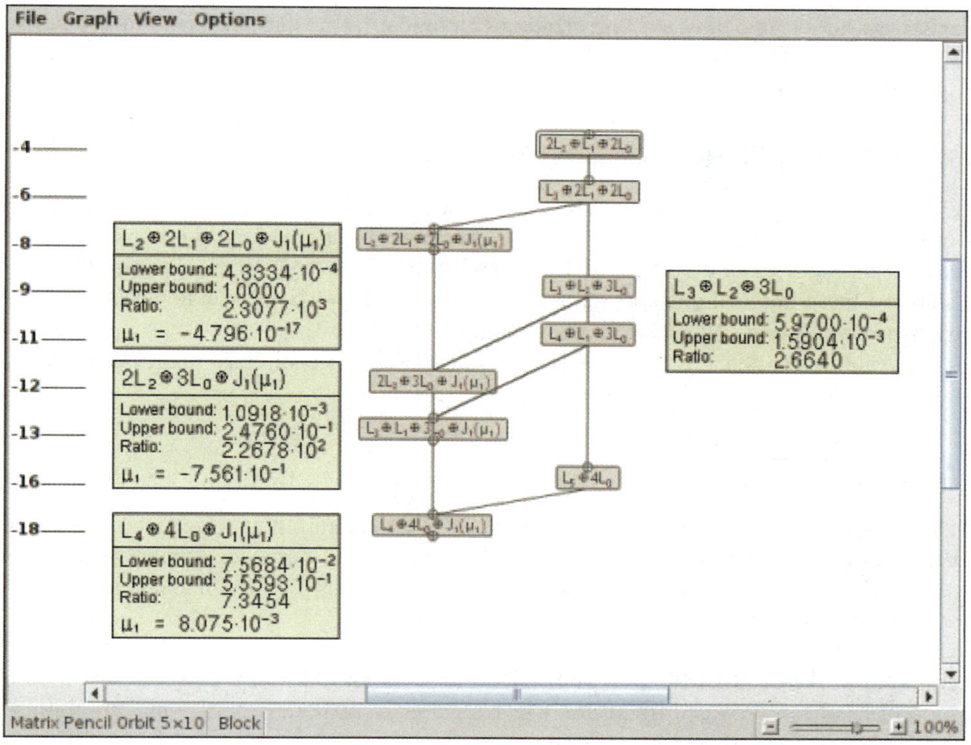

Figure 5.11. Subgraph of the orbit closure hierarchy of a 5×10 matrix pair (A, B). The node with double line frame is the active node corresponding to a Boeing 747 under flight ($\mathcal{O}(2L_2 \oplus L_1 \oplus 2L_0)$), and the symbol $\bigoplus$ on the frame indicates unexpanded edges in the graph. The bounds to a selection of orbits are shown together with the eigenvalue of the imposed structure.

In general, the stratification theory and covering relations for many of the problem setups listed below are open challenging problems. This also includes definitions of transformations and canonical forms which in turn define the orbits and bundles to be considered. Some related publications include relevant results on stratification theory [4, 12, 22, 23], canonical forms for descriptor systems [21, 34, 41], and staircase and condensed forms for descriptor systems [3, 33, 36].

State-Space Descriptor and Singular Systems

One goal is to cover generalized state-space systems (5.1) with full generality, including the matrix tuples (E, A, B), (E, A, B, C), and (E, A, B, C, D), where $A - \lambda E$ is regular or even singular. A follow-up extension to these problem setups would be to allow the matrices to be rectangular, which corresponds to various singular state-space systems.

Polynomial Matrices

In the case of full normal-rank polynomial matrices, the stratification can be done on the corresponding right linearization [29]. However, for general polynomial matrices there exists, as far as we know, no linearization (canonical form) that preserves the complete eigenstructure. This means that a perturbation in the linearization does not have a one-to-one correspondence with perturbations in the coefficients of the polynomial matrix.

System Pencils with Structure and Other Properties

Other problem setups of interest are systems with structure, e.g., fixed elements in the system matrices or with a specified block structure. In addition, it is common that the coefficient matrices of a system model are symmetric, skew-symmetric, Hamiltonian, etc. Many of these applications correspond to matrices and matrix pencils under congruence transformations (e.g., $\mathcal{O}(A) = \{S^T AS \,:\, \det(S) \neq 0\}$) [11, 13].

Another challenge is to consider the stratification of generalized matrix products including periodic eigenvalue problems (e.g., see [31, 39]).

Acknowledgments

The authors are grateful to Alan Edelman at MIT, Erik Elmroth at Umeå University, and Paul Van Dooren at UCL Louvain-la-Neuve, co-authors of our fundamental work partly reviewed in this chapter. We also thank Volker Mehrmann at TU Berlin and Andras Varga at DLR for suggesting interesting control applications.

This work was supported by the Swedish Foundation for Strategic Research under grant A3 02:128 and UMIT Research Laboratory via an EU Mål 2 project. In addition, support has been provided by eSSENCE, a strategic collaborative e-Science programme funded by the Swedish Research Council.

Bibliography

[1] V. I. ARNOLD, *On matrices depending on parameters*, Russian Math. Surveys, 26 (1971), pp. 29–43.

[2] K. BONGARTZ, *On degenerations and extensions of finite dimensional modules*, Adv. Math., 121 (1996), pp. 245–287.

[3] A. BUNSE-GERSTNER, R. BYERS, V. MEHRMANN, AND N. K. NICHOLS, *Feedback design for regularizing descriptor systems*, Linear Algebra Appl., 299 (1999), pp. 119–151.

[4] J. CLOTET AND M. D. MAGRET, *Dimension of orbits of linear time-invariant singular systems under restricted system equivalence*, Linear Algebra Appl., 429 (2008), pp. 1102–1113.

[5] L. DAI, *Singular Control Systems*, Lecture Notes in Control and Inform. Sci. 118, Springer-Verlag, Berlin, 1989.

[6] I. DE HOYOS, *Points of continuity of the Kronecker canonical form*, SIAM J. Matrix Anal. Appl., 11 (1990), pp. 278–300.

[7] J. DEMMEL AND A. EDELMAN, *The dimension of matrices (matrix pencils) with given Jordan (Kronecker) canonical forms*, Linear Algebra Appl., 230 (1995), pp. 61–87.

[8] J. DEMMEL AND B. KÅGSTRÖM, *Stably computing the Kronecker structure and reducing subspaces of singular pencils* $A - \lambda B$ *for uncertain data*, in Large Scale Eigenvalue Problems, J. Cullum and R. A. Willoughby, eds., North–Holland, Amsterdam, 1986, pp. 283–323.

[9] J. DEMMEL AND B. KÅGSTRÖM, *The generalized Schur decomposition of an arbitrary pencil* $A - \lambda B$*: Robust software with error bounds and applications. Part* I: *Theory and algorithms*, ACM Trans. Math. Software, 19 (1993), pp. 160–174.

[10] J. DEMMEL AND B. KÅGSTRÖM, *The generalized Schur decomposition of an arbitrary pencil* $A - \lambda B$*: Robust software with error bounds and applications. Part* II: *Software and applications*, ACM Trans. Math. Software, 19 (1993), pp. 175–201.

[11] F. DE TERÁN AND F. M. DOPICO, *The solution of the equation* $XA + AX^T = 0$ *and its application to the theory of orbits*, Linear Algebra Appl., 434 (2011), pp. 44–67.

[12] A. DÍAZ, M. I. GARCÍA-PLANAS, AND S. TARRAGONA, *Local perturbations of generalized systems under feedback and derivative feedback*, Comput. Math. Appl., 56 (2008), pp. 988–1000.

[13] A. DMYTRYSHYN, V. FUTORNY, AND V. V. SERGEICHUK, *Miniversal deformations of matrices of bilinear forms*, Linear Algebra Appl., 436 (2012), pp. 2670–2700.

[14] A. EDELMAN, E. ELMROTH, AND B. KÅGSTRÖM, *A geometric approach to perturbation theory of matrices and matrix pencils. Part* I: *Versal deformations*, SIAM J. Matrix Anal. Appl., 18 (1997), pp. 653–692.

[15] A. EDELMAN, E. ELMROTH, AND B. KÅGSTRÖM, *A geometric approach to perturbation theory of matrices and matrix pencils. Part* II: *A stratification-enhanced staircase algorithm*, SIAM J. Matrix Anal. Appl., 20 (1999), pp. 667–699.

[16] R. EISING, *Between controllable and uncontrollable*, Systems Control Lett., 4 (1984), pp. 263–264.

[17] E. ELMROTH, P. JOHANSSON, AND B. KÅGSTRÖM, *Computation and presentation of graph displaying closure hierarchies of Jordan and Kronecker structures*, Numer. Linear Algebra Appl., 8 (2001), pp. 381–399.

[18] E. ELMROTH, P. JOHANSSON, AND B. KÅGSTRÖM, *Bounds for the distance between nearby Jordan and Kronecker structures in a closure hierarchy*, J. Math. Sci. (N.Y.), 114 (2003), pp. 1765–1779.

[19] E. ELMROTH, S. JOHANSSON, AND B. KÅGSTRÖM, *Stratification of controllability and observability pairs—Theory and use in applications*, SIAM J. Matrix Anal. Appl., 31 (2009), pp. 203–226.

[20] F. GANTMACHER, *The Theory of Matrices*, Vols. I and II (transl.), Chelsea, New York, 1959.

[21] U. HELMKE AND M. A. SHAYMAN, *A canonical form for controllable singular systems*, Systems Control Lett., 12 (1989), pp. 111–122.

[22] D. HINRICHSEN AND J. O'HALLORAN, *A pencil approach to high gain feedback and generalized state space systems*, Kybernetika, 31 (1995), pp. 109–139.

[23] D. HINRICHSEN AND J. O'HALLORAN, *Limits of generalized state space systems under proportional and derivative feedback*, Math. Control Signals Systems, 10 (1997), pp. 97–124.

[24] P. JOHANSSON, *Matrix Canonical Structure Toolbox*, Tech. Rep. UMINF 06.15, Department of Computing Science, Umeå University, Umeå, Sweden, 2006.

[25] P. JOHANSSON, *Software Tools for Matrix Canonical Computations and Web-Based Software Library Environments*, Ph.D. thesis, Department of Computing Science, Umeå University, Umeå, Sweden, 2006.

[26] P. JOHANSSON, *StratiGraph homepage*, Department of Computing Science, Umeå University, Umeå, Sweden, 2011; http://www.cs.umu.se/english/research/groups/matrix-computations/stratigraph.

[27] S. JOHANSSON, *Reviewing the Closure Hierarchy of Orbits and Bundles of System Pencils and Their Canonical Forms*, Tech. Rep. UMINF 09.02, Department of Computing Science, Umeå University, Umeå, Sweden, 2009.

[28] S. JOHANSSON, *Tools for Control System Design—Stratification of Matrix Pairs and Periodic Riccati Differential Equation Solvers*, Ph.D. thesis, Report UMINF 09.04, Department of Computing Science, Umeå University, Umeå, Sweden, 2009.

[29] S. JOHANSSON, B. KÅGSTRÖM, AND P. VAN DOOREN, *Stratification of Full Rank Polynomial Matrices*, Tech. Rep. UMINF 11.11, Department of Computing Science, Umeå University, Umeå, Sweden, 2011.

[30] B. KÅGSTRÖM, *GUPTRI homepage*, Department of Computing Science, Umeå University, Umeå, Sweden, 2011; http://www8.cs.umu.se/research/nla/singular_pairs/guptri/.

[31] B. KÅGSTRÖM, L. KARLSSON, AND D. KRESSNER, *Computing codimensions and generic canonical forms for generalized matrix products*, Electron. J. Linear Algebra, 22 (2011), pp. 277–309.

[32] V. N. KUBLANOVSKAYA, *On a method of solving the complete eigenvalue problem for a degenerate matrix*, Zh. Vychisl. Mat. Fiz., 6 (1966), pp. 611–620 (in Russian). USSR Comput. Math. Phys., 6 (1968), pp. 1–16 (in English).

[33] P. KUNKEL, V. MEHRMANN, AND W. RATH, *Analysis and numerical solution of control problems in descriptor form*, Math. Control Signals Systems, 14 (2001), pp. 29–61.

[34] J. J. LOISEAU, K. ÖZÇALDIRAN, M. MALABRE, AND N. KARCANIAS, *Feedback canonical forms of singular systems*, Kybernetika, 27 (1991), pp. 289–305.

[35] E. MENGI, *Software for Robust Stability and Controllability Measures*, Computer Science Department, New York University, New York, NY, 2008; http://www.cs.nyu.edu/~mengi/robuststability.html.

[36] P. MISRA, P. VAN DOOREN, AND A. VARGA, *Computation of structural invariants of generalized state space systems*, Automatica, 30 (1994), pp. 1921–1936.

[37] C. PAIGE, *Properties of numerical algorithms related to computing controllability*, IEEE Trans. Automat. Control, AC-26 (1981), pp. 130–138.

[38] A. POKRZYWA, *On perturbations and the equivalence orbit of a matrix pencil*, Linear Algebra Appl., 82 (1986), pp. 99–121.

[39] V.V. SERGEICHUK, *Computation of canonical matrices for chains and cycles of linear mappings*, Linear Algebra Appl., 376 (2004), pp. 235–263.

[40] L. TREFETHEN AND M. EMBREE, *Spectra and Pseudospectra: The Behavior of Nonnormal Matrices and Operators*, Princeton University Press, Princeton, NJ, 2005.

[41] D. VAFIADIS AND N. KARCANIAS, *Canonical forms for descriptor systems under restricted system equivalence*, Automatica, 33 (1997), pp. 955–958.

[42] P. VAN DOOREN, *The computation of Kronecker's canonical form of a singular pencil*, Linear Algebra Appl., 27 (1979), pp. 103–141.

[43] P. VAN DOOREN, *The generalized eigenstructure problem in linear system theory*, IEEE Trans. Automat. Control, AC-26 (1981), pp. 111–129.

[44] P. VAN DOOREN, *Reducing subspaces: Definitions, properties and algorithms.*, in Matrix Pencils (Proc. Pite Havsbad, 1982), B. Kågström and A. Ruhe, eds., Lecture Notes in Math. 973, Springer-Verlag, Berlin, 1983, pp. 58–73.

[45] A. VARGA, *On designing least order residual generators for fault detection and isolation*, in Proceedings of the 16th International Conference on Control Systems and Computer Science (CSCS), Bucharest, Rumania, 2007, pp. pp. 323–330.

[46] G. C. VERGHESE, B. C. LÉVY, AND T. KAILATH, *A generalized state-space for singular systems*, IEEE Trans. Automat. Control, AC-26 (1981), pp. 811–830.

[47] E. YIP AND R. SINCOVEC, *Solvability, controllability and observability of continuous descriptor systems*, IEEE Trans. Automat. Control, AC-26 (1981), pp. 702–707.

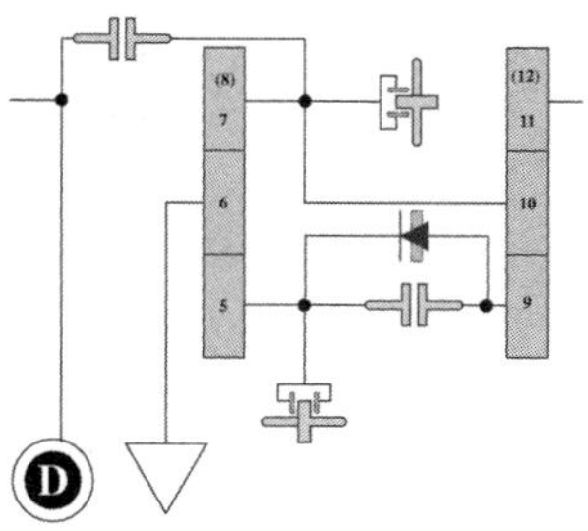

Chapter 6

Descriptor System Techniques in Solving $\mathcal{H}_{2/\infty}$-Optimal Fault Detection and Isolation Problems

Andreas Varga

This chapter illustrates the effectiveness of descriptor systems based algorithms in solving $\mathcal{H}_{2/\infty}$-optimal fault detection and isolation problems. The descriptor system based formulation allows the solution of these problems in the most general setting by eliminating all technical assumptions required when using standard approaches. The underlying numerical algorithms to compute rational nullspace bases, inner-outer factorizations, or proper coprime factorizations are based on descriptor system representations and rely on orthogonal matrix pencil reductions. The developed integrated computational approaches fully exploit the structural aspects at each solution step and produce fault detectors of least orders.

6.1 Introduction

We consider time-invariant linear *descriptor* system representations of the form

$$\begin{aligned} E\lambda x(t) &= Ax(t)+Bu(t), \\ y(t) &= Cx(t)+Du(t), \end{aligned} \tag{6.1}$$

where $x(t)$ is the system state vector, $u(t)$ is the system input vector, and $y(t)$ is the system output vectors, and where $\lambda x(t) = \dot{x}(t)$ for a continuous-time system and $\lambda x(t) = x(t+1)$ for a discrete-time system. All intervening matrices in (6.1) are real, and we assume E is a square matrix, which can be in general singular. A *standard* system corresponds to $E = I$. Continuous-time descriptor systems with singular E arise frequently from modeling interconnected systems with standard tools like Simulink or object-oriented modeling with Modelica. For example, descriptor models are commonly employed to model constrained mechanical systems (e.g., contact problems) [6]. Discrete-time descriptor representations are frequently used to model economic processes.

For our study, the importance of descriptor system representations lies primarily in solving computational problems involving the manipulation of rational matrices in a numerically reliable way. Note that polynomial matrices can be considered a particular case of rational matrices. Each real rational matrix $R(\lambda)$ can be interpreted as the *transfer-function matrix* (TFM) of a continuous- or discrete-time descriptor system of the form (6.1) from input u to output y, where the associated quadruple $(A - \lambda E, B, C, D)$, with $A - \lambda E$ assumed *regular*, satisfies

$$R(\lambda) = C(\lambda E - A)^{-1}B + D.$$

Here, λ stands for the *frequency variable*, which is either the complex variable s in the Laplace-transform or the complex variable z in the Z-transform for a continuous- or discrete-time realization, respectively. $R(\lambda)$ is a *proper* TFM if $R(\infty)$ is finite; otherwise it is called *improper*. $R(\lambda)$ is called *strictly proper* if $R(\infty) = 0$. A proper $R(\lambda)$ has a standard system realization with $E = I$. Many basic operations with improper and proper rational matrices (e.g., sum, difference, product, transposition, etc.) can be immediately transcribed in equivalent descriptor or standard system realizations, respectively. However, there are some operations on proper TFMs like conjugation or inversion, which cannot generally be expressed as standard system realizations, but can always be expressed as descriptor system realizations.

Many numerically reliable computational methods for descriptor systems are based on matrix pencil manipulation using orthogonal transformation matrices. Examples where pencil reduction techniques play a crucial role are the computation of controllability/observability staircase forms [20, 23], determination of infinite poles/zeros structure and minimal indices [14], performing additive spectral separations like finite-infinite or stable-unstable splitting [12], conversions between descriptor representations and rational matrices [22], and determination of various inverses [27]. We must say that, virtually, matrix pencil techniques can be used for every linear system analysis and design computation! Interestingly, the underlying pencil reduction tools are useful even in solving several difficult control problems for standard systems, as, for example, the solution of discrete-time Riccati equations [21], computation of infinite zeros [14], determination of the Kronecker structure of the system pencil [2, 19], computation of coprime factorizations [17, 25], and computation of inner-outer and spectral factorizations [16, 18].

Based on the above-mentioned algorithms, a DESCRIPTOR SYSTEMS toolbox for MATLAB has been developed by the author [11, 26]. Initially, the toolbox was mainly intended to provide an extended functionality for the CONTROL TOOLBOX of MATLAB, which formally supported descriptor systems, but only with nonsingular E. Consequently, some functions in the DESCRIPTOR SYSTEMS toolbox simply represent extensions of functions already present in the CONTROL TOOLBOX. Many functions are merely interfaces to powerful structure exploiting computational routines implemented as MEX-functions which call Fortran routines from the high performance linear algebra numerical library LAPACK [1] and control library SLICOT [11]. The implementations of all functions exploit the best of MATLAB and Fortran programming, by trying to balance the matrix manipulation power of MATLAB with the intrinsic high efficiency of carefully implemented structure-exploiting Fortran codes. For the contents of the current version of the toolbox (presently 1.06), see the associated web site.[2]

[2]http://www.robotic.de/fileadmin/control/varga/Desc_contents_01.m

The existence of the powerful descriptor systems algorithms and of the associated software tools in the DESCRIPTOR SYSTEMS toolbox allow one to formulate and implement a new generation of numerically reliable algorithms to solve *fault detection* (FD) and *fault detection and isolation* (FDI) problems [29, 30, 32, 33, 34, 35, 36, 38, 39]. The new computational methods have several distinctive features. The solution methods are *general*, being applicable to both continuous- and discrete-time systems in standard or descriptor system forms. Therefore, a solution can always be computed whenever one exists and no technical assumptions are necessary for the computation of the solution. The new synthesis algorithms belong to the so-called family of *integrated algorithms*, where the successive computational steps are strongly connected such that all structural information at the end of a computational step is fully exploited in the next step. For this, state-space representation based updating techniques of the intermediary solutions are employed, where the underlying descriptor system based formulation is of primordial importance. A collection of software tools for the synthesis of residual generators for fault detection has been implemented into a FAULT DETECTION toolbox by the author [31, 37].

In this chapter, we present computational procedures for the optimal synthesis of fault detection filters (or residual generators) for solving approximately FDI problems and illustrate how descriptor system techniques are instrumental for solving these problems in the most general setting and in a numerically reliable way. For illustration purposes, we describe two recently proposed algorithms for the synthesis of residual generators to solve $\mathcal{H}_2$- and $\mathcal{H}_\infty$-optimal FDI problems [35, 36]. After formulating the approximate synthesis problem, we provide a high-level description of the synthesis algorithms based on rational matrix manipulation techniques. Then we discuss an integrated approach based entirely on state-space computational techniques. Two academic examples illustrate the applicability of the algorithms to solve robust FDI problems.

6.2 The FDI Problem

Consider additive fault models described by input-output representations of the form

$$\mathbf{y}(\lambda) = G_u(\lambda)\mathbf{u}(\lambda) + G_d(\lambda)\mathbf{d}(\lambda) + G_w(\lambda)\mathbf{w}(\lambda) + G_f(\lambda)\mathbf{f}(\lambda), \tag{6.2}$$

where $\mathbf{y}(\lambda)$, $\mathbf{u}(\lambda)$, $\mathbf{d}(\lambda)$, $\mathbf{w}(\lambda)$, and $\mathbf{f}(\lambda)$ are Laplace- or Z-transformed vectors of the p-dimensional system output vector $y(t)$, m_u-dimensional control input vector $u(t)$, m_d-dimensional disturbance vector $d(t)$, m_w-dimensional noise vector $w(t)$, and m_f-dimensional fault vector $f(t)$, respectively, and where $G_u(\lambda)$, $G_d(\lambda)$, $G_w(\lambda)$, and $G_f(\lambda)$ are the TFMs from the control inputs to outputs, disturbance inputs to outputs, noise inputs to outputs, and fault inputs to outputs, respectively. For complete generality of our problem setting, we will allow that these TFMs are general rational matrices (proper or improper) for which we will not a priori assume any further properties.

A linear residual generator (or fault detection filter, or simply fault detector) processes the measurable system outputs $y(t)$ and control inputs $u(t)$ and generates the residual signals $r(t)$ which serve for decision making on the presence or absence of faults. The input-output form of this filter is

$$\mathbf{r}(\lambda) = Q(\lambda)\begin{bmatrix} \mathbf{y}(\lambda) \\ \mathbf{u}(\lambda) \end{bmatrix}, \tag{6.3}$$

where $Q(\lambda)$ is the TFM of the filter. For a physically realizable filter, $Q(\lambda)$ must be *proper* (i.e., only with finite poles) and *stable* (i.e., only with poles having negative real parts for a continuous-time system or magnitudes less than one for a discrete-time system). The (dynamic) *order* of $Q(\lambda)$ (also known as *McMillan degree*) is the dimension of the state vector of a minimal state-space realization of $Q(\lambda)$. The dimension q of the residual vector $r(t)$ depends on the FD problem to be solved.

The residual signal $r(t)$ in (6.3) generally depends on the system outputs $y(t)$ of all system inputs $u(t)$, $d(t)$, $w(t)$, and $f(t)$. The residual generation system, obtained by replacing in (6.3) $\mathbf{y}(\lambda)$ by its expression in (6.2), is given by

$$r(\lambda) = R_u(\lambda)\mathbf{u}(\lambda) + R_d(\lambda)\mathbf{d}(\lambda) + R_w(\lambda)\mathbf{w}(\lambda) + R_f(\lambda)\mathbf{f}(\lambda), \tag{6.4}$$

where

$$[\,R_u(\lambda) \mid R_d(\lambda) \mid R_w(\lambda) \mid R_f(\lambda)\,] := Q(\lambda)\left[\begin{array}{c|c|c|c} G_u(\lambda) & G_d(\lambda) & G_w(\lambda) & G_f(\lambda) \\ I_{m_u} & 0 & 0 & 0 \end{array}\right]. \tag{6.5}$$

For a successfully designed filter $Q(\lambda)$, the corresponding residual generation system is proper and stable and achieves specific fault detection requirements.

For the solution of FD problems it is always possible to completely decouple the control input $u(t)$ from the residuals $r(t)$ by requiring $R_u(\lambda) = 0$. Regarding the disturbance input $d(t)$ and noise input $w(t)$ we aim to impose a similar condition on the disturbances input $d(t)$ by requiring $R_d(\lambda) = 0$, while minimizing simultaneously the effect of noise input $w(t)$ on the residual (e.g., by minimizing the norm of $R_w(\lambda)$). Thus, from a practical synthesis point of view, the distinction between $d(t)$ and $w(t)$ lies solely in the way these signals are treated when solving the residual generator synthesis problem.

Let $M_r(\lambda)$ be a suitably chosen reference model (i.e., stable, proper, diagonal, and invertible) representing the desired TFM from the faults to residuals. We want to achieve that $\mathbf{r}(\lambda) \approx M_r(\lambda)\mathbf{f}(\lambda)$; that is, each residual $r_i(t)$ is influenced mainly by fault $f_i(t)$. Our formulation of the *approximate fault detection and isolation problem* (AFDIP) extends the formulation of the model-matching approach of [5, 3] by requiring us to determine a stable and proper filter $Q(\lambda)$ such that the following conditions are fulfilled:

$$\begin{array}{ll} \text{(i)} & R_u(\lambda) = 0, \\ \text{(ii)} & R_d(\lambda) = 0, \\ \text{(iii)} & R_f(\lambda) \approx M_r(\lambda), \ \text{ with } R_f(\lambda) \text{ stable;} \\ \text{(iv)} & R_w(\lambda) \approx 0, \ \text{ with } R_w(\lambda) \text{ stable.} \end{array} \tag{6.6}$$

The *exact fault detection and isolation problem* (EFDIP) requiring $R_f(\lambda) = M_r(\lambda)$ is included in this formulation and corresponds to $m_w = 0$, while the formulation of the AFDIP in [5, 3] corresponds to $m_d = 0$.

It is straightforward to show that for the solution of the AFDIP, the solvability conditions are those for the solvability of the EFDIP stated in [9] (see [35] for a proof).

Theorem 6.1. *For the system* (6.2) *there exists a stable, diagonal, proper, and invertible* $M_r(\lambda)$ *such that the AFIDP is solvable if and only if*

$$\operatorname{rank}[\,G_f(\lambda)\; G_d(\lambda)\,] = m_f + \operatorname{rank} G_d(\lambda). \tag{6.7}$$

Generically, the condition (6.7) is fulfilled if $p \geq m_f + m_d$, which implies that the system must have a sufficiently large number of measurements. For the case $m_d = 0$ considered in [3, 5], this condition reduces to the simple left invertibility condition

$$\operatorname{rank} G_f(\lambda) = m_f. \tag{6.8}$$

In the next section we describe the solution of the AFDIP by solving an approximate model-matching problem using $\mathcal{H}_{2/\infty}$-norm minimization techniques.

6.3 The Optimal Model-Matching Approach

Assume $m_u + m_d > 0$ and consider $Q(\lambda)$ in a factored form

$$Q(\lambda) = Q_1(\lambda)N_l(\lambda), \tag{6.9}$$

where $N_l(\lambda)$ is a proper left rational nullspace basis satisfying $N_l(\lambda)G(\lambda) = 0$, where $G(\lambda)$ is defined as

$$G(\lambda) = \begin{bmatrix} G_d(\lambda) & G_u(\lambda) \\ 0 & I_{m_u} \end{bmatrix} \tag{6.10}$$

and $Q_1(\lambda)$ is a factor to be further determined. With this choice it follows that $Q(\lambda)$ automatically fulfills the first two conditions in (6.6). The existence of $N_l(\lambda)$ is guaranteed provided condition (6.7) is fulfilled. The resulting $N_l(\lambda)$ has maximal row rank $p - r_d$, where $r_d = \operatorname{rank} G_d(\lambda)$. Moreover, we can choose $N_l(\lambda)$ stable and such that both $N_f(\lambda)$ and $N_w(\lambda)$, defined as

$$[\, N_f(\lambda)\ N_w(\lambda)\,] := N_l(\lambda)\begin{bmatrix} G_f(\lambda) & G_w(\lambda) \\ 0 & 0 \end{bmatrix}, \tag{6.11}$$

are proper and stable TFMs [33]. If $m_u = m_d = 0$, we can determine $N_l(\lambda)$ in (6.9) simply from a proper and stable left coprime factorization

$$[\, G_f(\lambda)\ G_w(\lambda)\,] = N_l^{-1}(\lambda)[\, N_f(\lambda)\ N_w(\lambda)\,].$$

To fulfill the last two conditions in (6.6) we can solve an $\mathcal{H}_{2/\infty}$-norm minimization problem to determine a stable and proper $Q_1(\lambda)$ such that

$$\big\| [\, Q_1(\lambda)N_f(\lambda) - M_r(\lambda) \ \ Q_1(\lambda)N_w(\lambda)\,] \big\|_{2/\infty} = \min.$$

This $\mathcal{H}_{2/\infty}$ model-matching problem can be easily reformulated as a standard $\mathcal{H}_{2/\infty}$-norm minimization based "controller" synthesis problem [40] as shown in Figure 6.1. Here, the underlying equations are

$$\begin{aligned}
\mathbf{e}(\lambda) &= \mathbf{r}(\lambda) - M_r(\lambda)\mathbf{f}(\lambda), \\
\bar{\mathbf{y}}(\lambda) &= N_f(\lambda)\mathbf{f}(\lambda) + N_w(\lambda)\mathbf{w}(\lambda), \\
\mathbf{r}(\lambda) &= Q_1(\lambda)\bar{\mathbf{y}}(\lambda)
\end{aligned}$$

and lead to the following definition of the generalized plant:

$$P(\lambda) = \left[\begin{array}{c|c} P_{11}(\lambda) & P_{12}(\lambda) \\ \hline P_{21}(\lambda) & P_{22}(\lambda) \end{array}\right] := \left[\begin{array}{cc|c} -M_r(\lambda) & 0 & I \\ \hline N_f(\lambda) & N_w(\lambda) & 0 \end{array}\right].$$

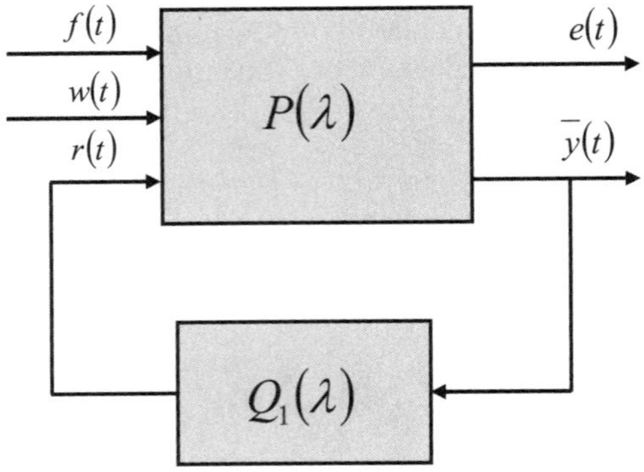

Figure 6.1. Standard $\mathcal{H}_{2/\infty}$ synthesis setting [35].

The minimization of the $\mathcal{H}_{2/\infty}$-norm of the TFM from $[\,f^T(t)\; w^T(t)]^T$ to $e(t)$ via an optimal $Q_1(\lambda)$ is thus formally a standard $\mathcal{H}_{2/\infty}$-synthesis problem for which software tools exist, as, for example, the functions `h2syn/hinfsyn` available in the MATLAB ROBUST CONTROL Toolbox. The main problem when employing standard tools like `h2syn/hinfsyn` is that, although a stable and proper solution of the AFDIP may exist, this solution cannot be computed because of an inappropriate choice of $M_r(\lambda)$ or of the presence of technical assumptions which must be fulfilled.

The first aspect, related to the choice of $M_r(\lambda)$, has already been discussed in [15]. The difficulty is that the determination of the solution $Q_1(\lambda)$ cannot be done simultaneously with the choice of $M_r(\lambda)$, because this would lead to the trivial (optimal) solution $M_r(\lambda) = 0$ and $Q_1(\lambda) = 0$. Therefore, the procedure suggested in [15] is to choose first $M_r(\lambda)$ and compute the corresponding $Q_1(\lambda)$, then solve the optimization problem for a new $M_r(\lambda)$ with fixed $Q_1(\lambda)$, and continue in this way until convergence to a satisfactory solution is achieved. However, it is not clear how to choose the initial $M_r(\lambda)$ or how to keep the dynamical order of $M_r(\lambda)$ as low as possible. Furthermore, there is not even a guarantee for the convergence of this process.

The second aspect, related to the presence of technical assumptions, is the need to fulfill by the underlying generalized plant model certain requirements when solving the $\mathcal{H}_{2/\infty}$-synthesis problem using available standard tools. For example, to obtain a proper solution, the TFM $P_{21}(\lambda)$ must not have zeros on the extended imaginary axis in the continuous-time case or on the unit circle. If this condition is not fulfilled, the synthesis fails, although an appropriate choice of $M_r(\lambda)$ would lead to the cancelation of these zeros in the final detector.

To face the above limitations, it is necessary to develop general synthesis procedures for which no such limitations exist. The key parameter to guarantee the stability and properness of the detector is $M_r(\lambda)$, the desired TFM relating the faults to the residuals. As already mentioned, the choice of $M_r(\lambda)$ is not obvious and can even be the object of an optimization based choice [15]. Often good candidates for $M_r(\lambda)$ result from an exact synthesis (for $m_w = 0$) [29]. However, in [30] a procedure has been proposed, where the choice of suitable $M_r(\lambda)$ is part of the solution. This procedure has been refined in [35, 36], by developing an integrated approach to the detector synthesis. An important feature of these computational approaches is that they rely on repeated updating of an initial fault detector. The underlying state-space representation based computations employ explicit least order realizations of the successive detectors, and thus a least order of the final detector is guaranteed. Since the successive steps are strongly connected, all structural features of the computed intermediary results can be exploited in the next steps. This leads to an integrated computational procedure based on highly efficient structure exploiting computations.

6.4 Enhanced Optimal Model-Matching Procedure

In this section we present enhanced versions of the algorithms of [30], where we exploit the additional structure in the model (6.2) owing to the separation of the unknown inputs in two components $d(t)$ and $w(t)$. Moreover, by using a new parameterization of the detector, we derive integrated computational procedures based on detector updating techniques. We describe in what follows the main stages of the overall computational procedures, pointing out the commonalities and differences between the $\mathcal{H}_2$- and $\mathcal{H}_\infty$-norm based optimization approaches. The enhanced synthesis procedures will be first presented as high-level computations in terms of TFMs. However, the actual computations have to be performed employing reliable numerical methods relying on descriptor system state-space representations. The corresponding computational methods are described in the next section.

Stage 1:

This stage employs the nullspace-based parameterization of the detector introduced in the previous section, which allows us to reduce the initial synthesis problem to a standard $\mathcal{H}_{2/\infty}$-norm optimization problem. The importance of this step is two-fold. First, the control inputs u and disturbance inputs d are decoupled from the residual via this parameterization. This decoupling can be performed whether the system is proper or not. Second, this step is common to both exact and approximate synthesis based approaches, and thus the exact synthesis procedure can be easily embedded into the approximate solution.

We employ the parameterization (6.9) of the detector in the factored form $Q(\lambda) = Q_1(\lambda)N_l(\lambda)$, where $N_l(\lambda)$ is a left proper rational nullspace basis satisfying (6.10), and $Q_1(\lambda)$ is a factor to be further determined. As already mentioned, the resulting $N_l(\lambda)$ has maximal row rank $p-r_d$, where $r_d = \operatorname{rank} G_d(\lambda)$. Moreover, we can choose $N_l(\lambda)$ stable and such that both proper TFMs $N_f(\lambda)$ and $N_w(\lambda)$ defined in (6.11) are also stable [33]. However, as it will be later apparent, enforcing the stability and even properness conditions are not necessary at this stage. We can easily check now the solvability of the AFDIP by verifying that

$$\operatorname{rank} N_f(\lambda) = m_f. \tag{6.12}$$

To fulfill the last two conditions in (6.6) we can solve an $\mathcal{H}_{2/\infty}$-norm minimization problem for $\|\widetilde{\mathcal{R}}(\lambda)\|_{2/\infty}$ to determine $Q_1(\lambda)$, where

$$\widetilde{\mathcal{R}}(\lambda) := \widetilde{F}(\lambda) - Q_1(\lambda)\widetilde{G}(\lambda), \tag{6.13}$$

with $\widetilde{G}(\lambda) = [\, N_f(\lambda)\; N_w(\lambda)\,]$ and $\widetilde{F}(\lambda) = [\, M_r(\lambda)\; O\,]$. Here, $M_r(\lambda)$ is the TFM of a given reference model (i.e., stable, proper, diagonal, invertible).

Let ℓ be the rank of the $(p-r_d)\times(m_f+m_w)$ TFM $\widetilde{G}(\lambda)$. If $\ell < p-r_d$ (i.e., $\widetilde{G}(\lambda)$ has no full row rank), we can take instead $N_l(\lambda)$, ℓ linear combinations of basis vectors of the form $W(\lambda)N_l(\lambda)$, which ensures that $W(\lambda)\widetilde{G}(\lambda)$ has full row rank ℓ. A suitable choice of the $\ell\times(p-r_d)$ TFM $W(\lambda)$ which also minimizes the McMillan degree of $W(\lambda)N_l(\lambda)$ is described in [33, 38].

Stage 2:

The second stage is standard in solving $\mathcal{H}_{2/\infty}$-norm optimization problems and consists in compressing the full row rank TFM $\widetilde{G}(\lambda)$ to a full column rank (thus invertible) TFM. For this, we compute a quasi-co-outer–inner factorization

$$\widetilde{G}(\lambda) = [\, G_{o,1}(\lambda)\; 0\,] \begin{bmatrix} G_{i,1}(\lambda) \\ G_{i,2}(\lambda) \end{bmatrix} := G_o(\lambda)G_i(\lambda), \tag{6.14}$$

where $G_i(\lambda)$ is an $(m_f+m_w)\times(m_f+m_w)$ inner TFM and $G_{o,1}(\lambda)$ is an $\ell\times\ell$ invertible TFM. Recall that a square TFM $G_i(\lambda)$ is *inner* (and simultaneously *co-inner*) if it has only stable poles and satisfies $G_i(\lambda)G_i^*(\lambda) = I$, where $G_i^*(s) := G_i^T(-s)$ in a continuous-time setting and $G_i^*(z) := G_i^T(1/z)$ in a discrete-time setting. The *quasi-co-outer* factor $G_o(\lambda)$ may have besides stable zeros also zeros which lie on the boundary of the stability domain.

We can refine the parameterization (6.9) of the detector by choosing $Q_1(\lambda)$ of the form

$$Q_1(\lambda) = Q_2(\lambda)G_{o,1}^{-1}(\lambda), \tag{6.15}$$

where $Q_2(\lambda)$ is to be determined. Using (6.14) and (6.15), we can express $\widetilde{\mathcal{R}}(\lambda)$ in (6.13) as $\widetilde{\mathcal{R}}(\lambda) = \overline{\mathcal{R}}(\lambda)G_i(\lambda)$, with

$$\overline{\mathcal{R}}(\lambda) = \left[\; \overline{F}_1(\lambda) - Q_2(\lambda) \;\middle|\; \overline{F}_2(\lambda) \;\right], \tag{6.16}$$

where $\overline{F}_1(\lambda) := \widetilde{F}(\lambda)G_{i,1}^*(\lambda)$ and $\overline{F}_2(\lambda) := \widetilde{F}(\lambda)G_{i,2}^*(\lambda)$. Since $G_i(\lambda)$ is an inner TFM, we have $\|\widetilde{\mathcal{R}}(\lambda)\|_{2/\infty} = \|\overline{\mathcal{R}}(\lambda)\|_{2/\infty}$. The determination of a stable and proper $Q_2(\lambda)$ which minimizes $\|\overline{\mathcal{R}}(\lambda)\|_{2/\infty}$ is a least distance problem. Depending on the employed norm, different solutions have to be computed.

From (6.9) and (6.15), the overall detection filter $Q(\lambda)$ has the product form

$$Q(\lambda) = Q_2(\lambda)G_{o,1}^{-1}(\lambda)N_l(\lambda) := Q_2(\lambda)\overline{Q}(\lambda),$$

where $\overline{Q}(\lambda)$ can be interpreted as an updated partial detector, to which correspond the updated quantities

$$[\,\overline{N}_f(\lambda)\;\overline{N}_w(\lambda)\,] := G_{o,1}^{-1}(\lambda)[\,N_f(\lambda)\;N_w(\lambda)\,] = [\,I_\ell\;\;0\,]\begin{bmatrix} G_{i,1}(\lambda) \\ G_{i,2}(\lambda) \end{bmatrix} = G_{i,1}(\lambda). \tag{6.17}$$

In the next section we show that pole-zero cancelations occur in forming $G_{o,1}^{-1}(\lambda)N_l(\lambda)$, and thus $\overline{Q}(\lambda)$ has as its only poles the zeros of $G_{o,1}(\lambda)$. These poles are stable, except possible poles corresponding to zeros of $G_{o,1}(\lambda)$ on the boundary of the stability domain (i.e., on the extended imaginary axis in the continuous-time case, or on the unit circle in discrete-time case).

Note: The solution of the EFDIP (for $m_w = 0$) corresponds to choosing $G_{o,1}(\lambda) = N_f(\lambda)$ and $G_i(\lambda) = I_{m_f}$. It follows that $Q_2(\lambda) = M_r(\lambda)$ and the next stage can be skipped.

Stage 3: $\mathcal{H}_2$-Optimal Synthesis

The solution of the least distance problem in the case of $\mathcal{H}_2$-norm is straightforward and involves computing the stable projection

$$Q_2(\lambda) = \{\overline{F}_1(\lambda)\}_+.$$

Here, $\{\cdot\}_+$ denotes the stable part of the underlying TFM including the direct feedthrough term, while $\{\cdot\}_-$ denotes the unstable part. With the above choice, we achieved the minimum $\mathcal{H}_2$-norm of $\mathcal{R}(\lambda)$, which can be computed as

$$\|\widetilde{\mathcal{R}}(\lambda)\|_2 = \|\overline{\mathcal{R}}(\lambda)\|_2 = \|[\,\{\overline{F}_1(\lambda)\}_- \;\; \overline{F}_2(\lambda)\,]\|_2.$$

Since the underlying TFMs are unstable, the $\mathcal{L}_2$-norm is used in the last equation.

Stage 3: $\mathcal{H}_\infty$-Optimal Synthesis

We determine a stable $Q_2(\lambda)$ as the solution of the suboptimal two-block least distance problem

$$\|[\,\overline{F}_1(\lambda) - Q_2(\lambda) \;\; \overline{F}_2(\lambda)\,]\|_\infty < \gamma, \tag{6.18}$$

where $\gamma_{opt} < \gamma \leq \gamma_{opt} + \varepsilon$, with ε an arbitrary user specified (accuracy) tolerance for the least achievable value γ_{opt} of γ. With the following lower and upper bounds for γ_{opt}

$$\gamma_l = \|\overline{F}_2(\lambda)\|_\infty, \quad \gamma_u = \left\|\left[\; \overline{F}_1(\lambda) \quad \overline{F}_2(\lambda) \;\right]\right\|_\infty \tag{6.19}$$

such a γ-suboptimal solution $Q_2(\lambda)$ can be computed using the bisection-based γ-iteration approach of [8], whose main steps are presented in what follows.

For a given $\gamma > \gamma_l$, we compute first the spectral factorization [40]

$$\gamma^2 I - \overline{F}_2(\lambda)\overline{F}_2^*(\lambda) = V(\lambda)V^*(\lambda), \tag{6.20}$$

where $V(\lambda)$ is biproper, stable, and minimum-phase. Further, we compute the additive decomposition

$$L_s(\lambda) + L_u(\lambda) = V^{-1}(\lambda)\overline{F}_1(\lambda), \tag{6.21}$$

where $L_s(\lambda)$ is the stable part and $L_u(\lambda)$ is the unstable part. If $\gamma > \gamma_{opt}$, the two-block problem (6.18) is equivalent to the one-block problem [8]

$$\left\| V^{-1}(\lambda)\left(\overline{F}_1(\lambda) - Q_2(\lambda)\right)\right\|_\infty \leq 1 \tag{6.22}$$

and $\gamma_H := \|L_u^*(\lambda)\|_H < 1$ ($\|\cdot\|_H$ denotes the Hankel norm of a stable TFM). In this case we readjust $\gamma_u = \gamma$. If $\gamma_H \geq 1$, we readjust $\gamma_l = \gamma$. Then, for $\gamma = (\gamma_l + \gamma_u)/2$ we redo the factorization (6.20) and decomposition (6.21). This process is repeated until $\gamma_u - \gamma_l \leq \varepsilon$.

If $\gamma_u \geq \gamma > \gamma_{opt}$, the stable solution of (6.22) is

$$Q_2(\lambda) = V(\lambda)(L_s(\lambda) + Q_{2,s}(\lambda)), \tag{6.23}$$

where, for any γ_1 satisfying $1 \geq \gamma_1 > \gamma_H$, $Q_{2,s}(\lambda)$ is the stable solution of the γ_1-suboptimal Nehari problem

$$\left\| L_u(\lambda) - Q_{2,s}(\lambda)\right\|_\infty < \gamma_1. \tag{6.24}$$

Stage 4:

Given the stable solution $Q_2(\lambda)$ computed at previous stage, the resulting $\mathcal{H}_{2/\infty}$-optimal detector can be expressed as

$$\widehat{Q}(\lambda) = Q_2(\lambda)G_{o,1}^{-1}(\lambda)N_l(\lambda). \tag{6.25}$$

Taking into account the pole-zero cancelations when forming $G_{o,1}^{-1}(\lambda)N_l(\lambda)$, the optimal detector $\widehat{Q}(\lambda)$ is stable (and also proper) only if $G_{o,1}$ is minimum-phase. However, since $G_{o,1}(\lambda)$ is in general a quasi-co-outer factor, it can still have unstable zeros on the boundary of the stability domain. Thus, these zeros may appear as poles of $\widehat{Q}(\lambda)$ if they are not canceled via the zeros of $Q_2(\lambda)$. To ensure that the final detector is stable (and thus also proper), the resulting $Q(\lambda)$ can be determined as

$$Q(\lambda) = M(\lambda)\widehat{Q}(\lambda),$$

where $M(\lambda)$ is chosen such that $Q(\lambda)$ is proper and stable. Since the matrix $\widehat{Q}_1(\lambda) := M(\lambda)Q_2(\lambda)G_{o,1}^{-1}(\lambda)$ can be seen as an approximation of the solution of the weighted minimization problem

$$\|M(\lambda)\widetilde{\mathcal{R}}(\lambda)\|_{2/\infty} = \min,$$

it follows that $M(\lambda)M_r(\lambda)$ can be interpreted as an updated reference model.

The computation of appropriate $M(\lambda)$ can be done using the stable and proper coprime factorization algorithm of [25]. The choice of $M(\lambda)$ can be done such that

$$\|M(\lambda)\widetilde{\mathcal{R}}(\lambda)\|_{2/\infty} \approx \|\widetilde{\mathcal{R}}(\lambda)\|_{2/\infty}$$

and $M(\lambda)$ has the least possible McMillan degree. For example, to ensure properness or strict properness, $M(\lambda)$ can be chosen diagonal with the diagonal terms $m_j(\lambda)$, $j = 1,\dots,m_f$, having the form

$$m_j(\lambda) = \frac{1}{(\tau s+1)^{k_j}} \quad \text{or} \quad m_j(z) = \frac{1}{z^{k_j}}$$

for continuous- or discrete-time settings, respectively. Both above factors have unit $\mathcal{H}_\infty$-norm. In the case of solving the EFDIP, the choice of $M(\lambda)$ can enforce arbitrary dynamics of the detector. In this case, the choice of a diagonal $M(\lambda)$ allows one to obtain an updated diagonal specification $M(\lambda)M_r(\lambda)$ for $R_f(\lambda)$.

6.5 Computational Issues

For computations we employ an equivalent *descriptor* state-space realization of the input-output model (6.2)

$$\begin{aligned} E\lambda x(t) &= Ax(t)+B_u u(t)+B_d d(t)+B_w w(t)+B_f f(t),\\ y(t) &= Cx(t)+D_u u(t)+D_d d(t)+D_w w(t)+D_f f(t) \end{aligned} \tag{6.26}$$

with the n-dimensional state vector $x(t)$. Recall that $\lambda x(t) = \dot{x}(t)$ or $\lambda x(t) = x(t+1)$ depending on the type of system, continuous or discrete, respectively. In general, we can assume that the representation (6.26) is *minimal*; that is, the descriptor pair $(A-\lambda E, C)$ is *observable* and the pair $(A-\lambda E, [\,B_u\; B_d\; B_w\; B_f\,])$ is *controllable*. The corresponding TFMs of the model in (6.2) are

$$\begin{aligned} G_u(\lambda) &= C(\lambda E-A)^{-1}B_u + D_u,\\ G_d(\lambda) &= C(\lambda E-A)^{-1}B_d + D_d,\\ G_w(\lambda) &= C(\lambda E-A)^{-1}B_w + D_w,\\ G_f(\lambda) &= C(\lambda E-A)^{-1}B_f + D_f \end{aligned} \tag{6.27}$$

or in an equivalent notation

$$\left[\begin{array}{cccc} G_u(\lambda) & G_d(\lambda) & G_w(\lambda) & G_f(\lambda) \end{array}\right] := \left[\begin{array}{c|cccc} A-\lambda E & B_u & B_d & B_w & B_f \\ \hline C & D_u & D_d & D_w & D_f \end{array}\right].$$

In what follows, we present the state-space representation based algorithms to be used to perform the successive stages of the synthesis procedures.

Stage 1:

If $m_u + m_d > 0$, we employ the orthogonal pencil reduction based algorithms described in [33, 38, 39] to compute a least order state-space representation of the $(p - r_d) \times (m_u + p)$ left proper rational nullspace $N_l(\lambda)$ satisfying $N_l(\lambda)G(\lambda) = 0$, with $G(\lambda)$ defined in (6.10). The corresponding $N_f(\lambda)$ and $N_w(\lambda)$ can be simultaneously obtained with realizations of the form

$$\left[\begin{array}{ccc} N_l(\lambda) & N_f(\lambda) & N_w(\lambda) \end{array}\right] = \left[\begin{array}{c|ccc} \widetilde{A} - \lambda\widetilde{E} & \widetilde{B}_{yu} & \widetilde{B}_f & \widetilde{B}_w \\ \hline \widetilde{C} & \widetilde{D}_{yu} & \widetilde{D}_f & \widetilde{D}_w \end{array}\right], \tag{6.28}$$

where the $n_l \times n_l$ matrix $\widetilde{E}$ is invertible (thus all TFMs are proper) and the pair $(\widetilde{A}, \widetilde{E})$ has only finite generalized eigenvalues which can be arbitrarily placed. The computation of the matrices of the state-space realizations in (6.28) is detailed in the above references and involves only an orthogonal reduction of the system matrix of the state-space realization of $G(\lambda)$ defined in (6.10) to a suitable Kronecker-like form which reveals the left Kronecker structure of $G(\lambda)$. The resulting Kronecker-like form allows one to read out, without further computations, a realization of a proper left nullspace basis $N_l(\lambda)$. The most important aspect, as shown in [33, 38], is that the realizations of $N_f(\lambda)$ and $N_w(\lambda)$ can be determined such that they share the same observable pair $(\widetilde{A} - \lambda\widetilde{E}, \widetilde{C})$ with the realization of $N_l(\lambda)$. This property can be preserved in the subsequent computation (see below) and exploited at the next computational stage. If $m_u = m_d = 0$, a realization of the form (6.28) can still be employed with the trivial choice $n_l = n$, $N_l(\lambda) = I_p$, $N_f(\lambda) = G_f(\lambda)$, and $N_w(\lambda) = G_w(\lambda)$. As will be apparent at the next stage, the presence of possibly singular $\widetilde{E}$ can be easily accommodated by the employed factorization algorithm.

The checking of condition (6.12) can be done by verifying that there exists a λ_0 (e.g., a randomly generated value) such that

$$\operatorname{rank}\left[\begin{array}{cc} \widetilde{A} - \lambda_0\widetilde{E} & \widetilde{B}_f \\ \widetilde{C} & \widetilde{D}_f \end{array}\right] = n_l + m_f. \tag{6.29}$$

If $\ell := \operatorname{rank}[\,N_f(\lambda)\ N_w(\lambda)\,] < p - r_d$, we can determine, by using minimum dynamic cover techniques, a suitable $\ell \times (p - r_d)$ prefilter $W(\lambda)$ such that $W(\lambda)N_l(\lambda)$ has the least possible McMillan degree and $W(\lambda)[\,N_f(\lambda)\ N_w(\lambda)\,]$ has full row rank ℓ. The state-space realization of $W(\lambda)\left[\,N_l(\lambda)\ \ N_f(\lambda)\ \ N_w(\lambda)\,\right]$ still has the form (6.28) and can be obtained by using updating techniques described in [33, 38].

Stage 2:

For the computation of the quasi-co-outer–inner factorization of $\widetilde{G}(\lambda) := [\,N_f(\lambda)\ N_w(\lambda)\,]$ in (6.14), we employ the dual of the algorithm of [18] for the continuous-time case and the dual of the algorithm of [16] for the discrete-time case. In both cases, the quasi-co-outer factor $G_{o,1}(\lambda)$ is obtained in the form

$$G_{o,1}(\lambda) = \left[\begin{array}{c|c} \widetilde{A} - \lambda\widetilde{E} & \overline{B}_o \\ \hline \widetilde{C} & \overline{D}_o \end{array}\right], \tag{6.30}$$

where $\overline{B}_o$ and $\overline{D}_o$ are matrices with ℓ columns. The system with the TFM $G_{o,1}(\lambda)$ may have besides the stable zeros (partly resulted from the column compression) also zeros on the imaginary axis (including infinity) in the continuous-time case or on the unit circle in the discrete-time case. The $(m_f + m_w) \times (m_f + m_w)$ TFM of the inner factor is proper and stable, and assume that its inverse (i.e., its conjugated TFM) has a state-space realization of the form

$$G_i^*(\lambda) = [\, G_{i,1}^*(\lambda)\; G_{i,2}^*(\lambda)\,] = \left[\begin{array}{c|cc} A_i - \lambda E_i & B_{i,1} & B_{i,2} \\ \hline C_i & D_{i,1} & D_{i,2} \end{array}\right], \tag{6.31}$$

where all generalized eigenvalues of the pair (A_i, E_i) are unstable. For a continuous-time system $E_i = I$, but this cannot be assumed, in general, for a discrete-time system with $\lambda = z$, unless $G_i(z)$ *does not have* poles in the origin. If $G_i(z)$ has poles in the origin, the resulting E_i is singular and thus the pair (A_i, E_i) also has infinite (unstable) generalized eigenvalues.

To compute the updated partial detector $\overline{Q}(\lambda) := G_{o,1}^{-1}(\lambda)N_l(\lambda)$ as well as $\overline{N}_f(\lambda) := G_{o,1}^{-1}(\lambda)R_f(\lambda)$ and $\overline{N}_w(\lambda) := G_{o,1}^{-1}(\lambda)R_w(\lambda)$, we can solve the linear rational system of equations

$$G_{o,1}(\lambda)\left[\; \overline{Q}(\lambda) \quad \overline{N}_f(\lambda) \quad \overline{N}_w(\lambda) \;\right] = \left[\; N_l(\lambda) \quad N_f(\lambda) \quad N_w(\lambda) \;\right]. \tag{6.32}$$

Observe that $G_{o,1}(\lambda)$, $N_l(\lambda)$, $N_f(\lambda)$, and $N_w(\lambda)$ have descriptor realizations which share the same state, descriptor, and output matrices. Using these state-space realizations, the linear rational equation (6.32) can be equivalently solved (see [28]) by computing first the solution $X(\lambda)$ of

$$\begin{bmatrix} \widetilde{A} - \lambda\widetilde{E} & \overline{B}_o \\ \widetilde{C} & \overline{D}_o \end{bmatrix} X(\lambda) = \begin{bmatrix} \widetilde{B}_{yu} & \widetilde{B}_f & \widetilde{B}_w \\ \widetilde{D}_{yu} & \widetilde{D}_f & \widetilde{D}_w \end{bmatrix}$$

and then

$$\left[\; \overline{Q}(\lambda) \quad \overline{N}_f(\lambda) \quad \overline{N}_w(\lambda) \;\right] = \left[\; 0 \quad I_\ell \;\right] X(\lambda).$$

With the invertible system matrix

$$S_o(\lambda) = \begin{bmatrix} \widetilde{A} - \lambda\widetilde{E} & \overline{B}_o \\ \widetilde{C} & \overline{D}_o \end{bmatrix} \tag{6.33}$$

we obtain

$$\left[\; \overline{Q}(\lambda) \quad \overline{N}_f(\lambda) \quad \overline{N}_w(\lambda) \;\right] = \left[\; 0 \quad I_\ell \;\right] S_o^{-1}(\lambda) \begin{bmatrix} \widetilde{B}_{yu} & \widetilde{B}_f & \widetilde{B}_w \\ \widetilde{D}_{yu} & \widetilde{D}_f & \widetilde{D}_w \end{bmatrix}.$$

Alternatively, explicit realizations of $\overline{N}_f(\lambda)$ and $\overline{N}_w(\lambda)$ can be determined using (6.17).

Let the quadruple (A_r, B_r, C_r, D_r) describe the state-space realization of $[\,M_r(\lambda)\;0\,]$. Then, using (6.31), the state-space realization of $[\,\overline{F}_1(\lambda)\;\overline{F}_2(\lambda)\,]$ has the form

$$[\,\overline{F}_1(\lambda)\;\overline{F}_2(\lambda)\,] = \left[\begin{array}{cc|cc} A_r - \lambda I & B_r C_i & B_r D_{i,1} & B_r D_{i,2} \\ 0 & A_i - \lambda E_i & B_{i,1} & B_{i,2} \\ \hline C_r & D_r C_i & D_r D_{i,1} & D_r D_{i,2} \end{array}\right], \qquad (6.34)$$

where A_r has only stable eigenvalues, while the pair (A_i, E_i) has only unstable generalized eigenvalues. The partial detector $\overline{Q}(\lambda)$ has an explicit descriptor realization of the form

$$\overline{Q}(\lambda) = \left[\begin{array}{cc|c} \widetilde{A} - \lambda \widetilde{E} & \overline{B}_o & \widetilde{B}_{yu} \\ \widetilde{C} & \overline{D}_o & \widetilde{D}_{yu} \\ \hline 0 & -I_\ell & 0 \end{array}\right]. \qquad (6.35)$$

Stage 3: $\mathcal{H}_2$-Optimal Synthesis

To compute the additive spectral separation

$$\overline{F}_1(\lambda) = \{\overline{F}_1(\lambda)\}_+ + \{\overline{F}_1(\lambda)\}_-$$

the general algorithm of [12] for descriptor systems can be employed. However, the structure of the realization of $\overline{F}_1(\lambda)$ in (6.34) can be easily exploited, taking into account that the first block $A_r - \lambda I$ has only stable eigenvalues, while the second block $A_i - \lambda E_i$ has only unstable eigenvalues. Thus, the computation of $\{\overline{F}_1(\lambda)\}_+$ involves only the solution of a Sylvester equation (or system) (see [12] for details).

Stage 3: $\mathcal{H}_\infty$-Optimal Synthesis

At this stage we need to perform the γ-iteration to solve the suboptimal two-block minimum distance problem (6.18). To start, we have to compute the $\mathcal{L}_\infty$-norms in (6.19) to obtain γ_l and γ_u. For this purpose, efficient algorithms can be employed based on extensions of the method of [4] (for which standard numerical tools are available in MATLAB). Note that for computing γ_u we can exploit that $\gamma_u = \|M_r(\lambda)\|_\infty$ as a consequence of the all-pass property of $G_i^*(\lambda)$.

The computation of the spectral factorization (6.20) involves two steps. First, we compute a right coprime factorization of $\overline{F}_2(\lambda)$ with inner denominator such that $\overline{F}_2(\lambda) = \overline{N}_2(\lambda)\overline{M}_2^{-1}(\lambda)$, where $\overline{M}_2(\lambda)$ is inner. It follows that $\overline{F}_2(\lambda)\overline{F}_2^*(\lambda) = \overline{N}_2(\lambda)\overline{N}_2^*(\lambda)$. This computation needs to be performed only once, and suitable algorithms for this purpose have been proposed in [25] for standard systems or in [17] for discrete-time descriptor systems. In both cases, the resulting $\overline{N}_2(\lambda)$ has a standard system realization of the form

$$\overline{N}_2(\lambda) = \left[\begin{array}{cc|c} A_r - \lambda I & B_r \overline{C}_i & B_r \overline{D}_{i,2} \\ 0 & \overline{A}_i - \lambda I & \overline{B}_{i,2} \\ \hline C_r & D_r \overline{C}_i & D_r \overline{D}_{i,2} \end{array}\right].$$

Second, we solve the spectral factorization problem

$$\gamma^2 I - \overline{N}_2(\lambda)\overline{N}_2^*(\lambda) = V(\lambda)V^*(\lambda)$$

using methods described in [40] to obtain a realization of $V^{-1}(\lambda)\overline{F}_1(\lambda)$ of the form

$$V^{-1}(\lambda)\overline{F}_1(\lambda) = \left[\begin{array}{cc|c} \overline{A}_{11} - \lambda I & \overline{A}_{12} & \overline{B}_1 \\ 0 & A_i - \lambda E_i & \overline{B}_2 \\ \hline \overline{C}_1 & \overline{C}_2 & \overline{D} \end{array}\right],$$

where $\overline{A}_{11}$ has only stable eigenvalues, while the pair (A_i, E_i) has only unstable eigenvalues. This computation involves the solution of an algebraic Riccati equation at each iteration. For details see [40].

To compute the spectral separation (6.21) we perform a similarity transformation to obtain the transformed pole pencil

$$\begin{bmatrix} I & X \\ 0 & I \end{bmatrix} \begin{bmatrix} \overline{A}_{11} - \lambda I & \overline{A}_{12} \\ 0 & A_i - \lambda E_i \end{bmatrix} \begin{bmatrix} I & Y \\ 0 & I \end{bmatrix}$$

in a block diagonal form by annihilating its (1,2) block. This comes down to solve the Sylvester system of matrix equations

$$\begin{aligned} 0 &= XA_i + \overline{A}_{11}Y + \overline{A}_{12}, \\ 0 &= XE_i + Y. \end{aligned}$$

After applying the transformations to the input and output matrices we obtain

$$\begin{bmatrix} I & X \\ 0 & I \end{bmatrix} \begin{bmatrix} \overline{B}_1 \\ \overline{B}_2 \end{bmatrix} = \begin{bmatrix} \overline{B}_1 + X\overline{B}_2 \\ \overline{B}_2 \end{bmatrix}, \qquad [\overline{C}_1 \; \overline{C}_2] \begin{bmatrix} I & Y \\ 0 & I \end{bmatrix} = [\overline{C}_1 \; \overline{C}_1 Y + \overline{C}_2].$$

The stable and unstable terms are given by

$$L_s(\lambda) = \left[\begin{array}{c|c} \overline{A}_{11} - \lambda I & \overline{B}_1 + X\overline{B}_2 \\ \hline \overline{C}_1 & \overline{D} \end{array}\right], \quad L_u(\lambda) = \left[\begin{array}{c|c} A_i - \lambda E_i & \overline{B}_2 \\ \hline \overline{C}_1 Y + \overline{C}_2 & 0 \end{array}\right].$$

The computation of the Hankel norm $\gamma_H = \|L_u^*(\lambda)\|_H$ can be performed using standard algorithms for proper and stable systems.

The computation of the Nehari approximation can be done using the algorithm of [10] for continuous-time systems. For discrete-time systems, the same algorithm is applicable after performing a bilinear transformation to map the exterior of the unit circle to the open right half plane. A suitable transformation and its inverse transformation are $z = \frac{1+s}{1-s}$ and $s = \frac{z-1}{z+1}$, respectively. Note that in the case of an improper $L_u(z)$, all infinite poles go to $s = 1$ and therefore the "equivalent" continuous-time system will be proper.

In light of the cancelation theory for continuous-time two-block problems of [13], pole-zero cancelations occur when forming $Q_2(s)$ in (6.23). In accordance with this theory, the expected order of $Q_2(s)$ is $n_r + n_i - 1$, where n_r and n_i are the McMillan degrees of $M_r(s)$ and $G_i(s)$, respectively. It is conjectured that similar cancelations will occur also for discrete-time systems, where a cancelation theory for two-block problems is still missing. Although we were not able to derive an explicit minimal state-space realization of $Q_2(\lambda)$, we can safely employ minimal realization procedures which exploit that the resulting $Q_2(\lambda)$ is stable. Balancing related methods are especially well suited for this computation, as, for example, the square-root balancing-free method of [24, Algorithm MR6].

Stage 4:

The resulting detector $\widehat{Q}(\lambda)$ in (6.25) may be improper and/or unstable, and therefore we need to determine a stable $M(\lambda)$ having the least McMillan degree such that $M(\lambda)\widehat{Q}(\lambda)$ is proper and stable. Suitable state-space based factorization algorithms are described in [25]. To choose $M(\lambda)$ diagonal, the same algorithms can be applied to the individual rows of $\widehat{Q}(\lambda)$, for which we can build minimal descriptor state-space realizations using the algorithm of [23].

6.6 Illustrative Example

We consider the robust actuator fault detection and isolation example of [7]. The fault system (6.2) has a standard state-space realization (6.26) with $E = I$ and

$$A(\delta_1,\delta_2) = \begin{bmatrix} -0.8 & 0 & 0 \\ 0 & -0.5(1+\delta_1) & 0.6(1+\delta_2) \\ 0 & -0.6(1+\delta_2) & -0.5(1+\delta_1) \end{bmatrix},$$

$$B_u = \begin{bmatrix} 1 & 1 \\ 1 & 0 \\ 0 & 1 \end{bmatrix}, \quad B_d = 0, \quad B_f = \begin{bmatrix} 1 & 1 \\ 1 & 0 \\ 0 & 1 \end{bmatrix}, \quad C = \begin{bmatrix} 0 & 1 & 1 \\ 1 & 1 & 0 \end{bmatrix},$$

$$D_u = 0, \quad D_d = 0, \quad D_f = 0.$$

In the expression of $A(\delta_1,\delta_2)$, δ_1 and δ_2 are uncertainties in the real and imaginary parts of the two complex conjugated eigenvalues $\lambda_{1,2} = -0.5 \pm j0.6$ of the nominal value $A(0,0)$. The fault detection filter is aimed to provide robust fault detection and isolation of actuator faults in the presence of these parametric uncertainties.

We reformulate the problem by assimilating δ_1 and δ_2 with fictitious noise inputs. We take A in (6.26) simply as the nominal value $A(0,0)$ and additionally define

$$B_w = \begin{bmatrix} 0 & 0 \\ 0 & 1 \\ 1 & 0 \end{bmatrix}, \quad D_w = 0.$$

At Stage 1 we choose as nullspace basis

$$N_l(s) = [\, I \;\; -G_u(s)\,] = \left[\begin{array}{c|cc} A - sI & 0 & -B_u \\ \hline C & I & -D_u \end{array}\right]$$

and the corresponding $N_f(s)$ and $N_w(s)$ are simply $N_f(s) = G_f(s)$ and $N_w(s) = G_w(s)$. The solvability condition (6.29) is fulfilled, and thus the AFDIP is solvable. Note that $N_f(s)$ is invertible.

At Stage 2 we compute the quasi-co-outer–inner factorization of

$$\widetilde{G}(s) = [\, N_f(s) \; N_w(s)\,]$$

in (6.14). The resulting realization of $G_{o,1}(s)$ in (6.30) has the matrices

$$\overline{B}_0 = \begin{bmatrix} -1.2181 & 0.3638 \\ -0.9828 & 0.7115 \\ -0.9913 & -1.0321 \end{bmatrix}, \quad \overline{D}_0 = 0.$$

Since $\widetilde{G}(s)$ has two zeros at infinity, $G_{o,1}(s)$ will also have these two zeros at infinity and an additional stable zero at -1.1336. This stable zero is also the only pole of the 4×4 inner factor $G_i(s)$.

The descriptor realization of the updated $\overline{Q}(s)$ in (6.35) is

$$\overline{Q}(s) = \left[\begin{array}{cc|cc} A - sI & \overline{B}_o & 0 & -B_u \\ C & \overline{D}_o & I & -D_u \\ \hline 0 & -I & 0 & 0 \end{array}\right].$$

While the updated detector $\overline{Q}(s)$ is improper (having two infinite poles), the updated $\overline{N}_f(s)$ and $\overline{N}_w(s)$ can alternatively be expressed as in (6.17) and therefore have minimal realizations which are stable standard systems (as parts of the inner factor).

With $M_r(s) = I_2$, we compute $\overline{F}_1(s)$ and $\overline{F}_2(s)$ as

$$[\,\overline{F}_1(s)\ \overline{F}_2(s)\,] = [\,I\ \ 0\,][\,G^*_{i,1}(s)\ \ G^*_{i,2}(s)\,] = \left[\begin{array}{c|cc} \overline{A} - sI & \overline{B}_1 & \overline{B}_2 \\ \hline \overline{C} & \overline{D}_1 & \overline{D}_2 \end{array}\right],$$

where

$$\overline{A} = 1.134, \qquad \overline{B}_1 = [\ -0.2830\ \ -0.41822\ ], \quad \overline{B}_2 = [\ 0.8756\ \ 1.4586\ ],$$

$$\overline{C} = \begin{bmatrix} -0.0623 \\ -0.7413 \end{bmatrix}, \quad \overline{D}_1 = \begin{bmatrix} -0.8314 & 0.2423 \\ -0.3914 & -0.3112 \end{bmatrix}, \quad \overline{D}_2 = \begin{bmatrix} 0.4477 & -0.2226 \\ -0.7625 & -0.4105 \end{bmatrix}.$$

Both $\overline{F}_1(s)$ and $\overline{F}_2(s)$ are first order systems with an unstable eigenvalue at 1.1336.

$\mathcal{H}_2$-Optimal Synthesis

With $\{\overline{F}_1(s)\}_+ = \overline{D}_1$ at Stage 3, we determine at Stage 4 the final detector as

$$Q(s) = M(s)\overline{D}_1 G_{o,1}^{-1}(s)[\,I\ -G_u(s)\,],$$

with $M(s) = \frac{10}{s+10}K$ chosen to ensure the properness and stability of the detector. The 2×2 scaling matrix K was determined to ensure that the resulting DC-gain of the TFM from faults to residuals is the identity matrix. The detector has a standard system realization of order 3. Note that the orders of the realizations of the individual factors are, respectively, 2, 0, 5, and 3, which sum together to 10. The resulting low order (in fact the least possible order) clearly illustrates the advantage of the integrated algorithm, which allows us, via explicitly computable realizations, to obtain at each step least order representations of the detector.

For completeness, we give the matrices of the state-space representation of the resulting detector

$$Q(s) = \left[\begin{array}{c|c} A_Q - sI & B_Q \\ \hline C_Q & D_Q \end{array}\right], \tag{6.36}$$

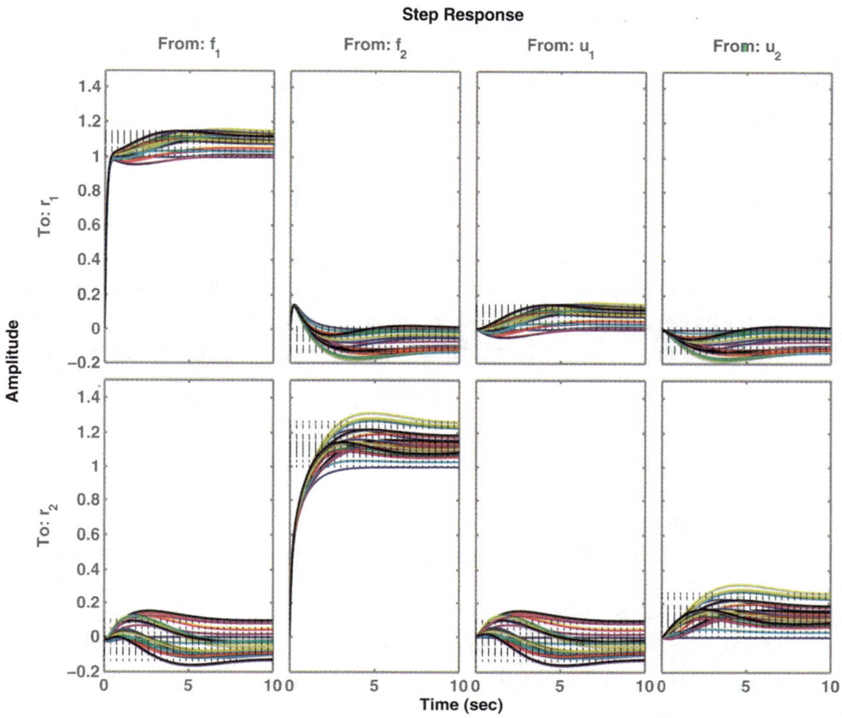

Figure 6.2. Parametric step response analysis for $\mathcal{H}_2$-synthesis [35].

with

$$A_Q = \begin{bmatrix} -10.0147 & -0.4346 & -3.0643 \\ -0.0057 & -10.1691 & -1.1925 \\ 0.0433 & 1.2836 & -0.9498 \end{bmatrix},$$

$$B_Q = \begin{bmatrix} -2.4464 & 0.4409 & -0.1912 & -0.3260 \\ -1.6712 & 3.6086 & 0.5794 & 0.1533 \\ -0.1336 & -0.3443 & -0.0512 & 0.3378 \end{bmatrix},$$

$$C_Q = \begin{bmatrix} -9.6340 & -21.1930 & -5.3805 \\ 36.3688 & 13.7607 & 12.5546 \end{bmatrix},$$

$$D_Q = \begin{bmatrix} -6.3099 & 8.2359 & 0 & 0 \\ 11.3898 & -5.8839 & 0 & 0 \end{bmatrix}.$$

The corresponding residual norm is $\|M(s)\widetilde{\mathcal{R}}(s)\|_2 = 7.9203$.

In Figure 6.2 we present the results of a Monte Carlo analysis of step responses of the parameter dependent residual generation system (of the form (6.4)) from the fault and control inputs for 20 random samples of δ_1 and δ_2 in the range $[-0.25, 0.25]$. The simulations have been performed using the original parameter uncertain state-space model.

The step responses corresponding to a randomly generated parameter combination have the same color. As it can be observed, with an appropriate choice of the detection threshold, the detection and isolation of constant faults can be reliably performed in the presence of parametric uncertainties.

$\mathcal{H}_\infty$-Optimal Synthesis

At Stage 3, the γ-iteration starts with $\gamma_l = 0.9239$ and $\gamma_u = 1$ and ends with $\gamma = 0.9239$ ($= \gamma_l$) for which the corresponding $\gamma_H = 0.5233$. The optimal Nehari approximation of the unstable part $L_u(s)$ has order zero, and the corresponding norm $\left\| V^{-1}(s)\left(\overline{F}_1(s) - Y(s)\right)\right\|_\infty = \gamma_H$. Full cancelation takes place when forming $Q_2(s)$ in (6.23), which thus results as a constant gain

$$Q_2(s) = \begin{bmatrix} -0.6389 & 0.5705 \\ -0.5081 & -0.1017 \end{bmatrix}.$$

The zero McMillan degree of $Q_2(s)$ fully agrees with the degree theory of [13].

Finally, at Stage 4 we choose $M(s) = \frac{10}{s+10}K$ to make $Q(s)$ proper and stable. The 2×2 scaling matrix K was determined to ensure that the resulting TFM from faults to residuals is the identity matrix. The expression of the detector $Q(s)$ can be written down explicitly as $Q(s) = M(s)Q_2(s)G_{o,1}^{-1}(s)[\,I\ - G_u(s)\,]$, which has a standard system realization of order 3. Note that the orders of the realizations of the individual factors are, respectively, 2, 0, 5, and 3, which, once again, sum together to 10.

The resulting state-space representation of the detector (6.36) has the matrices

$$A_Q = \begin{bmatrix} 10.0284 & 0.4410 & 3.2395 \\ -0.0070 & -10.1089 & -0.7999 \\ -0.0790 & -1.2256 & 0.9963 \end{bmatrix},$$

$$B_Q = \begin{bmatrix} 2.6035 & -0.5571 & 0.1846 & 0.3369 \\ -1.3316 & 3.4294 & 0.5806 & 0.1852 \\ 0.1780 & 0.3022 & 0.0473 & -0.3354 \end{bmatrix},$$

$$C_Q = \begin{bmatrix} -11.5159 & -21.6025 & -5.3846 \\ 36.0717 & 13.1418 & 12.5640 \end{bmatrix},$$

$$D_Q = \begin{bmatrix} -6.3099 & 8.2359 & 0 & 0 \\ 11.3898 & -5.8839 & 0 & 0 \end{bmatrix}.$$

The corresponding residual norm is $\|M(s)\widetilde{\mathcal{R}}(s)\|_\infty = 3.3973$.

We evaluated the step responses of the parameter dependent residual generation system (of the form (6.4)) from the faults and control inputs on a 5×5 uniform grid for δ_1 and δ_2 in the range $[-0.25, 0.25]$. The resulting parametric step responses can be seen in Figure 6.3. As can be observed, with an appropriate choice of the detection threshold, the detection and isolation of constant faults can be reliably performed in the presence of parametric uncertainties.

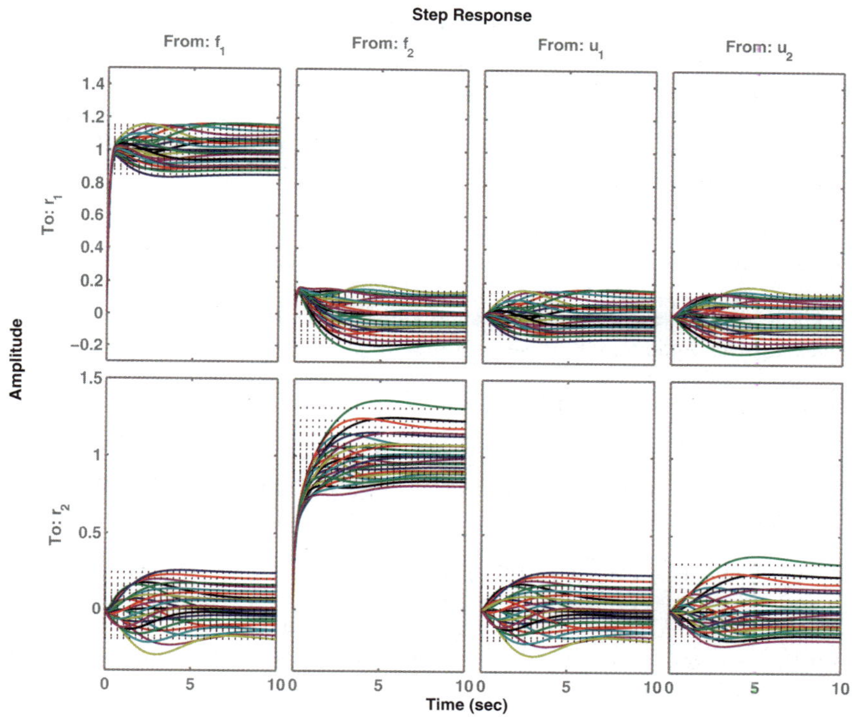

Figure 6.3. Parametric step response analysis for $\mathcal{H}_\infty$-synthesis [36].

6.7 Conclusions

We presented general computational approaches to solve the $\mathcal{H}_{2/\infty}$-norm optimal FDI filter design problems. The new approaches reformulate the filter design problems as equivalent model-matching problems for which an integrated algorithm is proposed which is able to solve this problem in the most general setting. In this way, the technical difficulties often encountered by the existing methods when trying to reduce the approximation problems to standard $\mathcal{H}_{2/\infty}$-norm synthesis problems are completely avoided. For example, the presence of zeros or poles on the boundary of stability domains or problems with nonfull rank and even improper transfer-function matrices can be easily handled. The underlying main computational algorithms are based on descriptor system representations and rely on orthogonal matrix pencil reductions. For all basic computations, reliable numerical software tools are available for MATLAB in the DESCRIPTOR SYSTEMS Toolbox [26] and in the current version of the FAULT DETECTION Toolbox [31, 37]. The described algorithms represent integrated alternative approaches to the exact synthesis method proposed in [29].

Bibliography

[1] E. ANDERSON, Z. BAI, C. BISCHOF, J. DEMMEL, J. DU CROZ, A. GREENBAUM, S. HAMMARLING, A. MCKENNEY, S. OSTROUCHOV, AND D. SORENSEN, *LAPACK User's Guide, Second Edition*, SIAM, Philadelphia, 1995.

[2] T. BEELEN AND P. VAN DOOREN, *An improved algorithm for the computation of Kronecker's canonical form of a singular pencil*, Linear Algebra Appl., 105 (1988), pp. 9–65.

[3] M. BLANKE, M. KINNAERT, J. LUNZE, AND M. STAROSWIECKI, *Diagnosis and Fault-Tolerant Control*, Springer-Verlag, Berlin, 2003.

[4] N. A. BRUINSMA AND M. STEINBUCH, *A fast algorithm to compute the H_∞-norm of a transfer function*, Systems Control Letters, 14 (1990), pp. 287–293.

[5] J. CHEN AND R. J. PATTON, *Robust Model-Based Fault Diagnosis for Dynamic Systems*, Kluwer Academic Publishers, London, 1999.

[6] G.-R. DUAN, *Analysis and Design of Descriptor Linear Systems*, Springer, New York, 2010.

[7] A. EDELMAYER AND J. BOKOR, *Optimal $\mathcal{H}_\infty$ scaling for sensitivity optimization of detection filters*, Internat. J. Robust Nonlinear Control, 12 (2002), pp. 749–760.

[8] B. A. FRANCIS, *A Course in H^∞ Theory*, Springer-Verlag, New York, 1987.

[9] P. M. FRANK AND X. DING, *Frequency domain approach to optimally robust residual generation and evaluation for model-based fault diagnosis*, Automatica, 30 (1994), pp. 789–804.

[10] K. GLOVER, *All optimal Hankel-norm approximations of linear multivariable systems and their L^∞-error bounds*, Internat. J. Control, 39 (1984), pp. 1115–1193.

[11] S. VAN HUFFEL, V. SIMA, A. VARGA, S. HAMMARLING, AND F. DELEBECQUE, *High-performance numerical software for control*, Control Systems Magazine, 24 (2004), pp. 60–76.

[12] B. KÅGSTRÖM AND P. VAN DOOREN, *Additive decomposition of a transfer function with respect to a specified region*, in Proceedings of the MTNS Symposium, Brussels, 1989.

[13] D. J. N. LIMEBEER AND G. D. HALIKIAS, *A controller degree bound for $\mathcal{H}^\infty$-optimal control problems of the second kind*, SIAM J. Control Optim., 26 (1988), pp. 646–677.

[14] P. MISRA, P. VAN DOOREN, AND A. VARGA, *Computation of structural invariants of generalized state-space systems*, Automatica, 30 (1994), pp. 1921–1936.

[15] H. NIEMANN AND J. STOUSTRUP, *Design of fault detectors using H_∞ optimization*, in Proceedings of CDC'00, Sydney, Australia, 2000, pp. 4237–4238.

[16] C. OARĂ, *Constructive solutions to spectral and inner-outer factorizations with respect to the disk*, Automatica, 41 (2005), pp. 1855–1866.

[17] C. OARĂ AND A. VARGA, *Minimal degree coprime factorization of rational matrices*, SIAM J. Matrix Anal. Appl., 21 (1999), pp. 245–278.

[18] C. OARĂ AND A. VARGA, *Computation of general inner-outer and spectral factorizations*, IEEE Trans. Automat. Control, 45 (2000), pp. 2307–2325.

[19] P. VAN DOOREN, *The computation of Kronecker's canonical form of a singular pencil*, Linear Algebra Appl., 27 (1979), pp. 103–141.

[20] P. VAN DOOREN, *The generalized eigenstructure problem in linear systems theory*, IEEE Trans. Automat. Control, 26 (1981), pp. 111–129.

[21] P. VAN DOOREN, *A generalized eigenvalue approach for solving Riccati equations*, SIAM J. Sci. Statist. Comput., 2 (1981), pp. 121–135.

[22] A. VARGA, *Computation of transfer function matrices of generalized state-space models*, Internat. J. Control, 50 (1989), pp. 2543–2561.

[23] A. VARGA, *Computation of irreducible generalized state-space realizations*, Kybernetika, 26 (1990), pp. 89–106.

[24] A. VARGA, *Minimal realization procedures based on balancing and related techniques*, in Computer Aided Systems Theory - EUROCAST'91, F. Pichler and R. Moreno Diaz, eds., Lecture Notes in Comput. Sci. 585, Springer Verlag, Berlin, 1992, pp. 733–761.

[25] A. VARGA, *Computation of coprime factorizations of rational matrices*, Linear Algebra Appl., 271 (1998), pp. 83–115.

[26] A. VARGA, *A* DESCRIPTOR SYSTEMS *toolbox for* MATLAB, in Proceedings of CACSD'2000 Symposium, Anchorage, AK, 2000.

[27] A. VARGA, *Computing generalized inverse systems using matrix pencil methods*, Int. J. Appl. Math. Comput. Sci., 11 (2001), pp. 1055–1068.

[28] A. VARGA, *Computation of least order solutions of linear rational equations*, in Proceedings of MTNS'04, Leuven, Belgium, 2004.

[29] A. VARGA, *New computational approach for the design of fault detection and isolation filters*, in Advances in Automatic Control, M. Voicu, ed., Kluwer Internat. Ser. Engrg. Comput. Sci. 754, Kluwer Academic Publishers, Boston, 2004, pp. 367–381.

[30] A. VARGA, *Numerically reliable methods for optimal design of fault detection filters*, in Proceedings of CDC'05, Seville, Spain, 2005.

[31] A. VARGA, *A* FAULT DETECTION *toolbox for* MATLAB, in Proceedings of CACSD'06, Munich, Germany, 2006.

[32] A. VARGA, *On designing least order residual generators for fault detection and isolation*, in Proceedings of the 16th International Conference on Control Systems and Computer Science, Bucharest, Romania, 2007, pp. 323–330.

[33] A. VARGA, *On computing nullspace bases – a fault detection perspective*, in Proceedings of the IFAC 2008 World Congress, Seoul, Korea, 2008, pp. 6295–6300.

[34] A. VARGA, *General computational approach for optimal fault detection*, in Proceedings of SAFEPROCESS'2009, Barcelona, Spain, 2009.

[35] A. VARGA, *Integrated algorithm for solving $\mathcal{H}_2$-optimal fault detection and isolation problems*, in Proceedings of SYSTOL'2010, Nice, France, 2010.

[36] A. VARGA, *Integrated computational algorithm for solving $\mathcal{H}_\infty$-optimal FDI problems*, in Proceedings of IFAC 2011 World Congress, Milano, Italy, 2011.

[37] A. VARGA, *Linear FDI-Techniques and Software Tools*, FAULT DETECTION Toolbox V1.0 – Technical Documentation, German Aerospace Center (DLR), Institute of Robotics and Mechatronics, 2011.

[38] A. VARGA, *On computing minimal proper nullspace bases with applications in fault detection*, in Numerical Linear Algebra in Signals, Systems and Control, P. Van Dooren, S. P. Bhattacharyya, R. H. Chan, V. Olshevsky, and A. Routray, eds., Lecture Notes in Electrical Engineering 80, Springer-Verlag, Berlin, 2011, pp. 433–465.

[39] A. VARGA AND B. D. O. ANDERSON, *Accuracy-enhancing methods for balancing-related frequency-weighted model and controller reduction*, Automatica, 39 (2003), pp. 919–927.

[40] K. ZHOU, J. C. DOYLE, AND K. GLOVER, *Robust and Optimal Control*, Prentice–Hall, Englewood Cliffs, NJ, 1996.

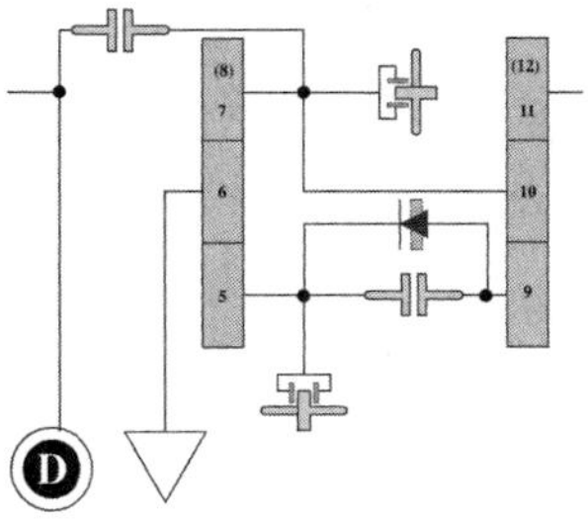

Chapter 7

Normal Forms, High-Gain, and Funnel Control for Linear Differential-Algebraic Systems

Thomas Berger, Achim Ilchmann, and Timo Reis

We consider linear differential-algebraic m-input m-output systems with positive strict relative degree or proper inverse transfer function; in the single-input single-output case these two disjoint classes make the whole of all linear DAEs without feedthrough term. Structural properties—such as normal forms (i.e., the counterpart to the Byrnes–Isidori form for ODE systems), zero dynamics, and high-gain stabilizability—are analyzed for two purposes: first, to gain insight into the system classes and, second, to solve the output regulation problem by funnel control. The funnel controller achieves tracking of a class of reference signals within a prespecified funnel; this means, in particular, the transient behavior of the output error can be specified and the funnel controller does not incorporate any internal model for the reference signals or any identification mechanism. It is simple in its design. The results are illuminated by position and velocity control of a mechanical system encompassing springs, masses, and dampers.

Nomenclature

$\mathbf{N}, \mathbf{N}_0$		set of natural numbers, $\mathbf{N}_0 = \mathbf{N} \cup \{0\}$, set of all integers, resp.
$\mathbf{R}_{\geq 0}$	$=$	$[0, \infty)$
$\mathbf{C}_+ (\overline{\mathbf{C}}_+)$		open (closed) set of complex numbers with positive real part, resp.
$\mathbf{C}_- (\overline{\mathbf{C}}_-)$		open (closed) set of complex numbers with negative real part, resp.
$\mathbf{Gl}_n(\mathbf{R})$		the group of invertible real $n \times n$ matrices
$\mathbf{R}[s]$		the ring of polynomials with coefficients in $\mathbf{R}$
$\mathbf{R}(s)$		the quotient field of $\mathbf{R}[s]$
$R^{n,m}$		the set of $n \times m$ matrices with entries in a ring R
$\lVert x \rVert$	$=$	$\sqrt{x^\top x}$, the Euclidean norm of $x \in \mathbf{R}^n$
$\lVert M \rVert$	$=$	$\max\left\{ \lVert Mx \rVert \,\middle|\, x \in \mathbf{R}^m, \lVert x \rVert = 1 \right\}$, induced norm of $M \in \mathbf{R}^{n,m}$
$\mathcal{C}^\ell(\mathcal{T}; \mathbf{R}^n)$		the set of ℓ-times continuously differentiable functions $f : \mathcal{T} \to \mathbf{R}^n$
$\mathcal{B}^\ell(\mathcal{T}; \mathbf{R}^n) =$		$\left\{ f \in \mathcal{C}^\ell(\mathcal{T}; \mathbf{R}^n) \,\middle|\, \frac{d^i}{dt^i} f \text{ is bounded for } i = 0, \ldots, \ell \right\}$

7.1 Introduction

We consider linear differential-algebraic systems of the form

$$\begin{aligned} E\dot{x}(t) &= Ax(t) + Bu(t), \\ y(t) &= Cx(t), \end{aligned} \tag{7.1}$$

where $E, A \in \mathbf{R}^{n,n}$, $B, C^\top \in \mathbf{R}^{n,m}$ are such that the pencil $sE - A \in \mathbf{R}[s]^{n,n}$ is *regular*, i.e., $\det(sE - A) \in \mathbf{R}[s] \setminus \{0\}$; the set of these square (i.e., same number of inputs and outputs) systems is denoted by $\Sigma_{n,m}$, and we write $[E, A, B, C] \in \Sigma_{n,m}$.

The functions $u, y : \mathbf{R} \to \mathbf{R}^m$ are called *input* and *output* of the system, respectively. A trajectory $(x, u, y) : \mathbf{R} \to \mathbf{R}^n \times \mathbf{R}^m \times \mathbf{R}^m$ is said to be a *solution* of (7.1) if and only if it belongs to the *behavior* of (7.1):

$$\mathfrak{B}_{(7.1)} := \left\{ (x, u, y) \in \mathcal{C}^1(\mathbf{R}; \mathbf{R}^n) \times \mathcal{C}(\mathbf{R}; \mathbf{R}^m) \times \mathcal{C}(\mathbf{R}; \mathbf{R}^m) \,\middle|\, \begin{array}{l} (x, u, y) \text{ solves (7.1)} \\ \text{for all } t \in \mathbf{R} \end{array} \right\}.$$

Regularity of the pencil guarantees that for each consistent initial value $x(0) = x^0 \in \mathbf{R}^n$ there exists a unique and global solution of $E\dot{x} = Ax$; see, for example, [15, Section 2.1]. More smoothness for u and y is required for some results.

We also derive frequency domain results for $[E, A, B, C] \in \Sigma_{n,m}$ and its *transfer function*, defined by

$$G(s) = C(sE - A)^{-1} B \in \mathbf{R}(s)^{m,m}.$$

Roughly speaking, and explained in detail in Section 7.1.3, we divide, in the single-input single-output case, the system class $\Sigma_{n,1}$ into the disjoint sets of systems with strictly proper transfer function and with nonstrictly proper transfer function $g(s) = C(sE - A)^{-1}B \in \mathbf{R}(s)$, respectively. Our generalization for multi-input, multi-output systems $\Sigma_{n,m}$ will treat systems with positive strict relative degree and systems with proper inverse transfer function (see Section 7.1.2 for the definitions), respectively, However, these two disjoint sets do not unify to whole $\Sigma_{n,m}$, as, for example, systems with strictly proper transfer function with a nonconstant vector relative degree are not encompassed.

7.1.1 System Equivalence

We will derive normal forms of systems $[E, A, B, C] \in \Sigma_{n,m}$ belonging to certain classes specified in Section 7.1.3. To this end, recall the following.

Two systems $[E_i, A_i, B_i, C_i] \in \Sigma_{n,m}$, $i = 1, 2$, are called *system equivalent* if and only if

$$\exists\, W, T \in \mathbf{Gl}_n(\mathbf{R}) \,:\, \begin{bmatrix} sE_1 - A_1 & B_1 \\ C_1 & 0 \end{bmatrix} = \begin{bmatrix} W & 0 \\ 0 & I_m \end{bmatrix} \begin{bmatrix} sE_2 - A_2 & B_2 \\ C_2 & 0 \end{bmatrix} \begin{bmatrix} T & 0 \\ 0 & I_m \end{bmatrix};$$

we write

$$[E_1, A_1, B_1, C_1] \overset{W,T}{\sim} [E_2, A_2, B_2, C_2]. \tag{7.2}$$

It is easy to see that system equivalence is an equivalence relation on $\Sigma_{n,m}$ and that the transfer function is invariant under system equivalence.

We quote the well-known result by Weierstraß for regular pencils $sE - A$.

Proposition 7.1 (Weierstraß form [6, Theorem XII.3]). *For any regular matrix pencil $sE - A \in \mathbf{R}[s]^{n,n}$, there exist $W, T \in \mathbf{Gl}_n(\mathbf{R})$ such that*

$$sE - A = W \begin{bmatrix} sI_{n_s} - A_s & 0 \\ 0 & sN - I_{n_f} \end{bmatrix} T \tag{7.3}$$

for some $A_s \in \mathbf{R}^{n_s,n_s}$ and nilpotent $N \in \mathbf{R}^{n_f,n_f}$. The dimensions $n_s, n_f \in \mathbf{N}_0$ are unique, and the matrices A_s and N are unique up to similarity.

The *index of nilpotency* of a nilpotent matrix $N \in \mathbf{R}^{k,k}$ is defined to be the smallest $\nu \in \mathbf{N}$ such that $N^\nu = 0$. It can be shown (see, e.g., [15, Lemma 2.10]) that the index of nilpotency ν of N in (7.3) is uniquely defined by the regular pencil $sE - A$; ν is therefore called the *index* of the pencil $sE - A$ if the nilpotent block is present and the index is $\nu = 0$ if the nilpotent block is absent (i.e., $n_f = 0$).

An immediate result from Proposition 7.1 for systems $[E, A, B, C]$ is the following.

Corollary 7.2 (Decoupled DAE). *Let $[E, A, B, C] \in \Sigma_{n,m}$. Then there exist $W, T \in \mathbf{Gl}_n(\mathbf{R})$ such that*

$$[E, A, B, C] \overset{W,T}{\sim} \left[\begin{bmatrix} I_{n_s} & 0 \\ 0 & N \end{bmatrix}, \begin{bmatrix} A_s & 0 \\ 0 & I_{n_f} \end{bmatrix}, \begin{bmatrix} B_s \\ B_f \end{bmatrix}, \begin{bmatrix} C_s & C_f \end{bmatrix} \right] \tag{7.4}$$

for some $B_s \in \mathbf{R}^{n_f,m}$, $B_f \in \mathbf{R}^{n_f,m}$, $C_s \in \mathbf{R}^{m,n_s}$, $C_f \in \mathbf{R}^{m,n_s}$, $A_s \in \mathbf{R}^{n_s,n_s}$ and nilpotent $N \in \mathbf{R}^{n_f,n_f}$.

The form (7.4) is interpreted, in terms of the DAE (7.1), as follows:

$$(x,u,y) \in \mathfrak{B}_{(7.1)} \qquad \text{if and only if} \qquad \begin{pmatrix} x_s(\cdot) \\ x_f(\cdot) \end{pmatrix} := Tx(\cdot)$$

solves the decoupled DAEs

$$\begin{aligned} \dot{x}_s(t) &= A_s x_s(t) + B_s u(t), & N\dot{x}_f(t) &= x_f(t) + B_f u(t), \\ y_s(t) &= C_s x_s(t), & y_f(t) &= C_f x_f(t), \\ y(t) &= y_s(t) + y_f(t). \end{aligned} \tag{7.5}$$

If $(x,u,y) \in \mathfrak{B}_{(7.1)}$ and in addition $u \in \mathcal{C}^{\nu-1}(\mathbf{R};\mathbf{R}^m)$, then by repeated multiplication of $N\dot{x}_f(t) = x_f(t) + B_f u(t)$ by N from the left, differentiation, and using the identity $N^\nu = 0$, it is easy to see that the solution satisfies

$$x_f(\cdot) = -\sum_{k=0}^{\nu-1} N^k B_f u^{(k)}(\cdot). \tag{7.6}$$

7.1.2 System Properties

Zero Dynamics

An essential concept in this chapter is the *zero dynamics* of system (7.1):

$$\mathcal{ZD}_{(7.1)} := \left\{ (x,u,y) \in \mathfrak{B}_{(7.1)} \;\middle|\; y = 0 \right\}.$$

By linearity of (7.1), $\mathcal{ZD}_{(7.1)}$ is a real vector space.

The zero dynamics of system (7.1) is called *asymptotically stable* if and only if

$$\forall \varepsilon > 0 \; \exists \delta > 0 \; \forall (x,u,y) \in \mathcal{ZD}_{(7.1)} \text{ s.t. } \|(x(0),u(0))\| < \delta \; \forall t \geq 0 \; : \; \|(x(t),u(t))\| < \varepsilon$$

and

$$\forall (x,u,y) \in \mathcal{ZD}_{(7.1)} : \lim_{t\to\infty} \big(x(t),u(t)\big) = 0.$$

(Asymptotically stable) zero dynamics is the vector space of those trajectories of the system which are, loosely spoken, not visible at the output (and tend to zero).

Relative Degree

Another fundamental concept is the *relative degree* of the transfer function $G(s) = C(sE - A)^{-1}B$ of (7.1).

A rational matrix function $G(s) \in \mathbf{R}(s)^{m,m}$ is called

$$\begin{aligned} &\textit{proper} &&\iff && \lim_{s\to\infty} G(s) = D \;\text{ for some }\; D \in \mathbf{R}^{m,m}, \\ &\textit{strictly proper} &&\iff && \lim_{s\to\infty} G(s) = 0, \end{aligned}$$

and we say that the square rational matrix $G(s) \in \mathbf{R}(s)^{m,m}$ has *strict relative degree* $\rho \in \mathbf{Z}$

if and only if

$$\rho \;=\; \operatorname{sr\,deg} G(s) \;:=\; \sup\left\{\, k \in \mathbf{Z} \;\middle|\; \lim_{s\to\infty} s^k\, G(s) \in \mathbf{Gl}_m(\mathbf{R}) \text{ and } \lim_{s\to\infty} s^{k-1}\, G(s) = 0 \,\right\}$$

exists. Note that for any $G(s) \in \mathbf{R}(s)^{m,m}$ we have (consider the entries)

$$\lim_{s\to\infty} s^{k-1}\, G(s) = 0 \quad \text{for some } k \in \mathbf{Z} \qquad \Longrightarrow \qquad \lim_{s\to\infty} s^{k-i}\, G(s) = 0 \quad \text{for all } i \in \mathbf{N}.$$

For convenience, we also say that $[E,A,B,C] \in \Sigma_{n,m}$ has strict relative degree ρ if $G(s) = C(sE-A)^{-1}B$ has strict relative degree ρ. Then the matrix

$$\Gamma := \lim_{s\to\infty} s^{\rho}\, G(s) \in \mathbf{Gl}_m(\mathbf{R})$$

is called *high-frequency gain matrix*.

If $g(s) = p(s)/q(s)$, for $p(s) \in \mathbf{R}[s]$ and $q(s) \in \mathbf{R}[s]\backslash\{0\}$, is a scalar rational function, then the strict relative degree always exists and coincides with the well-known definition of relative degree:

$$\operatorname{sr\,deg} g(s) \;=\; \deg q(s) - \deg p(s).$$

Stabilizability and Detectability

We recall different concepts of stabilizability and detectability for DAEs (7.1). For brevity, we do not define the concepts in system theoretic terms but recall algebraic characterizations in Proposition 7.3; the latter will be used in our proofs. For system theoretic notions of the concepts see [5, Sections 2 and 3].

Proposition 7.3 (Stabilizability and detectability). *A system* (7.1) *is*

(i) *stabilizable* $\Longleftrightarrow$ $\operatorname{rank}[sE-A,B] = n$ *for all* $s \in \overline{\mathbf{C}}_+$,

(ii) *detectable* $\Longleftrightarrow$ $\operatorname{rank}[sE^\top - A^\top, C^\top] = n$ *for all* $s \in \overline{\mathbf{C}}_+$.

Both properties are invariant under system equivalence.

Poles and Zeros

Finally, we recall the definition of transmission zeros and poles of a transfer function; see, for example, [14, Section 6.5].

Let $G(s) \in \mathbf{R}(s)^{m,m}$ with *Smith–McMillan form*

$$U^{-1}(s)G(s)V^{-1}(s) = \operatorname{diag}\left(\frac{\varepsilon_1(s)}{\psi_1(s)},\ldots,\frac{\varepsilon_r(s)}{\psi_r(s)},0,\ldots,0\right) \in \mathbf{R}(s)^{m,m},$$

where $U(s)$, $V(s) \in \mathbf{R}[s]^{m,m}$ are unimodular (i.e., invertible over $\mathbf{R}[s]^{m,m}$), $\operatorname{rank} G(s) = r$ and $\varepsilon_i(s), \psi_i(s) \in \mathbf{R}[s]$ are monic and coprime and satisfy $\varepsilon_i(s) \mid \varepsilon_{i+1}(s)$, $\psi_{i+1}(s) \mid \psi_i(s)$ for all $i = 1,\ldots,r-1$. $s_0 \in \mathbf{C}$ is called *transmission zero* of $G(s)$ if $\varepsilon_r(s_0) = 0$ and a *pole* of $G(s)$ if $\psi_1(s_0) = 0$.

Note that by the transfer function being invariant under system equivalence, this holds true for transmission zeros and poles as well.

7.1.3 System Classes

We introduce the two main system classes investigated in this chapter. In both of them, an explicit feedthrough term is not allowed. However, this assumption is made only for technical reasons. We also explain how systems with feedthrough are related to the two system classes under consideration.

Systems with Positive Strict Relative Degree

Consider the class of DAE systems (7.1) such that $[E,A,B,C] \in \Sigma_{n,m}$ has positive strict relative degree.

If $[E,A,B,C]$ is single-input, single-output, i.e., $m=1$, then its transfer function

$$g(s) = C(sE-A)^{-1}B = p(s)/q(s)$$

has positive strict relative degree if and only if $\deg q(s) > \deg p(s)$ or, equivalently, $g(s)$ is strictly proper.

Note that for any $G(s) \in \mathbf{R}(s)^{m,m}$ we have

$$G(s) \text{ has positive strict relative degree} \quad \underset{\text{i.g.}}{\overset{\Longrightarrow}{\not\Longleftarrow}} \quad G(s) \text{ is strictly proper.}$$

An ODE system $[E,A,B,C] = [I,A,B,C]$ has transfer function

$$G(s) \;=\; C(sI-A)^{-1}B \;=\; CBs^{-1} + CABs^{-2} + CA^2Bs^{-3} + \cdots$$

and therefore strict relative degree $\rho \in \mathbf{N}$ if and only if

$$\det(CA^{\rho-1}B) \neq 0 \quad \text{and,} \quad \text{if } \rho > 1, \quad \forall\, k = 0,\ldots,\rho-2 \,:\, CA^kB = 0.$$

Systems with Proper Inverse Transfer Function

The second class of DAE systems (7.1) considered in this chapter are those DAEs $[E,A,B,C] \in \Sigma_{n,m}$ whose transfer function $G(s) = C(sE-A)^{-1}B \in \mathbf{R}(s)^{m,m}$ has *proper inverse* over $\mathbf{R}(s)^{m,m}$; i.e., $G(s)$ is invertible over $\mathbf{R}(s)$ and $\lim_{s\to\infty} G^{-1}(s) \in \mathbf{R}^{m,m}$.

In [2, Proposition 1.2] we have shown that for any $G(s) \in \mathbf{R}(s)^{m,m}$ the properties of proper inverse transfer function and strict relative degree are related as follows:

$$\operatorname{sr\,deg} G(s) \leq 0 \quad \underset{\text{i.g.}}{\overset{\Longrightarrow}{\not\Longleftarrow}} \quad G(s) \text{ has proper inverse.}$$

A zero dynamics form for systems $[E,A,B,C] \in \Sigma_{n,m}$ with transfer function $G(s) \in \mathbf{R}(s)^{m,m}$ whose inverse $G^{-1}(s)$ is proper is studied in Theorem 7.8. Conversely, if a proper transfer function $G^{-1}(s) \in \mathbf{R}(s)^{m,m}$ is given, then this may be realized as an ODE $[I,A,B,C,D]$ with feedthrough term D and it is easily verified that the regular DAE

$$\left[\begin{bmatrix} 0 & 0 \\ 0 & I \end{bmatrix}, \begin{bmatrix} -D & -C \\ B & A \end{bmatrix}, \begin{bmatrix} I \\ 0 \end{bmatrix}, [I,0]\right] \tag{7.7}$$

without feedthrough term is a realization of $G(s)$.

If $[E, A, B, C]$ is single-input, single-output, then its transfer function $g(s) = C(sE - A)^{-1}B = p(s)/q(s)$ has proper inverse if and only if $\deg q(s) \leq \deg p(s)$ or, equivalently, $g(s)$ is not strictly proper. In this case, there does not exist a realization of $g(s)$ as an ODE system $[I, A, B, C]$ without feedthrough term. The class of single-input single-output DAE systems $\Sigma_{n,1}$ can be decomposed into the disjoint sets of systems with strictly proper transfer function and with nonstrictly proper transfer function $g(s) = C(sE - A)^{-1}B \in \mathbf{R}(s)$, respectively:

$$\begin{aligned}\Sigma_{n,m} &= \left\{[E,A,B,C] \in \Sigma_{n,1} \,\middle|\, g(s) = \tfrac{p(s)}{q(s)},\ \deg q(s) > \deg p(s)\right\} \\ &\quad \dot{\cup} \left\{[E,A,B,C] \in \Sigma_{n,1} \,\middle|\, g(s) = \tfrac{p(s)}{q(s)},\ \deg q(s) \leq \deg p(s)\right\}.\end{aligned}$$

For multi-input, multi-output systems $\Sigma_{n,m}$, generalization to systems with positive strict relative degree and systems with proper inverse transfer function does not allow for such a disjoint union. This is the topic of the following section.

Systems Which Have neither Positive Strict Relative Degree nor Proper Inverse Transfer Function

The class of multi-input, multi-output systems $[E, A, B, C] \in \Sigma_{n,m}$ with transfer function $G(s) = C(sE - A)^{-1}B \in \mathbf{R}(s)^{m,m}$ can be decomposed in the disjoint union of the following three sets:

$$\begin{aligned}\Sigma_{n,m} = &\left\{[E,A,B,C] \in \Sigma_{n,m} \,\middle|\, G(s) \text{ has positive strict relative degree}\right\} \\ &\dot{\cup} \left\{[E,A,B,C] \in \Sigma_{n,m} \,\middle|\, G(s) \text{ has proper inverse transfer function}\right\} \\ &\dot{\cup} \left\{[E,A,B,C] \in \Sigma_{n,m} \,\middle|\, \begin{array}{l} G(s) \text{ has neither positive strict relative degree} \\ \text{nor proper inverse transfer function} \end{array}\right\}.\end{aligned}$$

Note that the latter set is, for multi-input, multi-output systems, not empty. It contains in particular systems with a nonconstant vector relative degree and it contains, for example for $m = 2$, the system

$$[E, A, B, C] := \left[\begin{bmatrix} 0 & -1 & 0 \\ 0 & 0 & 0 \\ 0 & 0 & 1 \end{bmatrix}, \begin{bmatrix} 1 & 0 & 0 \\ 0 & 1 & 0 \\ 0 & 0 & 0 \end{bmatrix}, \begin{bmatrix} 0 & 0 \\ 1 & 0 \\ 0 & 1 \end{bmatrix}, \begin{bmatrix} 1 & 0 & 0 \\ 0 & 0 & 1 \end{bmatrix}\right]$$

with transfer function $G(s) = \operatorname{diag}(s, s^{-1}) \in \mathbf{R}(s)^{2,2}$. We stress that $G(s)$ has neither a positive strict relative degree nor a proper inverse. The study of normal forms for this general system class would be very involved, and it is not clear whether funnel control would be possible. Hence it is omitted.

Systems with Feedthrough

So far we have not considered systems with nonzero feedthrough term. In this section we will show that such systems can, by augmenting the state by the feedthrough term, be rewritten as a system of the form (7.1); the transfer function of the new system coincides with the transfer function of the original system. Furthermore, it is shown that a system with proper transfer function can be rewritten as an ODE with feedthrough.

Remark 7.4 (DAE with feedthrough $\longrightarrow$ DAE without feedthrough). *Consider a DAE system with feedthrough of the form*

$$\begin{aligned} E\dot{x}(t) &= Ax(t) + Bu(t), \\ y(t) &= Cx(t) + Du(t), \end{aligned} \tag{7.8}$$

where $E, A \in \mathbf{R}^{n,n}$, $B, C^\top \in \mathbf{R}^{n,m}$, $D \in \mathbf{R}^{m,m}$. *Then the behavior* $\mathfrak{B}_{(7.8)}$, *the zero dynamics* $\mathcal{ZD}_{(7.8)}$, *and its asymptotic stability can be defined completely analogously as for* (7.1); *this is omitted.*

By introducing an "augmented state" of $x(t)$ *and* $w(t) = Du(t)$, *system* (7.8) *can be rewritten as a DAE system without feedthrough term, namely*

$$\begin{aligned} \begin{bmatrix} E & 0 \\ 0 & 0 \end{bmatrix} \frac{d}{dt} \begin{pmatrix} x(t) \\ w(t) \end{pmatrix} &= \begin{bmatrix} A & 0 \\ 0 & -I \end{bmatrix} \begin{pmatrix} x(t) \\ w(t) \end{pmatrix} + \begin{bmatrix} B \\ D \end{bmatrix} u(t), \\ y(t) &= \begin{bmatrix} C & I \end{bmatrix} \begin{pmatrix} x(t) \\ w(t) \end{pmatrix}. \end{aligned} \tag{7.9}$$

Clearly, the behaviors and zero dynamics are related in the following way:

$$\begin{aligned} \mathfrak{B}_{(7.9)} &= \left\{ ((x^\top, w^\top)^\top, u, y) \;\middle|\; (x,u,y) \in \mathfrak{B}_{(7.8)},\ w = Du \right\}, \\ \mathcal{ZD}_{(7.9)} &= \left\{ ((x^\top, w^\top)^\top, u, y) \;\middle|\; (x,u,y) \in \mathcal{ZD}_{(7.8)},\ w = Du \right\}. \end{aligned}$$

The following observations can now be made:

(a) *The transfer functions of* (7.8) *and* (7.9) *coincide:*

$$C(sE - A)^{-1}B + D = \begin{bmatrix} C & I \end{bmatrix} \left(s \begin{bmatrix} E & 0 \\ 0 & 0 \end{bmatrix} - \begin{bmatrix} A & 0 \\ 0 & -I \end{bmatrix} \right)^{-1} \begin{bmatrix} B \\ D \end{bmatrix}.$$

(b) *The zero dynamics* $\mathcal{ZD}_{(7.8)}$ *is asymptotically stable if and only if the zero dynamics* $\mathcal{ZD}_{(7.9)}$ *is asymptotically stable.*

(c) *The pencil* $s\left[\begin{smallmatrix} E & 0 \\ 0 & 0 \end{smallmatrix}\right] - \left[\begin{smallmatrix} A & 0 \\ 0 & -I \end{smallmatrix}\right]$ *is regular if and only if* $sE - A$ *is regular.*

(d) *We have*

$$\begin{aligned} \det \left[\begin{array}{cc|c} sE - A & 0 & B \\ 0 & I & D \\ \hline C & I & 0 \end{array}\right] &= \det \begin{bmatrix} sE - A & 0 & B \\ 0 & I & D \\ C & 0 & -D \end{bmatrix} = \det \begin{bmatrix} sE - A & 0 & B \\ 0 & I & 0 \\ C & 0 & -D \end{bmatrix} \\ &= \det \begin{bmatrix} sE - A & B \\ C & -D \end{bmatrix}. \end{aligned}$$

(e) *Since, for all $s \in \mathbf{C}$,*

$$\operatorname{rank}\begin{bmatrix} sE-A & 0 & B \\ 0 & I & D \end{bmatrix} = \operatorname{rank}\begin{bmatrix} sE-A & B \end{bmatrix} + m,$$
$$\operatorname{rank}\begin{bmatrix} sE-A & 0 \\ 0 & I \\ C & I \end{bmatrix} = \operatorname{rank}\begin{bmatrix} sE-A \\ C \end{bmatrix} + m,$$

it is easily seen that system (7.8) *is stabilizable (detectable) if and only if system* (7.9) *is stabilizable (detectable).*

(f) *If $E = I$, then* (7.8) *is an ODE with feedthrough and can be rewritten as a DAE with regular matrix pencil and transfer function $G(s) = C(sI - A)^{-1}B + D$, which coincides with the transfer function of the original ODE system.*

Remark 7.5 (DAE with proper transfer function $\longrightarrow$ ODE with feedthrough). *For any DAE* (7.1) *with regular pencil $sE - A \in \mathbf{R}[s]^{n,n}$ and $\lim_{s\to\infty} C(sE-A)^{-1}B \in \mathbf{R}^{m,m}$ there exists an ODE system with feedthrough*

$$\begin{aligned} \dot{z}(t) &= A_s z(t) + B_s u(t), \\ y(t) &= C_s z(t) - C_f B_f u(t) \end{aligned} \tag{7.10}$$

for some $B_s \in \mathbf{R}^{n_f,m}$, $B_f \in \mathbf{R}^{n_f,m}$, $C_s \in \mathbf{R}^{p,n_s}$, $C_f \in \mathbf{R}^{p,n_s}$, $A_s \in \mathbf{R}^{n_s,n_s}$ such that the transfer functions of (7.1) *and* (7.10) *coincide:*

$$C(sE-A)^{-1}B = C_s(sI - A_s)^{-1}B_s + C_f B_f .$$

This can be seen as follows: Without restriction of generality, one may assume that (7.1) *is in Weierstraß form (see Proposition* 7.1*) and given by equations in* (7.5)*. Then its transfer function satisfies*

$$\begin{aligned} C(sE-A)^{-1}B &= C_f(sN - I_{n_f})^{-1}B_f + C_s(sI_{n_s} - A_s)^{-1}B_s \\ &= -\sum_{i=0}^{\nu-1} C_f N^i B_f\, s^i + \sum_{i\geq 1} C_s A_s^{i-1} B_s\, s^{-i} \end{aligned}$$

and the assumption $\lim_{s\to\infty} C(sE-A)^{-1}B \in \mathbf{R}^{m,m}$ yields $C_f N^i B_f = 0$ for all $i \geq 1$. Now invoking (7.6) *and setting $z = x_s$ in* (7.10) *proves the claim.*

7.1.4 Control Objectives

We consider two control strategies with respect to the two system classes: for DAE systems with positive strict relative degree, a constant proportional derivative output feedback is suggested; for DAE systems with proper inverse transfer function, a constant proportional output feedback is considered. These controllers achieve high-gain stabilization and exploit the underlying system properties. However, since no information on the system entries will be required it is unknown how "high" the feedback gain has to be chosen. To resolve this, we allow for funnel control, that is, a proportional time-varying high-gain error feedback in combination with a filter (the filter "adjusts" the higher relative degree) for DAE systems with positive strict relative degree, and a proportional time-varying high-gain error feedback for DAE systems with proper inverse transfer function.

Constant High-Gain Control

First, constant high-gain proportional output feedback is given by

$$u(t) = -k \cdot p(\tfrac{\mathrm{d}}{\mathrm{d}t})\, y(t) \qquad \begin{array}{l}\text{for } [E,A,B,C] \in \Sigma_{n,m} \\ \text{with positive strict relative degree,}\end{array} \tag{7.11}$$

$$u(t) = \ k \cdot y(t) \qquad \begin{array}{l}\text{for } [E,A,B,C] \in \Sigma_{n,m} \\ \text{with proper inverse transfer function,}\end{array} \tag{7.12}$$

where $k > 0$ and $p(s) \in \mathbf{R}[s]$ is Hurwitz in (7.11), and $k \in \mathbf{R}$ in (7.12). However, these results are of more theoretical nature to show the consequences of the underlying system properties such as asymptotically stable zero dynamics.

The control objective is output feedback regulation in the sense that the *high-gain controller*, i.e., the time-invariant proportional output-derivative feedback (7.11) or (7.12) respectively, yields an asymptotically stable closed-loop system. Here k is the high-gain parameter and, in general, k has to be large to achieve that the state $x(t)$ decays to zero.

In Section 7.4 we will show that the assumption of asymptotically stable zero dynamics of a system (7.1) which has either positive strict relative degree or proper inverse transfer function implies high-gain stabilizability in the following sense: If the system has positive strict relative degree, $k \gg 1$, and $p(s)$ is Hurwitz with positive leading coefficient, then the high-gain feedback (7.11) or (7.12), respectively, applied to (7.1) forces the output "very quickly" to zero. If the system has proper inverse transfer function, then it is sufficient to assume $|k| \gg 1$ and $p(s) \equiv 1$ to achieve this goal.

Funnel Control

The *control objective* is output feedback regulation in the sense that the *funnel controller*, applied to any system $[E,A,B,C] \in \Sigma_{n,m}$, achieves tracking of the output of any reference signal $y_{\text{ref}} \in \mathcal{B}^{\nu+1}(\mathbf{R}_{\geq 0}; \mathbf{R}^m)$, where ν is the index of the $sE - A$, with prespecified transient behavior:

$$\forall\, t > 0: \ \|e(t)\| < 1/\varphi(t).$$

The transient behavior is prespecified by a funnel boundary $1/\varphi(\cdot)$ given by

$$\varphi \in \Phi^{\mu} := \left\{ \varphi \in \mathcal{B}^{\mu}(\mathbf{R}_{\geq 0}, \mathbf{R}) \,\middle|\, \varphi(0) = 0, \ \ \varphi(s) > 0 \ \forall s > 0 \text{ and } \liminf_{s \to \infty} \varphi(s) > 0 \right\},$$

with which we associate, see Figure 7.1, the *performance funnel*

$$\mathcal{F}_{\varphi} := \left\{ (t,e) \in \mathbf{R}_{\geq 0} \times \mathbf{R}^m \,\middle|\, \varphi(t)\|e\| < 1 \right\}. \tag{7.13}$$

Note that there is no need that the funnel be monotone. However, the funnel is not allowed to "shrink" to zero as $t \to \infty$. But the minimal distance (at infinity) between

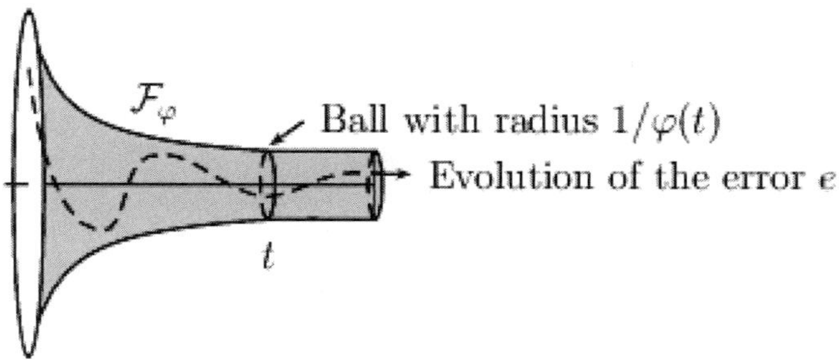

Figure 7.1. Error evolution in the funnel $\mathcal{F}_\varphi$ with "width ∞" at $t = 0$, i.e., $\varphi(0) = 0$.

the funnel boundary and zero, i.e., $\lambda = \liminf_{s\to\infty} \varphi(s)^{-1}$, must be positive albeit can be arbitrarily small; from a practical point of view, this is not a limitation.

The funnel controller takes two forms. For DAE systems $[E,A,B,C] \in \Sigma_{n,m}$ with positive strict relative degree the higher degree is an obstacle; in (7.11) we have used derivative feedback while now we will incorporate a filter. This idea goes back to ODEs; it is shown in [11] that funnel control is feasible if a filter is incorporated in the feedback. This filter is constructed as follows;

$$\dot{z}(t) = \underbrace{\begin{bmatrix} -I_m & I_m & 0 & \cdots & 0 & 0 \\ 0 & -I_m & I_m & \cdots & 0 & 0 \\ 0 & 0 & -I_m & \cdots & 0 & 0 \\ \vdots & \vdots & \vdots & \ddots & \vdots & \vdots \\ 0 & 0 & 0 & \cdots & -I_m & I_m \\ 0 & 0 & 0 & \cdots & 0 & -I_m \end{bmatrix}}_{=:F_\rho} z(t) + \underbrace{\begin{bmatrix} 0 \\ 0 \\ 0 \\ \vdots \\ 0 \\ I_m \end{bmatrix}}_{=:G_\rho} u(t), \quad z(0) = z^0 \tag{7.14}$$

with initial data $z^0 \in \mathbf{R}^{(\rho-1)m}$. The feedback law is defined recursively by the C^∞-functions

$$\begin{aligned} \gamma_1 : \quad \mathbf{R} \times \mathbf{R}^m &\to \mathbf{R}^m, \\ (k,e) &\mapsto ke, \\ \gamma_2 : \quad \mathbf{R} \times \mathbf{R}^m \times \mathbf{R}^m &\to \mathbf{R}^m, \\ (k,e,z_1) &\mapsto \gamma_1(k,e) + \|D\gamma_1(k,e)\|^2 k^4 (1 + \|z_1\|^2)(z_1 + \gamma_1(k,e)) \end{aligned}$$

and, for $i = 3, \ldots, \rho$,

$$\begin{aligned} &\gamma_i : \mathbf{R} \times \mathbf{R}^m \times \mathbf{R}^{(i-1)m} \to \mathbf{R}^m, \quad (k,e,(z_1,\ldots,z_{i-1})) \\ &\mapsto \gamma_{i-1}(k,e,(z_1,\ldots,z_{i-2})) + \|D\gamma_{i-1}(k,e,(z_1,\ldots,z_{i-2}))\|^2 k^4 (1 + \|(k,e,(z_1,\ldots,z_{i-1}))\|^2) \\ &\qquad \times \big(z_{i-1} + \gamma_{i-1}(k,e,(z_1,\ldots,z_{i-2}))\big), \end{aligned}$$

where D denotes the derivative (Jacobian matrix). For a lengthy discussion of the intuition for the filter see [11]. Now the funnel controller (with filter (7.14)) for systems

$[E,A,B,C] \in \Sigma_{n,m}$ with positive strict relative degree $\rho \in \mathbf{N}$ takes the form

$$\boxed{\begin{array}{rclr} u(t) & = & -\gamma_\rho\big(k(t),e(t),z(t)\big), & e(t) = y(t) - y_{\mathrm{ref}}(t), \\ \dot z(t) & = & F_\rho z(t) + G_\rho u(t), & k(t) = \frac{1}{1-\varphi(t)^2\|e(t)\|^2}. \end{array}} \tag{7.15}$$

The funnel controller for systems $[E,A,B,C] \in \Sigma_{n,m}$ with proper inverse transfer function is much simpler: For appropriately chosen gain parameter $\hat{k} \in \mathbf{R} \setminus \{0\}$ it is

$$\boxed{\begin{array}{rclr} u(t) & = & -k(t)e(t), & e(t) = y(t) - y_{\mathrm{ref}}(t), \\ k(t) & = & \frac{\hat{k}}{1-\varphi(t)^2\|e(t)\|^2}. & \end{array}} \tag{7.16}$$

In Section 7.5 we will show that the assumption of asymptotically stable zero dynamics of a system (7.1) which has either positive strict relative degree or proper inverse transfer function implies feasibility of the funnel controller. In view of the fact that such systems are high-gain stabilizable (see Section 7.4), intuitively we may believe that if $\|e(t)\|$ is close to the funnel boundary $\varphi(t)^{-1}$, then the high-gain $k(t)$ forces $\|e(t)\|$ away from the funnel boundary. This is the essential property to allow for funnel control of these systems: $k(t)$ is designed in such a way that it is large if the error $\|e(t)\|$ is close to the funnel boundary $\varphi(t)^{-1}$, hence avoiding contact.

The funnel controller is a high-gain adaptive controller which is not high-gain and not adaptive: Certainly, the gain $k(t)$ in (7.15) and (7.16) takes high values if $k(t)$ is "close" to the funnel boundary; but it is by no means monotone and exploits the high-gain property of the system class (i.e., asymptotically stable zero dynamics) only if necessary. The gain $k(t)$ is adapted by the output, more precisely by the distance of the norm of the output error to the funnel boundary.

7.1.5 Literature and Outline

This chapter is based on the following precursors: The Byrnes–Isidori form for ODE systems with strictly proper transfer function is derived in [12]; see also [13, Section 5.1]. The zero dynamics form for DAE systems with proper inverse transfer function is derived in [2]. The concept of funnel control is introduced in [10]; for a further discussion of funnel control see the survey article [9].

In this chapter we consider two classes of linear multi-input, multi-output DAE systems: systems with positive strict relative degree and systems with proper inverse transfer function; in the single-input single-output case, these two disjoint classes are the whole of all linear DAEs without feedthrough term. In Section 7.2, zero dynamics forms for these system classes are derived; they are the counterpart to the Byrnes–Isidori form for ODE systems. The normal forms may be interesting in their own right and are here essential for later proofs of funnel control. In Section 7.3, parameterization of the zero dynamics and characterization of asymptotically zero dynamics in terms of a matrix condition, the zero dynamics form, and the transfer function are derived. In Section 7.4, it is shown that a system with asymptotically stable zero dynamics is high-gain stabilizable with constant gain (in case of proper inverse transfer function) and constant gain in combination with a filter (in case of positive strict relative degree of the transfer function). In Section 7.5, we are ready to present two funnel controllers for the two system classes. These funnel controllers achieve tracking of a reference signal (belonging to a rather large class) within a prespecified funnel; this means in particular that transient behavior of the output error can be prespecified and the funnel controller does not incorporate any internal model but is simple in its design. These results are illustrated in Section 7.6 by the position control and velocity control of a mechanical system consisting of springs, masses, and dampers. The proofs are delegated to Section 7.7.

7.2 Normal Forms

The notion "normal form" means that it is weaker than "canonical form": Recall that if a set (of systems) may be divided into equivalence classes and if each orbit (i.e., equivalence class) has a unique representative, then the mapping from the set to each representative within the orbit is called a *canonical form*. The Byrnes–Isidori form for ODE state-space systems with strictly proper transfer function (see [9] and also [13, Section 5.1]) is not a canonical form. However, the freedom left within the nonzero entries of the Byrnes–Isidori form can be specified and is not significant; this justifies the notion "normal form."

In the present section, certain normal forms, the so-called zero dynamics forms, are studied for two classes of DAE systems with positive strict relative degree and with proper inverse transfer function. They are interesting in their own right since they give insight into the properties of the systems class; we will use them to derive results on zero dynamics and on high-gain output feedback and funnel control.

We first present a zero dynamics form (7.17) for descriptor systems with positive strict relative degree; see Section 7.1.3.

Theorem 7.6 (Zero dynamics form for systems with positive strict relative degree). *Suppose* $[E,A,B,C] \in \Sigma_{n,m}$ *has strict relative degree* $\rho \in \mathbf{N}$. *Then there exist* $W,T \in \mathbf{Gl}_n(\mathbf{R})$ *such that*

$$[E,A,B,C] \overset{W,T}{\sim} [\hat{E},\hat{A},\hat{B},\hat{C}],$$

with

$$\begin{aligned}
& s\hat{E}-\hat{A} \\
&= \left[\begin{array}{cccccc|cc}
sI_m & -I_m & 0 & \cdots & 0 & 0 & 0 & 0 \\
0 & sI_m & -I_m & 0 & & 0 & 0 & 0 \\
\vdots & & \ddots & \ddots & \ddots & \vdots & \vdots & \vdots \\
0 & 0 & \cdots & sI_m & -I_m & 0 & 0 & 0 \\
-R_1 & -R_2 & \cdots & -R_{\rho-1} & sI_m-R_\rho & -S & 0 & 0 \\
-P & 0 & \cdots & 0 & 0 & sI_\mu-Q & 0 & 0 \\
\hline
0 & 0 & \cdots & 0 & 0 & 0 & sN_c-I_{n_c} & sN_{c\bar{c}} \\
0 & 0 & \cdots & 0 & 0 & 0 & 0 & sN_{\bar{c}}-I_{n_{\bar{c}}}
\end{array}\right], \\
& \hat{B} = \left[\begin{array}{cccccc|cc} 0 & 0 & \cdots & 0 & \Gamma^\top & 0 & B_c^\top & 0 \end{array}\right]^\top, \\
& \hat{C} = \left[\begin{array}{cccccc|cc} I_m & 0 & \cdots & 0 & 0 & 0 & 0 & C_{\bar{c}} \end{array}\right],
\end{aligned} \tag{7.17}$$

where, for some $n_c, n_{\overline{c}} \in \mathbf{N}_0$, *and* $\mu = n - n_c - n_{\overline{c}} - \rho m$,

$$\begin{aligned} &\Gamma = \lim_{s\to\infty} s^\rho\, C(sE-A)^{-1}B \in \mathbf{Gl}_m(\mathbf{R}) \text{ is the high frequency gain matrix,}\\ &S \in \mathbf{R}^{m,\mu},\ P \in \mathbf{R}^{\mu,m},\ Q \in \mathbf{R}^{\mu,\mu},\ \left[R_1,\dots,R_\rho\right] \in \mathbf{R}^{m,\rho m},\\ &B_c \in \mathbf{R}^{n_c,m},\ C_{\overline{c}} \in \mathbf{R}^{m,n_{\overline{c}}},\ \mathbf{N}_{c\overline{c}} \in \mathbf{R}^{n_c,n_{\overline{c}}},\\ &N_c \in \mathbf{R}^{n_c,n_c},\ \mathbf{N}_{\overline{c}} \in \mathbf{R}^{n_{\overline{c}},n_{\overline{c}}} \text{ are nilpotent, and } \operatorname{rank}[\,N_c\,,\,B_c] = n_c. \end{aligned} \tag{7.18}$$

The entries $R_1,\dots,R_\rho$ *are unique, system* $[I,Q,P,S]$ *is unique up to system equivalence* $\overset{\hat{T}^{-1},\hat{T}}{\sim}$, *and the matrices* N_c *and* $N_{\overline{c}}$ *are unique up to similarity. If* E *is invertible, then* $n_c = n_{\overline{c}} = 0$; *this means only the upper left block in* $\hat{E}$ *is present. The form* (7.17) *diagrammed in Figure* 7.2 *is called* zero dynamics form *of* (7.1). *The transfer function satisfies, where* $R_{\rho+1} := -I$,

$$C(sE-A)^{-1}B \;=\; -\left[\sum_{i=1}^{\rho+1} R_i s^{i-1} + S(sI_{n-n_c-n_{\overline{c}}-\rho m} - Q)^{-1}P\right]^{-1} \Gamma. \tag{7.19}$$

The proof is given in Section 7.7.

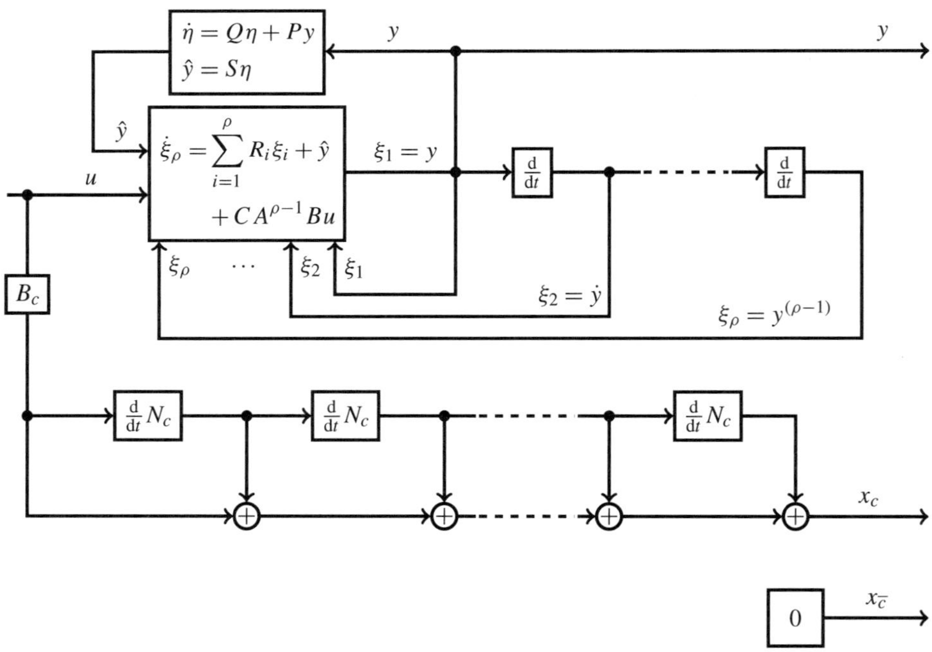

Figure 7.2. Zero dynamics form for systems with positive strict relative degree.

Remark 7.7 (Zero dynamics form for DAEs). *An immediate consequence of Theorem* 7.6 *is the simplified representation of system* $[E,A,B,C] \in \Sigma_{n,m}$: *If* ν *denotes the index of* $sE-A$, *then a trajectory satisfies*

$$(x,u,y) \in \mathfrak{B}_{(7.1)} \cap \left(\mathcal{C}^1(\mathbf{R};\mathbf{R}^n) \times \mathcal{C}^{\nu-1}(\mathbf{R};\mathbf{R}^m) \times \mathcal{C}^\rho(\mathbf{R};\mathbf{R}^m)\right)$$

if and only if $Tx = \left(y^\top, \dot{y}^\top, \ldots, y^{(\rho-1)\top}, \eta^\top, x_c^\top, x_{\overline{c}}^\top\right)^\top \in \mathcal{C}^1(\mathbf{R};\mathbf{R}^n)$ *fulfills*

$$\boxed{\begin{array}{rcl} y^{(\rho)}(t) &=& \sum_{i=1}^{\rho} R_i y^{(i-1)}(t) + S\eta(t) + \Gamma u(t), \\ \dot{\eta}(t) &=& P y(t) + Q\eta(t), \\ x_c(t) &=& -\sum_{i=0}^{\nu-1} N_c^i B_c u^{(i)}(t), \\ x_{\overline{c}}(t) &=& 0. \end{array}} \tag{7.20}$$

The second zero dynamics form is for systems $[E,A,B,C] \in \Sigma_{n,m}$, where the transfer function $C(sE-A)^{-1}B$ has proper inverse; see Section 7.1.3. This is the counterpart to (7.17).

Theorem 7.8 (Zero dynamics form for systems with proper inverse transfer function [2, Theorem 2.3]). *Suppose* $[E,A,B,C] \in \Sigma_{n,m}$ *is such that* $C(sE-A)^{-1}B$ *has proper inverse. Then there exist* $W, T \in \mathbf{Gl}_n(\mathbf{R})$ *such that*

$$[E,A,B,C] \overset{W,T}{\sim} [\hat{E},\hat{A},\hat{B},\hat{C}],$$

with

$$s\hat{E}-\hat{A} = \begin{bmatrix} -A_{11} & -A_{12} & 0 & sE_{\overline{c}} \\ -A_{21} & sI_{n_2}-Q & 0 & 0 \\ sE_c & 0 & sN_c - I_{n_c} & sN_{c\overline{c}} \\ 0 & 0 & 0 & sN_{\overline{c}} - I_{n_{\overline{c}}} \end{bmatrix}, \quad \hat{B} = \hat{C}^\top = \begin{bmatrix} I_m \\ 0_{n_2,m} \\ 0_{n_c,m} \\ 0_{n_{\overline{c}},m} \end{bmatrix}, \tag{7.21}$$

where, for some $n_2, n_c, n_{\overline{c}} \in \mathbf{N}_0$,

$$\begin{array}{l} A_{11} \in \mathbf{R}^{m,m},\ A_{12} \in \mathbf{R}^{m,n_2},\ A_{21} \in \mathbf{R}^{n_2,m},\ Q \in \mathbf{R}^{n_2,n_2}, \\ E_c \in \mathbf{R}^{n_c,m},\ E_{\overline{c}} \in \mathbf{R}^{m,n_{\overline{c}}},\ N_{c\overline{c}} \in \mathbf{R}^{n_c,n_{\overline{c}}}, \\ N_c \in \mathbf{R}^{n_c,n_c},\ N_{\overline{c}} \in \mathbf{R}^{n_{\overline{c}},n_{\overline{c}}} \text{ are nilpotent, and } \operatorname{rank}[N_c, E_c] = n_c. \end{array} \tag{7.22}$$

The form (7.21) *diagrammed in Figure* 7.3 *is called* zero dynamics form *of* (7.1). *Furthermore, the following hold:*

(i) $N_c^\nu = 0$ *and* $N_{\overline{c}}^\nu = 0$, *where* ν *denotes the index of the pencil* $sE-A$.

(ii) *The transfer function satisfies*

$$C(sE-A)^{-1}B = -\left(A_{11} + A_{12}(sI_{n_2}-Q)^{-1}A_{21}\right)^{-1}.$$

(iii) $\operatorname{sr deg}\left(C(sE-A)^{-1}B\right) = 0$ *if and only if* $A_{11} \in \mathbf{Gl}_m(\mathbf{R})$.

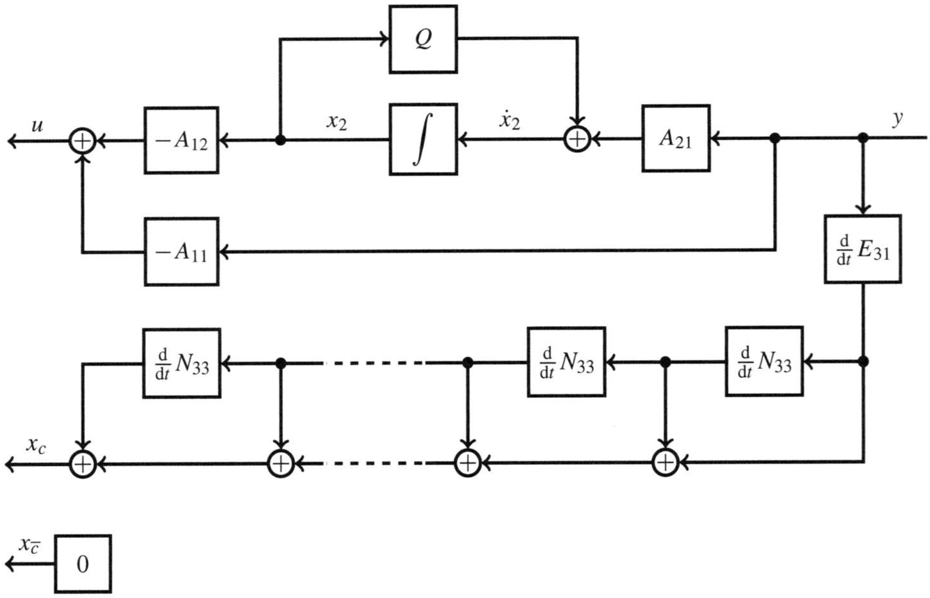

Figure 7.3. Zero dynamics form for systems with proper inverse transfer function.

(iv) $\operatorname{sr deg}\big(C(sE-A)^{-1}B\big) = \rho < 0$ *if and only if* $A_{11} = 0$ *and* $\operatorname{sr deg}\big(A_{12}(sI_{n_2}-Q)^{-1}A_{21}\big) = -\rho$.

For uniqueness we have the following:

(v) *If* $[E,A,B,C]$, $[\hat{E},\hat{A},\hat{B},\hat{C}] \in \Sigma_{n,m,m}$ *are both in zero dynamics form* (7.21) *and*

$$[\hat{E},\hat{A},\hat{B},\hat{C}] \overset{W,T}{\sim} [E,A,B,C] \qquad \text{for some} \quad W,T \in \mathbf{Gl}_n(\mathbf{R}), \tag{7.23}$$

then there exist $W_{ii} \in \mathbf{Gl}_{n_i}(\mathbf{R})$, $i \in \{2,c,\overline{c}\}$, $W_{c\overline{c}} \in \mathbf{R}^{n_c,n_{\overline{c}}}$ *such that*

$$W = T^{-1} = \begin{bmatrix} I_m & 0 & 0 & 0 \\ 0 & W_{22} & 0 & 0 \\ 0 & 0 & W_{cc} & W_{c\overline{c}} \\ 0 & 0 & 0 & W_{\overline{c}\overline{c}} \end{bmatrix}. \tag{7.24}$$

(vi) *The dimensions* $n_2, n_c, n_{\overline{c}}$ *are unique; the matrices* N_c, $N_{\overline{c}}$, Q *are unique up to similarity, so in particular the spectrum of* Q *is unique; and* $A_{11} = -\lim_{s\to\infty}\big(C(sE-A)^{-1}B\big)^{-1}$.

The proof is given in [2, Theorem 2.3].

Remark 7.9 (Zero dynamics form for DAE). *An immediate consequence of Theorem* 7.8 *is the representation as a DAE: If* ν *is the index of* $sE-A$, *then a trajectory satisfies*

$$(x,u,y) \in \mathfrak{B}_{(7.1)} \cap \Big(\mathcal{C}^1(\mathbf{R};\mathbf{R}^n) \times \mathcal{C}^1(\mathbf{R};\mathbf{R}^m) \times \mathcal{C}^\nu(\mathbf{R};\mathbf{R}^m)\Big)$$

if and only if $Tx = \left(y, x_2^\top, x_c^\top, x_{\overline{c}}^\top\right)^\top \in \mathcal{C}^1(\mathbf{R};\mathbf{R}^n)$ *fulfills*

$$\boxed{\begin{array}{rcl} 0 & = & A_{11}\, y(t) + A_{12} x_2(t) + u(t), \\ \dot{x}_2(t) & = & A_{21}\, y(t) + Q\, x_2(t), \\ x_c(t) & = & \sum_{i=0}^{\nu-1} N_c^i E_c\, y^{(i+1)}(t), \\ x_{\overline{c}}(t) & = & 0. \end{array}} \tag{7.25}$$

Remark 7.10 (ODE systems with feedthrough). *For ODE systems with feedthrough, i.e.,* (7.8) *with* $E = I$*, and strict relative degree zero, we have* $D \in \mathbf{Gl}_m(\mathbf{R})$*. In this case, by using the transcription of* (7.8) *as a DAE system* (7.9)*, the zero dynamics form* (7.25) *can be achieved without changing the coordinates of the state, namely*

$$\begin{array}{rcl} 0 & = & -D^{-1} y(t) + D^{-1} C x(t) + u(t), \\ \dot{x}(t) & = & B D^{-1} y(t) + (A - B D^{-1} C) x(t). \end{array} \tag{7.26}$$

In the notation of (7.25)*, the quantities* n_c *and* $n_{\overline{c}}$ *are then trivial, while* $n_2 = n$ *and* $A_{11} = -D^{-1}$, $A_{12} = D^{-1}C$, $A_{21} = A - BD^{-1}C$*, and* $Q = A - BD^{-1}C$*.*

7.3 Zero Dynamics

Before we characterize asymptotically stable zero dynamics, we show that both zero dynamics forms allow us to parameterize the zero dynamics.

Remark 7.11 (Parameterization of zero dynamics).

(i) *Suppose that* $[E, A, B, C] \in \Sigma_{n,m}$ *has positive strict relative degree and let* ν *be the index of* $sE - A$*. Then the zero dynamics are given, in terms of the matrices in* (7.17)*, by*

$$\mathcal{ZD}_{(7.1)} = \left\{ \left(T^{-1} \begin{pmatrix} 0_{\rho m} \\ e^{Q\cdot}\eta^0 \\ \sum\limits_{i=0}^{\nu-1} N_c^i B_c \Gamma^{-1} S Q^i e^{Q\cdot}\eta^0 \\ 0_{n_{\overline{c}}} \end{pmatrix}, -\Gamma^{-1} S e^{Q\cdot}\eta^0,\, 0 \right) \;\middle|\; \eta^0 \in \mathbf{R}^\mu \right\}. \tag{7.27}$$

(ii) *Suppose that* $[E, A, B, C] \in \Sigma_{n,m}$ *has proper inverse transfer function. Then the zero dynamics are given, in terms of* (7.21)*, by*

$$\mathcal{ZD}_{(7.1)} = \left\{ \left(T^{-1} \begin{pmatrix} 0_m \\ e^{Q\cdot}x_2^0 \\ 0_{n_c} \\ 0_{n_{\overline{c}}} \end{pmatrix}, -A_{12} e^{Q\cdot} x_2^0,\, 0 \right) \;\middle|\; x_2^0 \in \mathbf{R}^{n_z} \right\}. \tag{7.28}$$

We are now in a position to characterize asymptotically stable zero dynamics of systems $[E, A, B, C] \in \Sigma_{n,m}$ with positive strict relative degree or proper inverse transfer function.

Theorem 7.12 (Asymptotically stable zero dynamics). *Consider a system* $[E, A, B, C] \in \Sigma_{n,m}$ *and suppose it has positive strict relative degree or proper inverse transfer function. Then the following statements are, in terms of the system properties defined in Section* 7.1.2, *equivalent:*

(i) *the zero dynamics of system* (7.1) *are asymptotically stable ;*

(ii)
$$\forall s \in \overline{\mathbb{C}}_+ : \det \begin{bmatrix} sE - A & B \\ C & 0 \end{bmatrix} \neq 0;$$

(iii) *system* (7.1) *is* minimum phase, *i.e.,*

(a) (7.1) *is stabilizable,*

(b) (7.1) *is detectable,*

(c) $C(sE - A)^{-1}B$ *does not have any transmission zeros in* $\overline{\mathbb{C}}_+$*;*

(iv) $\sigma(Q) \subseteq \mathbb{C}_-$, *where* Q *is in the zero dynamics form* (7.17) *or* (7.21), *respectively.*

The proof is given in Section 7.7.

Condition (ii) in Theorem 7.12 is a convenient mathematical frequency domain characterization which leads to condition (iv); the latter is the important property to show constant high-gain stabilizability in Section 7.4. Condition (iii) is the traditional definition of a minimum phase system. Conditions (ii) and (iii) are very similar to what is known for asymptotically stable zero dynamics of ODE systems; see [13, Section 4.3].

The following two remarks, although not exploited in the next section, may be worth knowing in their own right.

Remark 7.13 (Zero dynamics and behavior).

(i) *Suppose that* $[E, A, B, C] \in \Sigma_{n,m}$ *has positive strict relative degree. Then the behavior* $\mathfrak{B}_{(7.1)}$ *can be decomposed, in terms of the transformation matrix* T *from Theorem* 7.6, *into a direct sum of the zero dynamics and a summand as*

$$\mathfrak{B}_{(7.1)} = \mathcal{ZD}_{(7.1)} \oplus \mathcal{R},$$

where

$$\mathcal{R} := \left\{ \begin{array}{c} (x,u,y) \in \mathcal{C}^1(\mathbb{R};\mathbb{R}^n) \\ \times \mathcal{C}(\mathbb{R};\mathbb{R}^m) \times \mathcal{C}(\mathbb{R};\mathbb{R}^m) \end{array} \middle| \begin{array}{l} (x,u,y) \text{ solves } (7.1) \text{ and} \\ [0_{\mu,\rho m}, I_\mu, 0_{\mu,n_c}, 0_{\mu,n_{\overline{c}}}]Tx(0) = 0 \end{array} \right\}.$$

In terms of (7.27)*, the representation is immediate from*

$$\big(x(\cdot),u(\cdot),y(\cdot)\big) = \left(T^{-1}\begin{pmatrix} 0_{\rho m,\rho m} \\ e^{Q\cdot}\eta^0 \\ \sum_{i=0}^{\nu-1} N_c^i B_c \Gamma^{-1} S Q^i e^{Q\cdot}\eta^0 \\ 0_{n_{\overline{c}},n_{\overline{c}}} \end{pmatrix},\ -\Gamma^{-1} S e^{Q\cdot}\eta^0,\ 0 \right)$$

$$+ \left(x(\cdot) - T^{-1}\begin{pmatrix} 0_{\rho m,\rho m} \\ e^{Q\cdot}\eta^0 \\ \sum_{i=0}^{\nu-1} N_c^i B_c \Gamma^{-1} S Q^i e^{Q\cdot}\eta^0 \\ 0_{n_{\overline{c}},n_{\overline{c}}} \end{pmatrix},\ u(\cdot)+\Gamma^{-1} S e^{Q\cdot}\eta^0,\ y(\cdot) \right)$$

for any $(x,u,y) \in \mathfrak{B}_{(7.1)}$*, where* $\eta^0 = [0_{\mu,\rho m}, I_\mu, 0_{\mu,n_c}, 0_{\mu,n_{\overline{c}}}]Tx(0)$.

(ii) *Suppose that* $[E,A,B,C] \in \Sigma_{n,m}$ *has proper inverse transfer function. Then the behavior* $\mathfrak{B}_{(7.1)}$ *can be decomposed, in terms of the transformation matrix* T *from Theorem* 7.8*, into a direct sum of the zero dynamics and a summand as*

$$\mathfrak{B}_{(7.1)} = \mathcal{ZD}_{(7.1)} \oplus \mathcal{R},$$

where

$$\mathcal{R} := \left\{ \begin{array}{l|l} (x,u,y) \in \mathcal{C}^1(\mathbf{R};\mathbf{R}^n) & (x,u,y) \text{ solves } (7.1) \text{ and} \\ \quad \times\mathcal{C}(\mathbf{R};\mathbf{R}^m) \times \mathcal{C}(\mathbf{R};\mathbf{R}^m) & [0_{n_2,m}, I_{n_2}, 0_{n_2,n_c}, 0_{n_2,n_{\overline{c}}}]Tx(0) = 0 \end{array} \right\}.$$

This fact is shown in [2, Remark 3.3].

Finally, we show that the zero dynamics carries, in a certain sense, the structure of a dynamical system.

Remark 7.14 (Zero dynamics are a dynamical system). *Let* ν *be the index of the pencil* $sE-A$*. The transition map of system* (7.1) *is defined, in terms of Proposition* 7.1*, as*

$$\begin{aligned} \varphi : \mathbf{R}\times\mathbf{R}\times\mathbf{R}^n\times\mathcal{C}^{\nu-1}(\mathbf{R};\mathbf{R}^m) &\to \mathbf{R}^n \\ (t,t_0,x^0,u(\cdot)) &\mapsto T^{-1}\begin{bmatrix} e^{A_s(t-t_0)} & 0 \\ 0 & 0 \end{bmatrix} Tx^0 \\ &\quad + \int_{t_0}^{t} T^{-1}\begin{bmatrix} e^{A_s(t-\tau)} & 0 \\ 0 & 0 \end{bmatrix} W^{-1}Bu(\tau)\,d\tau \\ &\quad - \sum_{k=0}^{\nu-1} T^{-1}\begin{bmatrix} 0 & 0 \\ 0 & N^k \end{bmatrix} W^{-1}Bu^{(k)}(t). \end{aligned}$$

It has been shown in [3, Proposition 2.20] *that for any* $(t_0, x^0, u) \in \mathbf{R} \times \mathbf{R}^n \times \mathcal{C}^{\nu-1}(\mathbf{R};\mathbf{R}^m)$ *the map* $t \mapsto x(t) := \varphi(t, t_0, x^0, u(\cdot))$ *solves the IVP*

$$E\dot{x}(t) = Ax(t) + Bu(t), \qquad x(t_0) = x^0 \tag{7.29}$$

if and only if

$$x^0 \in \mathcal{V}_{t_0,u(\cdot)} := \left\{ x \in \mathbf{R}^n \;\middle|\; x + \sum_{i=0}^{\nu-1} T^{-1} \begin{bmatrix} 0 & 0 \\ 0 & N^i \end{bmatrix} W^{-1} B u^{(i)}(t_0) \in \operatorname{im} T^{-1} \begin{bmatrix} I_{n_s} \\ 0 \end{bmatrix} \right\}.$$

Therefore, consistency of the initial value x^0 *depends on the initial time* t_0 *and the input* $u(\cdot)$. *The output map of system* (7.1) *is defined by*

$$\eta : \mathbf{R} \times \mathbf{R}^n \times \mathbf{R}^m \to \mathbf{R}^m, \quad (t, x, u) \mapsto Cx.$$

It is readily verified that the structure $(\mathbf{R}, \mathbf{R}^m, \mathcal{C}^{\nu-1}(\mathbf{R};\mathbf{R}^m), \mathbf{R}^n, \mathbf{R}^m, \varphi, \eta)$, *where* $\varphi : \mathcal{D}_\varphi \to \mathbf{R}^n$ *is the restriction of the transition map (by abuse of notation we write the same symbol) on*

$$\mathcal{D}_\varphi := \left\{ (t, t_0, x^0, u) \in \mathbf{R} \times \mathbf{R} \times \mathbf{R}^n \times \mathcal{C}^{\nu-1}(\mathbf{R};\mathbf{R}^m) \;\middle|\; \begin{array}{l} x^0 \in \mathcal{V}_{t_0,u(\cdot)}, \\ C\varphi(\cdot; t_0, x^0, u(\cdot)) = 0 \end{array} \right\},$$

is an **R***-linear time-invariant dynamical system as defined in* [7, Definitions 2.1.1, 2.1.24, 2.1.26].

Next let $u \in \mathcal{C}^{\nu-1}(\mathbf{R};\mathbf{R}^m)$. *As a consequence of uniqueness and global existence of the solution of the IVP* (7.29) *for* $x^0 \in \mathcal{V}_{0,u(\cdot)}$ *(see again* [3, Proposition 2.20]*), the map*

$$\Psi : \mathcal{D}_{\varphi,0} \to \mathcal{ZD}_{(7.1)}, \quad (0, 0, x^0, u(\cdot)) \mapsto \big(\varphi(\cdot; 0, x^0, u(\cdot)), u(\cdot), C\varphi(\cdot; 0, x^0, u(\cdot))\big)$$

is well defined, where

$$\mathcal{D}_{\varphi,0} := \left\{ (0, 0, x^0, u) \in \mathcal{D}_\varphi \right\} \subset \mathcal{D}_\varphi.$$

Most importantly, if $[E, A, B, C]$ *has positive strict relative degree or proper inverse transfer function, then* Ψ *is an isomorphism: it is surjective since a preimage of* $(x, u, 0) \in \mathcal{ZD}_{(7.1)}$ *is* $(0, 0, x(0), u(\cdot)) \in \mathcal{D}_{\varphi,0}$ *(note that* $u \in \mathcal{C}^{\nu-1}(\mathbf{R};\mathbf{R}^m)$ *by Remark* 7.11*), it is injective by uniqueness of the solution of the IVP* (7.29)*. In this sense, we may say that* $\mathcal{ZD}_{(7.1)}$ *is a dynamical system.*

7.4 Stabilization by Constant High-Gain Output Feedback

Next we investigate whether asymptotically stable zero dynamics is a sufficient (or even necessary) condition for high-gain stabilizability in the sense that the feedback (7.11) or (7.12), respectively, applied to (7.1) yields an asymptotically stable closed-loop system.

Theorem 7.15 (High-gain stabilizability). *Consider a system* $[E, A, B, C] \in \Sigma_{n,m}$. *Then the following statements hold:*

(i) *Suppose* $[E, A, B, C]$ *has strict relative degree* $\rho \in \mathbf{N}$ *and positive definite high frequency gain matrix* $\Gamma \in \mathbf{R}^{m,m}$ *(cf.* (7.18)*). Let* $p(s) = \sum_{i=0}^{\rho-1} p_i s^i \in \mathbf{R}[s]$ *be Hurwitz* *and* $p_{\rho-1} > 0$. *Then*

$$\mathcal{ZD}_{(7.1)} \text{ is asympt. stable} \quad \underset{\text{i.g.}}{\overset{\Longrightarrow}{\not\Longleftarrow}} \quad \begin{cases} \exists k^* \geq 0\ \forall k \geq k^* : \\ \text{“(7.1) \& (7.11)” is asympt. stable.} \end{cases}$$

(ii) *Suppose* $[E,A,B,C]$ *has proper inverse transfer function. Then*

$$\mathcal{ZD}_{(7.1)} \text{ is asympt. stable} \quad \underset{\text{i.g.}}{\overset{\Longrightarrow}{\not\Longleftarrow}} \quad \begin{cases} \exists k^* \geq 0\ \forall k \in \mathbf{R}, |k| \geq k^*\ \forall \text{sln. } x(\cdot) \text{ of} \\ \text{“(7.1) \& (7.12)”} : \lim_{t\to\infty} x(t) = 0. \end{cases}$$

The proof is given in Section 7.7.

Remark 7.16 (High-gain stabilizability).

(i) *In case of strict relative degree one, the feedback law* (7.11) *reduces to the proportional output feedback* $u(t) = -k\,y(t)$. *If the system has higher relative degree,* (7.11) *incorporates a compensator* $p(s)$ *(and thus derivative feedback) to achieve a relative degree one system.*

For ODE systems, the result is proven in [4] *for relative degree one. By using the form* (7.17)*, this result can be generalized to differential-algebraic systems with positive strict relative degree by using the same techniques.*

(ii) *Consider* $[E,A,B,C]$ *with proper inverse transfer function. It may be surprising that the sign of* k *in* (7.12) *is irrelevant and, furthermore, no compensator is required.*

In view of Remark 7.9*, the closed-loop system “*$u(t) = ky(t)$ *&* (7.1)*” is equivalent to*

$$\begin{aligned} -(A_{11} + kI_m)\,y(t) &= A_{12}x_2(t), \\ \dot{x}_2(t) &= Qx_2(t) + A_{21}y(t), \\ x_c(t) &= \sum_{i=0}^{\nu-1} N_c^i\, E_c\, y^{(i+1)}(t) \end{aligned}$$

and, if $|k| > \|A_{11}\|$*, equivalent to*

$$\begin{aligned} y(t) &= -(A_{11} + kI_m)^{-1} A_{12}x_2(t), \\ \dot{x}_2(t) &= \left[Q - A_{21}(A_{11} + kI_m)^{-1}A_{12}\right] x_2(t), \\ x_c(t) &= \sum_{i=0}^{\nu-1} N_c^i\, E_c\, y^{(i+1)}(t). \end{aligned} \tag{7.30}$$

Note that $x_2 \in \mathcal{C}^\infty(\mathbf{R};\mathbf{R}^{n_2})$ *yields* $y \in \mathcal{C}^\infty(\mathbf{R};\mathbf{R}^m)$ *and so the algebraic equation for* x_c *is well defined. Note also that asymptotic stability of the zero dynamics is equivalent to* $\sigma(Q) \subseteq \mathbf{C}_-$ *by Theorem* 7.12*. Since*

$$\lim_{k\to\pm\infty} \sigma\left(Q - A_{21}(A_{11} + kI_m)^{-1}A_{12}\right) = \sigma(Q),$$

the assumptions "$|k|$ sufficiently large" and "(7.1) has asymptotically stable zero dynamics" yield exponential decay of $x_2(\cdot)$, and therefore $x_c(\cdot)$ and $y(\cdot)$ decay exponentially, too.

7.5 Funnel Control

In this section we will show that the two funnel controllers (7.15) and (7.16) achieve output tracking of a rather general reference signal within a prespecified funnel.

Theorem 7.17 (Funnel control). *Suppose that $[E,A,B,C] \in \Sigma_{n,m}$ has asymptotically stable zero dynamics, and let ν be the index of $sE - A$. Let $\varphi \in \Phi^{\nu+1}$ define a performance funnel $\mathcal{F}_\varphi$. The initial data are an arbitrary consistent initial value $x^0 \in \mathbf{R}^n$ and a reference signal $y_{\text{ref}} \in \mathcal{B}^{\nu+1}(\mathbf{R}_{\geq 0};\mathbf{R}^m)$.*

(i) *If $[E,A,B,C]$ has positive strict relative degree $\rho \in \mathbf{N}$ and positive definite high frequency gain matrix $\Gamma = \lim_{s\to\infty} s^\rho C(sE-A)^{-1}B$, then the application of the funnel controller* (7.15) *to* (7.1) *yields a closed-loop IVP with precisely one maximal continuously differentiable solution $x\colon [0,\omega) \to \mathbf{R}^n$ and this solution is global (i.e., $\omega = \infty$), and all functions x,z,k,u are bounded.*

(ii) *If $[E,A,B,C]$ has proper inverse transfer function and the initial gain $k(0) = \hat{k} \in \mathbf{R}$ satisfies $|\hat{k}| > \lim_{s\to\infty} \|G^{-1}(s)\|$, then the application of the funnel controller* (7.16) *to* (7.1) *yields a closed-loop IVP with precisely one maximal continuously differentiable solution $x\colon [0,\omega) \to \mathbf{R}^n$ and this solution is global (i.e., $\omega = \infty$), and all functions x,k,u are bounded.*

Most importantly, in both cases (i) *and* (ii), *the tracking error $e(\cdot) = Cx(\cdot) - y_{\text{ref}}(\cdot)$ satisfies*

$$\exists \varepsilon > 0\ \forall t > 0 : \ \|e(t)\| \leq \varphi(t)^{-1} - \varepsilon \tag{7.31}$$

(that means $e(\cdot)$ evolves within the performance funnel $\mathcal{F}_\varphi$ and is uniformly bounded away from the boundary) and for the same ε the gain is bounded by

$$\forall t_0 > 0 : \ \sup_{t \geq t_0} |k(t)| \leq \frac{|\hat{k}|}{1 - (1 - \varepsilon\lambda_{t_0})^2}, \tag{7.32}$$

where $\lambda_{t_0} := \inf_{t\geq t_0} \varphi(t) > 0$ for all $t_0 > 0$. $\hat{k} = 1$ in case of (i).

The proof is given in Section 7.7.

Remark 7.18.

(i) *Minimum-phase systems (7.8), with transfer function $G(s) = C(sE-A)^{-1}B + D$ that has either a proper inverse or positive strict relative degree, also allow funnel control in the sense of Theorem 7.17. This follows from the artificial extension (7.9).*

(ii) *Consistency of the initial value $x^0 \in \mathbf{R}^n$ means that the closed-loop system (7.1), (7.15) or (7.1), (7.16), respectively, with $x(0) = x^0$ has a solution $x : [0,\omega) \to \mathbf{R}^n$ for some $\omega \in (0,\infty]$. In practice, consistency of the initial state of the "unknown" system should be satisfied as far as the DAE $[E,A,B,C]$ is the correct model.*

(iii) *In case of Theorem* 7.17(ii), *the assumption on the initial gain* $k(0) = \hat{k}$ *reduces to* $\hat{k} > 0$ *if* $\lim_{s\to\infty} G^{-1}(s) = 0$; *the latter means that the transfer function cannot be realized by an ODE system with feedthrough; see* [2, Remark 6.3].

(iv) *In case of ODE systems* (7.8), $E = I$, *with invertible feedthrough term, the transfer function* $G(s) = C(sI - A)^{-1}B + D$ *has proper inverse and the assumption on the initial gain* $k(0) = \hat{k}$ *is equivalent to* $|\hat{k}| > \|D^{-1}\|$.

(v) *Theorem* 7.17 *is on system class* $\Sigma_{n,m}$. *A careful inspection of the proof of the theorem shows that one actually needs only that the nominal system can be brought into the form* (7.20) *or* (7.25), *respectively, Thus in particular, the pencil* $sE - A$ *does not have to be regular; what is important is that* $\sigma(Q) \subseteq \mathbb{C}_-$.

7.6 Application: Position and Velocity Control of a Mechanical System with Springs, Masses, and Dampers

7.6.1 The Mechanical Model

For purposes of illustration, we consider a mechanical system, see Figure 7.4, with springs, masses, and dampers with single-input spatial distance between the two masses and single-output position of one mass. We are indebted to Professor P.C. Müller (BU Wuppertal) for suggesting this example to us. The masses m_1, m_2, damping constants d_1, d_2, and spring constants c_1, c_2 are all assumed to be positive. As output $y(t) = z_2(t)$ we take the position of the mass m_2. In the first example, the input $u(t) = z_2(t) - z_1(t)$ is the spatial distance between the masses m_1 and m_2, whereas, in the second example, the input is the relative velocity between the masses m_1 and m_2, i.e., $u(t) = \dot{z}_2(t) - \dot{z}_1(t)$.

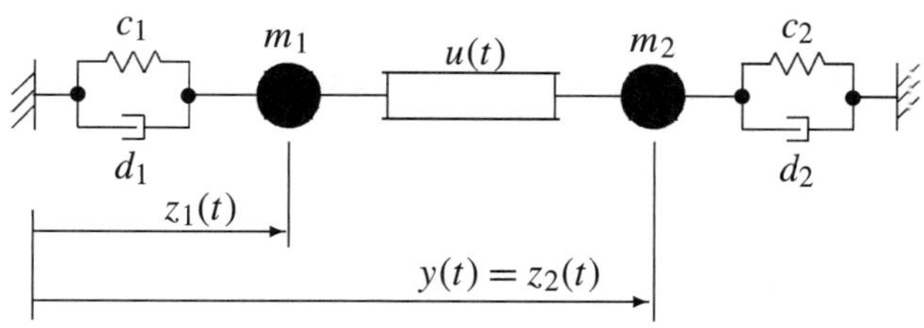

Figure 7.4. Mass-spring-damper system.

Position Control

With input $u(t) = z_2(t) - z_1(t)$, the mechanical system in Figure 7.4 may be modeled by the second order DAE

$$\begin{aligned} m_1\ddot{z}_1(t) + d_1\dot{z}_1(t) + c_1 z_1(t) - \lambda(t) &= 0, \\ m_2\ddot{z}_2(t) + d_2\dot{z}_2(t) + c_2 z_2(t) + \lambda(t) &= 0, \\ z_2(t) - z_1(t) &= u(t), \\ y(t) &= z_2(t), \end{aligned} \tag{7.33}$$

where $\lambda(\cdot)$ is a constraint force viewed as a variable. Defining the state

$$x(t) = (z_1(t), \dot{z}_1(t), z_2(t), \dot{z}_2(t), \lambda(t))^\top, \tag{7.34}$$

model (7.33) may be rewritten as the linear differential-algebraic input-output system (7.1) for

$$sE - A = \begin{bmatrix} s & -1 & 0 & 0 & 0 \\ c_1 & sm_1 + d_1 & 0 & 0 & -1 \\ 0 & 0 & s & -1 & 0 \\ 0 & 0 & c_2 & sm_2 + d_2 & 1 \\ -1 & 0 & 1 & 0 & 0 \end{bmatrix}, \quad B = \begin{bmatrix} 0 \\ 0 \\ 0 \\ 0 \\ 1 \end{bmatrix}, \quad C = \begin{bmatrix} 0 \\ 0 \\ 1 \\ 0 \\ 0 \end{bmatrix}^\top. \tag{7.35}$$

We may immediately see that the pencil $sE - A$ is regular and has index $\nu = 3$. The transfer function

$$G(s) = C(sE - A)^{-1}B = \frac{m_1 s^2 + d_1 s + c_1}{(m_1 + m_2)s^2 + (d_1 + d_2)s + (c_1 + c_2)}$$

has proper inverse: $\lim_{s\to\infty} G^{-1}(s) = (m_1 + m_2)/m_1$. The zero dynamics of (7.35) is asymptotically stable: setting $y(\cdot) = 0$ in (7.33) yields $z_2(\cdot) = 0$, $\lambda(\cdot) = 0$, $z_1(\cdot) = -u(\cdot)$ and $m_1 z_1(t) + d_1 \dot{z}_1(t) + c_1 z_1(t) = 0$ for all $t \geq 0$; positivity of m_1, d_1, and c_1 then gives $\lim_{t\to\infty} \dot{z}_1(t) = \lim_{t\to\infty} z_1(t) = 0$. Summarizing, system (7.35) satisfies the assumptions of Theorem 7.17(ii).

Velocity Control

With input $u(t) = \dot{z}_2(t) - \dot{z}_1(t)$, the mechanical system in Figure 7.4 may, analogous to position control, be modeled by the second order DAE

$$\begin{aligned} m_1 \ddot{z}_1(t) + d_1 \dot{z}_1(t) + c_1 z_1(t) - \lambda(t) &= 0, \\ m_2 \ddot{z}_2(t) + d_2 \dot{z}_2(t) + c_2 z_2(t) + \lambda(t) &= 0, \\ \dot{z}_2(t) - \dot{z}_1(t) &= u(t), \\ y(t) &= z_2(t). \end{aligned} \tag{7.36}$$

Defining the state as in (7.34), the model (7.36) may be rewritten as the linear differential-algebraic input-output system (7.1) for

$$A = \begin{bmatrix} 0 & 1 & 0 & 0 & 0 \\ -c_1 & -d_1 & 0 & 0 & 1 \\ 0 & 0 & 0 & 1 & 0 \\ 0 & 0 & -c_2 & -d_2 & -1 \\ 0 & 1 & 0 & -1 & 0 \end{bmatrix} \quad \text{and} \quad E, B, C \text{ as in (7.35)}. \tag{7.37}$$

We may immediately see that the pencil $sE - A$ is regular and has index $\nu = 2$. The transfer function

$$G(s) = C(sE - A)^{-1}B = \frac{m_1 s^2 + d_1 s + c_1}{(m_1 + m_2)s^3 + (d_1 + d_2)s^2 + (c_1 + c_2)s}$$

has strict relative degree one: $\lim_{s\to\infty} s \cdot G(s) = m_1/(m_1 + m_2)$. Similar to Section 7.6.1, we may see that the zero dynamics of (7.37) is asymptotically stable, whence we are in the situation of Theorem 7.17(i).

7.6.2 Simulations

In both examples, as reference signal $y_{\text{ref}} : \mathbf{R}_{\geq 0} \to \mathbf{R}$, we take the first component of the solution of the following IVP for the Lorenz system:

$$\begin{array}{rcll} \dot{\xi}_1(t) &=& 10(\xi_2(t) - \xi_1(t)), & \xi_1(0) = 5, \\ \dot{\xi}_2(t) &=& 28\,\xi_1(t) - \xi_1(t)\xi_3(t) - \xi_2(t), & \xi_2(0) = 5, \\ \dot{\xi}_3(t) &=& \xi_1(t)\xi_2(t) - \frac{8}{3}\,\xi_3(t), & \xi_3(0) = 5. \end{array} \tag{7.38}$$

This may be viewed as a rather academic choice; however, it is well known (see, for example, [16, Appendix C]) that the Lorenz system is chaotic (and thus the reference signal is rather "wild"), and the unique global solution of (7.38) is bounded with bounded derivative on the positive real axis (and thus our assumptions on the class of reference signals are satisfied). The solution of (7.38) is depicted in Figure 7.5. The funnel $\mathcal{F}_\varphi$ is determined by the function

$$\varphi : \mathbf{R}_{\geq 0} \to \mathbf{R}_{\geq 0}, \quad t \mapsto 0.5\, te^{-t} + 2 \arctan t. \tag{7.39}$$

Note that this prescribes an exponentially (exponent 1) decaying funnel in the transient phase $[0, T]$, where $T \approx 3$, and a tracking accuracy quantified by $\lambda = 1/\pi$ thereafter; see, e.g., Figure 7.6(d).

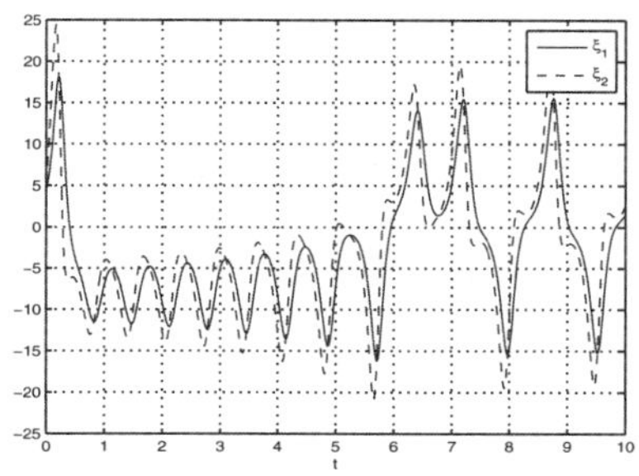

Figure 7.5. Solution of the Lorenz system (7.38).

Spring and damping constants, masses, and their initial positions are chosen, for the simulations, as

$$\begin{array}{l} m_1 = 1, \quad m_2 = 3, \quad c_1 = 2, \quad c_2 = 1, \quad d_1 = 3, \quad d_2 = 5, \\ z_1(0) = 101, \quad z_2(0) = 21, \quad \text{and} \quad \hat{k} = 5. \end{array} \tag{7.40}$$

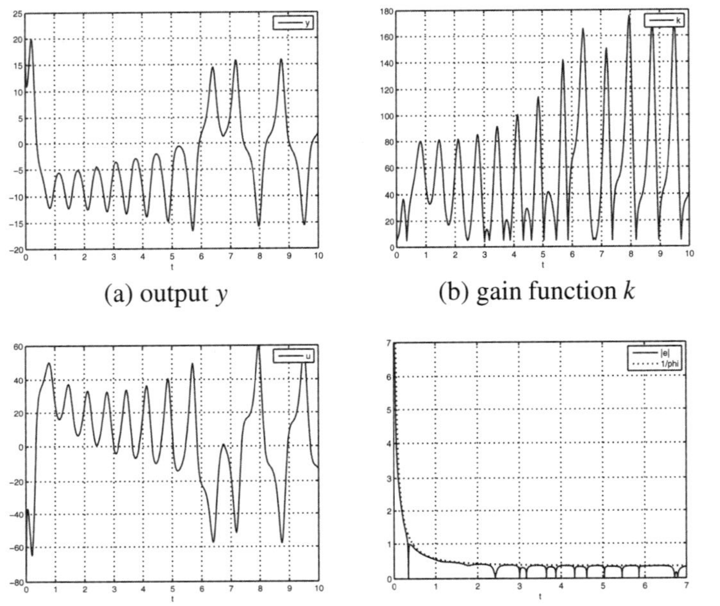

(a) output y (b) gain function k

(c) input u (d) norm of error $|e(t)|$ and funnel

Figure 7.6. Position control: Simulation of the funnel controller (7.16) with funnel boundary specified in (7.39) and reference signal $y_{\text{ref}}(\cdot) = \xi_1(\cdot)$ given in (7.38) applied to the mechanical model (7.33) with data (7.40).

For position control, straightforward calculations show that the closed-loop system (7.16), (7.33) has uniquely determined initial velocities $\dot{z}_1(0)$, $\dot{z}_2(0)$ as well as initial constraint force $\lambda(0)$ and that the initialization is consistent.

In the case of velocity control we further choose the initial velocities

$$\dot{z}_1(0) = -11, \quad \dot{z}_2(0) = -3 \tag{7.41}$$

and clearly there is a unique initial constraint force $\lambda(0)$ and the initialization of (7.15), (7.37) is consistent.

All numerical simulations are performed by MATLAB.

Position Control

Since

$$\hat{k} = 5 > 4 = \lim_{s\to\infty} G^{-1}(s),$$

all assumptions of Theorem 7.17(ii) are satisfied and we may apply the funnel controller (7.16) with funnel boundary specified in (7.39) and reference signal $y_{\text{ref}}(\cdot) = \xi_1(\cdot)$ given in (7.38).

The simulations over the time interval $[0,10]$ are depicted in Figure 7.6: Figure 7.6(a) shows the output $y(\cdot)$ tracking the rather "vivid" reference signal $y_{\text{ref}}(\cdot)$ within the funnel shown in Figure 7.6(d). Note that the input $u(\cdot)$ in Figure 7.6(c) as well as the gain function $k(\cdot)$ in Figure 7.6(b) have spikes at those times t when the norm of the error $\|e(t)\|$ is "close" to the funnel boundary $\varphi(t)^{-1}$; this is due to rapid change of the reference signal. We stress that the gain function $k(\cdot)$ is nonmonotone.

Velocity Control

Since the system has relative degree one with positive high frequency gain $\Gamma = m_1/(m_1 + m_2) = 1/4$, all assumptions of Theorem 7.17(i) are satisfied and we may apply the funnel controller (7.15) with funnel boundary specified in (7.39) and reference signal $y_{\mathrm{ref}}(\cdot) = \xi_1(\cdot)$ given in (7.38).

The simulations over the time interval $[0, 10]$ are depicted in Figure 7.7: Figure 7.7(a) shows the output $y(\cdot)$ tracking the reference signal $y_{\mathrm{ref}}(\cdot)$; the error within the funnel is depicted in Figure 7.7(d). Note that, due to the rather "academic choice" of the example, the input $u(\cdot)$ (in Figure 7.7(c)) and the gain function $k(\cdot)$ (in Figure 7.7(b)) both take considerable larger values than for position control. Another reason for this behavior is that we have kept the funnel as tight as for position control simulated in Figure 7.6, and the velocity exhibits a very "vivid" behavior which causes the error to approach the funnel boundary faster, resulting in the high values of the gain function.

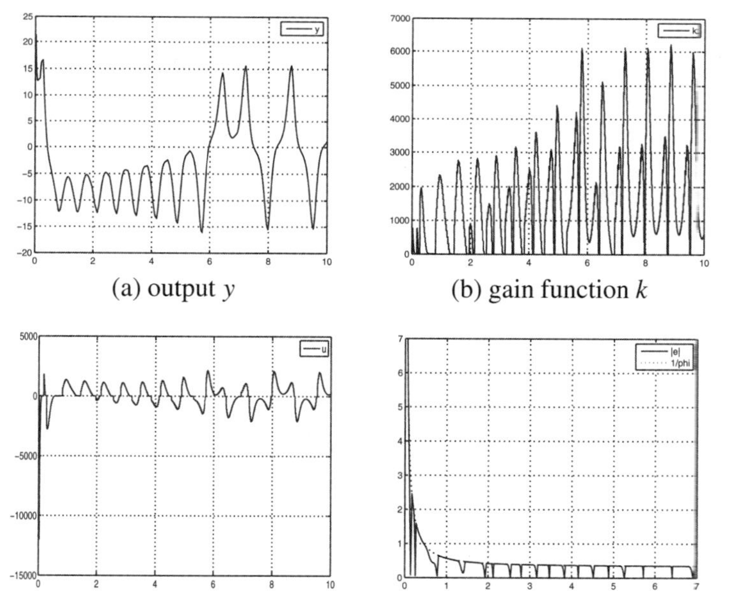

(a) output y (b) gain function k

(c) input u (d) norm of error $|e(t)|$ and funnel

Figure 7.7. Velocity control: Simulation of the funnel controller (7.15) with funnel boundary specified in (7.39) and reference signal $y_{\mathrm{ref}}(\cdot) = \xi_1(\cdot)$ given in (7.38) applied to the mechanical model (7.36) with data (7.40), (7.41).

7.7 Appendix: Proofs

This section contains the proofs of the results presented in Sections 7.1–7.5. First we prove a preliminary lemma.

Lemma 7.19. *Suppose $[E,A,B,C] \in \Sigma_{n,m}$ has strictly proper transfer function $G(s) = C(sE-A)^{-1}B$. Then there exist $W,T \in \mathbf{Gl}_n(\mathbf{R})$ such that*

$$[E,A,B,C] \overset{W,T}{\sim} \left[\begin{bmatrix} I_{n_s} & 0 & 0 \\ 0 & N_c & N_{c\bar{c}} \\ 0 & 0 & N_{\bar{c}} \end{bmatrix}, \begin{bmatrix} A_s & 0 & 0 \\ 0 & I_{n_{fc}} & 0 \\ 0 & 0 & I_{n_{f\bar{c}}} \end{bmatrix}, \begin{bmatrix} B_s \\ B_{fc} \\ 0 \end{bmatrix}, \begin{bmatrix} C_s & 0 & C_{f\bar{c}} \end{bmatrix} \right] \tag{7.42}$$

for some $A_s \in \mathbf{R}^{n_s,n_s}$, $B_s \in \mathbf{R}^{n_s,m}$, $C_s \in \mathbf{R}^{m,n_s}$, $N_c \in \mathbf{R}^{n_{fc},n_{fc}}$, $N_{c\bar{c}} \in \mathbf{R}^{n_{fc},n_{f\bar{c}}}$, $N_{\bar{c}} \in \mathbf{R}^{n_{f\bar{c}},n_{f\bar{c}}}$, $B_{fc} \in \mathbf{R}^{n_{fc},m}$, and $C_{f\bar{c}} \in \mathbf{R}^{m,n_{f\bar{c}}}$, where $N_c, N_{\bar{c}}$ are nilpotent and $\operatorname{rank}[N_c, B_{fc}] = n_{fc}$. *The dimensions $n_s, n_{fc}, n_{f\bar{c}} \in \mathbf{N}_0$ are unique, and the matrices A_s, N_c, and $N_{\bar{c}}$ are unique up to similarity. Furthermore, the system $[E,A,B,C]$ has strict relative degree $\rho > 0$ if and only if*

$$\det(C_s A_s^{\rho-1} B_s) \neq 0 \quad \text{and, if } \rho > 1, \quad \forall\, k = 0,\dots,\rho-2:\; C_s A_s^k B_s = 0. \tag{7.43}$$

Proof. We proceed in several steps.

Step 1: We show that there exist $W,T \in \mathbf{Gl}_n(\mathbf{R})$ such that (7.42) holds true. Corollary 7.2 yields (7.4) for some $W_1, T_1 \in \mathbf{Gl}_n(\mathbf{R})$. It follows from [5, Section 2-1] that the system $[N, I_{n_f}, B_f, C_f]$ may be decomposed into controllability form so that, for some $T_2 \in \mathbf{Gl}_{n_f}(\mathbf{R})$,

$$[N, I_{n_f}, B_f, C_f] \overset{T_2^{-1},T_2}{\sim} \left[\begin{bmatrix} N_c & N_{c\bar{c}} \\ 0 & N_{\bar{c}} \end{bmatrix}, \begin{bmatrix} I_{n_{fc}} & 0 \\ 0 & I_{n_{f\bar{c}}} \end{bmatrix}, \begin{bmatrix} B_{fc} \\ 0 \end{bmatrix}, \begin{bmatrix} C_{fc}, & C_{f\bar{c}} \end{bmatrix} \right],$$

where $N_c \in \mathbf{R}^{n_{fc},n_{fc}}$, $N_{\bar{c}} \in \mathbf{R}^{n_{f\bar{c}},n_{f\bar{c}}}$, $N_{12} \in \mathbf{R}^{n_{fc},n_{f\bar{c}}}$, $B_{fc} \in \mathbf{R}^{n_{fc},m}$, $C_{fc} \in \mathbf{R}^{m,n_{fc}}$, and $C_{f\bar{c}} \in \mathbf{R}^{m,n_{f\bar{c}}}$ such that N_c, $N_{\bar{c}}$ are nilpotent and $\operatorname{rank}[N_c, B_{fc}] = n_{fc}$.

We show that $C_{fc} = 0$: Since the transfer function is invariant under system equivalence we have, using $(sN_c - I_{n_f})^{-1} = -I_{n_f} - sN - s^2N^2 - \dots - s^{\nu-1}N^{\nu-1}$,

$$G(s) = C(sE-A)^{-1}B = C_s(sI_{n_s} - A_s)^{-1}B_s - \sum_{k=0}^{\nu-1} s^k C_{fc} N_c^k B_{fc},$$

and since $G(s)$ is strictly proper, it follows that $C_{fc}N_c^i B_{fc} = 0$ for $i = 1,\dots,\nu-1$. The nilpotency of N_c gives $C_{fc}N_c^{\nu-1}[N_c, B_{fc}] = 0$, whence $C_{fc}N_c^{\nu-1} = 0$. Repeating this argument $\nu - 1$ times, we obtain $C_{fc} = 0$. Setting $W := W_1 \left[\begin{smallmatrix} I & 0 \\ 0 & T_2^{-1} \end{smallmatrix}\right]$ and $T := \left[\begin{smallmatrix} I & 0 \\ 0 & T_2 \end{smallmatrix}\right] T_1$, we obtain (7.42).

Step 2: We show that the dimensions $n_s, n_{fc}, n_{f\bar{c}} \in \mathbf{N}_0$ are unique and that the matrices A_s, N_c, and $N_{\bar{c}}$ are unique up to similarity: Assume that

$$\left[\begin{bmatrix} I_{n_{s1}} & 0 & 0 \\ 0 & N_{c1} & N_{c\bar{c}1} \\ 0 & 0 & N_{\bar{c}1} \end{bmatrix}, \begin{bmatrix} A_{s1} & 0 & 0 \\ 0 & I_{n_{fc1}} & 0 \\ 0 & 0 & I_{n_{f\bar{c}1}} \end{bmatrix}, \begin{bmatrix} B_{s1} \\ B_{fc1} \\ 0 \end{bmatrix}, \begin{bmatrix} C_{s1} & 0 & C_{f\bar{c}1} \end{bmatrix}\right]$$
$$\overset{W,T}{\sim} \left[\begin{bmatrix} I_{n_{s2}} & 0 & 0 \\ 0 & N_{c2} & N_{c\bar{c}2} \\ 0 & 0 & N_{\bar{c}2} \end{bmatrix}, \begin{bmatrix} A_{s2} & 0 & 0 \\ 0 & I_{n_{fc2}} & 0 \\ 0 & 0 & I_{n_{f\bar{c}2}} \end{bmatrix}, \begin{bmatrix} B_{s2} \\ B_{fc2} \\ 0 \end{bmatrix}, \begin{bmatrix} C_{s2} & 0 & C_{f\bar{c}2} \end{bmatrix}\right].$$

Proposition 7.1 implies that $n_{s1} = n_{s2}$ as well as the similarity of A_{s1} and A_{s2}. Proposition 7.1 also yields the existence of $T_{11} \in \mathbf{R}^{n_{fc1}, n_{fc2}}$, $T_{12} \in \mathbf{R}^{n_{fc1}, n_{f\bar{c}2}}$, $T_{21} \in \mathbf{R}^{n_{f\bar{c}1}, n_{fc2}}$, and $T_{22} \in \mathbf{R}^{n_{f\bar{c}1}, n_{f\bar{c}2}}$ such that

$$T = \begin{bmatrix} T_{11} & T_{12} \\ T_{21} & T_{22} \end{bmatrix} \in \mathbf{Gl}_{n_f}(\mathbf{R}),$$
$$\begin{bmatrix} N_{c1} & N_{c\bar{c}1} \\ 0 & N_{\bar{c}1} \end{bmatrix} \begin{bmatrix} T_{11} & T_{12} \\ T_{21} & T_{22} \end{bmatrix} = \begin{bmatrix} T_{11} & T_{12} \\ T_{21} & T_{22} \end{bmatrix} \begin{bmatrix} N_{c2} & N_{c\bar{c}2} \\ 0 & N_{\bar{c}2} \end{bmatrix}, \begin{bmatrix} B_{fc1} \\ 0 \end{bmatrix} = \begin{bmatrix} T_{11} & T_{12} \\ T_{21} & T_{22} \end{bmatrix} \begin{bmatrix} B_{fc2} \\ 0 \end{bmatrix}.$$

Therefore, $0 = T_{21} B_{fc2}$ and $T_{21} N_{c2} = N_{\bar{c}1} T_{21}$.

Hence, for $k = 1, \ldots, \nu - 1$, we have $T_{21} N_{c2}^k B_{fc2} = N_{\bar{c}1}^k T_{21} B_{fc2} = 0$, and so $T_{21} N_{c2}^{\nu-1} [\, N_{c2} \; B_{fc2} \,] = 0$, whence $T_{21} N_{c2}^{\nu-1} = 0$. Repeating this argument $\nu - 1$ times, we obtain $T_{21} = 0$. Then $T \in \mathbf{Gl}_{n_f}(\mathbf{R})$ yields $n_{fc2} \le n_{fc1}$. By reversing the roles of the above matrices, we analogously obtain $n_{fc1} \le n_{fc2}$ and thus $n_{fc1} = n_{fc2}$, $n_{f\bar{c}1} = n_{f\bar{c}2}$. This shows that T_{11} and T_{22} are square. Together with $T \in \mathbf{Gl}_{nf}$, we obtain $T_{11} \in \mathbf{Gl}_{n_{fc}}(\mathbf{R})$ and $T_{22} \in \mathbf{Gl}_{n_{f\bar{c}}}(\mathbf{R})$. Hence N_{c1}, N_{c2} and $N_{\bar{c}1}, N_{\bar{c}2}$ are similar, respectively.

Step 3: We show that $[E, A, B, C]$ has strict relative degree $\rho > 0$ if and only if (7.43) holds.

This is an immediate consequence of the fact that, due to Step 1, the transfer function has the representation $G(s) = C(sE - A)^{-1} B = C_s (s I_{n_s} - A_s)^{-1} B_s$. This completes the proof of the lemma. □

Proof of Theorem 7.6. We proceed in several steps.

Step 1: We show that there exist $W, T \in \mathbf{Gl}_n(\mathbf{R})$ such that $[E, A, B, C] \overset{W,T}{\sim} [\hat{E}, \hat{A}, \hat{B}, \hat{C}]$ for $[\hat{E}, \hat{A}, \hat{B}, \hat{C}]$ as in (7.17).

Since a positive strict relative degree implies that $G(s)$ is strictly proper, we may apply Lemma 7.19 to obtain (7.42) for some $W_1, T_1 \in \mathbf{Gl}_n(\mathbf{R})$. Furthermore, (7.43) holds and hence we may transform $[I, A_s, B_s, C_s]$ into Byrnes–Isidori form (see [12, Lemma

3.5]); i.e., there exists $T_2 \in \mathbf{Gl}_{n_s}(\mathbf{R})$ such that

$$[I, A_s, B_s, C_s] \overset{T_2^{-1},T_2}{\sim} \left[I, \begin{bmatrix} 0 & I_m & 0 & \cdots & 0 & 0 \\ 0 & 0 & I_m & 0 & & 0 \\ \vdots & & \ddots & \ddots & \ddots & \vdots \\ 0 & 0 & \cdots & 0 & I_m & 0 \\ R_1 & R_2 & \cdots & R_{\rho-1} & R_\rho & S \\ P & 0 & \cdots & 0 & 0 & Q \end{bmatrix}, \begin{bmatrix} 0 \\ 0 \\ \vdots \\ 0 \\ \Gamma \\ 0 \end{bmatrix}, \begin{bmatrix} I_m \\ 0 \\ \vdots \\ 0 \end{bmatrix}^\top \right]. \tag{7.44}$$

Set $W := W_1 \begin{bmatrix} I & 0 \\ 0 & T_2^{-1} \end{bmatrix}$ and $T := \begin{bmatrix} I & 0 \\ 0 & T_2 \end{bmatrix} T_1$. Since $N_c, N_{\bar{c}}$ are nilpotent and $\operatorname{rank}[N_c, B_{fc}] = n_{fc}$, the claim follows.

Step 2: For the proof of the uniqueness statements see [1, Theorem 2.5] in combination with Lemma 7.19. In particular, Γ is uniquely determined.

Step 3: It remains to prove (7.19) and that $\Gamma = \lim_{s\to\infty} s^\rho C(sE - A)^{-1}B$.

We prove (7.19): Determine the solution $X(s)$ of the linear equation

$$\begin{bmatrix} sI_m & -I_m & & & \\ & \ddots & \ddots & & \\ & & sI_m & -I_m & \\ -R_1 & \cdots & -R_{\rho-1} & (sI_m - R_\rho) & -S \\ -P & 0 & \cdots & 0 & sI_{n-n_{fc}-n_{f\bar{c}}-\rho m} - Q \end{bmatrix} \begin{bmatrix} X_1(s) \\ X_2(s) \\ \vdots \\ X_\rho(s) \\ X_{\rho+1}(s) \end{bmatrix} = \begin{bmatrix} 0 \\ \vdots \\ 0 \\ \Gamma \\ 0 \end{bmatrix}. \tag{7.45}$$

A simple iterative calculation yields

$$\begin{aligned} sX_i(s) &= X_{i+1}(s) \qquad \text{for } i = 1, \ldots, \rho - 1, \\ -\sum_{i=1}^{\rho-1} R_i X_i(s) + (sI_m - R_\rho) X_\rho(s) - S X_{\rho+1}(s) &= \Gamma, \\ -P X_1(s) + (sI_{n-n_{fc}-n_{f\bar{c}}-\rho m} - Q) X_{\rho+1}(s) &= 0, \end{aligned}$$

and this is equivalent to

$$\begin{aligned} X(s) &= \left(X_1(s)^\top, sX_1(s)^\top, \ldots, s^{\rho-1} X_1(s)^\top, X_{\rho+1}(s)^\top \right)^\top, \\ \Gamma &= -\sum_{i=1}^{\rho-1} R_i s^{i-1} X_1(s) + (sI_m - R_\rho) s^{\rho-1} X_1(s) - S X_{\rho+1}(s), \\ X_{\rho+1}(s) &= (sI_{n-\rho m} - Q)^{-1} P X_1(s). \end{aligned} \tag{7.46}$$

Since the transfer function is invariant under system equivalence, letting $\bar{n} = n - n_{fc} - n_{f\bar{c}}$ we have

$$
\begin{aligned}
C(sE-A)^{-1}B &= \hat{C}(s\hat{E}-\hat{A})^{-1}\hat{B} \\
&= \begin{bmatrix} I_m & 0 & \cdots & 0 \end{bmatrix} \left(sI_{\bar{n}} - \begin{bmatrix} 0 & I_m & 0 & \cdots & 0 & 0 \\ 0 & 0 & I_m & 0 & & 0 \\ \vdots & & \ddots & \ddots & \ddots & \vdots \\ 0 & 0 & \cdots & 0 & I_m & 0 \\ R_1 & R_2 & \cdots & R_{\rho-1} & R_\rho & S \\ P & 0 & \cdots & 0 & 0 & Q \end{bmatrix} \right)^{-1} \begin{bmatrix} 0 \\ \vdots \\ 0 \\ \Gamma \\ 0 \end{bmatrix} \\
&\quad + \begin{bmatrix} 0 & C_{f\bar{c}} \end{bmatrix} \begin{bmatrix} sN_c - I_{n_{fc}} & sN_{c\bar{c}} \\ 0 & sN_{\bar{c}} - I_{n_{f\bar{c}}} \end{bmatrix}^{-1} \begin{bmatrix} B_{fc} \\ 0 \end{bmatrix} \\
&= \begin{bmatrix} I_m & 0 & \cdots & 0 \end{bmatrix} \left(sI_{\bar{n}} - \begin{bmatrix} 0 & I_m & 0 & \cdots & 0 & 0 \\ 0 & 0 & I_m & 0 & & 0 \\ \vdots & & \ddots & \ddots & \ddots & \vdots \\ 0 & 0 & \cdots & 0 & I_m & 0 \\ R_1 & R_2 & \cdots & R_{\rho-1} & R_\rho & S \\ P & 0 & \cdots & 0 & 0 & Q \end{bmatrix} \right)^{-1} \begin{bmatrix} 0 \\ \vdots \\ 0 \\ \Gamma \\ 0 \end{bmatrix} \\
&\overset{(7.45)}{=} X_1(s) \overset{(7.46), R_{\rho+1}=-I}{=} - \left[\sum_{i=1}^{\rho+1} R_i s^{i-1} + S(sI_{n-n_{fc}-n_{f\bar{c}}-\rho m} - Q)^{-1} P \right]^{-1} \Gamma.
\end{aligned}
$$

This proves (7.19). Finally,

$$
\begin{aligned}
\Gamma &= \lim_{s\to\infty} - \left[\sum_{i=1}^{\rho+1} R_i s^{i-1} + S(sI_{n-n_{fc}-n_{f\bar{c}}-\rho m} - Q)^{-1} P \right] G(s) \\
&= - \sum_{i=1}^{\rho} R_i \lim_{s\to\infty} s^{i-1} G(s) + \lim_{s\to\infty} s^\rho G(s) - \lim_{s\to\infty} S(sI_{n-n_{fc}-n_{f\bar{c}}-\rho m} - Q)^{-1} P\, G(s) \\
&= \lim_{s\to\infty} s^\rho G(s),
\end{aligned}
$$

and the proof of the theorem is complete. □

Proof of Theorem 7.12. (i) ⇔ (iv): This is immediate from Remark 7.11.

For the remainder of the proof we consider only systems with positive strict relative degree; the proof for systems which have a proper inverse transfer function is in [2, Theorem 5.4]. In view of invariance of the transfer function under system equivalence we may assume, without loss of generality, that $[E, A, B, C]$ is in the form (7.17).

(ii) $\Leftrightarrow$ (iv): First note that Schur's complement formula (see, for example, [7, Lemma A.1.17]) yields, for all $X \in \mathbf{R}^{k,m}$, $Y \in \mathbf{R}^{k,k}$, $Z \in \mathbf{R}^{m,m}$,

$$\det\begin{bmatrix} X & Y \\ Z & 0 \end{bmatrix} = \det\begin{bmatrix} 0 & I_k \\ I_m & 0 \end{bmatrix} \cdot \det\begin{bmatrix} Z & 0 \\ X & Y \end{bmatrix} = (-1)^{km} \det Z \cdot \det Y. \tag{7.47}$$

This and the fact that $\det(sN - I_k) = (-1)^k$ for any nilpotent $N \in \mathbf{R}^{k,k}$ yield

$$\begin{aligned}
&\det\begin{bmatrix} sE - A & B \\ C & 0 \end{bmatrix} \\
&= \det\left[\begin{array}{ccccccc|c}
sI_m & -I_m & & & & & & 0 \\
& \ddots & \ddots & & & & & \vdots \\
& & sI_m & -I_m & & & & 0 \\
-R_1 & \cdots & -R_{\rho-1} & sI_m - R_\rho & -S & & & \Gamma \\
-P & 0 & \cdots & 0 & sI_\mu - Q & & & 0 \\
0 & 0 & & \cdots & 0 & sN_c - I_{n_c} & sN_{c\bar{c}} & B_c \\
0 & 0 & & \cdots & 0 & 0 & sN_{\bar{c}} - I_{n_{\bar{c}}} & 0 \\
\hline
I_m & 0 & & \cdots & 0 & 0 & C_{\bar{c}} & 0
\end{array}\right] \\
&= (-1)^m \det\begin{bmatrix}
0 & -I_m & & & & & & sI_m \\
\vdots & \ddots & \ddots & & & & & \\
0 & & sI_m & -I_m & & & & \\
\Gamma & \cdots & -R_{\rho-1} & sI_m - R_\rho & -S & & & -R_1 \\
0 & 0 & \cdots & 0 & sI_\mu - Q & & & -P \\
B_c & 0 & 0 & \cdots & 0 & sN_c - I_{n_c} & sN_{c\bar{c}} & 0 \\
0 & 0 & 0 & \cdots & 0 & 0 & sN_{\bar{c}} - I_{n_{\bar{c}}} & 0 \\
0 & 0 & 0 & \cdots & 0 & 0 & C_{\bar{c}} & I_m
\end{bmatrix} \\
&= (-1)^m \det\begin{bmatrix} sN_{\bar{c}} - I_{n_{\bar{c}}} & 0 \\ C_{\bar{c}} & I_m \end{bmatrix} \cdot \det(sN_c - I_{n_c}) \cdot \det(sI_\mu - Q) \\
&\qquad \cdot \det\begin{bmatrix}
0 & -I_m & & & \\
0 & sI_m & -I_m & & \\
\vdots & & \ddots & \ddots & \\
0 & & & sI_m & -I_m \\
\Gamma & -R_2 & \cdots & -R_{\rho-1} & sI_m - R_\rho
\end{bmatrix} \\
&\overset{(7.47)}{=} (-1)^{m+n_c+n_{\bar{c}}+(\rho-1)(m^2+m)} \det(sI_\mu - Q) \det\Gamma.
\end{aligned}$$

This proves the claim.

(iii) ⇔ (iv): Set

$$[A_s, B_s, C_s] = \left[\begin{bmatrix} 0 & I_m & 0 & \cdots & 0 & 0 \\ 0 & 0 & I_m & 0 & & 0 \\ \vdots & & \ddots & \ddots & \ddots & \vdots \\ 0 & 0 & \cdots & 0 & I_m & 0 \\ R_1 & R_2 & \cdots & R_{\rho-1} & R_\rho & S \\ P & 0 & \cdots & 0 & 0 & Q \end{bmatrix}, \begin{bmatrix} 0 \\ 0 \\ \vdots \\ 0 \\ \Gamma \\ 0 \end{bmatrix}, \begin{bmatrix} I_m & 0 & \cdots & 0 \end{bmatrix} \right].$$

We immediately see that

(a) $\forall s \in \mathbf{C}$: $\operatorname{rank}[sE - A, B] = \operatorname{rank}[sI_{n_s} - A_s, B_s] + n_c + n_{\overline{c}}$,

(b) $\forall s \in \mathbf{C}$: $\operatorname{rank}[sE^\top - A^\top, C^\top] = \operatorname{rank}[sI_{n_s} - A_s^\top, C_s^\top] + n_c + n_{\overline{c}}$,

(c) $G(s) = C(sE - A)^{-1}B = C_s(sI_{n_s} - A_s)^{-1}B_s$,

where (c) is proved similar to Step 3 in the proof of Theorem 7.6. Therefore, the equivalence of (iii) and (iv) has been reduced to the ODE system $[I_{n_s}, A_s, B_s, C_s]$. Now the claim follows from [8, Proposition 2.1.2]. This completes the proof of the theorem. □

For the proof of Theorem 7.15 a lemma is required.

Lemma 7.20 (High-gain lemma). *Consider, for $\Gamma \in \mathbf{Gl}_m(\mathbf{C})$, $\tilde{R} \in \mathbf{C}^{m,m}$, $\tilde{S}^\top, \tilde{P} \in \mathbf{C}^{n-m,m}$, $\tilde{Q} \in \mathbf{C}^{n-m,n-m}$, the parameterized matrix*

$$A_\kappa := \begin{bmatrix} \tilde{R} - \kappa\Gamma & \tilde{S} \\ \tilde{P} & \tilde{Q} \end{bmatrix}, \qquad \kappa \geq 0.$$

Denote the spectra of Γ and $\tilde{Q}$ by

$$\sigma(\Gamma) = \{\gamma_1, \dots, \gamma_m\} \subseteq \mathbf{C} \setminus \{0\} \quad \textit{and} \quad \sigma(\tilde{Q}) = \{q_{m+1}, \dots, q_n\} \subseteq \mathbf{C}.$$

Then there exist $z_1, \dots, z_m \in \mathbf{C}$ and $\hat{\theta} > 0$ with the following property: For all $\varepsilon > 0$ and all $\theta \in (0, \hat{\theta})$ there exist $r \geq 0$ and $\kappa^* \geq 1$ such that, with a suitable enumeration of the eigenvalues $\lambda_1(A_\kappa), \dots, \lambda_n(A_\kappa)$ of A_κ, we have, for all $\kappa \geq \kappa^*$,

(i) $B(z_i - \kappa\gamma_i, r + \kappa\theta) \cap B(0, 1/\varepsilon) = \emptyset$ *for* $i = 1, \dots, m$,

(ii) $\lambda_i(A_\kappa) \in \bigcup_{j=1}^m B\big((z_j - \kappa\gamma_j, r + \kappa\theta)\big)$ *for* $i = 1, \dots, m$,

(iii) $\lambda_i(A_\kappa) \in \bigcup_{j=m+1}^n B\big((q_j, \varepsilon)\big)$ *for* $i = m+1, \dots, n$,

where $B(z, \varepsilon) = \{\, w \in \mathbf{C} \mid |z - w| < \varepsilon \,\}$ denotes the ball of radius ε around z in $\mathbf{C}$.

Proof. Let

$$\hat{\theta} := \tfrac{1}{4} \min\big\{|\gamma_1|, \dots, |\gamma_m|\big\} > 0 \tag{7.48}$$

and choose arbitrary $\theta \in (0,\hat{\theta})$. Choose $U_1 \in \mathbf{Gl}_m(\mathbf{C})$, $U_2 \in \mathbf{Gl}_{n-m}(\mathbf{C})$ such that, for appropriately chosen $\delta_1,\dots,\delta_{n-1} \in \{0,1\}$, we have Jordan forms

$$U_1 \Gamma U_1^{-1} = \begin{bmatrix} \gamma_1 & \delta_1 & & \\ & \ddots & \ddots & \\ & & \gamma_{m-1} & \delta_{m-1} \\ & & & \gamma_m \end{bmatrix}, \qquad U_2 \tilde{Q} U_2^{-1} = \begin{bmatrix} q_{m+1} & \delta_{m+1} & & \\ & \ddots & \ddots & \\ & & q_{n-1} & \delta_{n-1} \\ & & & q_n \end{bmatrix}.$$

Set

$$T_\theta := \operatorname{diag}(\theta,\theta^2,\dots,\theta^m) \quad \text{and} \quad T_\alpha := \operatorname{diag}(\alpha,\alpha^2,\dots,\alpha^{n-m}) \qquad \text{for } \alpha > 0$$

and transform A_κ to the similar matrix

$$M(\kappa,\theta,\alpha) := \begin{bmatrix} T_\theta^{-1}U_1^{-1} & 0 \\ 0 & T_\alpha U_2^{-1} \end{bmatrix} \begin{bmatrix} \tilde{R} - \kappa\Gamma & \tilde{S} \\ \tilde{P} & \tilde{Q} \end{bmatrix} \begin{bmatrix} U_1 T_\theta & 0 \\ 0 & U_2 T_\alpha^{-1} \end{bmatrix}$$

$$= \begin{bmatrix} T_\theta^{-1}U_1^{-1}\tilde{R}U_1 T_\theta - \kappa \begin{bmatrix} \gamma_1 & \theta\delta_1 & & \\ & \ddots & \ddots & \\ & & \gamma_{m-1} & \theta\delta_{m-1} \\ & & & \gamma_m \end{bmatrix} & T_\theta^{-1}U_1^{-1}\tilde{S}U_2 T_\alpha^{-1} \\ T_\alpha U_2^{-1}\tilde{P}U_1 T_\theta & \begin{bmatrix} q_{m+1} & \delta_{m+1}/\alpha & & \\ & \ddots & \ddots & \\ & & q_{n-1} & \delta_{n-1}/\alpha \\ & & & q_n \end{bmatrix} \end{bmatrix}$$

and the off-diagonal column sums

$$\rho_j(\kappa,\theta,\alpha) := \sum_{\substack{i=1 \\ i\neq j}}^{n} \left| M(\kappa,\theta,\alpha)_{ij} \right|, \qquad j = 1,\dots,n.$$

Fix $\varepsilon > 0$. We may now choose $\alpha > 0$ sufficiently large so that the effect of the scaling matrix T_α^{-1} in the last $n-m$ columns of $M(\kappa,\theta,\alpha)$ is

$$\forall i = m+1,\dots,n \; : \; \rho_j(\kappa,\theta,\alpha) = \rho_j(\alpha) \in [0,\varepsilon).$$

Consider next the first m columns of $M(\kappa,\theta,\alpha)$. Noting that every summand in $\rho_i(\kappa,\theta,\alpha)$ which involves κ must be a product of κ and θ, we find that there exists $r = r(\alpha,\theta) \geq 0$ such that

$$\forall i = 1,\dots,m \quad \forall \kappa \geq 0 \; : \; \rho_i(\kappa,\theta,\alpha) \leq r + \kappa\theta. \tag{7.49}$$

Define the diagonal entries

$$z_i := (U_1^{-1}\tilde{R}U_1)_{ii} = (T_\theta^{-1}U_1^{-1}\tilde{R}U_1 T_\theta)_{ii}, \qquad i = 1,\dots,m.$$

We now show that the center of the balls $B(z_i - \kappa\gamma_i, r+\kappa\theta)$, $i=1,\dots,m$, tends, as $\kappa\to\infty$, to infinity at a faster pace than its radius. To this end note that using (7.48) gives

$$|z_i - \kappa\gamma_i| \ge \big||z_i| - 4\kappa\hat{\theta}\big|$$

and for $\kappa > (r+|z_i|)/\hat{\theta}$ we have $|z_i| < \kappa\hat{\theta}$ and hence

$$|z_i - \kappa\gamma_i| > 3\kappa\hat{\theta} - |z_i| > 2\kappa\hat{\theta} + r > \kappa\hat{\theta} + (r+\kappa\theta).$$

Therefore,

$$|z_i - \kappa\gamma_i| - (\kappa\theta + r) > \kappa\hat{\theta},$$

which implies that $B(z_i - \kappa\gamma_i, r+\kappa\theta) \cap B(0,\kappa\hat{\theta}) = \emptyset$. Choosing

$$\kappa^* > \max\left\{1/(\varepsilon\hat{\theta}), (r+|z_1|)/\hat{\theta}, \dots, (r+|z_m|)/\hat{\theta}\right\}$$

we obtain assertion (i). Since $\gamma_i \ne 0$ for all $i=1,\dots,m$, we may now choose $\kappa^* \ge 1$ sufficiently large so that

$$\forall \kappa \ge \kappa^* : \bigcup_{i=1}^{m} B(z_i - \kappa\gamma_i, r+\kappa\theta) \cap \bigcup_{j=m+1}^{n} B(q_j, \varepsilon) = \emptyset.$$

We are now in a position to apply Gershgorin's disks, see [7, Theorem 4.2.19], to deduce (ii) and (iii). This completes the proof of the lemma. □

Proof of Theorem 7.15.
(i) We prove "⇒". By Theorem 7.6, $[E,A,B,C]$ is equivalent to a system in the form (7.20). We introduce the "new states"

$$\xi := \frac{1}{p_{\rho-1}} \cdot p(\tfrac{\mathrm{d}}{\mathrm{d}t})y, \quad \chi := \left(y^\top, \dot{y}^\top, \dots, \left(y^{(\rho-2)}\right)^\top, \eta^\top\right)^\top$$

and observe that

$$\begin{aligned} \dot{\xi}(t) &= \tilde{R}\xi(t) + \tilde{S}\chi(t) + \Gamma u(t), \\ \dot{\chi}(t) &= \tilde{P}\xi(t) + \tilde{Q}\chi(t), \end{aligned} \tag{7.50}$$

where $\tilde{R}, \tilde{S}$ are matrices of appropriate size and

$$\tilde{P} = [0,\dots,0,I_m,0]^\top, \quad \tilde{Q} = \begin{bmatrix} \hat{A} & 0 \\ \hat{P} & Q \end{bmatrix},$$

$$\text{with} \quad \hat{A} = \begin{bmatrix} 0 & I & & \\ & \ddots & \ddots & \\ & & 0 & I \\ -\frac{p_0}{p_{\rho-1}}I & \cdots & -\frac{p_{\rho-3}}{p_{\rho-1}}I & -\frac{p_{\rho-2}}{p_{\rho-1}}I \end{bmatrix}, \quad \hat{P} = [P,0,\dots,0].$$

Note that $\hat{A}$ is Hurwitz since $p(s)$ is Hurwitz. The feedback (7.15) reads in the new coordinates as

$$u(t) = -k\, p(\tfrac{\mathrm{d}}{\mathrm{d}t})\, y(t) = -k\, p_{\rho-1}\xi(t), \tag{7.51}$$

and therefore the application of (7.15) to $[E,A,B,C]$ results, in terms of the new system (7.50), in the closed-loop system

$$\frac{\mathrm{d}}{\mathrm{d}t}\begin{pmatrix}\xi(t)\\ \chi(t)\end{pmatrix} = \underbrace{\begin{bmatrix}\tilde{R}-kp_{\rho-1}\Gamma & \tilde{S}\\ \tilde{P} & \tilde{Q}\end{bmatrix}}_{=:A_k}\begin{pmatrix}\xi(t)\\ \chi(t)\end{pmatrix}. \tag{7.52}$$

Note that the closed-loop system (7.1), (7.11) is asymptotically stable if (7.52) is asymptotically stable. This is due to the fact that $p(s)$ is Hurwitz and if y decays exponentially so do all derivatives of y and, by (7.20), also u and all derivatives of u, which finally gives that x_c decays exponentially.

We show that (7.52) is asymptotically stable. Note that $\tilde{Q}$ is Hurwitz since Q is Hurwitz. Therefore, by $p_{\rho-1} > 0$, Γ positive definite and $\sigma(\tilde{Q}) \subseteq \mathbb{C}_-$, we may apply Lemma 7.20 to conclude that

$$\exists k^* \geq 0\, \forall k \geq k^* \; : \; \sigma(A_k) \subseteq \mathbb{C}_-.$$

This proves the claim.

To see that "$\Leftarrow$" does not hold in general, consider system (7.1) for

$$E = I, \quad A = \begin{bmatrix}0 & 1\\ -1 & 0\end{bmatrix}, \quad B = \begin{bmatrix}1\\ 0\end{bmatrix}, \quad C = [1,0], \tag{7.53}$$

which is in zero dynamics form (7.17) and has strict relative degree one. We may observe that $Q = 0$ and therefore the zero dynamics of (7.53) are not asymptotically stable. However, the closed-loop system (7.53), (7.11) where $p(s) \equiv p_0 > 0$ takes the form

$$\dot{x}(t) = A_k x(t) = \begin{bmatrix}-kp_0 & 1\\ -1 & 0\end{bmatrix} x(t),$$

which is, since $\sigma(A_k) = \{-p_0k/2 \pm \sqrt{(p_0k)^2/4 - 1}\}$, asymptotically stable for all $k > 0$. This shows that $[E,A,B,C]$ is high-gain stabilizable.

(ii) The proof of "$\Rightarrow$" has been carried out in Remark 7.16(ii). It remains to show that "$\Leftarrow$" does not hold in general. Consider system (7.1) for

$$E = \begin{bmatrix}0&0&0&0\\0&0&0&0\\0&0&1&0\\0&0&0&1\end{bmatrix}, \; A = \begin{bmatrix}0&0&1&0\\0&0&0&1\\0&-1&0&0\\1&0&0&-1\end{bmatrix}, \; B = \begin{bmatrix}1&0\\0&1\\0&0\\0&0\end{bmatrix}, \; C = \begin{bmatrix}1&0&0&0\\0&1&0&0\end{bmatrix}, \tag{7.54}$$

which is in zero dynamics form (7.21) with $n_1 = n_2 = 2$, $n_c = n_{\overline{c}} = 0$ and

$$A_{11} = \begin{bmatrix}0&0\\0&0\end{bmatrix}, \quad A_{12} = \begin{bmatrix}1&0\\0&1\end{bmatrix}, \quad A_{21} = \begin{bmatrix}0&1\\-1&0\end{bmatrix}, \quad Q = \begin{bmatrix}0&0\\0&-1\end{bmatrix}.$$

Since Q has an eigenvalue at the origin, Theorem 7.12 yields that the zero dynamics are not asymptotically stable. The asymptotic stability of the closed-loop system "(7.54) & (7.12)" is determined by the spectrum of the matrix

$$Q - A_{21}(kI_2 + A_{11})^{-1}A_{12} = \begin{bmatrix} 0 & 1/k \\ -1/k & -1 \end{bmatrix},$$

and this satisfies, for all $k \in \mathbf{R}\setminus\{0\}$,

$$\sigma(Q - A_{21}(kI_2 - A_{11})^{-1}A_{12}) = \left\{-1/2 \pm \sqrt{1/4 - 1/k^2},\right\} \subseteq \mathbf{C}_-.$$

This shows that $[E, A, B, C]$ is high-gain stabilizable and completes the proof of the theorem. □

Proof of Theorem 7.17.
(i) Without restriction of generality, one may consider $[E, A, B, C]$ in the form (7.20). Ignoring the bottom two algebraic equations in (7.20), the claim in (i) and the bound on e follow from [11, Theorem 2]. Since γ_ρ is a $\mathcal{C}^\infty$-function, it is easy to see that u is $(\nu - 1)$-times continuously differentiable and all of these derivatives are bounded functions. Therefore, x_c and $\bar{x}_c$ in (7.20) are bounded functions. It remains to show the bound on k in (7.32): This follows from the following inequalities, which hold for all $t > 0$:

$$k(t) = \hat{k} + k(t)\varphi(t)^2\|e(t)\|^2 \overset{(7.31)}{\leq} \hat{k} + k(t)\varphi(t)^2(\varphi(t)^{-1} - \varepsilon)^2 = \hat{k} + k(t)(1 - \varphi(t)\varepsilon)^2.$$

(ii) This result is shown in [2, Theorem 6.2]. □

Acknowledgments

We are indebted to our colleague Fabian Wirth (University of Würzburg) for constructive discussions.

Bibliography

[1] T. BERGER AND A. ILCHMANN, *Zero Dynamics of Time-Varying Linear Systems*, Preprint available online, Institute for Mathematics, Ilmenau University of Technology, Preprint 10-05, 2010.

[2] T. BERGER, A. ILCHMANN, AND T. REIS, *Zero Dynamics and Funnel Control of Linear Differential-Algebraic Systems*, Preprint available online, Institute for Mathematics, Ilmenau University of Technology, Preprint 10-15, submitted to Mathematics of Control, Signals, and Systems, 2010.

[3] T. BERGER, A. ILCHMANN, AND S. TRENN, *The quasi-Weierstraß form for regular matrix pencils*, Linear Algebra Appl., to appear; preprint available online, Institute for Mathematics, Ilmenau University of Technology, Preprint 09-21.

[4] C. I. BYRNES AND J. C. WILLEMS, *Adaptive stabilization of multivariable linear systems*, in Proceedings of the 23rd IEEE Conference on Decision and Control, 1984, pp. 1574–1577.

[5] L. DAI, *Singular Control Systems*, Lecture Notes in Control and Inform. Sci. 118, Springer-Verlag, Berlin, 1989.

[6] F. R. GANTMACHER, *The Theory of Matrices*, Vol. II, Chelsea, New York, 1959.

[7] D. HINRICHSEN AND A. J. PRITCHARD, *Mathematical Systems Theory* I. *Modelling, State Space Analysis, Stability and Robustness*, Texts Appl. Math. 48, Springer-Verlag, Berlin, 2005.

[8] A. ILCHMANN, *Non-identifier-Based High-Gain Adaptive Control*, Lecture Notes in Control and Inform. Sci. 189, Springer-Verlag, London, 1993.

[9] A. ILCHMANN AND E. P. RYAN, *High-gain control without identification: A survey*, GAMM Mitt., 31 (2008), pp. 115–125.

[10] A. ILCHMANN, E. P. RYAN, AND C. J. SANGWIN, *Tracking with prescribed transient behaviour*, ESAIM Control Optim. Cal. Var., 7 (2002), pp. 471–493.

[11] A. ILCHMANN, E. P. RYAN, AND P. TOWNSEND, *Tracking control with prescribed transient behaviour for systems of known relative degree*, Systems Control Lett., 55 (2006), pp. 396–406.

[12] A. ILCHMANN, E. P. RYAN, AND P. TOWNSEND, *Tracking with prescribed transient behavior for nonlinear systems of known relative degree*, SIAM J. Control Optim., 46 (2007), pp. 210–230.

[13] A. ISIDORI, *Nonlinear Control Systems*, 3rd ed., Comm. Control Engrg. Ser., Springer-Verlag, Berlin, 1995.

[14] T. KAILATH, *Linear Systems*, Prentice–Hall, Englewood Cliffs, NJ, 1980.

[15] P. KUNKEL AND V. MEHRMANN, *Differential-Algebraic Equations. Analysis and Numerical Solution*, EMS Publishing House, Zürich, Switzerland, 2006.

[16] C. SPARROW, *The Lorenz Equations: Bifurcations, Chaos, and Strange Attractors*, Appl. Math. Sci. 41, Springer-Verlag, New York, 1982.

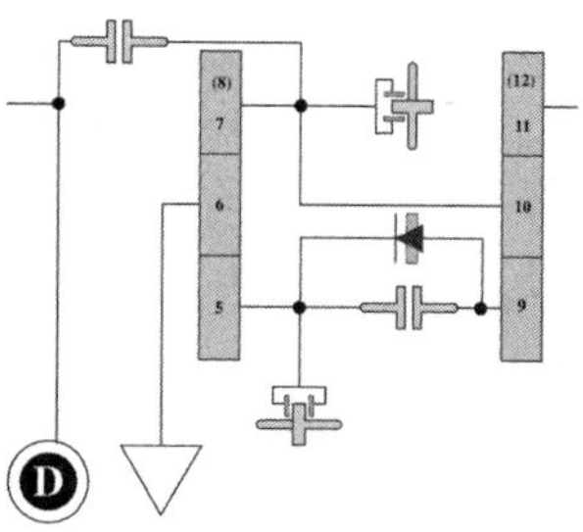

Chapter 8

Linear-Quadratic Optimal Control Problems with Switch Points and a Small Parameter

Galina Kurina

Results are presented on two types of linear-quadratic optimal control problems for DAEs. The first type of problems has discontinuous coefficients and the state equations are switched systems. In the second type of problems, the cost functional depends on values of the state variable at intermediate points. We also present a procedure for obtaining asymptotic solutions of problems with singularly perturbed switched state equations and problems with intermediate points and a small parameter in the cost functional. The procedure is called the direct scheme. *This approach enables us to prove that the minimized functional is nonincreasing if higher order approximations to the optimal control are used. Also, we present a procedure for obtaining asymptotic solutions of singularly perturbed problems with discontinuous coefficients, which exploits a feedback control.*

8.1 Introduction

This chapter deals with linear-quadratic optimal control problems for systems with differential-algebraic constraints, so-called *descriptor systems*, having intermediate points and a small parameter in the problem statement.

DAEs as well as discontinuous (switched) and singularly perturbed systems have numerous applications in physics, chemical engineering, circuit simulation, mathematical biology, technology, and engineering (see the works cited below and also the references therein). For example, if a DAE in a switched system describes a robot motion with a small mass [44] or an electric circuit with a small capacity [41], then we arrive at a singularly perturbed problem.

The literature on optimization problems for descriptor systems without intermediate points is very extensive (see, for instance, [5, 7, 14, 20, 21, 24, 26, 27, 29, 32, 33, 34,

43, 44]). There are quite a lot of studies on control problems with intermediate points, in particular, on problems with switched systems (see, e.g.. [4, 9, 15, 40, 45, 46]). The above list of publications is by no means complete, and we refer the reader to the references therein for further sources.

Feedback solutions via matrix Riccati differential equations are a known and proven tool for solving linear-quadratic optimal control problems (see, e.g., [39]). In contrast to the control optimality conditions obtained in e.g., [4, 18, 19] for various control problems for discontinuous systems, this approach allows for linear-quadratic problems to avoid solving auxiliary boundary value problems (BVPs) for obtaining optimal controls. The matrix Riccati differential equation has been considered in many publications (see, for instance, the monographs [1, 48] and the references therein). In the case of problems with DAE constraints, nonstandard operator Riccati equations arise. Several classes of such equations have been obtained in [11, 23, 25, 30, 34]. A survey of Riccati DAEs is given in [34].

The mentioned works motivate us to search for similar results for problems with intermediate points. In this chapter, using discontinuous solutions of a special implicit operator Riccati equation, we obtain a feedback optimal control for two types of problems. The first one has discontinuous coefficients, and the state equations are switched systems. The second type of problems are those with the cost functional depending on values of a state variable at intermediate points.

It should be noted that the possibility for finding optimal feedback controls for linear-quadratic problems with DAE constraints substantially depends on a choice of a suitable Riccati equation. For example, it was shown in [25] that the Riccati DAE of the form [25, equation (44)] can be used only in very special cases. As was demonstrated in [34], the Riccati DAE system from [34] (see [34, equation (1.15)]) is more appropriate. However, there are linear-quadratic problems which cannot be solved using the Riccati DAE system from [34] (see, e.g., Example 3.16 in [5]).

Sometimes control problems with a small parameter become simpler if the small parameter is set to 0. It is interesting to estimate the proximity between the solutions of the perturbed problem and the degenerate case, when the small parameter is set to zero, and to construct an asymptotic expansion for the solution of the perturbed problem solving simpler problems than the original one.

In most publications devoted to optimal control problems with a small parameter, the asymptotic analysis of solutions is based on the asymptotic solution of BVPs arising from control optimality conditions (see, e.g., [42, p. 139]). Another approach to constructing asymptotic solutions consists of an immediate substitution of a postulated asymptotic expansion of a solution into the problem condition. Then a family of control problems to find asymptotic expansion terms is defined. This approach is called the *direct scheme* [10]. It enables us to prove that the value of the minimized functional is nonincreasing if higher order approximations to the optimal control are used. Due to the direct scheme we can use program packages for solving optimal control problems in order to find asymptotic expansion terms. Several applications of the direct scheme can be found in [17].

There is one more approach to construct asymptotic solutions for linear-quadratic problems. It is based on the feedback form for an optimal control and on an asymptotic solution of the corresponding matrix differential Riccati equation. We refer the reader to [17] for a survey of various methods of studying optimal control problems with a small parameter.

Application of control optimality conditions to singularly perturbed problems may lead to systems having solutions, which contain fast and slow components. If fast time scales are present, then numerical methods may be restricted to using very small steps. These are the so-called stiff problems (see, e.g., [3]). For them, asymptotic methods, which allow one to reduce the solution of the perturbed problems to the solution of simpler ones, are of great importance. Asymptotic solutions can be used as initial approximations in numerical methods. The latter motivates the interest in procedures for obtaining asymptotic solutions of the problems considered in this chapter.

The chapter is organized as follows. In Section 8.2 we present control optimality conditions for problems with differential-algebraic constraints and intermediate points. In Section 8.3 we consider problems with intermediate points and a small parameter, for which some qualitative properties of the problem are changed if the small parameter is set to 0. Namely, the state equation in subsections 8.3.1 and 8.3.2 is a singularly perturbed switched system. In subsection 8.3.3 a small parameter is placed in the cost functional, which depends on values of the state variable at intermediate points. Using the direct scheme in subsections 8.3.1 and 8.3.3, we construct asymptotic solutions of the problems under consideration. Subsection 8.3.2 deals with feedback control. The correspondence between the asymptotic solution constructed by means of the direct scheme and the exact solution is illustrated by Example 8.7.

8.2 Control Optimality Conditions for Problems with Intermediate Points

In this section we consider two classes of linear-quadratic problems with DAE constraints. First, we present control optimality conditions for problems with discontinuous coefficients; the state equations in these problems are switched systems. Then we obtain control optimality conditions for problems with intermediate points in the cost functional.

We use the following notation. X, Y, Z, U stand for real finite-dimensional Euclidean spaces; $\langle \cdot, \cdot \rangle$ is the standard inner product in appropriate finite-dimensional spaces; the superscript ' denotes the transposition; I is the identity operator.

Let E be a linear bounded operator acting in X, $E \in L(X)$. Assuming that $\mathrm{Ker}E \neq X$, we denote by P the orthogonal projection onto $\mathrm{Ker}E$ with respect to the decomposition $X = \mathrm{Ker}E \oplus \mathrm{Im}E'$ and by Q the orthogonal projection onto $\mathrm{Ker}E'$ with respect to the decomposition $X = \mathrm{Ker}E' \oplus \mathrm{Im}E$. By E_+ we denote the inverse to the operator $(I - Q)E(I-P) : \mathrm{Im}E' \to \mathrm{Im}E$. We say that an operator E is positive definite (semidefinite) if $\langle Ex, x \rangle > 0$ (≥ 0) for all $x \in X \setminus \{0\}$.

In subsections 8.2.1 and 8.2.2, admissible controls are assumed to be piecewise continuous functions guaranteeing the existence of a solution of the state equation, which is a DAE, with given additional conditions. We consider the left-hand side of the state equation in the form $\frac{d}{dt}(Ex(t))$, i.e., the differentiability for $Px(\cdot)$ is not required. A solution of a DAE is a continuous function $x(\cdot)$ satisfying this equation almost everywhere. Note that IVPs for linear DAEs with various conditions have been studied in many publications (see, e.g., [8, 12, 13, 26]).

Let us emphasize here that we are concerned with sufficient control optimality conditions but not with necessary conditions. In this subsection, we do not suppose that an index is defined for differential-algebraic constraints. Index criteria for DAEs arising from linear-quadratic optimal control problems were studied in [7]. Note also the papers [6, 28] devoted to such DAEs.

8.2.1 Problems with Discontinuous Coefficients

Consider the optimal control problem of minimizing the functional

$$J(u,x) = \frac{1}{2}\langle x_2(T), Fx_2(T)\rangle + \frac{1}{2}\sum_{j=1}^{2}\int_{t_{j-1}}^{t_j} \{\langle x_j(t), W_j(t)x_j(t)\rangle + \langle u_j(t), R_j(t)u_j(t)\rangle\}dt \tag{8.1}$$

on trajectories of the system

$$\frac{d}{dt}(Ex_j(t)) = A_j(t)x_j(t) + B_j(t)u_j(t), \quad t \in [t_{j-1}, t_j], \quad j = 1,2, \tag{8.2}$$

$$Ex_1(0) = x^0, \quad Ex_2(t_1) = Ex_1(t_1). \tag{8.3}$$

Here $0 = t_0 < t_1 < t_2 = T$; t_1, T and $x^0 \in \operatorname{Im}E$ are fixed; $x_j = x_j(t) \in X$, $u_j = u_j(t) \in U$; the operators $W_j(t) \in L(X)$, $R_j(t) \in L(U)$, $A_j(t) \in L(X)$, $B_j(t) \in L(U,X)$ and admissible controls are continuous with respect to t for all $t \in [t_{j-1}, t_j]$. The operators $F \in L(X)$, $W_j(t)$, and $R_j(t)$ are symmetric, F and $W_j(t)$ are positive semidefinite, and $R_j(t)$ is positive definite for all $t \in [t_{j-1}, t_j]$, $j = 1,2$.

Problems of the form (8.1)–(8.3) arise in the description of the motion of a multibody system by DAEs [44] when this system is changed at time t_1. For instance, industrial robots are typical examples of multibody systems. The object structure is very often subjected to change in the situation of multistage space vehicle flights [22].

We present the theorem on the feedback solution of the optimal control problem (8.1)–(8.3).

Theorem 8.1. *Let the operator-function $K : [0,T] \to L(X)$, consisting of $K_j : [t_{j-1}, t_j] \to L(X)$, $j = 1,2$, be a solution of the problem*

$$\begin{aligned} &\frac{d}{dt}(E'K_j) = -K'_jA_j - A'_jK_j + K'_jS_jK_j - W_j, \\ &S_j = B_jR_j^{-1}B_j', \quad t \in [t_{j-1}, t_j], \quad j = 1,2, \end{aligned} \tag{8.4}$$

$$E'K_2(T) = F, \quad E'K_1(t_1) = E'K_2(t_1), \tag{8.5}$$

and let $x_* : [0,T] \to X$, consisting of $x_{j*} : [t_{j-1}, t_j] \to X$, $j = 1,2$, be a solution of the problem

$$\frac{d}{dt}(Ex_{j*}) = \left(A_j - S_jK_j\right)x_{j*}, \quad t \in [t_{j-1}, t_j], \quad j = 1,2, \tag{8.6}$$

$$Ex_{1*}(0) = x^0, \quad Ex_{2*}(t_1) = Ex_{1*}(t_1). \tag{8.7}$$

Then the function u_*, consisting of

$$u_{j*} = -R_j^{-1}B_j'K_jx_{j*}, \quad t \in [t_{j-1}, t_j], \quad j = 1,2, \tag{8.8}$$

is an optimal control for the problem (8.1)–(8.3) *and the minimal value of the functional* (8.1) *is*

$$J(u_*, x_*) = \frac{1}{2}\left\langle x^0, K_1(0)E_+x^0 \right\rangle.$$

Theorem 8.1 can be easily proved using the scheme of the proof of Theorem 8.2 (see below).

Setting in (8.8)

$$\psi_j = -K_j x_{j*}, \quad j = 1,2,$$

we get the functions defining the optimal control in the following form:

$$u_{j*}(t) = R_j(t)^{-1} B_j(t)' \psi_j(t), \quad t \in [t_{j-1}, t_j].$$

Taking into account (8.4)–(8.7) and noting that $E'K_j(t)$ is symmetric, we obtain the following BVP for solving problem (8.1)–(8.3):

$$\frac{d}{dt}(Ex_{j*}(t)) = A_j(t)x_{j*}(t) + S_j(t)\psi_j(t),$$

$$\frac{d}{dt}(E'\psi_j(t)) = W_j(t)x_{j*}(t) - A_j(t)'\psi_j(t), \quad t \in [t_{j-1}, t_j], \quad j = 1,2,$$

$$Ex_{1*}(0) = x^0, \quad Ex_{2*}(t_1) = Ex_{1*}(t_1),$$

$$E'\psi_2(T) = -Fx_{2*}(T), \quad E'\psi_1(t_1) = E'\psi_2(t_1).$$

8.2.2 Intermediate Points in Cost Functional

In this subsection, we consider the nonstandard quadratic cost functional

$$J(u,x) = \frac{1}{2}\sum_{j=1}^{N+1} \left\langle x(t_j) - \xi_j, F_j(x(t_j) - \xi_j) \right\rangle$$
$$+ \frac{1}{2}\int_0^T \left\langle \begin{pmatrix} x(t) \\ u(t) \end{pmatrix}, \begin{pmatrix} W(t) & V(t) \\ V(t)' & R(t) \end{pmatrix} \begin{pmatrix} x(t) \\ u(t) \end{pmatrix} \right\rangle dt \tag{8.9}$$

to be minimized with respect to trajectories of the system

$$\frac{d}{dt}(Ex(t)) = A(t)x(t) + B(t)u(t) + f(t), \tag{8.10}$$

$$Ex(0) = x^0. \tag{8.11}$$

We assume that $t \in [0,T]$, $0 = t_0 < t_1 < \cdots < t_N < t_{N+1} = T$; $t_j, \xi_j \in X$, $j = 1, \ldots, N+1$, and $x_0 \in \operatorname{Im} E$ are fixed; $W(t)$, $R(t)$ are symmetric; $\begin{pmatrix} W(t) & V(t) \\ V(t)' & R(t) \end{pmatrix}$ is positive semidefinite and $R(t)$ is positive definite for all $t \in [0,T]$; the operator-functions $W(\cdot)$, $V(\cdot)$, $R(\cdot)$, $A(\cdot)$, $B(\cdot)$ and the function $f(\cdot)$ are continuous on $[0,T]$.

In the following, we use the notation $\varphi(t_j \pm 0) = \lim_{t \to \pm 0} \varphi(t)$.

Theorem 8.2. *Let the operator-function* $K:[0,T]\to L(X)$ *be a solution of the problem*

$$\frac{d}{dt}(E'K(t)) = -K(t)'A(t) - A(t)'K(t) \\ + (V(t)+K(t)'B(t))R(t)^{-1}(V(t)+K(t)'B(t))' - W(t), \quad t\neq t_j, \tag{8.12}$$

$$E'(K(t_j-0)-K(t_j+0)) = F_j, \quad j=1,\dots,N, \quad E'K(T)=F_{N+1}, \tag{8.13}$$

and let the function $\varphi:[0,T]\to X$ *be a solution of the problem*

$$\frac{d}{dt}(E'\varphi(t)) = -(A(t)-B(t)R(t)^{-1}(B(t)'K(t)+V(t)'))'\varphi(t) - K(t)'f(t), \quad t\neq t_j, \tag{8.14}$$

$$E'(\varphi(t_j+0)-\varphi(t_j-0)) = F_j\xi_j, \quad j=1,\dots,N, \quad E'\varphi(T) = -F_{N+1}\xi_{N+1}. \tag{8.15}$$

Let also $x_*:[0,T]\to X$ *be a solution of the problem*

$$\frac{d}{dt}(Ex(t)) = (A(t)-B(t)R(t)^{-1}(V(t)+K(t)'B(t))')x(t) - B(t)R(t)^{-1}B(t)'\varphi(t) + f(t), \tag{8.16}$$

$$Ex(0) = x^0. \tag{8.17}$$

Then

$$u_*(t) = -R(t)^{-1}((V(t)+K(t)'B(t))'x_*(t) + B(t)'\varphi(t)) \tag{8.18}$$

is an optimal control for problem (8.9)–(8.11) *and the minimal value of the functional* (8.9) *is given by*

$$J(u_*,x_*) = \frac{1}{2}\sum_{j=1}^{N+1}\langle \xi_j, F_j\xi_j\rangle + \left\langle x^0, \varphi(0) + \frac{1}{2}K(0)E_+x^0\right\rangle \\ + \frac{1}{2}\int_0^T \left\langle \varphi(t), 2f(t) - B(t)R(t)^{-1}B(t)'\varphi(t)\right\rangle dt. \tag{8.19}$$

Let us emphasize that the operator-function K, the function φ, and the optimal control u_* are, generally speaking, discontinuous, though the coefficients in problem (8.9)–(8.11) are continuous.

Proof. Since $W(t)$, $R(t)$, and F_j are symmetric, using (8.12), (8.13), we obtain

$$E'K(t) = K(t)'E, \qquad t\neq t_j, \quad j=1,\dots,N. \tag{8.20}$$

Let $u(\cdot)$ be an admissible control and let $x(\cdot)$ be the corresponding solution of problem (8.10)–(8.11). Using (8.20), we derive the expression

$$\frac{d}{dt}\Big(\langle x, E'Kx\rangle + 2\langle x, E'\varphi\rangle\Big) = \left\langle x, \frac{d}{dt}(E'K)x\right\rangle \\ +2\left(\left\langle \frac{d}{dt}(Ex), Kx+\varphi\right\rangle + \left\langle x, \frac{d}{dt}(E'\varphi)\right\rangle\right), \quad t\neq t_j. \tag{8.21}$$

Taking into account (8.21) and (8.10), (8.12), (8.14), after simple transformations we obtain the relation

$$\frac{1}{2}\frac{d}{dt}\Big(\langle x, E'Kx\rangle + 2\langle x, E'\varphi\rangle\Big) = -\frac{1}{2}\langle x, Wx\rangle - \frac{1}{2}\langle u, Ru\rangle - \langle x, Vu\rangle$$

$$+\frac{1}{2}\langle u + R^{-1}((V' + B'K)x + B'\varphi), R(u + R^{-1}((V' + B'K)x + B'\varphi))\rangle$$

$$+\frac{1}{2}\langle \varphi, 2f - BR^{-1}B'\varphi\rangle, \quad t \neq t_j.$$

Integrating and using the form of $J(u,x)$, we find

$$\begin{aligned} J(u,x) = {} & \frac{1}{2}\sum_{j=1}^{N+1}\langle x(t_j) - \xi_j, F_j(x(t_j) - \xi_j)\rangle \\ & - \frac{1}{2}\int_0^T \frac{d}{dt}\big(\langle x(t), E'K(t)x(t)\rangle + 2\langle x(t), E'\varphi(t)\rangle\big)\,dt \\ & + \frac{1}{2}\int_0^T \langle \varphi(t), 2f(t) - B(t)R(t)^{-1}B(t)'\varphi(t)\rangle\,dt + M(u,x), \end{aligned}$$

where

$$M(u,x) = \frac{1}{2}\int_0^T \langle u + R^{-1}((V' + B'K)x + B'\varphi), R(u + R^{-1}((V' + B'K)x + B'\varphi))\rangle\,dt.$$

The positivity of $R(t)$ and (8.18) clearly yields $M(u,x) \geq 0$ and $M(u_*, x_*) = 0$.

Taking into account the symmetry condition (8.20) and the continuity of $Ex(\cdot)$, we further compute

$$\begin{aligned} J(u,x) = {} & \frac{1}{2}\sum_{j=1}^{N+1}\langle x(t_j) - \xi_j, F_j(x(t_j) - \xi_j)\rangle - \frac{1}{2}\sum_{j=1}^{N}\langle x(t_j), E'(K(t_j - 0) - K(t_j + 0))x(t_j)\rangle \\ & - \sum_{j=1}^{N}\langle x(t_j), E'(\varphi(t_j - 0) - \varphi(t_j + 0))\rangle - \frac{1}{2}\langle x(T), E'K(T)x(T)\rangle - \langle x(T), E'\varphi(T)\rangle \\ & + \frac{1}{2}\langle x(0), E'K(0)x(0)\rangle + \langle x(0), E'\varphi(0)\rangle \\ & + \frac{1}{2}\int_0^T \langle \varphi(t), 2f(t) - B(t)R(t)^{-1}B(t)'\varphi(t)\rangle dt + M(u,x). \end{aligned}$$

Using (8.13), (8.15), (8.17), we derive

$$\begin{aligned} J(u,x) = {} & \frac{1}{2}\sum_{j=1}^{N+1}\langle \xi_j, F_j\xi_j\rangle + \frac{1}{2}\langle x^0, K(0)x(0)\rangle + \langle x^0, \varphi(0)\rangle \\ & + \frac{1}{2}\int_0^T \langle \varphi(t), 2f(t) - B(t)R(t)^{-1}B(t)'\varphi(t)\rangle dt + M(u,x). \end{aligned}$$

Since $\langle x^0, K(0)x(0)\rangle = \langle Ex(0), K(0)x(0)\rangle = \langle (I-P)x(0), K(0)'Ex(0)\rangle = \langle E_+x^0, K(0)'x^0\rangle$, in view of the relation $M(u_*, x_*) = 0$, we conclude that

$$J(u,x) \geq \frac{1}{2}\sum_{j=1}^{N+1}\langle \xi_j, F_j\xi_j\rangle + \frac{1}{2}\langle x^0, K(0)E_+x^0\rangle + \langle x^0, \varphi(0)\rangle$$
$$+\frac{1}{2}\int_0^T \langle \varphi(t), 2f(t) - B(t)R(t)^{-1}B(t)'\varphi(t)\rangle dt = J(u_*, x_*).$$

Hence, u_* is an optimal control and the minimal value of the minimized functional is given by (8.19). □

Theorem 8.2 allows us to avoid solving BVPs for obtaining the solution of problem (8.9)–(8.11). Namely, to find the solution of this problem we need to solve sequentially three problems: (8.12)–(8.13), (8.14)–(8.15), and (8.16)–(8.17).

Concerning the solvability of problems from Theorem 8.2, let us mention the following. Problem (8.16)–(8.17) is the IVP for a linear DAE. This type of problems has been successively studied by many authors (see, e.g., [8, 12, 13, 26]). Problem (8.14)–(8.15) can be considered as an ordered family of terminal problems for linear DAEs on the corresponding segments $[t_{j-1}, t_j]$, $j = N, \dots, 1$. Problem (8.12)–(8.13) is an ordered family of terminal problems for Riccati DAEs on the segments $[t_{j-1}, t_j]$, $j = N, \dots, 1$. Under some mild assumptions (see [31] and also [34]), (8.12) can be reduced to three matrix equations: (1) the standard algebraic Riccati equation, (2) a linear equation, and (3) the standard differential Riccati equation.

Setting $\psi = -(\varphi + Kx_*)$, we obtain the optimal control in the form

$$u_*(t) = R(t)^{-1}(B(t)'\psi(t) - V(t)'x_*(t)).$$

Taking into account (8.12)–(8.17) and using the symmetry of $E'K(t)$, we obtain the following multipoint BVP for solving problem (8.9)–(8.11):

$$\frac{d}{dt}(Ex_*(t)) = (A(t) - B(t)R(t)^{-1}V(t)')x_*(t) + B(t)R(t)^{-1}B(t)'\psi(t) + f(t),$$

$$\frac{d}{dt}(E'\psi(t)) = (W(t) - V(t)R(t)^{-1}V(t)')x_*(t) - (A(t) - B(t)R(t)^{-1}V(t)')'\psi(t), \quad t \neq t_j,$$

$$Ex(0) = x^0,$$

$$E'(\psi(t_j - 0) - \psi(t_j + 0)) = F_j(\xi_j - x_*(t_j)), \quad j = 1, \dots, N, \quad E'\psi(T) = F_{N+1}(\xi_{N+1} - x_*(T)).$$

Note that this BVP can be solved using the solutions of problems (8.12)–(8.13), (8.14)–(8.15), and (8.16)–(8.17).

8.3 Asymptotic Solutions of Problems with a Small Parameter

This section deals with asymptotic solutions of the linear-quadratic problems with intermediate points and a small parameter $\varepsilon \geq 0$. The estimates for the proximity of the approximate asymptotic solutions to the exact one are given in terms of the control, the trajectory, and the functional.

The first two parts are devoted to singularly perturbed problems with discontinuous coefficients; state equations in these problems are switched systems. If the small parameter is set to 0, then we have a problem with DAE constraints. First, exploiting the direct scheme, we construct an asymptotic expansion of the solution. A numerical example illustrating our approach is given. Second, using an optimal feedback control and an asymptotic expansion for a solution of the corresponding matrix differential Riccati equation with discontinuous coefficients, we present a procedure for constructing an asymptotic expansion of the solution.

In the last part of this section, we present the results on asymptotic solution of the problem with DAE constraints, which contains intermediate points and a small parameter in the cost functional. These results are obtained by means of the direct scheme.

In the following, we sometimes denote $D(t,0)$ by $D_0(t)$.

8.3.1 Singularly Perturbed Problems for Discontinuous Systems

ODEs with a small parameter standing before the derivative belong to the class of singularly perturbed problems. They are of common occurrence in all branches of applied mathematics and engineering. Various applied problems formulated in terms of singularly perturbed equations have been considered in, e.g., [42, 47]. One of the effective asymptotic methods in the theory of singular perturbations is the boundary function method (for further details see, e.g., [47]).

In this subsection, we construct asymptotic solutions of singularly perturbed linear-quadratic optimal control problems with discontinuous coefficients using the *direct scheme* based on the boundary function method. Namely, we substitute a postulated asymptotic expansion of the boundary layer type for a solution into the cost functional, the state equation, and the given conditions for the state variable. Then we define four types of optimal control problems in order to find asymptotic expansion terms.

We consider the following linear-quadratic optimal control problem P_ε: the cost functional

$$J_\varepsilon(u) = \frac{1}{2}\sum_{j=1}^{2}\int_{t_{j-1}}^{t_j}\Big\{\big\langle x_j(t,\varepsilon), W_j(t,\varepsilon)x_j(t,\varepsilon)\big\rangle + \big\langle u_j(t,\varepsilon), R_j(t,\varepsilon)u_j(t,\varepsilon)\big\rangle\Big\}dt \tag{8.22}$$

to be minimized on trajectories of the system

$$\begin{aligned} E(\varepsilon)\frac{d}{dt}x_j(t,\varepsilon) = {}& A_j(t,\varepsilon)x_j(t,\varepsilon) + B_j(t,\varepsilon)u_j(t,\varepsilon),\ t\in[t_{j-1},t_j],\ j=1,2,\\ & x_1(0,\varepsilon) = x^0, \quad x_2(t_1,\varepsilon) = x_1(t_1,\varepsilon). \end{aligned} \tag{8.23}$$

Here $0 = t_0 < t_1 < t_2 = T$; t_1, T, and $x^0 \in \mathbf{R}^{n+m}$ are fixed; $E(\varepsilon) = \operatorname{diag}(I_n, \varepsilon I_m)$, where I_n and I_m are identity operators in $\mathbf{R}^n$ and $\mathbf{R}^m$, respectively; $x_j = ({y'}_j, {z'}_j)'$, $y_j = y_j(t,\varepsilon) \in \mathbf{R}^n$, $z_j = z_j(t,\varepsilon) \in \mathbf{R}^m$, $u_j = u_j(t,\varepsilon) \in \mathbf{R}^r$. We suppose that matrices $W_j(t,\varepsilon)$, $R_j(t,\varepsilon)$, $A_j(t,\varepsilon)$, $B_j(t,\varepsilon)$ and admissible controls are sufficiently smooth for all $t \in [t_{j-1}, t_j]$ and $\varepsilon \geq 0$; matrices $W_j(t,\varepsilon)$ and $R_j(t,\varepsilon)$ are symmetric, $W_j(t,0)$ and $R_j(t,0)$ are positive definite for all $t \in [t_{j-1}, t_j]$, $j = 1,2$. We can change the assumption that $W_j(t,0)$ is positive definite by $W_j(t,\varepsilon)$ is positive semidefinite for all sufficiently small $\varepsilon > 0$.

If $\varepsilon = 0$, then the order of (8.23) drops and we obtain a DAE instead of a differential one. In general, the solution of the state equation does not satisfy the given initial conditions and the fast variable consisting of $z_1(t,\varepsilon)$ and $z_2(t,\varepsilon)$ may be a discontinuous function though trajectories of the perturbed problem are continuous.

Let $\begin{pmatrix} A_{j1}(t,\varepsilon) & A_{j2}(t,\varepsilon) \\ A_{j3}(t,\varepsilon) & A_{j4}(t,\varepsilon) \end{pmatrix}$ be a block-matrix representation of the operator $A_j(t,\varepsilon)$ with respect to the direct sum $\mathbf{R}^n \dot{+} \mathbf{R}^m$, i.e., for $x = (y', z')' = \binom{y}{z}$, where $y \in \mathbf{R}^n$, $z \in \mathbf{R}^m$, we have $A_j(t,\varepsilon)x = \begin{pmatrix} A_{j1}(t,\varepsilon)y + A_{j2}(t,\varepsilon)z \\ A_{j3}(t,\varepsilon)y + A_{j4}(t,\varepsilon)z \end{pmatrix}$.

We suppose that the operator $A_{j4}(t,0)$ is stable for every specified $t \in [t_{j-1}, t_j]$, $j = 1,2$; i.e., eigenvalues of $A_{j4}(t,0)$ have negative real parts.

We will seek a solution of the perturbed problem $v = (u', x')'$ in the asymptotic series form

$$v_j(t,\varepsilon) = \sum_{i\geq 0} \varepsilon^i v_{ji}(t,\varepsilon) = \sum_{i\geq 0} \varepsilon^i (\overline{v}_{ji}(t) + \Pi_i v_j(\tau_{j-1}) + Q_i v_j(\tau_j)). \tag{8.24}$$

Here $\tau_j = (t - t_j)/\varepsilon$, $j = 0,1,2$; by symbols Πv_1 and Πv_2 (Qv_1 and Qv_2) we denote boundary functions of the exponential type in neighborhoods of the left (right) ends of the segments $[0,t_1]$ and $[t_1,T]$, respectively. Such notation for boundary functions is often used; see, for instance, [47, p. 42].

Let us emphasize that the asymptotic series (8.24) may not converge to $v_j(t,\varepsilon)$ and may even diverge. An example of a divergent asymptotic series can be found in [47, p. 7].

We substitute expansions (8.24) into (8.22), (8.23) and then rewrite the right-hand side of the state equation and the integrand in asymptotic series form with respect to powers of ε with coefficients depending on $t, \tau_0, \tau_1, \tau_2$. Equating in the state equation the coefficients of equal powers of ε separately depending on $t, \tau_0, \tau_1, \tau_2$, we obtain a family of equations for coefficients in the expansion (8.24). Then the functional to be minimized admits the representation

$$J_\varepsilon(u) = \sum_{j\geq 0} \varepsilon^j J_j. \tag{8.25}$$

Let us determine a family of optimal control problems in order to find the coefficients in the expansion (8.24). We introduce the following notation. For the expansion $h = h(\varepsilon) = \sum_{i\geq 0} \varepsilon^i h_i$, we set

$$[h]_n := h_n, \quad \{h\}_{n-1} := \sum_{i=0}^{n-1} \varepsilon^i h_i.$$

By $\widetilde{v}_n(t,\varepsilon)$ we denote the function consisting of $\widetilde{v}_{1n}(t,\varepsilon) = \{v_1(t,\varepsilon)\}_n$, $t \in [0,t_1]$, and $\widetilde{v}_{2n}(t,\varepsilon) = \{v_2(t,\varepsilon)\}_n$, $t \in [t_1,T]$, and

$$\widetilde{\overline{v}}_{jn}(t,\varepsilon) := \sum_{i=0}^{n} \varepsilon^i \overline{v}_{ji}(t), \quad \widetilde{\Pi}_n v_j(\tau_{j-1},\varepsilon) := \sum_{i=0}^{n} \varepsilon^i \Pi_i v_j(\tau_{j-1}),$$

$$\widetilde{Q}_n v_j(\tau_j,\varepsilon) := \sum_{i=0}^{n} \varepsilon^i Q_i v_j(\tau_j), \quad f_j(x_j,u_j,t,\varepsilon) := A_j(t,\varepsilon)x_j + B_j(t,\varepsilon)u_j,$$

$$q_j(x_j,\omega_j,t,\varepsilon) := W_j(t,\varepsilon)x_j - A_j(t,\varepsilon)'\omega_j, \quad g_j(u_j,\omega_j,t,\varepsilon) := R_j(t,\varepsilon)u_j - B_j(t,\varepsilon)'\omega_j.$$

Thus we arrive at the following systems for coefficients in the expansion (8.24):

$$\left[E(\varepsilon)\frac{d}{dt}\overline{x}_j(t,\varepsilon)\right]_i = A_{j0}(t)\overline{x}_{ji} + B_{j0}(t)\overline{u}_{ji} + [\widehat{f}_{j(i-1)}]_i,\ t_{j-1} \le t \le t_j,\ j = 1,2, \quad (8.26)$$

$$\begin{aligned}\frac{d}{d\tau_{j-1}}\Pi_i x_j &= E_2(A_{j0}(t_{j-1})\Pi_i x_j + B_{j0}(t_{j-1})\Pi_i u_j)\\ &\quad + E_1[\widehat{\Pi}_{i-1} f_j]_{i-1} + E_2[\widehat{\Pi}_{i-1} f_j]_i,\ \tau_{j-1} \ge 0,\end{aligned} \quad (8.27)$$

$$\frac{d}{d\tau_j}Q_i x_j = E_2(A_{j0}(t_j)Q_i x_j + B_{j0}(t_j)Q_i u_j) + E_1[\widehat{Q}_{i-1} f_j]_{i-1} + E_2[\widehat{Q}_{i-1} f_j]_i,\ \tau_j \le 0. \quad (8.28)$$

The notation $\widehat{f}_{j(i-1)}$, $\widehat{\Pi}_{i-1} f_j$, and $\widehat{Q}_{i-1} f_j$ means the values of the function f under $v_j = \widetilde{\overline{v}}_{j(i-1)}(t,\varepsilon)$; $v_j = \widetilde{\Pi}_{i-1} v_j(\tau_{j-1},\varepsilon)$, $t = t_{j-1} + \varepsilon\tau_{j-1}$, and $v_j = \widetilde{Q}_{i-1} v_j(\tau_j,\varepsilon)$, $t = t_j + \varepsilon\tau_j$, respectively; $E_1 = \mathrm{diag}(I_n, 0)$, $E_2 = \mathrm{diag}(0, I_m)$.

The functions $\overline{u}_{j0}$, $\overline{x}_{j0}$ can be found from the degenerate problem $\overline{P}_0$ (set $\varepsilon = 0$ in (8.22)–(8.23)):

$$\begin{aligned}\overline{J}_0(\overline{u}_0) &= \frac{1}{2}\sum_{j=1}^{2}\int_{t_{j-1}}^{t_j}\left(\langle\overline{x}_{j0}, W_{j0}(t)\overline{x}_{j0}\rangle + \langle\overline{u}_{j0}, R_{j0}(t)\overline{u}_{j0}\rangle\right)dt \to \min,\\ \frac{d}{dt}(E_1\overline{x}_{j0}) &= A_{j0}(t)\overline{x}_{j0} + B_{j0}(t)\overline{u}_{j0},\ t \in [t_{j-1}, t_j],\ j = 1,2,\\ E_1\overline{x}_{10}(0) &= E_1 x^0, \quad E_1\overline{x}_{20}(t_1) = E_1\overline{x}_{10}(t_1).\end{aligned}$$

In view of the stability of the operator $A_{j4}(t,0)$ for every specified $t \in [t_{j-1}, t_j]$, $j = 1,2$, the state equation in the degenerate problem $\overline{P}_0$ with the state variable $\overline{x}_0 = (\overline{y}_0', \overline{z}_0')'$ is of index one. Note that in general the trajectory $\overline{z}_0(\cdot)$ is discontinuous at t_1. The problem $\overline{P}_0$ can be reduced to a classical linear-quadratic optimal control problem with the state variable $\overline{y}_0$. Therefore it has a unique solution. Functions defining the adjoint variable for the problem $\overline{P}_0$ will be denoted by $\overline{\omega}_{j0}$, $j = 1,2$.

Using the control optimality conditions for the problem $\overline{P}_0$, we transform the expression for J_1 in (8.25). Removing summands, which are known after solving problem $\overline{P}_0$, we obtain the sum (see [35]) $J_0^1 + J_0^2 + J_0^3$, where

$$\begin{aligned}J_0^1 = J_0^1(\Pi_0 u_1) &= \frac{1}{2}\int_0^{+\infty}\{\langle\Pi_0 x_1, W_{10}(0)\Pi_0 x_1\rangle + \langle\Pi_0 u_1, R_{10}(0)\Pi_0 u_1\rangle\}d\tau_0,\\ J_0^2 = J_0^2(Q_0 u_1, \Pi_0 u_2) &= \langle Q_0 x_1(0), E_2'\overline{\omega}_{10}(t_1)\rangle - \langle\Pi_0 x_2(0), E_2'\overline{\omega}_{20}(t_1)\rangle\\ &\quad + \frac{1}{2}\int_{-\infty}^{0}\{\langle Q_0 x_1, W_{10}(t_1)Q_0 x_1\rangle + \langle Q_0 u_1, R_{10}(t_1)Q_0 u_1\rangle\}d\tau_1\\ &\quad + \frac{1}{2}\int_0^{+\infty}\{\langle\Pi_0 x_2, W_{20}(t_1)\Pi_0 x_2\rangle + \langle\Pi_0 u_2, R_{20}(t_1)\Pi_0 u_2\rangle\}d\tau_1,\\ J_0^3 = J_0^3(Q_0 u_2) &= \langle Q_0 x_2(0), E_2'\overline{\omega}_{20}(T)\rangle\\ &\quad + \frac{1}{2}\int_{-\infty}^{0}\{\langle Q_0 x_2, W_{20}(T)Q_0 x_2\rangle + \langle Q_0 u_2, R_{20}(T)Q_0 u_2\rangle\}d\tau_2.\end{aligned}$$

Equations (8.27) and (8.28) depend only on one type of boundary functions among all four considered types. Therefore, coefficients of boundary layer series in (8.24) with $i = 0$ (the zero approximation) can be found from the following three problems:

$$P_0^1: \quad J_0^1(\Pi_0 u_1) \to \min,$$
$$\frac{d}{d\tau_0}\Pi_0 x_1 = E_2(A_{10}(0)\Pi_0 x_1 + B_{10}(0)\Pi_0 u_1), \quad \tau_0 \geq 0,$$
$$E_1\Pi_0 x_1(+\infty) = 0, \quad E_2\Pi_0 x_1(0) = E_2(x^0 - \overline{x}_{10}(0));$$

$$P_0^2: \quad J_0^2(Q_0 u_1, \Pi_0 u_2) \to \min,$$
$$\frac{d}{d\tau_1}Q_0 x_1 = E_2(A_{10}(t_1)Q_0 x_1 + B_{10}(t_1)Q_0 u_1), \quad \tau_1 \leq 0,$$
$$\frac{d}{d\tau_1}\Pi_0 x_2 = E_2(A_{20}(t_1)\Pi_0 x_2 + B_{20}(t_1)\Pi_0 u_2), \quad \tau_1 \geq 0,$$
$$Q_0 x_1(-\infty) = 0, \quad E_1\Pi_0 x_2(+\infty) = 0,$$
$$E_2\Pi_0 x_2(0) = E_2(Q_0 x_1(0) + \overline{x}_{10}(t_1) - \overline{x}_{20}(t_1));$$

$$P_0^3: \quad J_0^3(Q_0 u_2) \to \min,$$
$$\frac{d}{d\tau_2}Q_0 x_2 = E_2(A_{20}(T)Q_0 x_2 + B_{20}(T)Q_0 u_2), \quad \tau_2 \leq 0,$$
$$Q_0 x_2(-\infty) = 0.$$

It follows from [35] that these problems have unique solutions, which can be found from the control optimality conditions.

Let us introduce the recurrent formulas for finding expansion terms with positive indices. Suppose that the problems $\overline{P}_i$, P_i^1, P_i^2, P_i^3, $i = 0, 1, \ldots, n-1$, are already solved. We denote the adjoint variables for these problems by $(\overline{\omega}_{ji}(t),\ j = 1, 2)$, $\Pi_i\omega_1(\tau_0)$, $(Q_i\omega_1(\tau_1)$, $\Pi_i\omega_2(\tau_1))$, $Q_i\omega_2(\tau_2)$, respectively.

The functions $\overline{v}_n(\cdot)$ can be found from the problem $\overline{P}_n$ of minimizing the functional

$$\overline{J}_n(\overline{u}_n) = \sum_{j=1}^{2}\langle \overline{x}_{jn}(t_j), E_1' Q_{n-1}\omega_j(0)\rangle - \langle \overline{x}_{2n}(t_1), E_1'\Pi_{n-1}\omega_2(0)\rangle$$
$$+\sum_{j=1}^{2}\int_{t_{j-1}}^{t_j}\left\langle \overline{x}_{jn}(t), \frac{1}{2}W_{j0}(t)\overline{x}_{jn}(t) + [\,\widehat{q}_{j(n-1)}]_n - E_2'\frac{d}{dt}\overline{\omega}_{j(n-1)}(t)\right\rangle dt$$
$$+\sum_{j=1}^{2}\int_{t_{j-1}}^{t_j}\left\langle \overline{u}_{jn}(t), \frac{1}{2}R_{j0}(t)\overline{u}_{jn}(t) + [\widehat{g}_{j(n-1)}]_n\right\rangle dt$$

on trajectories of the system (8.26) with $i = n$ under the conditions

$$E_1\overline{x}_{1n}(0) = -E_1\Pi_n x_1(0), \quad E_1(\overline{x}_{2n}(t_1) - \overline{x}_{1n}(t_1)) = E_1(Q_n x_1(0) - \Pi_n x_2(0)).$$

Here the symbols $\widehat{q}_{j(n-1)}$ and $\widehat{g}_{j(n-1)}$ denote the values of the functions q_j and g_j with $v_j = \widetilde{\overline{v}}_{j(n-1)}$, $\omega_j = \widetilde{\overline{\omega}}_{j(n-1)}$, respectively.

The functions $\Pi_n v_1(\tau_0)$, $\tau_0 \in [0,+\infty)$, are obtained from the problem P_n^1 of minimizing the functional

$$J_n^1(\Pi_n u_1) = \int_0^{+\infty} \left\langle \Pi_n x_1(\tau_0), \frac{1}{2} W_{10}(0) + [\widehat{\Pi}_{n-1} q_1]_n \right\rangle d\tau_0 + \int_0^{+\infty} \left\langle \Pi_n u_1(\tau_0), \frac{1}{2} R_{10}(0) \Pi_n u_1(\tau_0) + [\widehat{\Pi}_{n-1} g_1]_n \right\rangle d\tau_0$$

on trajectories of the system (8.27) with $j = 1$ and $i = n$ under the conditions

$$E_1 \Pi_n x_1(+\infty) = 0, \quad E_2 \Pi_n x_1(0) = -E_2 \overline{x}_{1n}(0).$$

The functions $Q_n v_1(\tau_1)$, $\tau_1 \in (-\infty, 0]$, and $\Pi_n v_2(\tau_1)$, $\tau_1 \in [0,+\infty)$, can be found from the problem P_n^2 of minimizing the functional

$$\begin{aligned} J_n^2(Q_n u_1, \Pi_n u_2) = {} & \langle Q_n x_1(0), E_2' \overline{\omega}_{1n}(t_1) \rangle - \langle \Pi_n x_2(0), E_2' \overline{\omega}_{2n}(t_1) \rangle \\ & + \int_{-\infty}^0 \left\langle Q_n x_1(\tau_1), \frac{1}{2} W_{10}(t_1) Q_n x_1(\tau_1) + [\widehat{Q}_{n-1} q_1]_n \right\rangle d\tau_1 \\ & + \int_{-\infty}^0 \left\langle Q_n u_1(\tau_1), \frac{1}{2} R_{10}(t_1) Q_n u_1(\tau_1) + [\widehat{Q}_{n-1} g_1]_n \right\rangle d\tau_1 \\ & + \int_0^{+\infty} \left\langle \Pi_n x_2(\tau_1), \frac{1}{2} W_{20}(t_1) \Pi_n x_2(\tau_1) + [\widehat{\Pi}_{n-1} q_2]_n \right\rangle d\tau_1 \\ & + \int_0^{+\infty} \left\langle \Pi_n u_2(\tau_1), \frac{1}{2} R_{20}(t_1) \Pi_n u_2(\tau_1) + [\widehat{\Pi}_{n-1} g_2]_n \right\rangle d\tau_1 \end{aligned}$$

on trajectories of the system (8.28) with $j = 1$, $i = n$ and (8.27) with $j = 2$, $i = n$ under the conditions

$$Q_n x_1(-\infty) = 0, \quad E_1 \Pi_n x_2(+\infty) = 0, \quad E_2(\Pi_n x_2(0) - Q_n x_1(0)) = E_2(\overline{x}_{1n}(t_1) - \overline{x}_{2n}(t_1)).$$

The functions $Q_n v_2(\tau_2)$, $\tau_2 \in (-\infty, 0]$, are obtained from the problem P_n^3 of minimizing the functional

$$\begin{aligned} J_n^3(Q_n u_2) = {} & \langle Q_n x_2(0), E_2' \overline{\omega}_{2n}(T) \rangle \\ & + \int_{-\infty}^0 \left\langle Q_n x_2(\tau_2), \frac{1}{2} W_{20}(T) Q_n x_2(\tau_2) + [\widehat{Q}_{n-1} q_2]_n \right\rangle d\tau_2 \\ & + \int_{-\infty}^0 \left\langle Q_n u_2(\tau_2), \frac{1}{2} R_{20}(T) Q_n u_2(\tau_2) + [\widehat{Q}_{n-1} g_2]_n \right\rangle d\tau_2 \end{aligned}$$

on trajectories of the system (8.28) with $j = 2$ and $i = n$ under the condition

$$Q_n x_2(-\infty) = 0.$$

In problems P_n^1, P_n^2, P_n^3, the symbols $\widehat{\Pi}_{n-1}q_j$, $\widehat{\Pi}_{n-1}g_j$, $\widehat{Q}_{n-1}q_j$, $\widehat{Q}_{n-1}g_j$ denote the values of the functions q_j, g_j, respectively, with arguments depending on terms of the asymptotic expansion (8.24) of order less than n. The problems $\overline{P}_n$, P_n^1, P_n^2, P_n^3 have unique solutions, which can be found from the control optimality conditions (see [37]).

Due to the results of subsection 8.2.1, the optimal control for problem (8.22), (8.23) with $\varepsilon > 0$ has the form

$$u_j(t,\varepsilon) = R_j(t,\varepsilon)^{-1}B_j(t,\varepsilon)'\zeta_j(t,\varepsilon), \quad t \in [t_{j-1},t_j], \quad j = 1,2, \tag{8.29}$$

where the adjoint variable is a solution of the problem

$$\begin{gathered} E(\varepsilon)'\tfrac{d}{dt}\zeta_j(t,\varepsilon) = W_j(t,\varepsilon)x_j(t,\varepsilon) - A_j(t,\varepsilon)'\zeta_j(t,\varepsilon), \quad t \in [t_{j-1},t_j], \quad j = 1,2, \\ \zeta_2(T,\varepsilon) = 0, \quad \zeta_1(t_1,\varepsilon) = \zeta_2(t_1,\varepsilon). \end{gathered} \tag{8.30}$$

Let us denote the coefficients in an asymptotic expansion for $\zeta_j(t,\varepsilon)$ by $\overline{\zeta}_{ji}(t)$, $\Pi_i\zeta_j(\tau_{j-1})$, $Q_i\zeta_j(\tau_j)$.

The following two theorems explain the form of the problems for finding terms of the asymptotic expansion (8.24).

Theorem 8.3. *Equations for optimal trajectories and adjoint variables for the problems $\overline{P}_n$, P_n^1, P_n^2, and P_n^3 coincide with the equations for the functions $(\overline{v}_{jn}, \overline{\zeta}_{jn})$, $j = 1,2$, $(\Pi_n v_1, E_1'\Pi_{n+1}\zeta_1 + E_2'\Pi_n\zeta_1)$, $(Q_n v_1, E_1'Q_{n+1}\zeta_1 + E_2'Q_n\zeta_1, \Pi_n v_2, E_1'\Pi_{n+1}\zeta_2 + E_2'\Pi_n\zeta_2)$, and $(Q_n v_2, E_1'Q_{n+1}\zeta_2 + E_2'Q_n\zeta_2)$, respectively, from the asymptotic expansion of the solution of the problem* (8.23), (8.29), (8.30).

Theorem 8.4. *The cost functional $\overline{J}_n$ for the problem $\overline{P}_n$ is a result of a transformation of the coefficient J_{2n} in expansion* (8.25). *The sum of the cost functionals for the problems $P_n^i (i = 1,2,3)$: $J_n^1 + J_n^2 + J_n^3$ is a result of a transformation of the coefficient J_{2n+1} in expansion* (8.25) *after removing summands which are known if the problems $\overline{P}_k$, P_k^i, $i = 1,2,3$, $k < n$, have been solved.*

The proofs of Theorems 8.3 and 8.4 are based on immediate and simple but cumbersome transformations, which exploit the formulations of problems $\overline{P}_n$, P_n^i, $i = 1,2,3$, and also control optimality conditions for these problems. The transformation of coefficients in (8.25) in the proof of Theorem 8.4 is similar to the transformation of the coefficient J_1 in [35]. The detailed proofs are presented in [37].

Let us denote the block-matrix representation of the operators $B_j(t,\varepsilon)$, $W_j(t,\varepsilon)$, and $S_j(t,\varepsilon) = B_j(t,\varepsilon)R_j(t,\varepsilon)^{-1}B_j(t,\varepsilon)'$ with respect to the direct sum $\mathbf{R}^n \dotplus \mathbf{R}^m$ by $\begin{pmatrix} B_{j1} \\ B_{j2} \end{pmatrix}$, $\begin{pmatrix} W_{j1} & W_{j2} \\ W_{j2}' & W_{j3} \end{pmatrix}$, and $\begin{pmatrix} S_{j1} & S_{j2} \\ S_{j2}' & S_{j3} \end{pmatrix}$, respectively. By $\{\mu_{ji}(t), -\mu_{ji}(t)\}, i = 1,\dots,m$, we denote the eigenvalues of $G_j(t) = \begin{pmatrix} A_{j4}(t,0) & S_{j3}(t,0) \\ W_{j3}(t,0) & -A_{j4}(t,0)' \end{pmatrix}$. Here $t \in [t_{j-1},t_j]$, $j = 1,2$.

Assume that the following three conditions are satisfied:

(1) $\mu_{ji}(t) \neq \mu_{jk}(t)\,, i \neq k;\ i,k = 1,\dots,m,\ t \in [t_{j-1},t_j],\ j = 1,2.$

Then there exists an operator $D_j(t) = \begin{pmatrix} D_{j1}(t) & D_{j2}(t) \\ D_{j3}(t) & D_{j4}(t) \end{pmatrix}$, $j = 1,2$, which is sufficiently smooth with respect to t and it reduces $G_j(t)$ to the form diag$\{\mu_{j1}(t)$,

$\ldots, \mu_{jm}(t), \mu_{jm}(t), \ldots, -\mu_{j1}(t)\}$ (see [16, Chapter VI]). Note that in the general case the reduction to Jordan normal form is an unstable procedure. Another normal form is studied in [2] for holomorphic matrix-functions depending on complex variables.

(2) The inequalities $\mathrm{Re}\mu_{j1}(t) \leq \mathrm{Re}\mu_{j2}(t) \leq \cdots \leq \mathrm{Re}\mu_{jm}(t) < 0, t \in [t_{j-1}, t_j],\ j = 1,2$, hold.

(3) The operators $D_{j1}(t)$, $D_{j4}(t)$ at $t = t_{j-1}, t = t_j, j = 1,2$, and $\begin{pmatrix} D_{12}(t_1) & D_{21}(t_1) \\ D_{14}(t_1) & D_{23}(t_1) \end{pmatrix}$ are invertible.

Note that we need these restrictions only to simplify the proof of the next theorem.

Theorem 8.5. *Let $u_*(\cdot,\varepsilon)$, $x_*(\cdot,\varepsilon)$ be a solution of the perturbed problem P_ε and let $\widetilde{u}_n(\cdot,\varepsilon)$, $\widetilde{x}_n(\cdot,\varepsilon)$ be the corresponding approximate asymptotic solution. Then for all sufficiently small $\varepsilon > 0$ the following estimates*

$$\|\widetilde{v}_{jn}(t,\varepsilon) - v_{j*}(t,\varepsilon)\| \leq c\varepsilon^{n+1}, \quad t \in [t_{j-1}, t_j], \quad j = 1,2,$$
$$J_\varepsilon(\widetilde{u}_n) - J_\varepsilon(u_*) \leq c\varepsilon^{2(n+2)}$$

hold, where the constant c does not depend on t and ε.

The proof of this theorem is in two steps. First, using Green's functions, we reduce the BVP obtained from the control optimality condition for the perturbed problem (8.22), (8.23) to a system of integral equations. Next we apply the method of successive approximations. The details of the proof are given in [37, Theorem 5.1].

It follows from Theorem 8.5 that the sequence $\{\widetilde{u}_n(\cdot,\varepsilon)\}$ is a minimizing sequence for the functional (8.22).

Theorem 8.6. *For a sufficiently small $\varepsilon > 0$, we have*

$$J_\varepsilon(\widetilde{u}_i) \leq J_\varepsilon(\widetilde{u}_{i-1} + \varepsilon^i \overline{u}_i) \leq J_\varepsilon(\widetilde{u}_{i-1}), \quad i = 1, \ldots, n; \tag{8.31}$$

i.e., for any fixed and sufficiently small $\varepsilon > 0$, the sequence $\{J_\varepsilon(\widetilde{u}_i)\}$ is nonincreasing.

Proof. Let us construct the asymptotic solution of the problem (8.23) with the control $\widetilde{u}_{i-1}$:

$$x = \sum_{k=0}^{i-1} \varepsilon^k (\overline{x}_k + \Pi_k x + Q_k x) + O(\varepsilon^i), \qquad i \geq 1.$$

Note that the terms in this expansion coincide with the corresponding terms in the asymptotic expansion of the optimal trajectory for the perturbed problem. Further, we expand $J_\varepsilon(\widetilde{u}_{i-1})$ and $J_\varepsilon(\widetilde{u}_{i-1} + \varepsilon^i \overline{u}_i)$ into series with respect to powers of ε. Comparing the obtained expansions and using Theorems 8.3 and 8.4, we obtain the inequality $J_\varepsilon(\widetilde{u}_{i-1} + \varepsilon^i \overline{u}_i) \leq J_\varepsilon(\widetilde{u}_{i-1})$ for a sufficiently small $\varepsilon > 0$.

The proof of the second inequality in (8.31) can be obtained in the same way. $\square$

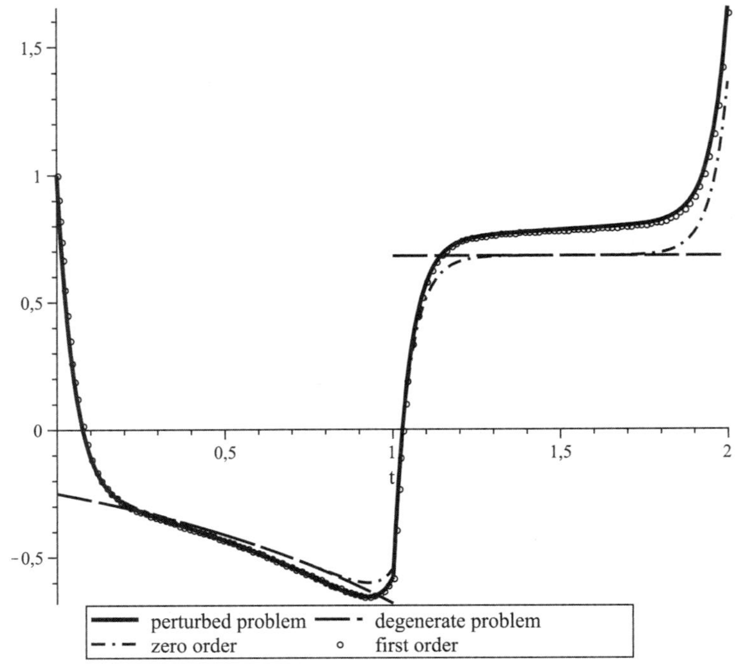

Figure 8.1. Fast component of optimal trajectory $z(t,\varepsilon)$ [37].

Example 8.7. Let us consider the problem from [35], where the zero order approximation for the solution has been obtained using the direct scheme (see also [37]):

$$P_\varepsilon: \quad J_\varepsilon(u) = \frac{1}{2}\int_0^1 \left((y_1)^2 + 2y_1 z_1 + 3(z_1)^2 + (u_1)^2\right) dt + \frac{1}{2}\int_1^2 \left((y_2)^2 + (z_2)^2 + \frac{1}{3}(u_2)^2\right) dt,$$

$$\frac{d}{dt}y_1 = (1+\varepsilon)y_1, \qquad \varepsilon\frac{d}{dt}z_1 = -z_1 + u_1, \quad t \in [0,1],$$

$$\frac{d}{dt}y_2 = \varepsilon y_2, \qquad \varepsilon\frac{d}{dt}z_2 = y_2 - z_2 + u_2, \quad t \in [1,2],$$

$$y_1(0,\varepsilon) = 1, \ y_2(1,\varepsilon) = y_1(1,\varepsilon), \ z_1(0,\varepsilon) = 1, \ z_2(1,\varepsilon) = z_1(1,\varepsilon).$$

The results of evaluations with $\varepsilon = 0.1$ for the fast optimal trajectory $z(t,\varepsilon)$ are shown for the solutions of the perturbed and degenerate problems and for the approximations of the zero and the first orders in Figure 8.1. It shows that the first order approximation almost coincides with the solution of the considered perturbed problem. ∎

8.3.2 Feedback Control

The optimal feedback control and the asymptotic expansion of a solution of a matrix singularly perturbed differential Riccati equation with discontinuous coefficients are used to construct asymptotic solutions of singularly perturbed linear-quadratic optimal control problems with discontinuous coefficients. Applying this method, we avoid solving BVPs.

Let us consider the problem P_ε of minimizing the functional

$$J_\varepsilon(u) = \frac{1}{2}\langle x_2(T,\varepsilon), F(\varepsilon)x_2(T,\varepsilon)\rangle + \frac{1}{2}\sum_{j=1}^{2}\int_{t_{j-1}}^{t_j}\{\langle x_j(t,\varepsilon), W_j(t,\varepsilon)x_j(t,\varepsilon)\rangle + \langle u_j(t,\varepsilon), R_j(t,\varepsilon)u_j(t,\varepsilon)\rangle\}\,dt \tag{8.32}$$

on trajectories of the system (8.23).

We assume that coefficients $W_j(t,\varepsilon)$, $R_j(t,\varepsilon)$, $A_j(t,\varepsilon)$, $B_j(t,\varepsilon)$ and admissible controls satisfy conditions from subsection 8.3.1. The sufficiently smooth in ε operator $F(\varepsilon) = \begin{pmatrix} F_1(\varepsilon) & \varepsilon F_2(\varepsilon) \\ \varepsilon F_2(\varepsilon)' & \varepsilon F_3(\varepsilon)\end{pmatrix}$ does not depend on t; it is symmetric positive semidefinite.

In this subsection, we assume that one of the following conditions is satisfied for every specified $t \in [t_{j-1}, t_j]$, $j = 1,2$:

(1^0) The pair $(A_{j4}(t,0), B_{j2}(t,0))$ is controllable, (see [39]), i.e.,

$$\text{rank}(B_{j2}(t,0), A_{j4}(t,0)B_{j2}(t,0), \ldots, (A_{j4}(t,0))^{m-1}B_{j2}(t,0)) = m.$$

(2^0) $B_{j2}(t,0) = 0$ and the operator $A_{j4}(t,0)$ is stable; i.e., eigenvalues of $A_{j4}(t,0)$ have negative real parts.

According to Theorem 8.1, to construct the asymptotic solution of the problem (8.32), (8.23) we need to find asymptotic expansions of solutions of the problems (8.4), (8.5) and (8.6), (8.7) with $E(\varepsilon)$, $F(\varepsilon)$, and $E(\varepsilon)x^0$ in place of E, F, and x^0, respectively. Using the method of boundary functions, asymptotic solutions of these problems are sought in the series form

$$\begin{aligned} K_j(t,\varepsilon) &= \sum_{k\ge 0}\varepsilon^k\left(\overline{K}_{jk}(t) + Q_kK_j(\tau_j)\right), \\ x_{j*}(t,\varepsilon) &= \sum_{k\ge 0}\varepsilon^k\left(\overline{x}_{jk}(t) + \Pi_kx_j(\tau_{j-1}) + Q_kx_j(\tau_j)\right). \end{aligned} \tag{8.33}$$

Then the components u_{j*} of the optimal control u_* (see (8.8)) admit the representation

$$u_{j*}(t,\varepsilon) = \sum_{k\ge 0}\varepsilon^k\left(\overline{u}_{jk}(t) + \Pi_ku_j(\tau_{j-1}) + Q_ku_j(\tau_j)\right). \tag{8.34}$$

The procedure for finding the terms in the expansions (8.33) and (8.34) as well as the proof of the estimates in the following two theorems are presented in [36].

Assume that the first $n+1$ terms in the expansions (8.33), (8.34) are already found. Set

$$w_{jk}(t,\varepsilon) := \overline{w}_{jk}(t) + \Pi_kw_j(\tau_{j-1}) + Q_kw_j(\tau_j), \quad k = 0,1,\ldots,n,$$

$$\widetilde{w}_{jn}(t,\varepsilon) := \sum_{k=0}^{n}\varepsilon^kw_{jk}(t,\varepsilon), \quad t\in[t_{j-1},t_j], \quad j = 1,2,$$

where w denotes one of three functions K, x_*, or u_*.

Theorem 8.8. *Let $w_j(t,\varepsilon)$, $j=1,2$, be the functions determining the solution of the problem* (8.4), (8.5) *or* (8.6), (8.7) *with $E(\varepsilon)$, $F(\varepsilon)$, and $E(\varepsilon)x^0$ in place of E, F, and x^0, respectively, or the functions determining the optimal control for the problem* (8.32), (8.23). *There is a constant $c>0$ independent of t and ε such that the inequality*

$$\|\widetilde{w}_{jn}(t,\varepsilon)-w_j(t,\varepsilon)\| \le c\varepsilon^{n+1}, \quad t\in[t_{j-1},t_j], \quad j=1,2,$$

holds for all sufficiently small $\varepsilon>0$.

Let us denote by $\widehat{x}_{jn}$, $j=1,2$, the solutions of the problem (8.6), (8.7) with $E=E(\varepsilon)$, $x^0=E(\varepsilon)x^0$, and $K_j=\widetilde{K}_{jn}$, $j=1,2$; i.e.,

$$\begin{gathered} E(\varepsilon)\frac{d}{dt}\widehat{x}_{jn}=(A_j-S_j\widetilde{K}_{jn})\widehat{x}_{jn}, \quad t\in[t_{j-1},t_j], \quad j=1,2, \\ \widehat{x}_{1n}(0,\varepsilon)=x^0, \quad \widehat{x}_{2n}(t_1,\varepsilon)=\widehat{x}_{1n}(t_1,\varepsilon). \end{gathered} \tag{8.35}$$

Further, set

$$\widehat{u}_{jn} := -R_j^{-1}B_j'\widetilde{K}_{jn}\widehat{x}_{jn}, \quad j=1,2. \tag{8.36}$$

Theorem 8.9. *The asymptotic approximation for the optimal control u_* for the problem* (8.32), (8.23) *can be constructed using the asymptotic expansions* (8.33). *Moreover, the estimates*

$$\begin{gathered} \|\widehat{u}_{jn}(t,\varepsilon)-u_{j*}(t,\varepsilon)\| \le c\varepsilon^{n+1}, \quad \|\widehat{x}_{jn}(t,\varepsilon)-x_{j*}(t,\varepsilon)\| \le c\varepsilon^{n+1}, \quad t\in[t_{j-1},t_j], \quad j=1,2, \\ J_\varepsilon(\widehat{u}_n)-J_\varepsilon(u_*) \le c\varepsilon^{2(n+1)}, \end{gathered}$$

where the constant c does not depend on t and ε, hold for all sufficiently small $\varepsilon>0$.

For the proof of Theorems 8.8 and 8.9 we refer the reader to [36]. Also, in [36] an example is given, illustrating the construction of asymptotic expansions using the feedback optimal control.

8.3.3 Problems with DAE Constraints

In this subsection we present some results from [38] on the asymptotic solution of a linear-quadratic optimal control problem P_ε with DAE constraints containing intermediate points and a small parameter in the cost functional. The problem is to minimize the functional

$$\begin{aligned} J_\varepsilon(u,x) &= \frac{1}{2}\langle x(T)-\xi_{N+1}, F_{N+1}(x(T)-\xi_{N+1})\rangle + \frac{\varepsilon}{2}\sum_{j=1}^{N}\big\langle x(t_j)-\xi_j, F_j(x(t_j)-\xi_j)\big\rangle \\ &+ \frac{1}{2}\int_0^T \left\langle \begin{pmatrix} x(t) \\ u(t) \end{pmatrix}, \begin{pmatrix} W(t) & V(t) \\ V(t)' & R(t) \end{pmatrix}\begin{pmatrix} x(t) \\ u(t) \end{pmatrix}\right\rangle dt \end{aligned} \tag{8.37}$$

on trajectories of the system

$$\frac{d}{dt}(Ex(t)) = A(t)x(t)+B(t)u(t)+f(t), \quad Ex(0)=x^0. \tag{8.38}$$

The conditions for the coefficients and admissible controls in this problem are the same as in subsection 8.2.2. In addition, we suppose that the operator

$$QA(t)P : \mathrm{Ker}E \to \mathrm{Ker}E' \tag{8.39}$$

is invertible, and hence the DAE (8.38) is of index one. We also assume that the operator $\begin{pmatrix} PW(t)P & PV(t) \\ V(t)'P & R(t) \end{pmatrix}$ is positive definite for all $t \in [0,T]$. Here, as above, P and Q are orthogonal, and hence symmetric, projections onto $\mathrm{Ker}E$ and $\mathrm{Ker}E'$, respectively.

Due to the direct scheme, a solution of the perturbed problem (8.37), (8.38) is sought in the series form

$$u(t,\varepsilon) = \sum_{j\geq 0} \varepsilon^j u_j(t), \quad x(t,\varepsilon) = \sum_{j\geq 0} \varepsilon^j x_j(t). \tag{8.40}$$

Then we substitute (8.40) into the problem conditions. Hence the minimized functional admits the representation

$$J_\varepsilon(u,x) = \sum_{j\geq 0} \varepsilon^j J_j. \tag{8.41}$$

Equating the coefficients of equal powers of ε in (8.38), we obtain equations for the terms in the expansions (8.40). We then determine a family of optimal control problems in order to find the coefficients in the expansions (8.40).

Setting $\varepsilon = 0$, we obtain from (8.37), (8.38) the degenerate problem without intermediate points. We denote it by P_0. Its solution is denoted by (u_0, x_0). Since the operator (8.39) is invertible, the problem P_0 can be reduced to a classical linear-quadratic problem with state variable $(I - P)x$. Therefore the optimal control for the problem P_0 is a continuous function.

To find the pair of functions (u_k, x_k) for $k \geq 1$, we consider the following problem with intermediate points:

$$P_k: \quad \widetilde{J}_k(u_k,x_k) = \frac{1}{2}\langle x_k(T), F_{N+1}x_k(T)\rangle + \sum_{j=1}^{N} \big\langle x_k(t_j), F_j(x_{k-1}(t_j) - \xi_{j,k-1})\big\rangle$$
$$+ \frac{1}{2}\int_0^T \left\langle \begin{pmatrix} x_k \\ u_k \end{pmatrix}, \begin{pmatrix} W & V \\ V' & R \end{pmatrix} \begin{pmatrix} x_k \\ u_k \end{pmatrix} \right\rangle dt \to \min,$$
$$\frac{d}{dt}(Ex_k) = Ax_k + Bu_k, \quad Ex_k(0) = 0,$$

where $\xi_{j,0} = \xi_j$ and $\xi_{j,k-1} = 0$ if $k > 1$.

The solution of the problem P_k can be found from the multipoint BVP from subsection 8.2.2. It should be noted that, in general, the optimal control for the problem P_k, $k \geq 1$, is a discontinuous function. At points of discontinuity, we define the values of functions so that the functions become right continuous.

The expressions for the coefficients in (8.41) take the form

$$J_{2k-1} = \langle x_{2k-1}(T), F_{N+1}(x_0(T) - \xi_{N+1})\rangle + \sum_{m=1}^{k-1} \langle x_m(T), F_{N+1}x_{2k-1-m}(T)\rangle$$

$$+\sum_{j=1}^{N}\left(\sum_{m=1}^{k-2}\big\langle x_m(t_j), F_j x_{2k-2-m}(t_j)\big\rangle + \big\langle x_{2k-2}(t_j), F_j(x_0(t_j) - \xi_j)\big\rangle + \frac{1}{2}\big\langle x_{k-1}(t_j), F_j x_{k-1}(t_j)\big\rangle\right)$$

$$+\sum_{m=0}^{k-1}\int_0^T \big(\langle x_{2k-1-m}, Wx_m + Vu_m\rangle + \big\langle u_{2k-1-m}, V'x_m + Ru_m\big\rangle\big)\,dt,$$

$$J_{2k} = \sum_{j=1}^{N}\left(\sum_{m=1}^{k-1}\big\langle x_m(t_j), F_j x_{2k-1-m}(t_j)\big\rangle + \big\langle x_{2k-1}(t_j), F_j(x_0(t_j) - \xi_j)\big\rangle\right)$$

$$+ \langle x_{2k}(T), F_{N+1}(x_0(T) - \xi_{N+1})\rangle + \sum_{m=1}^{k-1} \langle x_m(T), F_{N+1}x_{2k-m}(T)\rangle + \frac{1}{2}\langle x_k(T), F_{N+1}x_k(T)\rangle$$

$$+\int_0^T \left(\frac{1}{2}\left\langle \begin{pmatrix} x_k \\ u_k \end{pmatrix}, \begin{pmatrix} W & V \\ V' & R \end{pmatrix}\begin{pmatrix} x_k \\ u_k \end{pmatrix}\right\rangle\right.$$

$$\left. + \sum_{m=0}^{k-1}\big(\langle x_{2k-m}, Wx_m + Vu_m\rangle + \big\langle u_{2k-m}, V'x_m + Ru_m\big\rangle\big)\right) dt.$$

The structure of these coefficients is described in the following theorem.

Theorem 8.10. *The coefficient J_{2k-1} can be found solving the problem P_{k-1}. The cost functional $\widetilde{J}_k(u_k, x_k)$ in the problem P_k, $k \geq 1$, is the transformed expression for the coefficient J_{2k}.*

The proof of this theorem is based on transformations of the expressions for J_{2k-1} and J_{2k} using control optimality conditions for problems $P_j, j = 0, 1, \ldots, k-1$. We note that the assumption on the invertibility of the operator (8.39) is not used in the proof of Theorem 8.10.

Assume now that the solutions (u_j, x_j) of the problems $P_j, j = 0, 1, \ldots, n$, are already obtained. Let us estimate the approximate solution of the perturbed problem (8.37), (8.38):

$$\widetilde{u}_n(t,\varepsilon) = \sum_{j=0}^{n} \varepsilon^j u_j(t), \quad \widetilde{x}_n(t,\varepsilon) = \sum_{j=0}^{n} \varepsilon^j x_j(t).$$

It is not difficult to see that the function $\widetilde{x}_n(\cdot,\varepsilon)$ is a solution of the problem (8.38) with $u(t) = \widetilde{u}_n(t,\varepsilon)$.

Theorem 8.11. *For the solution* $(u_*(\cdot,\varepsilon), x_*(\cdot,\varepsilon))$ *of the problem* (8.37), (8.38), *the following estimates*

$$\|u_*(t,\varepsilon) - \widetilde{u}_n(t,\varepsilon)\| \le c\varepsilon^{n+1}, \quad \|x_*(t,\varepsilon) - \widetilde{x}_n(t,\varepsilon)\| \le c\varepsilon^{n+1},$$

$$J_\varepsilon(\widetilde{u}_n, \widetilde{x}_n) - J_\varepsilon(u_*, x_*) \le c\varepsilon^{2(n+1)}$$

hold for all $t \in [0,T]$ *and sufficiently small* $\varepsilon > 0$.

This theorem implies that the sequence $\{\widetilde{u}_n\}$ is a minimizing sequence for the functional (8.37).

Theorem 8.12. *For a sufficiently small* $\varepsilon > 0$, *we have*

$$J_\varepsilon(\widetilde{u}_i, \widetilde{x}_i) \le J_\varepsilon(\widetilde{u}_{i-1}, \widetilde{x}_{i-1}), \quad i = 1, \ldots, n. \tag{8.42}$$

If $u_i \neq 0$, *then the inequality in* (8.42) *is strict.*

Theorem 8.12 shows that the sequence $\{J_\varepsilon(\widetilde{u}_i, \widetilde{x}_i)\}$ is nonincreasing for any fixed sufficiently small $\varepsilon > 0$.

The detailed proofs of Theorems 8.10, 8.11, and 8.12 can be found in [38] (see Theorems 1–3, respectively). Also, [38] contains a numerical example with graphs for exact and approximate asymptotic solutions, which illustrates the direct scheme method for the problem (8.37), (8.38).

Acknowledgments

The author thanks the anonymous referee for numerous critical and constructive remarks that have helped to improve the text of the chapter.

Bibliography

[1] H. Abou-Kandil, G. Freiling, V. Ionescu, and G. Jank, *Matrix Riccati Equations in Control and Systems Theory*, Birkhäuser Verlag, Basel, Boston, Berlin, 2003.

[2] V. I. Arnold, *On matrices depending on parameters*, Uspehi Mat. Nauk, 26 (1971), pp. 101–114 (in Russian); Russian Math. Surveys, 26 (1971), pp. 29–43 (in English).

[3] U. M. Ascher, R. M. M. Mattheij, and R. D. Russell, *Numerical Solution of Boundary Value Problems for Ordinary Differential Equations*, Classics Appl. Math. 13, SIAM, Philadelphia, 1995.

[4] L. T. Ashchepkov, *Optimal Control for Discontinuous Systems*, Nauka, Novosibirsk, 1987 (in Russian).

[5] A. Backes, *Extremalbedingungen für Optimierungs–Probleme mit Algebro–Differentialgleichungen*, Dissertation, Mathematisch–Naturwissenschaftlichen Fakultät II, Humboldt–Universität zu Berlin, Berlin, Germany, 2006.

[6] K. BALLA, G. A. KURINA, AND R. MÄRZ, *On self-adjoint differential-algebraic equations in Hilbert space*, Trudy Matematicheskogo Centra Imeni N. I. Lobachevskogo. Teoriya Funkcii, Ee Prilozheniya i Smezhnye Voprosy, Kazan', Izdatel'stvo Kazanskogo Matematicheskogo Obshchestva, 19 (2003), pp. 26–28 (in Russian).

[7] K. BALLA, G. A. KURINA, AND R. MÄRZ, *Index criteria for differential algebraic equations arising from linear–quadratic optimal control problems*, J. Dyn. Control Syst., 12 (2006), pp. 289–311.

[8] K. BALLA AND R. MÄRZ, *A unified approach to linear differential algebraic equations and their adjoints*, Z. Anal. Anwendungen, 21 (2002), pp. 783–802.

[9] F. J. BEJARANO AND L. FRIDMAN, *State exact reconstruction for switched linear systems via a super–twisting algorithm*, Internat. J. Systems Sci., 42 (2011), pp. 717–724.

[10] S. V. BELOKOPYTOV AND M. G. DMITRIEV, *Direct scheme in optimal control problems with fast and slow motions*, Systems Control Lett., 8 (1986), pp. 129–135.

[11] D. J. BENDER AND A. J. LAUB, *The linear–quadratic optimal regulator for descriptor systems*, IEEE Trans. Automat. Control, AC-32 (1987), pp. 672–688.

[12] YU. E. BOYARINCEV, *Regular and Singular Systems of Linear Ordinary Differential Equations*, Nauka, Novosibirsk, 1980 (in Russian).

[13] K. E. BRENAN, S. L. CAMPBELL, AND L. R. PETZOLD, *Numerical Solution of Initial Value Problems in Differential-Algebraic Equations*, North–Holland, New York, 1989.

[14] S. L. CAMPBELL, *Singular Systems of Differential Equations* II, Pitman Advanced Publishing Program, San Francisco, London, Melbourne, 1982.

[15] D. CHENG, *Stabilization of planar switched systems*, Systems Control Lett., 51 (2004), pp. 79–88.

[16] E. A. CODDINGTON AND N. LEWINSON, *Theory of Ordinary Differential Equations*, McGraw–Hill, New York, Toronto, London, 1955.

[17] M. DMITRIEV AND G. KURINA, *Singular perturbations in control problems*, Avtomat. i Telemekh., 2006 (2006), pp. 3–51 (in Russian); Autom. Remote Control, 67 (2006), pp. 1–43 (in English).

[18] A. V. DMITRUK AND A. M. KAGANOVICH, *The hybrid maximum principle is a consequence of Pontryagin maximum principle*, Systems Control Lett., 57 (2008), pp. 964–970.

[19] A. V. DMITRUK AND A. M. KAGANOVICH, *Quadratic order conditions for an extended weak minimum in optimal control problems with intermediate and mixed constraints*, Discrete Contin. Dyn. Syst., 29 (2011), pp. 523–545.

[20] G. R. DUAN, *Analysis and Design of Descriptor Linear Systems,* Springer, New York, 2010.

[21] M. GERDTS, *Direct shooting method for the numerical solution of higher-index DAE optimal control problems*, J. Optim. Theory Appl., 117 (2003), pp. 267–294.

[22] G. L. GRODZOVSKII, YU. N. IVANOV, AND V. V. TOKAREV, *Mechanics of Space Flight with Low Thrust*, Nauka, Moscow, 1966 (in Russian).

[23] T. KATAYAMA AND K. MINAMINO, *Linear quadratic regulator and spectral factorization for continuous–time descriptor systems*, in Proceedings of the 31st Conference Decision and Control, Tucson, AZ, 1992, pp. 967–972.

[24] A. KUMAR AND P. DAOUTIDIS, *Control of Nonlinear Differential Algebraic Equation Systems,* CRC Press, Boca Raton, FL, 1999.

[25] P. KUNKEL AND V. MEHRMANN, *The linear quadratic optimal control problem for linear descriptor systems with variable coefficients*, Math. Control Signals Systems, 10 (1997), pp. 247–264.

[26] P. KUNKEL AND V. MEHRMANN, *Differential-Algebraic Equations: Analysis and Numerical Solution,* EMS Publishing House, Zürich, Switzerland, 2006.

[27] P. KUNKEL AND V. MEHRMANN, *Optimal control for unstructured nonlinear differential-algebraic equations of arbitrary index*, Math. Control Signals Systems, 20 (2008), pp. 227–269.

[28] P. KUNKEL AND V. MEHRMANN, *Formal adjoints of linear DAE operators and their role in optimal control*, Electron. J. Linear Algebra, 22 (2011), pp. 672–693.

[29] G. A. KURINA, *Application of the method of tents to an optimal control problem for a differential equation with a singular matrix multiplying the derivative*, Differencial'nye Uravneniya, 15 (1979), pp. 600–608 (in Russian); Diff. Eq., 15 (1979), pp. 417–423 (in English).

[30] G. A. KURINA, *Feedback control for linear systems unresolved with respect to derivative*, Avtomat. i Telemekh., 1984 (1984), pp. 37–41 (in Russian); Autom. Remote Control, 45 (1984), pp. 713–717 (in English).

[31] G. A. KURINA, *On operator Riccati equation unresolved with respect to derivative*, Differencial'nye Uravneniya, 22 (1986), pp. 1826–1829 (in Russian).

[32] G. A. KURINA, *Singular perturbations of control problems with equation of state not solved for the derivative (a survey)*, Izv. Ross. Akad. Nauk. Tekhn. Kibernet., 1992 (1992), pp. 20–48 (in Russian); J. Comput. Systems Sci. Internat., 31 (1993), pp. 17–45 (in English).

[33] G. A. KURINA AND R. MÄRZ, *On linear-quadratic optimal control problems for time-varying descriptor systems*, SIAM J. Control Optim., 42 (2004), pp. 2062–2077.

[34] G. A. KURINA AND R. MÄRZ, *Feedback solutions of optimal control problems with DAE constraints*, SIAM J. Control Optim., 46 (2007), pp. 1277–1298.

[35] G. A. KURINA AND T. H. NGUYEN, *Zero order approximation for asymptotic solution of singularly perturbed linear-quadratic control problem with discontinuous coefficients*, Modelirovanie i Analiz Informacionnyh Sistem, 17 (2010), pp. 93–116 (in Russian).

[36] G. A. KURINA AND T. H. NGUYEN, *Asymptotics of feedback optimal control for singularly perturbed linear-quadratic problem with discontinuous coefficients*, Vestnik Voronezhskogo gosudarstvennogo universiteta, Seriya: Fizika. Matematika, 2010 (2010), pp. 103–117 (in Russian).

[37] G. A. KURINA AND T. H. NGUYEN, *Asymptotic solution of singularly perturbed linear-quadratic optimal control problems with discontinuous coefficients*, Zhurnal Vychislitel'noi Matematiki i Matematicheskoi Fiziki, 52 (2012), to appear (in Russian).

[38] G. A. KURINA AND E. V. SMIRNOVA, *Asymptotic solution of linear-quadratic control problem with intermediate points and small parameter in performance index,* in Proceedings of the 47th IEEE Conference on Decision and Control, Cancun, Mexico, 2008, pp. 3688–3693.

[39] E. B. LEE AND L. MARKUS, *Foundations of Optimal Control Theory*, John Wiley & Sons, New York, London, Sydney, 1967.

[40] D. LIBERZON, *Switching in Systems and Control*, Birkhäuser, Boston, 2003.

[41] R. MÄRZ, *Differential algebraic systems with properly stated leading term with MNA equations*, in Modelling, Simulation and Optimization of Integrated Circuits, K. Antreich, R. Bulirsch, A. Gilg, and P. Rentrop, eds., Internat. Ser. Numer. Math. 146, Birkhäuser, Basel, 2003, pp. 135–151.

[42] R. E. O'MALLEY, JR., *Introduction to Singular Perturbations*, Academic Press, New York, London, 1974.

[43] M. DO R. DE PINHO AND R. B. VINTER, *Necessary conditions for optimal control problems involving nonlinear differential algebraic equations*, J. Math. Anal. Appl., 212 (1997), pp. 493–516.

[44] T. ROUBIČEK AND M. VALÁŠEK, *Optimal control of causal differential-algebraic systems*, J. Math. Anal. Appl., 269 (2002), pp. 616–641.

[45] S. SAGER, *Reformulations and algorithms for the optimization of switching decisions in nonlinear optimal control*, J. Process Control, 19 (2009), pp. 1238–1247.

[46] Z. SUN AND S. S. GE, *Analysis and synthesis of switched linear control systems*, Automatica, 41 (2005), pp. 181–195.

[47] A. B. VASIL'EVA, V. F. BUTUZOV, AND L. V. KALACHEV, *The Boundary Function Method for Singular Perturbation Problems*, SIAM, Philadelphia, 1995.

[48] M. I. ZELIKIN, *Homogeneous Spaces and the Riccati Equation in Calculus of Variations*, Factorial, Moscow, 1998 (in Russian); *Control Theory and Optimization* I., Springer, Berlin, 2000 (in English).

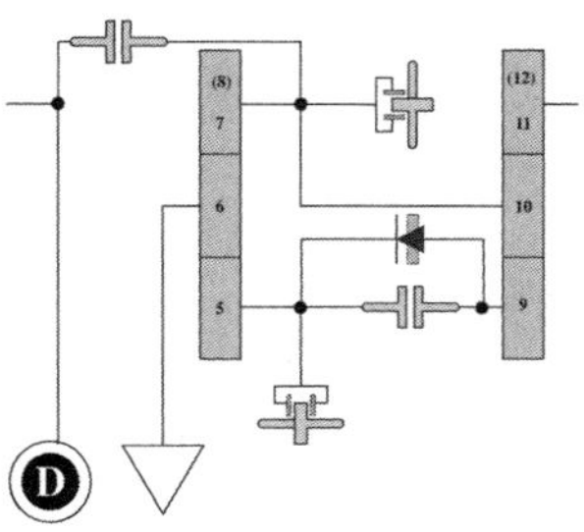

Chapter 9

Mixed-Integer DAE Optimal Control Problems: Necessary Conditions and Bounds

Matthias Gerdts and Sebastian Sager

We are interested in the optimal control of dynamic processes that can be described by DAEs and that include integer restrictions on some or all of the control functions. We assume the DAE system to be of index one. In our study we consider necessary conditions of optimality for this specific case of a hybrid system and results on lower bounds that are important in an algorithmic setting. Both results generalize previous work for the case of ODEs. Interestingly, the proofs for both analytical results are based on constructive elements to obtain integer controls from reformulations or relaxations to purely continuous control functions. These constructive elements can also be used for an efficient numerical calculation of optimal solutions. We illustrate the theoretical results by means of a mixed-integer nonlinear optimal control benchmark problem with algebraic variables.

9.1 Introduction

Technical or economical processes often involve *discrete* control variables, which are used to model finitely many decisions, discrete resources, or switching structures like gear shifts in a car or operating modes of a device. This leads to optimal control problems with nonconvex and partly discrete control set $\mathcal{U}$. More specifically, some of the control variables may still assume any real value within a given convex set with nonempty interior, which are called *continuous-valued control variables* in what follows, while other control variables which are restricted to a finite set of values are called *discrete control variables*.

An optimal control problem involving continuous-valued and discrete control variables is called *mixed-integer optimal control problem* (MIOCP). Mixed-integer optimal control is a field of increasing importance, and practical applications can be found in [11, 13, 15, 19, 34]. For a website of further benchmark problems please refer to [27] and the corresponding paper [30]. A typical example of discrete control functions and index-one DAEs are chemical processes that involve open/closed valves or on/off pumps.

An approach to solve MIOCPs is by exploiting necessary optimality conditions. A proof for index-one DAEs will be provided in Section 9.3. The proof exploits an idea of Dubovitskii and Milyutin, see [7, 8], [14, p. 95], [16, p. 148], who used a time transformation to transform the MIOCP into an equivalent optimal control problem without discrete control variables. Necessary conditions are then obtained by applying suitable local minimum principles to the transformed problem. The result are necessary conditions in terms of *global minimum principles*.

A global minimum principle for disjoint control sets and (noncontinuous) ODEs has been formulated and solved numerically via the newly developed method of *Competing Hamiltonians* in the work of Bock and Longman; see [2, 3, 20]. To our knowledge this was the first time that a global minimum principle has been applied to solve an MIOCP.

Global minimum principles for DAE optimal control problems can be found in [26] for Hessenberg DAE optimal control problems, in [5] for semi-explicit index-one DAEs, in [6] for implicit control systems, in [1] for quasi-linear DAEs, in [21] for nonlinear DAEs of arbitrary index, and in [36] for switched ODEs.

The global minimum principle can be exploited numerically using an indirect *first optimize, then discretize* approach, but a very good initial guess of the problem's switching structure is needed. Such an initial guess is often not available for practical applications.

The time transformation of Dubovitskii and Milyutin will be used in Section 9.3 as a theoretical tool to prove the global minimum principle. Interestingly, the same variable time transformation can be used numerically to solve MIOCPs, see [13, 23, 24, 34, 37], time optimal control problems, see [23], and singular optimal control problems, see [35]. A method for solving nonlinear mixed-integer programming problems based on a suitable formulation of an equivalent optimal control problem was introduced in [22].

An alternative approach based on a partial outer convexification, relaxation, and control grid adaptivity has been proposed in [28, 31, 33]. Extensions include the explicit treatment of combinatorial [32] and vanishing constraints [18]. Crucial ingredients are tight lower bounds. An extension of an important result to the DAE case is deduced in Section 9.4. Again, parts of the proof are constructive in the sense that they provide discrete control values.

We illustrate the global minimum principle for MIOCPs in DAE in Section 9.5 and close with a summary.

Throughout, $L^n_\infty(\mathcal{I})$ denotes the Banach space of essentially bounded n-vector functions on the compact interval $\mathcal{I} \subset \mathbf{R}$ and $W^n_{1,\infty}(\mathcal{I})$ denotes the Banach space of absolutely continuous n-vector functions on the compact interval $\mathcal{I}$ with essentially bounded first derivative.

9.2 Time Transformation

In this section we distinguish between continuous-valued controls u with values in the closed convex set $\mathcal{U} \subseteq \mathbf{R}^{n_u}$ with $\text{int}(\mathcal{U}) \neq \emptyset$ and discrete controls v with values in the discrete finite set

$$\mathcal{V} := \{v^1, \dots, v^{n_\omega}\}, \qquad v^i \in \mathbf{R}^{n_v},\ n_\omega \in \mathbf{N}. \tag{9.1}$$

We investigate the following problem.

Problem 9.1 (MIOCP). *Let $\mathcal{I} := [t_0, t_f]$ be a nonempty compact time interval with $t_0 < t_f$ fixed. Let*

$$\begin{aligned}
\varphi &: \mathbf{R}^{n_x} \times \mathbf{R}^{n_x} \longrightarrow \mathbf{R},\\
f &: \mathbf{R}^{n_x} \times \mathbf{R}^{n_y} \times \mathbf{R}^{n_u} \times \mathbf{R}^{n_v} \longrightarrow \mathbf{R}^{n_x},\\
g &: \mathbf{R}^{n_x} \times \mathbf{R}^{n_y} \times \mathbf{R}^{n_u} \times \mathbf{R}^{n_v} \longrightarrow \mathbf{R}^{n_y},\\
s &: \mathbf{R}^{n_x} \longrightarrow \mathbf{R}^{n_x},\\
\psi &: \mathbf{R}^{n_x} \times \mathbf{R}^{n_x} \longrightarrow \mathbf{R}^{n_\psi}
\end{aligned}$$

be sufficiently smooth functions, $\mathcal{U} \subseteq \mathbf{R}^{n_u}$ closed and convex with nonempty interior, and $\mathcal{V}$ as in (9.1).

Minimize the objective function

$$\varphi(x(t_0), x(t_f))$$

with respect to $x \in W^{n_x}_{1,\infty}(\mathcal{I})$, $y \in L^{n_y}_{\infty}(\mathcal{I})$, $u \in L^{n_u}_{\infty}(\mathcal{I})$, $v \in L^{n_v}_{\infty}(\mathcal{I})$ subject to the semi-explicit DAE

$$\begin{aligned}
\dot{x}(t) &= f(x(t), y(t), u(t), v(t)) \quad \text{a.e. in } \mathcal{I},\\
0_{\mathbf{R}^{n_y}} &= g(x(t), y(t), u(t), v(t)) \quad \text{a.e. in } \mathcal{I},
\end{aligned}$$

the state constraint

$$s(x(t)) \leq 0_{\mathbf{R}^{n_s}},$$

the boundary condition

$$\psi(x(t_0), x(t_f)) = 0_{\mathbf{R}^{n_\psi}},$$

and the set constraints

$$\begin{aligned}
u(t) &\in \mathcal{U} \quad \text{a.e. in } \mathcal{I},\\
v(t) &\in \mathcal{V} \quad \text{a.e. in } \mathcal{I}.
\end{aligned}$$

■

The variable-time transformation method is based on a discretization. For simplicity only equally spaced grids are discussed. Let the *major grid*

$$\mathbf{G}_N := \{t_i = t_0 + ih \mid i = 0, \ldots, N\}, \quad h = \frac{t_f - t_0}{N}$$

with $N \in \mathbf{N}$ intervals be given. Each major grid interval is subdivided into n_ω equally spaced subintervals, where n_ω denotes the number of values in the discrete control set $\mathcal{V}$ in (9.1). This leads to the *minor grid*

$$\mathbf{G}_{N,n_\omega} := \left\{ \tau_{i,j} = t_i + j\frac{h}{n_\omega} \mid j = 0, \ldots, n_\omega,\ i = 0, \ldots, N-1 \right\}.$$

On the minor grid define the fixed and piecewise constant function

$$v_{\mathbf{G}_{N,n_\omega}}(\tau) := v^j \quad \text{for } \tau \in [\tau_{i,j-1}, \tau_{i,j}),\ i = 0, \ldots, N-1,\ j = 1, \ldots, n_\omega. \tag{9.2}$$

Consider the time transformation

$$t(\tau) := t_0 + \int_{t_0}^{\tau} w(r)\,\mathrm{d}r, \quad \tau \in \mathcal{I},$$

with

$$\int_{t_0}^{t_\mathrm{f}} w(r)\,\mathrm{d}r = t_\mathrm{f} - t_0 \quad \text{and} \quad w(\tau) \geq 0 \quad \text{for almost every } \tau \in \mathcal{I}. \tag{9.3}$$

The inverse mapping is defined by

$$\tau(t) := \inf\{\tau \in \mathcal{I} \mid t(\tau) = t\}. \tag{9.4}$$

The time transformation controls the length of the intervals $[t(\tau_{i,j}), t(\tau_{i,j+1})]$ by proper choice of w according to

$$\int_{\tau_{i,j}}^{\tau_{i,j+1}} w(\tau)\,\mathrm{d}\tau = t(\tau_{i,j+1}) - t(\tau_{i,j}).$$

Note that the time transformation maps $\mathcal{I}$ onto itself but changes the speed of running through this interval. In particular, we have

$$\frac{\mathrm{d}t}{\mathrm{d}\tau}(\tau) = w(\tau) \quad \text{for } \tau \in \mathcal{I},$$

and the interval $[t(\tau_{i,j}), t(\tau_{i,j+1})]$ shrinks to the point $\{t(\tau_{i,j})\}$ if

$$w(\tau) = 0 \quad \text{in } [\tau_{i,j}, \tau_{i,j+1}).$$

State constraints $s(x(t)) \leq 0_{\mathbb{R}^{n_s}}$ will be evaluated at major grid points t_i, $i = 0, \ldots, N$, only, and hence we impose the additional constraints

$$\int_{t_i}^{t_{i+1}} w(\tau)\,\mathrm{d}\tau = t_{i+1} - t_i = h, \quad i = 0, \ldots, N-1, \tag{9.5}$$

which ensure that the transformed time points $t_i = t(t_i)$, $i = 1, \ldots, N$, are fixed points. Without these constraints, the time transformation tends to optimize the points $t(t_i)$ such that constraints can be fulfilled easily.

Joining the function $v_{\mathbf{G}_{N,n_\omega}}$ from (9.2) and any w satisfying the conditions (9.3) and (9.5) yields a feasible discrete control $v(t) \in \mathcal{V}$ defined by

$$v(t) := v_{\mathbf{G}_{N,n_\omega}}(\tau(t)), \qquad t \in [t_0, t_\mathrm{f}];$$

see Figure 9.1. Notice that minor intervals with $w(\tau) = 0$ do not contribute to $v(t)$.

Vice versa, every piecewise constant discrete control v on the major grid $\mathbf{G}_N$ can be described by $v_{\mathbf{G}_{N,n_\omega}}$ and some feasible w.

The preference of values given by the definition of the fixed function $v_{\mathbf{G}_{N,n_\omega}}$ on the minor grid $\mathbf{G}_{N,n_\omega}$ is arbitrary and any other order would be feasible as well. As any discrete control with finitely many jumps can be approximated arbitrarily close already on the major grid for h sufficiently small, a specific order on the minor grid is not required.

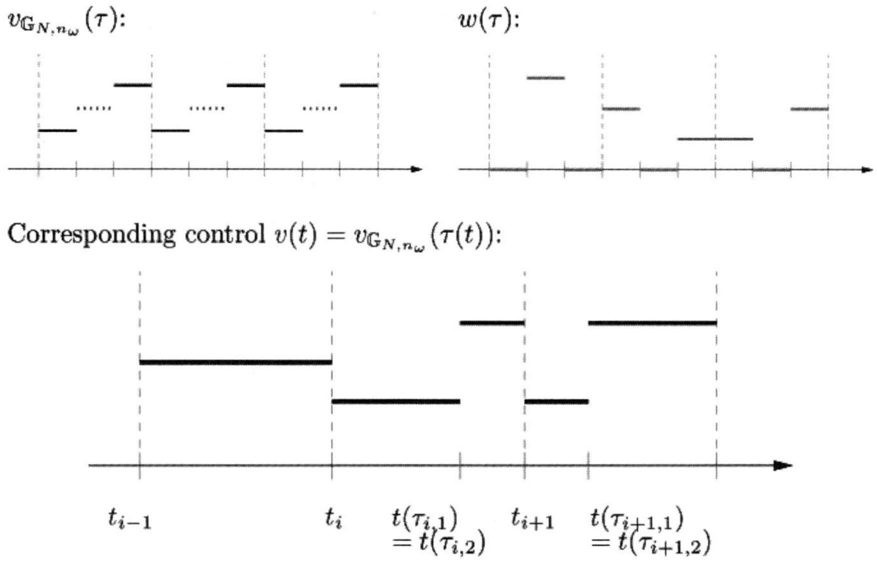

Figure 9.1. Back-transformation v (bottom) of variable time transformation for given w and fixed $v_{\mathbf{G}_{N,n_\omega}}$ (top) [13].

Summarizing, the time transformation leads to the following partly discretized optimal control problem.

Problem 9.2.

Minimize

$$\varphi(x(t_0), x(t_f))$$

with respect to $x \in W_{1,\infty}^{n_x}(\mathcal{I})$, $y \in L_\infty^{n_y}(\mathcal{I})$, $u \in L_\infty^{n_u}(\mathcal{I})$, $w \in L_\infty(\mathcal{I})$ subject to

$$\begin{aligned}
\dot{x}(\tau) &= w(\tau) f(x(\tau), y(\tau), u(\tau), v_{\mathbf{G}_{N,n_\omega}}(\tau)) && \text{a.e. in } \mathcal{I},\\
0_{\mathbb{R}^{n_y}} &= g(x(\tau), y(\tau), u(\tau), v_{\mathbf{G}_{N,n_\omega}}(\tau)) && \text{a.e. in } \mathcal{I},\\
s(x(\tau)) &\leq 0_{\mathbb{R}^{n_s}} && \text{in } \mathcal{I},\\
\psi(x(t_0), x(t_f)) &= 0_{\mathbb{R}^{n_\psi}}, &&\\
u(\tau) &\in \mathcal{U} && \text{a.e. in } \mathcal{I},\\
w &\in \mathcal{W}. &&
\end{aligned}$$

Herein, $\mathcal{W}$ is defined by

$$\mathcal{W} := \left\{ w \in L_\infty(\mathcal{I}) \;\middle|\; \begin{array}{l} w(\tau) \geq 0, \\ w \textit{ piecewise constant on } \mathbf{G}_{N,n_\omega}, \\ \displaystyle\int_{t_i}^{t_{i+1}} w(\tau)\,\mathrm{d}\tau = t_{i+1} - t_i,\ i = 0,\ldots,N \end{array} \right\}.$$

■

Problem 9.2 has only continuous-valued controls and can be solved by a direct discretization method (first discretize, then optimize). Application of the inverse time transformation

$$x(t) := \hat{x}(\tau(t)),\ y(t) := \hat{y}(\tau(t)),\ u(t) := \hat{u}(\tau(t)),\ v(t) := v_{G_{N,n_\omega}}(\tau(t)),$$

with $\tau(t)$ according to (9.4), to an optimal solution $(\hat{x}, \hat{y}, \hat{u}, \hat{w})$ of Problem 9.2 yields an approximate solution of Problem 9.1.

9.3 Necessary Conditions for Optimality

We exploit the time transformation in Section 9.2 in order to prove a global minimum principle. To this end we consider the following autonomous optimal control problem on a fixed time interval $[t_0, t_f]$ subject to an index-one DAE with general set constraints for the control.

Problem 9.3 (Optimal control problem). *Let $\mathcal{I} := [t_0, t_f]$ be a nonempty compact time interval with $t_0 < t_f$ fixed. Let*

$$\begin{aligned}
\varphi &: \mathbf{R}^{n_x} \times \mathbf{R}^{n_x} \longrightarrow \mathbf{R},\\
f_0 &: \mathbf{R}^{n_x} \times \mathbf{R}^{n_y} \times \mathbf{R}^{n_u} \longrightarrow \mathbf{R},\\
f &: \mathbf{R}^{n_x} \times \mathbf{R}^{n_y} \times \mathbf{R}^{n_u} \longrightarrow \mathbf{R}^{n_x},\\
g &: \mathbf{R}^{n_x} \times \mathbf{R}^{n_y} \times \mathbf{R}^{n_u} \longrightarrow \mathbf{R}^{n_y},\\
\psi &: \mathbf{R}^{n_x} \times \mathbf{R}^{n_x} \longrightarrow \mathbf{R}^{n_\psi}
\end{aligned}$$

be sufficiently smooth functions and $\mathcal{U} \subseteq \mathbf{R}^{n_u}$ a nonempty set.

Minimize the objective function

$$\varphi(x(t_0), x(t_f)) + \int_{t_0}^{t_f} f_0(x(t), y(t), u(t))\,dt$$

with respect to $x \in W_{1,\infty}^{n_x}(\mathcal{I})$, $y \in L_\infty^{n_y}(\mathcal{I})$, $u \in L_\infty^{n_u}(\mathcal{I})$, subject to the DAE

$$\begin{aligned}
\dot{x}(t) &= f(x(t), y(t), u(t)) \quad \text{a.e. in } \mathcal{I},\\
0_{\mathbf{R}^{n_y}} &= g(x(t), y(t), u(t)) \quad \text{a.e. in } \mathcal{I},
\end{aligned}$$

the boundary condition

$$\psi(x(t_0), x(t_f)) = 0_{\mathbf{R}^{n_\psi}},$$

and the set constraint

$$u(t) \in \mathcal{U} \quad \text{a.e. in } \mathcal{I}.$$

■

The set $\mathcal{U}$ in Problem 9.3 is supposed to be an arbitrary set. We particularly allow that $\mathcal{U}$ may only contain finitely many vectors so that Problem 9.3 contains problems with discrete controls.

In proving necessary conditions we cannot exploit a special structure of $\mathcal{U}$ as it was done in the local minimum principles in [12] and [10, p. 95] by assuming that the set was convex with nonempty interior. Hence, these necessary optimality conditions do not hold for Problem 9.3, but they are not worthless. A special time transformation similar to the one in Section 9.2 is used to transform Problem 9.3 into an equivalent problem with a nice convex control set with nonempty interior for which the local minimum principles are valid. This time transformation is due to Dubovitskii and Milyutin and the following proof techniques in the case of ODEs can be found in [16, p. 148] and [14, p. 95]. The results are extended to the DAE setting in Problem 9.3. To this end let

$$\mathcal{H}(x,y,u,\lambda_f,\lambda_g,\ell_0) := \ell_0 f_0(x,y,u) + \lambda_f^\top f(x,y,u) + \lambda_g^\top g(x,y,u)$$

denote the Hamilton function (also called Hamiltonian) for Problem 9.3, let

$$\tilde{\mathcal{H}}(x,y,u,\lambda_f,\ell_0) := \ell_0 f_0(x,y,u) + \lambda_f^\top f(x,y,u)$$

denote the reduced Hamilton function, and let $(\hat{x},\hat{y},\hat{u})$ be a solution of Problem 9.3. Moreover, let the DAE have index one.

Assumption 9.4. *Let the Jacobian matrix g'_y of g with respect to the component y be uniformly nonsingular in the solution $(\hat{x},\hat{y},\hat{u})$ and let the inverse matrix $(g'_y)^{-1}$ be essentially bounded.*

As in Section 9.2 we use the time transformation

$$t(\tau) := t_0 + \int_0^\tau w(r)\mathrm{d}r, \quad t(0) = t_0, \quad t(1) = t_\mathrm{f}, \quad w(\tau) \geq 0 \tag{9.6}$$

for $\tau \in [0,1]$. For any function $w \in L_\infty([0,1])$ satisfying (9.6) define

$$\begin{aligned}
\tilde{u}(\tau) &:= \begin{cases} \hat{u}(t(\tau)) & \text{for } \tau \in \Delta_w, \\ \text{arbitrary} & \text{for } \tau \in [0,1]\setminus\Delta_w, \end{cases} \\
\tilde{x}(\tau) &:= \hat{x}(t(\tau)), \\
\tilde{y}(\tau) &:= \begin{cases} \hat{y}(t(\tau)) & \text{for } \tau \in \Delta_w, \\ \text{suitable} & \text{for } \tau \in [0,1]\setminus\Delta_w, \end{cases}
\end{aligned}$$

where

$$\Delta_w := \{\tau \in [0,1] \mid w(\tau) > 0\}.$$

Suitable values for $\tilde{y}$ on $[0,1]\setminus\Delta_w$ will be provided later.

The functions $\tilde{x}$, $\tilde{y}$, $\tilde{u}$ are feasible for the following auxiliary DAE optimal control problem in which w is considered a *control* and $\tilde{u}$ a *fixed* function.

Problem 9.5 (Auxiliary DAE optimal control problem).

Minimize

$$\varphi(x(0),x(1)) + \int_0^1 w(\tau) f_0(x(\tau),y(\tau),\tilde{u}(\tau))\mathrm{d}\tau$$

with respect to $x \in W^{n_x}_{1,\infty}([0,1])$, $y \in L^{n_y}_{\infty}([0,1])$, $t \in W_{1,\infty}([0,1])$, $w \in L_\infty([0,1])$ subject to the constraints

$$\begin{aligned}
\dot{x}(\tau) &= w(\tau) f(x(\tau), y(\tau), \tilde{u}(\tau)) && \text{a.e. in } [0,1],\\
0_{\mathbf{R}^{n_y}} &= g(x(\tau), y(\tau), \tilde{u}(\tau)) && \text{a.e. in } [0,1],\\
\dot{t}(\tau) &= w(\tau) && \text{a.e. in } [0,1],\\
0_{\mathbf{R}^{n_\psi}} &= \psi(x(0), x(1)),\\
t(0) &= t_0,\\
t(1) &= t_{\mathrm{f}},\\
w(\tau) &\geq 0 && \text{a.e. in } [0,1]!
\end{aligned}$$

■

Note that the control w in Problem 9.5 is restricted only by the control constraint $w(\tau) \geq 0$. Consequently, the necessary optimality conditions in [10] hold for Problem 9.5. A formal proof similar to [16, pp. 149-156] shows that $\tilde{x}$ and $\tilde{y}$ are actually optimal for Problem 9.5 for any feasible control w.

Remark 9.6.

(a) *Note: If $w \equiv 0$ on some interval, then on this interval*

$$\begin{aligned}
\dot{x}(\tau) &\equiv 0 &&\Longrightarrow && x(\tau) \equiv \text{const},\\
\dot{t}(\tau) &\equiv 0 &&\Longrightarrow && t(\tau) \equiv \text{const}.
\end{aligned}$$

(b) *The necessary conditions in* [10] *require the functions*

$$\begin{aligned}
\tilde{f}_0(\tau, x, y, w) &:= w f_0(x, y, \tilde{u}(\tau)),\\
\tilde{f}(\tau, x, y, w) &:= w f(x, y, \tilde{u}(\tau)),\\
\tilde{g}(\tau, x, y) &:= g(x, y, \tilde{u}(\tau))
\end{aligned}$$

to be continuous with respect to the component τ and continuously differentiable with respect to x, y, and w. This assumption is not satisfied in general for Problem 9.5 *as $\tilde{u}$ is not continuous in general, but it can be relaxed appropriately for measurable τ functions.*

First order necessary optimality conditions for Problem 9.5 with the augmented Hamilton function

$$\begin{aligned}
&\tilde{\mathcal{H}}(\tau, x, y, t, w, \lambda_f, \lambda_g, \lambda_t, \eta, \ell_0)\\
&\quad := w\left(\ell_0 f_0(x, y, \tilde{u}(\tau)) + \lambda_f^\top f(x, y, \tilde{u}(\tau)) + \lambda_t - \eta\right) + \lambda_g^\top g(x, y, \tilde{u}(\tau))\\
&\quad = w\left(\mathcal{H}(x, y, \tilde{u}(\tau), \lambda_f, \ell_0) + \lambda_t - \eta\right) + \lambda_g^\top g(x, y, \tilde{u}(\tau))
\end{aligned}$$

read as follows: There exist multipliers $\tilde{\ell}_0 \in \mathbf{R}$, $\tilde{\lambda}_f \in W^{n_x}_{1,\infty}([0,1])$, $\tilde{\lambda}_g \in L^{n_y}_{\infty}([0,1])$, $\tilde{\lambda}_t \in W_{1,\infty}([0,1])$, $\tilde{\eta} \in L_\infty([0,1])$, and $\tilde{\sigma} \in \mathbf{R}^{n_\psi}$, not all zero, with the following:

(a) $\tilde{\ell}_0 \geq 0$.

(b) In $[0,1]$ we have the adjoint equation

$$\frac{d}{d\tau}\tilde{\lambda}_f(\tau) = -w(\tau)\tilde{\mathcal{H}}'_x(\tilde{x}(\tau),\tilde{y}(\tau),\tilde{u}(\tau),\tilde{\lambda}_f(\tau),\tilde{\ell}_0)^\top - g'_x(\tilde{x}(\tau),\tilde{y}(\tau),\tilde{u}(\tau))^\top\tilde{\lambda}_g(\tau),$$

$$0_{\mathbb{R}^{n_y}} = w(\tau)\tilde{\mathcal{H}}'_y(\tilde{x}(\tau),\tilde{y}(\tau),\tilde{u}(\tau),\tilde{\lambda}_f(\tau),\tilde{\ell}_0)^\top + g'_y(\tilde{x}(\tau),\tilde{y}(\tau),\tilde{u}(\tau))^\top\tilde{\lambda}_g(\tau),$$

$$\frac{d}{d\tau}\tilde{\lambda}_t(\tau) = 0.$$

In particular, $\tilde{\lambda}_t$ is constant.

(c) Transversality conditions:

$$\tilde{\lambda}_f(0)^\top = -\left(\tilde{\ell}_0\varphi'_{x_0} + \tilde{\sigma}^\top\psi'_{x_0}\right), \quad \tilde{\lambda}_f(1)^\top = \tilde{\ell}_0\varphi'_{x_f} + \tilde{\sigma}^\top\psi'_{x_f}.$$

(d) Almost everywhere in $[0,1]$ we have

$$0 = \tilde{\mathcal{H}}(\tilde{x}(\tau),\tilde{y}(\tau),\tilde{u}(\tau),\tilde{\lambda}_f(\tau),\tilde{\ell}_0) + \tilde{\lambda}_t(\tau) - \tilde{\eta}(\tau).$$

Owing to the complementarity condition in (e) we thus have

$$\tilde{\mathcal{H}}(\tilde{x}(\tau),\tilde{y}(\tau),\tilde{u}(\tau),\tilde{\lambda}_f(\tau),\tilde{\ell}_0) + \tilde{\lambda}_t(\tau)\begin{cases} = 0 & \text{if } \tau \in \Delta_w, \\ \geq 0 & \text{if } \tau \notin \Delta_w. \end{cases}$$

(e) Almost everywhere in $[0,1]$ we have that

$$\tilde{\eta}(\tau)w(\tau) = 0 \quad \text{and} \quad \tilde{\eta}(t) \geq 0.$$

Using the inverse time transformation defined in (9.4) we may define $\ell_0 := \tilde{\ell}_0$, $\sigma := \tilde{\sigma}$, $\lambda_f(t) := \tilde{\lambda}_f(\tau(t))$, $\lambda_t(t) := \tilde{\lambda}_t(\tau(t))$, and in addition

$$\lambda_g(t)^\top := -\tilde{\mathcal{H}}'_y(\hat{x}(t),\hat{y}(t),\hat{u}(t),\lambda_f(t),\ell_0)\left(g'_y(\hat{x}(t),\hat{y}(t),\hat{u}(t))\right)^{-1}.$$

Then, for almost every $\tau \in \Delta_w$ we see that

$$\lambda_f(t(\tau)) = \tilde{\lambda}_f(\tau), \quad \lambda_t(t(\tau)) = \tilde{\lambda}_t(\tau),$$

and λ_f and λ_g satisfy the adjoint equation

$$\dot{\lambda}_f(t) = -\mathcal{H}'_x(\hat{x}(t),\hat{y}(t),\hat{u}(t),\lambda_f(t),\lambda_g(t),\ell_0)^\top,$$

$$0_{\mathbb{R}^{n_y}} = \mathcal{H}'_y(\hat{x}(t),\hat{y}(t),\hat{u}(t),\lambda_f(t),\lambda_g(t),\ell_0)^\top,$$

and the transversality conditions

$$\lambda_f(t_0)^\top = -\left(\ell_0\varphi'_{x_0} + \sigma^\top\psi'_{x_0}\right), \quad \lambda_f(t_f)^\top = \ell_0\varphi'_{x_f} + \sigma^\top\psi'_{x_f}.$$

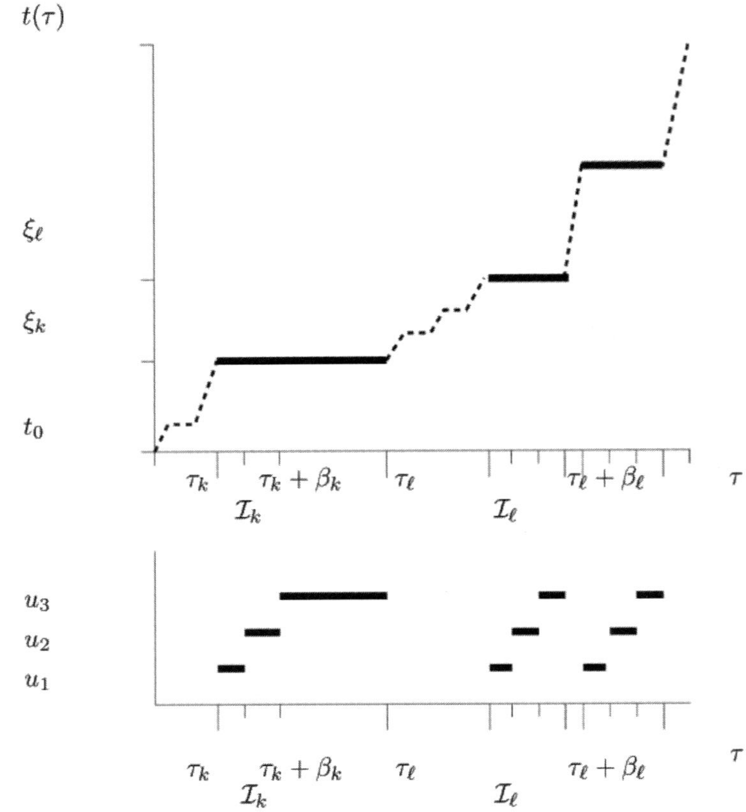

Figure 9.2. Construction principle of the new w.

The above conditions (a)–(e) hold for every w. Now, we will choose w in a special way in order to exploit this degree of freedom. The following construction follows [16, p. 157], compare Figure 9.2, and was exploited numerically in Section 9.2.

Let $w(\tau)$ vanish on the intervals $\mathcal{I}_k := (\tau_k, \tau_k + \beta_k]$, $k = 1, 2, \ldots$, which are to be constructed such that the image of $\bigcup_k \mathcal{I}_k$ under the mapping $\tau \mapsto t(\tau)$ is dense in $\mathcal{I}$. To this end let $\{\xi_1, \xi_2, \ldots\}$ be a countable dense subset of $\mathcal{I}$. Choose $\beta_k > 0$ with $\sum_k \beta_k = \frac{1}{2}$ and let

$$\tau_k := \frac{\xi_k - t_0}{2(t_f - t_0)} + \sum_{j:\xi_j < \xi_k} \beta_j.$$

Then, the intervals $\mathcal{I}_k = (\tau_k, \tau_k + \beta_k]$ are pairwise disjoint. Define

$$w(\tau) := \begin{cases} 0 & \text{if } \tau \in \bigcup_k \mathcal{I}_k, \\ 2(t_f - t_0) & \text{if } \tau \notin \bigcup_k \mathcal{I}_k. \end{cases}$$

We will show that $t(\tau) = \xi_k$ for any $\tau \in \mathcal{I}_k$. As $\{\xi_k\}$ was chosen to be dense in $\mathcal{I}$, so is the image of $\bigcup_k \mathcal{I}_k$ under the mapping $\tau \mapsto t(\tau)$.

We note that $\tau_j < \tau_k$ if and only if $\xi_j < \xi_k$ and $t(\tau) = t(\tau_k)$ for all $\tau \in \mathcal{I}_k$. For $\tau \in \mathcal{I}_k$ we find

$$\begin{aligned} t(\tau) &= t_0 + \int_0^\tau w(\xi) d\xi \\ &= t_0 + 2(t_\mathrm{f} - t_0) \left(\tau_k - \sum_{j:\tau_j < \tau_k} \beta_j \right) \\ &= t_0 + 2(t_\mathrm{f} - t_0) \left(\tau_k - \sum_{j:\xi_j < \xi_k} \beta_j \right) \\ &= t_0 + (t_\mathrm{f} - t_0) \frac{\xi_k - t_0}{t_\mathrm{f} - t_0} = \xi_k. \end{aligned}$$

Now let

(i) $\mathcal{I}_k = \bigcup_j \mathcal{I}_{kj}$, where $\mathcal{I}_{kj}$ are nonempty intervals that are closed from the right;

(ii) $\{u_1, u_2, \ldots\}$ be a countable dense subset of $\mathcal{U}$;

(iii) $\tilde{u}(\tau) := u_j$ if $\tau \in \mathcal{I}_{kj}$;

(iv) $\tilde{y}(\tau) := y_j$ with $0_{\mathbf{R}^{n_y}} = g(\tilde{x}(\tau), y_j, u_j)$ if $\tau \in \mathcal{I}_{kj}$.

According to (d) we have for almost every $\tau \in \bigcup_k \mathcal{I}_k$ the inequality

$$\tilde{\mathcal{H}}(\tilde{x}(\tau), \tilde{y}(\tau), \tilde{u}(\tau), \tilde{\lambda}_f(\tau), \tilde{\ell}_0) + \tilde{\lambda}_t(\tau) \geq 0.$$

As every interval $\mathcal{I}_{kj}$ has a positive measure, there exists $\tau \in \mathcal{I}_{kj}$ with $t(\tau) = \xi_k$ such that

$$\begin{aligned} &\tilde{\mathcal{H}}(\tilde{x}(\tau), \tilde{y}(\tau), \tilde{u}(\tau), \tilde{\lambda}_f(\tau), \tilde{\ell}_0) + \tilde{\lambda}_t(\tau) \\ &\quad = \tilde{\mathcal{H}}(\hat{x}(\xi_k), y_j, u_j, \lambda_f(\xi_k), \ell_0) + \lambda_t(\xi_k) \geq 0. \end{aligned}$$

Since the set $\{\xi_1, \xi_2, \ldots\}$ is dense in $\mathcal{I}$, $\{u_1, u_2, \ldots\}$ is dense in $\mathcal{U}$, and

$$h(t, y, u) := \tilde{\mathcal{H}}(\hat{x}(t), y, u, \lambda_f(t), \ell_0)$$

is continuous, it follows for almost all $t \in \mathcal{I}$ that

$$\tilde{\mathcal{H}}(\hat{x}(t), y, u, \lambda_f(t), \ell_0) + \lambda_t(t) \geq 0$$

for all

$$(u, y) \in M(\hat{x}(t)) := \{(u, y) \in \mathcal{U} \times \mathbf{R}^{n_y} \mid 0_{\mathbf{R}^{n_y}} = g(\hat{x}(t), y, u)\}.$$

Note that $\mathcal{H} \equiv \tilde{\mathcal{H}}$ whenever $(u, y) \in M(\hat{x}(t))$.

On the other hand, according to (d) for almost every $\tau \in \Delta_w$ and thus for almost every $t \in \mathcal{I}$ we have

$$\tilde{\mathcal{H}}(\hat{x}(t), \hat{y}(t), \hat{u}(t), \lambda_f(t), \ell_0) + \lambda_t(t) = 0.$$

Putting both relations together yields in the minimality of the reduced Hamilton function for almost every $t \in \mathcal{I}$

$$\begin{aligned}&\tilde{\mathcal{H}}(\hat{x}(t),\hat{y}(t),\hat{u}(t),\lambda_f(t),\ell_0)\\&\quad\leq \tilde{\mathcal{H}}(\hat{x}(t),y,u,\lambda_f(t),\ell_0)\quad \text{for all } (u,y)\in M(\hat{x}(t)).\end{aligned}$$

Moreover, since $\hat{u}$ is essentially bounded and $h(t,y,u)$ is continuous, it follows that

$$\tilde{\mathcal{H}}(\hat{x}(t),\hat{y}(t),\hat{u}(t),\lambda_f(t),\ell_0)+\lambda_t(t)\equiv 0$$

almost everywhere. Since λ_t is continuous and constant according to (b), so is $\tilde{\mathcal{H}}$ as a function of time. Exploiting $\tilde{\mathcal{H}} \equiv \mathcal{H}$ for every $(u,y)\in M(\hat{x}(t))$ we have thus proved the following global minimum principle.

Theorem 9.7 (Global minimum principle). *Let the following assumptions be fulfilled for the optimal control problem* 9.3*:*

(i) *Let the functions φ, f_0, f, g, ψ be continuous with respect to all arguments and continuously differentiable with respect to x and y.*

(ii) *Let $(\hat{x},\hat{y},\hat{u})$ be a strong local minimum of the optimal control problem* 9.3.

(iii) *Let Assumption* 9.4 *be valid.*

Then there exist multipliers $\ell_0 \in \mathbf{R}$, $\lambda_f \in W^{n_x}_{1,\infty}(\mathcal{I})$, $\lambda_g \in L^{n_y}_{\infty}(\mathcal{I})$, $\sigma \in \mathbf{R}^{n_\psi}$ such that the following conditions are satisfied:

(a) $\ell_0 \geq 0$, $(\ell_0,\sigma,\lambda_f,\lambda_g) \neq 0_{\mathbf{R}^{1+n_\psi+n_x+n_y}}$.

(b) Adjoint equations*: Almost everywhere in $\mathcal{I}$ such that*

$$\begin{aligned}\dot{\lambda}_f(t) &= -\mathcal{H}'_x(\hat{x}(t),\hat{y}(t),\hat{u}(t),\lambda_f(t),\lambda_g(t),\ell_0)^\top,\\ 0_{\mathbf{R}^{n_y}} &= \mathcal{H}'_y(\hat{x}(t),\hat{y}(t),\hat{u}(t),\lambda_f(t),\lambda_g(t),\ell_0)^\top.\end{aligned}$$

(c) Transversality conditions*:*

$$\begin{aligned}\lambda_f(t_0)^\top &= -\Big(\ell_0\varphi'_{x_0}(\hat{x}(t_0),\hat{x}(t_f))+\sigma^\top\psi'_{x_0}(\hat{x}(t_0),\hat{x}(t_f))\Big),\\ \lambda_f(t_f)^\top &= \ell_0\varphi'_{x_f}(\hat{x}(t_0),\hat{x}(t_f))+\sigma^\top\psi'_{x_f}(\hat{x}(t_0),\hat{x}(t_f)).\end{aligned}$$

(d) Optimality condition*: Almost everywhere in $\mathcal{I}$ such that*

$$\mathcal{H}(\hat{x}(t),\hat{y}(t),\hat{u}(t),\lambda_f(t),\lambda_g(t),\ell_0)\leq \mathcal{H}(\hat{x}(t),y,u,\lambda_f(t),\lambda_g(t),\ell_0)$$

for all $(u,y)\in M(\hat{x}(t))$, where

$$M(x) := \{(u,y)\in \mathcal{U}\times\mathbf{R}^{n_y} \mid g(x,y,u)=0_{\mathbf{R}^{n_y}}\}.$$

(e) *The Hamilton function is constant with respect to time:*

$$\mathcal{H}(\hat{x}(t),\hat{y}(t),\hat{u}(t),\lambda_f(t),\lambda_g(t),\ell_0)\equiv \text{const.}$$

Herein, $(\hat{x}, \hat{y}, \hat{u})$ is called *strong local minimum* if $(\hat{x}, \hat{y}, \hat{u})$ minimizes the objective function among all feasible functions with $\|x - \hat{x}\|_\infty < \varepsilon$ for some $\varepsilon > 0$.

Example 9.8 (See [26, p. 620]). Consider the following optimal control problem:
Minimize

$$\int_0^1 x(t)^2 + \alpha(y(t) - u(t))^2 \mathrm{d}t$$

subject to the constraints

$$\begin{aligned} \dot{x}(t) &= y(t) - u(t), \quad x(0) = 0, \\ 0 &= y(t) - u(t), \\ u(t) &\in \mathcal{U} := [-1, 1]. \end{aligned}$$

Apparently, *every* feasible control is optimal, as it yields the optimal objective function value 0. The Hamilton function is

$$\mathcal{H}(x, y, u, \lambda_f, \lambda_g, \ell_0) = \ell_0(x^2 + \alpha(y - u)^2) + \lambda_f(y - u) + \lambda_g(y - u).$$

Minimization of $\mathcal{H}$ with respect to

$$(u, y) \in M = \{(u, y) \in \mathcal{U} \times \mathbf{R} \mid y - u = 0\}$$

yields that every u satisfies the global minimum principle.

One should note that the simultaneous coupling between u and y by means of the set M is important in the global minimum principle. A wrong condition would be obtained, if the Hamilton function was first minimized with respect to u (assuming y to be fixed) and the consistent algebraic variable y corresponding to the minimizing u was determined afterwards. For instance consider the case $\alpha = -1$ and $\hat{u} \equiv 0$. The algebraic equation, the adjoint equation, and the transversality condition yield

$$\dot{\lambda}_f(t) = -2\ell_0 \hat{x}(t) = 0, \quad \lambda_f(1) = 0 \quad \Longrightarrow \quad \lambda_f(t) \equiv 0.$$

Moreover, $g(\hat{x}(t), \hat{y}(t), \hat{u}(t)) \equiv 0$ and hence assuming $\ell_0 = 1$,

$$\mathcal{H}(\hat{x}, \hat{y}, \hat{u}, \lambda_f, \lambda_g, \ell_0) = \alpha(\hat{y} - \hat{u})^2 = -(\hat{y} - \hat{u})^2.$$

Minimizing $\mathcal{H}$ with respect to $u \in [-1, 1]$ yields either $\hat{u} = +1$ or $\hat{u} = -1$ depending on $\hat{y} \in [-1, 1]$. This contradicts $\hat{u} \equiv 0$. ∎

The global minimum principle allows one to prove additional properties of the Hamilton function in the case of a free final time. In this case, the Hamilton function vanishes almost everywhere.

Theorem 9.9. *Let the assumptions of Theorem 9.7 hold and let the final time in Problem 9.3 be free. Then, $\mathcal{H}$ vanishes almost everywhere and*

$$\mathcal{H}(\hat{x}(t_f), \hat{y}(t_f), \hat{u}(t_f), \lambda_f(t_f), \lambda_g(t_f), \ell_0) = 0.$$

Proof. We use standard transformation techniques to transform the problem to an equivalent problem on a fixed time interval:

Minimize

$$\varphi(\bar{x}(0),\bar{x}(1)) + \int_0^1 (t_f(\tau) - t_0) f_0(\bar{x}(\tau),\bar{y}(\tau),\bar{u}(\tau))\,\mathrm{d}\tau$$

subject to the constraints

$$\begin{aligned}
\frac{d}{d\tau}\bar{x}(\tau) &= (t_f(\tau) - t_0) f(\bar{x}(\tau),\bar{y}(\tau),\bar{u}(\tau)) && \text{a.e. in } [0,1],\\
0_{\mathbf{R}^{n_y}} &= (t_f(\tau) - t_0) g(\bar{x}(\tau),\bar{y}(\tau),\bar{u}(\tau)) && \text{a.e. in } [0,1],\\
\frac{d}{d\tau} t_f(\tau) &= 0 && \text{in } [0,1],\\
\psi(\bar{x}(0),\bar{x}(1)) &= 0_{n_\psi},\\
\bar{u}(\tau) &\in \mathcal{U} && \text{a.e. in } [0,1].
\end{aligned}$$

The Hamilton function for the transformed problem reads as

$$\bar{\mathcal{H}}(\bar{x},\bar{y},t_\mathrm{f},\bar{u},\bar{\lambda}_f,\bar{\lambda}_g,\bar{\lambda}_t,\bar{\ell}_0) = (t_\mathrm{f} - t_0)\mathcal{H}(\bar{x},\bar{y},\bar{u},\bar{\lambda}_f,\bar{\lambda}_g,\bar{\ell}_0),$$

where $\mathcal{H}$ denotes the Hamilton function of the original problem. Theorem 9.7 yields the following adjoint equation and transversality conditions for the adjoint $\bar{\lambda}_t$:

$$\frac{d}{d\tau}\bar{\lambda}_t(\tau) = -\bar{\mathcal{H}}'_{t_\mathrm{f}}[\tau] = -\mathcal{H}[\tau], \quad \bar{\lambda}_t(0) = \bar{\lambda}_t(1) = 0. \tag{9.7}$$

According to part (e) of Theorem 9.7 the Hamilton function $\bar{\mathcal{H}}$ is constant almost everywhere and thus $\mathcal{H}$ is constant almost everywhere as well since t_f is constant. From (9.7) it follows that $\mathcal{H}[\tau] = 0$ almost everywhere and particularly after back-transformation $\mathcal{H}(\hat{x}(t_\mathrm{f}),\hat{y}(t_\mathrm{f}),\hat{u}(t_\mathrm{f}),\lambda_f(t_\mathrm{f}),\lambda_g(t_\mathrm{f}),\ell_0) = 0$. □

9.4 On the Relation between Relaxed and Integer Solutions

The necessary conditions of optimality that have been derived in the previous section are important, as they can be used from an algorithmical point of view and to gain analytical insight into solution structures. In practice, however, often *first discretize, then optimize* methods are used; compare [29] for an overview. Whenever integer programming is applied, lower bounds obtained from relaxations are crucial. In this context a *partial outer convexification* has been proposed for MIOCPs in ODEs [28]. We are going to extend this to the case of MIOCP in DAE as formulated in Problem 9.1.

Problem 9.10 (MIOCP after outer convexification). *Let $\mathcal{I},\varphi,f,g,s,\psi,\mathcal{U},\mathcal{V}$ be defined as in Problem* 9.1.

Let the matrix $g'_y(x,y,u,v^i)$ be uniformly nonsingular for all arguments and let the inverse matrix $(g'_y(x,y,u,v^i))^{-1}$ be essentially bounded for all $i = 1,\ldots,n_\omega$.

Minimize the objective function

$$\varphi(x(t_0),x(t_\mathrm{f}))$$

with respect to $x \in W_{1,\infty}^{n_x}(\mathcal{I})$, $u \in L_{\infty}^{n_u}(\mathcal{I})$, $\omega \in L_{\infty}^{n_\omega}(\mathcal{I})$ subject to the system of ODEs

$$\dot{x}(t) = \sum_{i=1}^{n_\omega} \omega_i(t)\, f\left(x(t), \theta^{i,t}(x(t),u(t)), u(t), v^i\right) \quad \text{a.e. in } \mathcal{I},$$

with continuous mappings $\theta^{i,t} : \mathbf{R}^{n_x} \times \mathbf{R}^{n_u} \longrightarrow \mathbf{R}^{n_y}$ for $i = 1,\dots,n_\omega$ and $t \in \mathcal{I}$ a.e., to the state constraint

$$s(x(t)) \leq 0_{\mathbf{R}^{n_s}},$$

the boundary conditions

$$\psi(x(t_0), x(t_f)) = 0_{\mathbf{R}^{n_\psi}},$$

the set constraints

$$\begin{aligned} u(t) &\in \mathcal{U} && \text{a.e. in } \mathcal{I}, \\ \omega(t) &\in \{0,1\}^{n_\omega} && \text{a.e. in } \mathcal{I}, \end{aligned}$$

and the special ordered set type 1 condition

$$\sum_{i=1}^{n_\omega} \omega_i(t) = 1 \quad \text{a.e. in } \mathcal{I}. \tag{9.8}$$

■

Theorem 9.11. *Let $(x^*, y^*, u^*, v^*)(\cdot)$ be an optimal solution of Problem* 9.1. *Define $\omega^* : \mathcal{I} \longrightarrow \{0,1\}^{n_\omega}$ as*

$$\omega_i^*(t) := \begin{cases} 1 & \text{if } v^*(t) = v^i, \\ 0 & \text{else.} \end{cases}$$

Then continuous mappings $\theta^{i,t} : \mathbf{R}^{n_x} \times \mathbf{R}^{n_u} \longrightarrow \mathbf{R}^{n_y}$ exist such that

$$y^*(t) = \sum_{i=1}^{n_\omega} \omega_i^*(t)\, \theta^{i,t}\left(x^*(t), u^*(t)\right) \quad \text{a.e. in } \mathcal{I} \tag{9.9}$$

and $(x^*, u^*, \omega^*)(\cdot)$ is an optimal solution of Problem 9.10.

If, conversely, $(x^*, u^*, \omega^*)(\cdot)$ is an optimal solution of Problem 9.10 *for mappings $\theta^{i,t}(\cdot)$ that fulfill*

$$0_{\mathbf{R}^{n_y}} = \sum_{i=1}^{n_\omega} \omega_i^*(t)\, g\left(x^*(t), \theta^{i,t}\left(x^*(t), u^*(t)\right), u^*(t), v^i\right) \tag{9.10}$$

a.e. in $\mathcal{I}$ and if we define

$$v^*(t) = v^i \text{ if and only if } \omega_i^*(t) = 1, \tag{9.11}$$

then also $(x^*, y^*, u^*, v^*)(\cdot)$ with $y^*(\cdot)$ given by (9.9) *is an optimal solution of Problem* 9.1.

Proof. Assume (x^*, y^*, u^*, v^*) to be a feasible solution of Problem 9.1. By construction we have

$$v^*(t) = \sum_{i=1}^{n_\omega} \omega_i^*(t)\, v^i \tag{9.12}$$

for a.e. t in $\mathcal{I}$. As (x^*, y^*, u^*, v^*) is feasible, in particular

$$0_{\mathbf{R}^{n_y}} = g(x^*(t), y^*(t), u^*(t), v^*(t))$$

is true. Because of the index-one assumption the matrices

$$\frac{\partial g(x^*(t), y^*(t), u^*(t), v^i)}{\partial y}$$

are invertible for all $i = 1, \ldots, n_\omega$. Hence the implicit function theorem states for all $i = 1, \ldots, n_\omega$ and $t \in \mathcal{I}$ a.e. the existence of open sets $\mathcal{S}^{1,i,t} \subset \mathbf{R}^{n_x} \times \mathbf{R}^{n_u}$ and $\mathcal{S}^{2,i,t} \subset \mathbf{R}^{n_y}$ with $s^{1,i,*}(t) := (x^*, u^*)(t) \in \mathcal{S}^{1,i,t}$ and $s^{2,i,*}(t) := y^*(t) \in \mathcal{S}^{2,i,t}$ and continuous mappings $\theta^{i,t} : \mathcal{S}^{1,i,t} \longrightarrow \mathcal{S}^{2,i,t}$ exist such that

$$0_{\mathbf{R}^{n_y}} = g(s^1, \theta^{i,t}(s^1), v^i) \tag{9.13}$$

for all $(s^1, s^2) \in \mathcal{S}^{1,i,t} \times \mathcal{S}^{2,i,t}$ with a slight abuse of notation in the order of the arguments. In particular, we have (9.10) for $s^1 = (x^*, u^*)(t)$.

Hence it is possible to pointwise replace $y^*(t)$ in Problem 9.1 by

$$y^*(t) = \sum_{i=1}^{n_\omega} \omega_i^*(t)\, \theta^{i,t}\left(x^*(t), u^*(t)\right). \tag{9.14}$$

Substituting (9.12) and (9.14) into the right-hand side function $f(\cdot)$ yields equivalence, i.e.,

$$\begin{aligned}
\dot{x}(t) &= f(x^*(t), y^*(t), u^*(t), v^*(t)) \\
&= f\left(x^*(t), \sum_{i=1}^{n_\omega} \omega_i^*(t)\, \theta^{i,t}\left(x^*(t), u^*(t)\right), u^*(t), \sum_{i=1}^{n_\omega} \omega_i^*(t)\, v^i\right) \\
&= \sum_{i=1}^{n_\omega} \omega_i^*(t)\, f\left(x^*(t), \theta^{i,t}\left(x^*(t), u^*(t)\right), u^*(t), v^i\right),
\end{aligned}$$

making use of $\omega_i^* \in \{0,1\}$ and $\sum_{i=1}^{n_\omega} \omega_i^*(t) = 1$. Hence (x^*, u^*, ω^*) is also a feasible solution of Problem 9.10. This solution is also optimal, because any better trajectory (x^*, u^*, ω^*) can be mapped via (9.11) and (9.14) to a feasible trajectory of Problem 9.1 with the same objective function value, contradicting the original optimality assumption.

Let us now assume (x^*, u^*, ω^*) to be an optimal solution of Problem 9.10 for mappings $\theta^{i,t}(\cdot)$ that fulfill (9.10) a.e. in $\mathcal{I}$. We define $v^*(\cdot)$ via (9.11) and $y^*(\cdot)$ via (9.14) and again have an equivalence of the right-hand sides of the differential equations, which results in identical feasibility and optimality. □

Note that the proof is similar to the one in [28] for the ODE case. The main extension is the equivalence of the ODE formulation to the DAE in Problem 9.1. It follows from the index-one assumption which allows us to apply the implicit function theorem.

Theorem 9.11 allows us to transfer formally a MIOCP in DAE in which the discrete controls $v(\cdot)$ enter in an arbitrary manner to a MIOCP in ODE in which the discrete controls $\omega(\cdot)$ enter linearly. This is important, as it allows to apply the integer gap theorem from [31] to Problem 9.10. It states that for any given tolerance ϵ a control discretization grid size can be determined, such that the difference of norms of the optimal trajectory of a relaxed ($\omega_i(t) \in [0,1]$) and an integer-valued ($\omega_i(t) \in \{0,1\}$) trajectory is smaller than ϵ. This property carries over by continuity arguments to the objective function and state constraints.

Corollary 9.12. *If $f(\cdot)$ and $g(\cdot)$ are sufficiently smooth, the optimal objective function value of Problem* 9.1 *is given by the optimal objective function value of Problem* 9.10, *where the requirements $\omega_i(t) \in \{0,1\}$ have been relaxed to $\omega_i(t) \in [0,1]$.*

The proof is a simple combination of Theorem 9.11 and [31, Corollary 8], in which also precise assumptions on the smoothness of $f(\cdot)$ are formulated.

It also allows to apply the Sum Up Rounding strategy to calculate integer controls from relaxed controls in linear time; compare also [31]. Interestingly, also the Sum Up Rounding strategy is a constructive element of a theoretical proof, similar to the time transformation in the previous sections.

From a practical point of view, however, it is not always desirable or even possible to explicitly calculate the mappings $\theta^{i,t}(\cdot)$. The question of how the algebraic equations can be altered to maintain a DAE formulation with a property similar to Corollary 9.12 is an open question, like the extension to DAE systems for which the index-one assumption does not hold.

9.5 Numerical Benchmark

We will illustrate the results from the previous sections with a MIOC benchmark example from the literature. Our focus is not on efficient numerical methods for MIOCPs, which have been described, e.g., in [13, 33]. Rather we give an example of a BVP that is deduced directly from the necessary conditions of optimality and follow a *first optimize, then discretize* approach.

We choose a control problem motivated by F-8 aircraft control. It is based on ODEs, but we reformulate it with the help of two artificial algebraic variables. This is not necessarily helpful from a computational point of view, but it will allow us to compare the results to published results for this benchmark problem [30], and we can be certain that for this test problem the index-one assumption (Assumption 9.4) is always valid.

The differential equations were introduced by Garrard and Jordan [9]. The differential states consist of $x_1(\cdot)$ as the angle of attack in radians, $x_2(\cdot)$ as the pitch angle, and $x_3(\cdot)$ as the pitch rate in rad/s. The F-8 aircraft control problem was introduced by Kaya and Noakes [17] and aims at controlling an aircraft in a time-optimal way from an initial state to a terminal state. The only control function $v(\cdot)$ is the tail deflection angle in radians which may attain only two different values. For $t \in [0, t_f]$ almost everywhere the MIOCP is given as follows.

Problem 9.13 (F-8 aircraft control with algebraic variables).

Minimize $\int_0^{t_f} 1 \, dt$

with respect to $x \in W^3_{1,\infty}(\mathcal{I})$, $y \in L^2_\infty(\mathcal{I})$, $v \in L^1_\infty(\mathcal{I})$ subject to

$$\dot{x}_1(t) = y_1(t)x_1(t) + x_3(t) - 0.019x_2(t)^2 - 0.215v(t) + 0.63v(t)^3, \tag{9.15}$$

$$\dot{x}_2(t) = x_3(t), \tag{9.16}$$

$$\begin{aligned} \dot{x}_3(t) = &-4.208x_1(t) - 0.396x_3(t) - 0.47x_1(t)^2 - 3.564x_1(t)^3 \\ &- 20.967v(t) + 6.265x_1(t)y_2(t) + 46y_2(t)v(t) + 61.4v(t)^3, \end{aligned} \tag{9.17}$$

$$\begin{aligned} 0 = &-y_1(t) - 0.877 - 0.088x_3(t) + 0.47x_1(t) - x_1(t)x_3(t) \\ &+ 3.846x_1(t)^2 + 0.28y_2(t) + 0.47v(t)^2, \end{aligned} \tag{9.18}$$

$$0 = -y_2(t) + x_1(t)v(t), \tag{9.19}$$

$$v(t) \in \{-0.05236, 0.05236\}, \tag{9.20}$$

$$x(0) = (0.4655, 0, 0)^\top, \quad x(t_f) = (0, 0, 0)^\top. \tag{9.21}$$

■

Note that the problem formulation in [17, 30] can be easily regained by substituting first $y_2(\cdot)$ and then $y_1(\cdot)$ back in the differential equations. In the following we leave off the time argument (t) for notational simplicity. We look at the necessary conditions of optimality for the case without time transformation, which we apply in our numerical scheme.

With $\ell_0 = 1$, the Hamiltonian of Problem 9.13 is given by

$$\begin{aligned} \mathcal{H}(\cdot) &= 1 + \lambda_f^\top f(x,y,u) + \lambda_g^\top g(x,y,u) \\ &= 1 + \lambda_{f1}(y_1x_1 + x_3 - 0.019x_2^2 - 0.215v + 0.63v^3) + \lambda_{f2}x_3 \\ &\quad + \lambda_{f3}(-4.208x_1 - 0.396x_3 - 0.47x_1^2 - 3.564x_1^3 \\ &\quad\quad - 20.967v + 6.265x_1y_2 + 46y_2v + 61.4v^3) \\ &\quad + \lambda_{g1}(-y_1 - 0.877 - 0.088x_3 + 0.47x_1 \\ &\quad\quad - x_1x_3 + 3.846x_1^2 + 0.28y_2 + 0.47v^2) \\ &\quad + \lambda_{g2}(-y_2 + x_1v). \end{aligned} \tag{9.22}$$

The adjoint equations read as

$$\begin{aligned} -\dot{\lambda}_{f1} = \frac{\partial \mathcal{H}(\cdot)}{\partial x_1} &= \lambda_{f3}(-4.208 - 2 \cdot 0.47x_1 - 3 \cdot 3.564x_1^2 + 6.265y_2) \\ &\quad + \lambda_{f1}y_1 + \lambda_{g1}(0.47 - x_3 + 2 \cdot 3.846x_1) + \lambda_{g2}v, \end{aligned} \tag{9.23}$$

$$-\dot{\lambda}_{f2} = \frac{\partial \mathcal{H}(\cdot)}{\partial x_2} = -0.038\lambda_{f1}x_2, \tag{9.24}$$

$$-\dot{\lambda}_{f3} = \frac{\partial \mathcal{H}(\cdot)}{\partial x_3} = \lambda_{f1} + \lambda_{f2} - 0.396\lambda_{f3} + \lambda_{g1}(-0.088 - x_1), \tag{9.25}$$

$$0 = \frac{\partial \mathcal{H}(\cdot)}{\partial y_1} = \lambda_{f1}x_1 - \lambda_{g1}, \tag{9.26}$$

$$0 = \frac{\partial \mathcal{H}(\cdot)}{\partial y_2} = \lambda_{f3}(6.265x_1 + 46v) + 0.28\lambda_{g1} - \lambda_{g2} \tag{9.27}$$

with the transversality conditions

$$\lambda_{fi}(t_0) = -\frac{\partial(\varphi(\cdot)+\sigma^\top\psi(\cdot))}{\partial x_i(t_0)} = \sigma_{0i}, \quad \lambda_{fi}(t_\mathrm{f}) = -\frac{\partial(\varphi(\cdot)+\sigma^\top\psi(\cdot))}{\partial x_i(t_\mathrm{f})} = \sigma_{fi} \quad (9.28)$$

for $i = 1,\dots,3$. In other words, the initial and terminal values of the differential adjoint states are free, because all initial and terminal values of the differential states $x_i(\cdot)$ are fixed. However, the value of the Hamiltonian at the free end time is fixed to 0,

$$\mathcal{H}(\cdot, t_\mathrm{f}) = 0. \quad (9.29)$$

The value of the optimal control $v(t)$ is determined according to the global minimum principle in Theorem 9.7 as the pointwise minimizer

$$v^*(t) = \arg\min\{\mathcal{H}(\cdot, v \equiv -0.05236), \mathcal{H}(\cdot, v \equiv 0.05236)\}. \quad (9.30)$$

Fast and reliable methods for MIOCPs have been developed recently [13, 33]. They are based, among others, on the transformations that have been described in the previous sections. For illustration and validation, however, we solve the boundary problem that results from the global minimum principle at this point. As a numerical solution we apply the method of Competing Hamiltonians that to our knowledge was first proposed in [3]. It enumerates the values of the Hamiltonian for all discrete control choices and chooses the pointwise minimizer, in our case as (9.30). Note that this approach is restricted to optimal control problems that have optimal bang-bang solutions for their relaxations, [33].

The resulting BVP consists of determining $(x^*, y^*, \lambda_f^*, \lambda_g^*, v^*)(\cdot)$ and a final time t_f such that (9.15)–(9.21), (9.23)–(9.27), (9.28)–(9.30) are fulfilled. Equations (9.26) and (9.27) trivially result in an extended DAE system of index one.

We need good initial guesses for the switching structure and for the differential and algebraic variables. We obtain them by running the MS MINTOC algorithm [28, 33]. It is based on a *first discretize, then optimize* approach, in particular on Bock's direct multiple shooting method [4]. We can directly read off the optimal switching structure and the values of differential and algebraic variables. For the adjoint variables $\lambda_f(\cdot)$ we take Lagrange multipliers of the matching conditions as a discrete approximation. This pretty accurate initialization allows us to solve the DAE BVP despite its well-known small region of convergence. We use the software MUSCOD-II, again based on direct multiple shooting [25]. Note that the F-8 control problem has several local minima; several are listed on [27]. We use the best known solution as initialization for our study. It results in an optimal integer control

$$v^*(t) = \begin{cases} 0.05236 & \text{if } t \in [0, \tau_1] \cup [\tau_2, \tau_3], \\ -0.05236 & \text{if } t \in [\tau_1, \tau_2] \cup [\tau_3, t_\mathrm{f}], \end{cases}$$

with $\tau_1 = 1.135007, \tau_2 = 1.482512, \tau_3 = 3.088809, t_\mathrm{f} = 3.780858$. Figures 9.3 and 9.4 show the optimal trajectories and the competing Hamiltonians.

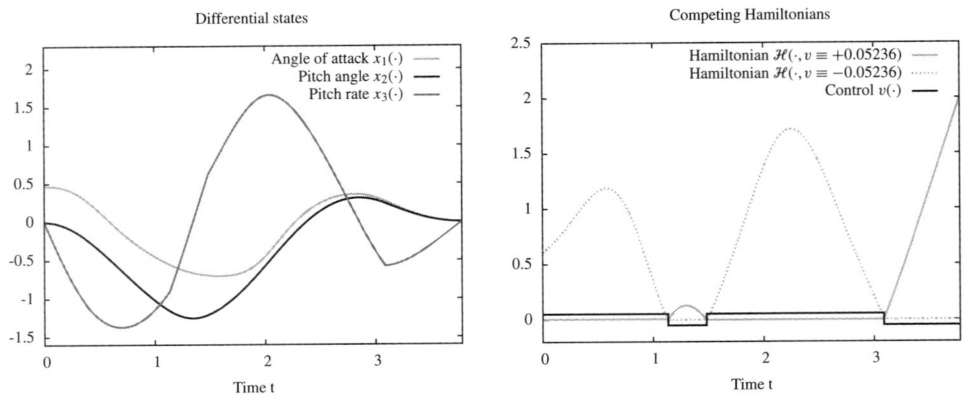

Figure 9.3. Left: differential states of the optimal trajectory. Right: the two competing Hamiltonians and the optimal control. Note that the minimizing Hamiltonian (9.30) is identical 0, in accordance to its end value (9.29) and the requirement to be constant, Theorem 9.7(e).

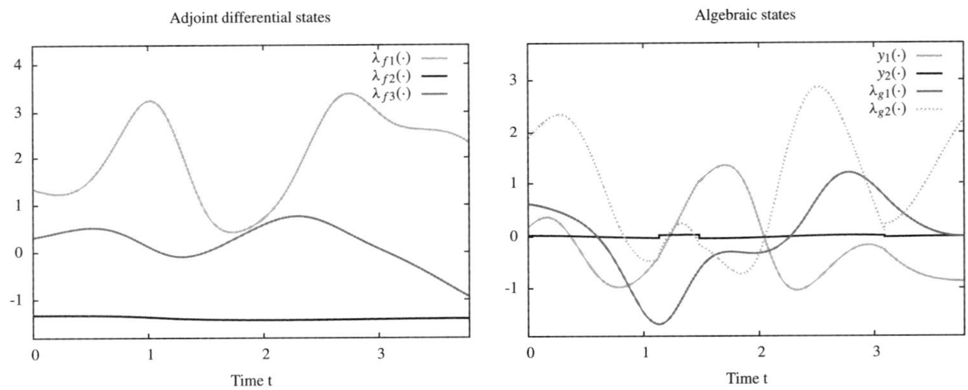

Figure 9.4. Left: the adjoint states plotted over time. Right: the algebraic states. The discontinuity in $v^*(\cdot)$ is accounted for by jumps in $y_2(\cdot)$.

It is interesting to note that outer convexification can be applied to Problem 9.13. With $\gamma = 0.05236$, the constraints (9.15)–(9.21) become

$$\begin{aligned}\dot{x}_1 = {} & -0.877\, x_1 + x_3 - 0.088\, x_1\, x_3 + 0.47\, x_1^2 - 0.019\, x_2^2 - x_1^2\, x_3 \\ & + 3.846\, x_1^3 + 0.215\, \gamma - 0.28\, x_1^2\, \gamma + 0.47\, x_1\, \gamma^2 - 0.63\, \gamma^3 \\ & - \left(0.215\, \gamma - 0.28\, x_1^2\, \gamma - 0.63\, \gamma^3\right) 2\omega, \end{aligned} \tag{9.31}$$

$$\dot{x}_2 = x_3, \tag{9.32}$$

$$\begin{aligned}\dot{x}_3 = {} & -4.208\, x_1 - 0.396\, x_3 - 0.47\, x_1^2 - 3.564\, x_1^3 \\ & + 20.967\, \gamma - 6.265\, x_1^2\, \gamma + 46\, x_1\, \gamma^2 - 61.4\, \gamma^3 \\ & - \left(20.967\, \gamma - 6.265\, x_1^2\, \gamma - 61.4\, \gamma^3\right) 2\omega, \end{aligned} \tag{9.33}$$

$$\omega(t) \in \{0,1\}, \tag{9.34}$$
$$x(0) = (0.4655, 0, 0)^\top, \quad x(t_f) = (0,0,0)^\top. \tag{9.35}$$

Any trajectory that minimizes t_f subject to (9.31)–(9.35) is also an optimal solution to Problem 9.13. According to Corollary 9.12 the objective function value is identical to the one of the relaxed version with $\omega(t) \in [0,1]$. In this particular case, there does not seem to be an integer gap to the nonlinear relaxation (Problem 9.13 with $v(t) \in [-0.05236, 0.05236]$). Examples for integer gaps can be found in [28, 31].

9.6 Summary

We discussed MIOCPs constrained by DAEs of index one. We derived a global minimum principle, making use of a time transformation technique. We also established theoretical results on the integer gap between the objective function value of the original MIOCP and of a relaxed, convexified version which is easier to solve with gradient-based methods. We illustrated the theoretical results by means of a challenging MIOCP benchmark problem. Future work is necessary for DAE systems that do not have index one.

Acknowledgments

Financial support of the Heidelberg Graduate School of Mathematical and Computational Methods for the Sciences and of the EU project EMBOCON under grant FP7-ICT-2009-4 248940 is gratefully acknowledged.

Bibliography

[1] A. Backes, *Extremalbedingungen für Optimierungs-Probleme mit Algebro-Differentialgleichungen*, Ph.D. thesis, Mathematisch-Naturwissenschaftliche Fakultät, Humboldt-Universität Berlin, Berlin, Germany, 2006.

[2] H. G. Bock and R. W. Longman, *Optimal control of velocity profiles for minimization of energy consumption in the New York subway system*, in Proceedings of the Second IFAC Workshop on Control Applications of Nonlinear Programming and Optimization, International Federation of Automatic Control, 1980, pp. 34–43.

[3] H. G. Bock and R. W. Longman, *Computation of optimal controls on disjoint control sets for minimum energy subway operation*, in Proceedings of the American Astronomical Society. Symposium on Engineering Science and Mechanics, Taiwan, 1982.

[4] H. G. Bock and K. J. Plitt, *A multiple shooting algorithm for direct solution of optimal control problems*, in Proceedings of the 9th IFAC World Congress, Budapest, 1984, Pergamon Press, Oxford, UK, 1995, pp. 242–247. Available at http://www.iwr.uni-heidelberg.de/groups/agbock/FILES/Bock1984.pdf.

[5] M. de Pinho and R. B. Vinter, *Necessary conditions for optimal control problems involving nonlinear differential algebraic equations*, J. Math. Anal. Appl., 212 (1997), pp. 493–516.

[6] E. N. DEVDARIANI AND YU. S. LEDYAEV, *Maximum principle for implicit control systems*, Appl. Math. Optim., 40 (1999), pp. 79–103.

[7] A. J. DUBOVITSKII AND A. A. MILYUTIN, *Extremum problems in the presence of restrictions.*, U.S.S.R. Comput. Math. Math. Phys., 5 (1965), pp. 1–80.

[8] A.YA. DUBOVITSKIJ AND A.A. MILYUTIN, *Extremum problems with constraints*, Soviet Math. Dokl., 4 (1963), pp. 452–455.

[9] W. L. GARRARD AND J. M. JORDAN, *Design of nonlinear automatic control systems*, Automatica, 13 (1977), pp. 497–505.

[10] M. GERDTS, *Local Minimum Principle for Optimal Control Problems Subject to Index One Differential-Algebraic Equations*, Tech. report, Institute of Mathematics and Applied Computing, Universität der Bundeswehr München, München, Germany; http://www.unibw.de/lrt1/gerdts/forschung/publikationen/report_index1.pdf, 2005.

[11] M. GERDTS, *Solving mixed-integer optimal control problems by branch & bound: A case study from automobile test-driving with gear shift*, Optimal Control Appl. Methods, 26 (2005), pp. 1–18.

[12] M. GERDTS, *Local minimum principle for optimal control problems subject to index-two differential-algebraic equations*, J. Optim. Theory Appl., 130 (2006), pp. 441–460.

[13] M. GERDTS, *A variable time transformation method for mixed-integer optimal control problems*, Optimal Control Appl. Methods, 27 (2006), pp. 169–182.

[14] I. V. GIRSANOV, *Lectures on Mathematical Theory of Extremum Problems*, Lecture Notes in Econom. Math. Systems 67, Springer, Berlin, Heidelberg, New York, 1972.

[15] P. HOWLETT, *Optimal strategies for the control of a train*, Automatica, 32 (1996), pp. 519–532.

[16] A. D. IOFFE AND V. M. TIHOMIROV, *Theory of Extremal Problems*, Stud. Math. Appl. 6, North–Holland, Amsterdam, New York, Oxford, 1979.

[17] C. Y. KAYA AND J. L. NOAKES, *A computational method for time-optimal control*, J. Optim. Theory Appl., 117 (2003), pp. 69–92.

[18] C. KIRCHES, *Fast Numerical Methods for Mixed–Integer Nonlinear Model–Predictive Control*, Ph.D. thesis, Ruprecht–Karls–Universität Heidelberg, 2010. Available at http://www.ub.uni-heidelberg.de/archiv/11636/.

[19] C. KIRCHES, S. SAGER, H.G. BOCK, AND J.P. SCHLÖDER, *Time-optimal control of automobile test drives with gear shifts*, Optimal Control Appl. Methods, 31 (2010), pp. 137–153.

[20] P. M. Krämer-Eis, H. G. Bock, R. W. Longman, and J. P. Schlöder, *Numerical determination of optimal feedback control in nonlinear problems with state/-control constraints*, Adv. Astronaut. Sci., 105 (2000), pp. 53–71.

[21] P. Kunkel and V. Mehrmann, *Optimal control for unstructured nonlinear differential-algebraic equations of arbitrary index.*, Math. Control Signals Systems, 20 (2008), pp. 227–269.

[22] H. W. J. Lee, K. L. Teo, and X. Q. Cai, *An optimal control approach to nonlinear mixed integer programming problems*, Comput. Math. Appl., 36 (1998), pp. 87–105.

[23] H. W. J. Lee, K. L. Teo, V. Rehbock, and L. S. Jennings, *Control parameterization enhancing technique for time optimal control problems*, Dynam. Systems Appl., 6 (1997), pp. 243–262.

[24] H. W. J. Lee, K. L. Teo, V. Rehbock, and L. S. Jennings, *Control parametrization enhancing technique for optimal discrete-valued control problems*, Automatica, 35 (1999), pp. 1401–1407.

[25] D. B. Leineweber, I. Bauer, H. G. Bock, and J. P. Schlöder, *An efficient multiple shooting based reduced SQP strategy for large-scale dynamic process optimization. Part* I: *Theoretical aspects*, Comput. Chem. Eng., 27 (2003), pp. 157–166.

[26] T. Roubicek and M. Valásek, *Optimal control of causal differential-algebraic systems*, J. Math. Anal. Appl., 269 (2002), pp. 616–641.

[27] S. Sager, *MIOCP Benchmark Site*, http://mintoc.de.

[28] S. Sager, *Numerical Methods for Mixed–Integer Optimal Control Problems*, Der andere Verlag, Tönning, Lübeck, Marburg, 2005.

[29] S. Sager, *Reformulations and algorithms for the optimization of switching decisions in nonlinear optimal control*, J. Process Control, 19 (2009), pp. 1238–1247.

[30] S. Sager, *A benchmark library of mixed-integer optimal control problems*, in Proceedings MINLP09, to appear.

[31] S. Sager, H. G. Bock, and M. Diehl, *The integer approximation error in mixed-integer optimal control*, Math. Program., to appear.

[32] S. Sager, M. Jung, and C. Kirches, *Combinatorial integral approximation*, Math. Methods Oper. Res., 73 (2011), pp. 363–380.

[33] S. Sager, G. Reinelt, and H. G. Bock, *Direct methods with maximal lower bound for mixed-integer optimal control problems*, Math. Program., 118 (2009), pp. 109–149.

[34] A. Siburian, *Numerical Methods for Robust, Singular and Discrete Valued Optimal Control Problems*, Ph.D. thesis, Curtin University of Technology, Perth, Australia, 2004.

[35] A. Siburian and V. Rehbock, *Numerical procedure for solving a class of singular optimal control problems*, Optim. Methods Softw., 19 (2004), pp. 413–426.

[36] N. TAUCHNITZ, *Das Pontrjaginsche Maximumprinzip für eine Klasse hybrider Steuerungsprobleme mit Zustandsbeschränkungen und seine Anwendung*, Ph.D. thesis, BTU Cottbus, Cottbus, Germany, 2010.

[37] K. L. TEO, L. S. JENNINGS, V. LEE, AND H. W. J. REHBOCK, *The control parameterization enhancing transform for constrained optimal control problems*, J. Aust. Math. Soc., 40 (1999), pp. 314–335.

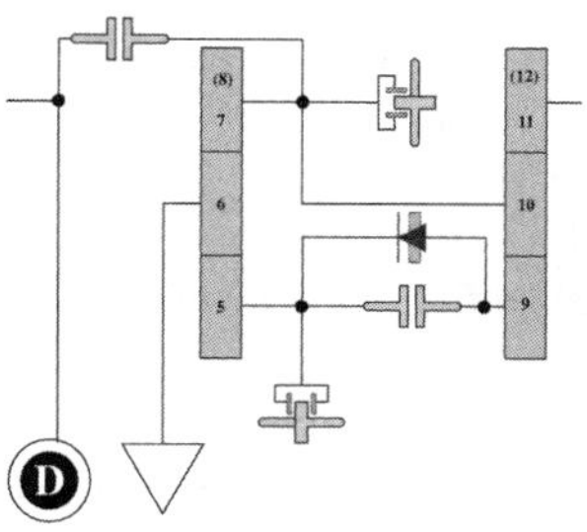

Chapter 10
Optimal Control of a Delay PDE

John T. Betts, Stephen L. Campbell, and Karmethia C. Thompson

Direct transcription is a popular method for the solution of optimal control problems in industry. Historically it has not been used on problems with delays. Progress on a direct transcription approach to optimal control problems with state delays is given and illustrated with the solution of nonlinear PDEs with delays.

10.1 Introduction

Differential equations are one of the most widespread mathematical tools used to describe physical phenomena. ODEs involve a single independent variable, and PDEs are used when there are derivatives of dependent variables with respect to more than one independent variable. It is not surprising that models used to describe a physical process often involve time as an independent variable. However, for a physical system, the current state usually depends on what happened at an earlier time. Predator-prey models describe the behavior of predators "now" feeding on populations of prey born "earlier." "In some situations, for example in real-time simulation, where time delays can be introduced by the computer time needed to compute an output after the input has been sampled, and where additional delays can be introduced by the operator-in-the-loop, differential equations with delays must be included in the model" [2].

The study of *delay-differential equations* (DDEs) as well as *delay-differential-algebraic equations* (DDAEs) has provided challenging analytic and numerical problems for over thirty years, and space precludes an exhaustive discussion of the many issues. Most widely used numerical methods for solving ODEs are based on a "stepwise" approach; that is, given values $\mathbf{y}_k$ at time t_k they construct new values $\mathbf{y}_{k+1}$ by stepping to the new time $t_k + h_k$ for a positive step h_k. Algorithmically, these methods can be implemented using information that is *local*. Single-step methods, such as the classical Runge–Kutta scheme, require information about the differential equations entirely within the step, i.e., for values $t_k \leq t \leq t_k + h_k$. Even multistep methods use only information over a few nearby steps. In contrast, solution of a DDE requires some type of "interpolation" for evaluating the delayed arguments at points well removed from the local step. At the very least, a stepwise algorithm must "save" enough information about early solution steps just to construct the current step. Thus one computational approach is to combine a method for solving an ODE with a technique for interpolation of previous steps. In essence, when solving a DDE, of necessity the method is "less local" or "more global" than single-step techniques widely used for solving ODEs. Paul [17] discusses a number of the issues that must be addressed when constructing DDE software and Shampine and Gahinet [18] present a MATLAB implementation. An example originally published in Russian by G. I. Marchuk is also cited by Hairer, Norsett, and Wanner [13, pp. 349–351]. Landry et al. [14] analyze the dynamics of a classical "inverted pendulum" problem when the control incorporates a time delay.

While many authors have considered stability and numerical simulation of differential equations with delays, research into methods for *optimization* of such systems is relatively new. Historically the first techniques developed for optimal control of systems without delay combined a numerical stepwise integrator with an optimization algorithm. Indeed these so-called indirect shooting methods have found wide applicability. The technique requires numerically solving the necessary conditions for optimality, i.e., the adjoint equations. Unfortunately, this approach becomes problematic, because the adjoint equations for systems with delay involve "advances," thus precluding a simple stepwise integration process. In fact, just deriving the necessary conditions for delay equations is often very difficult. In spite of these computational challenges, results have been published for some applications. Deshmukh, Ma, and Butcher [9] present a finite horizon optimal control model. The optimal control of a two-stage continuous stirred chemical tank reactor problem is posed in Büskens, Göllmann, and Maurer [6] and revisited in the thesis of Gretschko [12]. An optimal control model that describes the immune response of a pathogenic disease process is developed in Stengel et al. [19]. Maurer presents an example in [15] that incorporates a nonconvex model with state delay. Göllmann, Kern, and Maurer present an example in [11] that models optimal fishing, as well as another continuous stirred tank reactor system (CSTR) model. Batzel, Timischl-Teschl, and Kappel [3] describe a biomedical application of delay systems that considers a model of the human cardiovascular-respiratory control system with one and two transport delays in the state equations describing the respiratory system.

In contrast to stepwise methods, the direct transcription method [4] treats the problem in a global fashion. The entire problem is first discretized, and then a large scale optimization algorithm is used to compute the optimal solution of the discrete approximation. Accuracy of the discrete approximation is improved by a mesh refinement algorithm. Since this "discretize then optimize" approach is global it permits looking forward and/or backwards, and does not require explicit derivation of the adjoint equations. The method has demonstrated success for problems with ordinary differential-algebraic systems. After defining the delay problem with ODEs in Section 10.2, we extend the transcription method to include delays in Section 10.3. Section 10.4 introduces an example that has both delays and a PDE. In Section 10.5 we describe how the PDE can be converted to a system of ODEs using the *method of lines*. Sections 10.6–10.8 then present a series of examples that demonstrate the utility of the new method.

10.2 DDEs

Many dynamic systems can be represented by a DAE model of the form

$$\dot{\mathbf{y}} = \mathbf{f}(\mathbf{y}(t), \mathbf{u}(t), \mathbf{p}, t), \tag{10.1a}$$

$$\mathbf{0} = \mathbf{g}(\mathbf{y}(t), \mathbf{u}(t), \mathbf{p}, t) \tag{10.1b}$$

for $t_I \le t \le t_F$. The differential or state variables are denoted by $\mathbf{y}$, the algebraic or control variables by $\mathbf{u}$, and the (nondynamic) parameters by $\mathbf{p}$. For simplicity we consider equality path constraints (10.1b), although extending the results to inequalities is straightforward. The behavior of the system is quantified by the value of an *objective function* such as

$$F = \phi(\mathbf{y}(t_F), \mathbf{u}(t_F), \mathbf{p}, t_F) + \int_{t_I}^{t_F} L(\mathbf{y}(t), \mathbf{u}(t), \mathbf{p}, t)\, dt, \tag{10.2}$$

which must be minimized by the optimal choice for the control functions $\mathbf{u}(t)$ and parameters $\mathbf{p}$.

It should be noted that the terminology used here is consistent with how $\mathbb{SOCX}$ treats the various variables. However, this is slightly different from how a user might view the problem. For a typical user, they may view the problem as

$$\dot{\mathbf{y}}_1 = \mathbf{f}(\mathbf{y}_1(t), \mathbf{y}_2(t), \mathbf{u}_1(t), \mathbf{p}, t), \tag{10.3a}$$

$$\mathbf{0} = \mathbf{g}(\mathbf{y}_1(t), \mathbf{y}_2(t), \mathbf{u}_1(t), \mathbf{p}, t), \tag{10.3b}$$

where $\mathbf{y}_1$ is the differential part of the state, $\mathbf{y}_2(t)$ is the algebraic part of the state, and $\mathbf{u}_1(t)$ is the control. If the control was given, then we would need that $\partial \mathbf{g}/\partial \mathbf{y}_1$ is nonsingular to get an index-one system. However, $\mathbb{SOCX}$ treats $\mathbf{y}_2$ and $\mathbf{u}_1$ the same and what counts for successful solution is that there is some choice of control that results in an index-one system, and that choice need not be made explicitly [7, 10]. Thus we use the formulation (10.1) where $\mathbf{y}$ of (10.1) is $\mathbf{y}_1$ of (10.3) and $\mathbf{u}$ of (10.1) is $\mathbf{y}_1$ and $\mathbf{u}_1$. Initial conditions specified for $\mathbf{y}$ need to be consistent with an index-one formulation.

Let us define a set of ν *delay functions*, ω_k that are prescribed functions of time for $k = 1, \ldots, \nu$. Although we generically refer to the functions as "delay" functions, as in

$$\omega_1(t) = t - \tau$$

for a delay time $\tau > 0$, the approach can also be used to model "advances" such as

$$\omega_2(t) = t + 5\tau.$$

Although the function $\omega_k(t)$ is not necessarily linear, the behavior of $\omega_k(t)$ over the entire phase is restricted. To characterize delay we require

$$\omega_k(t) \le t \qquad \text{for } t_I \le t \le t_F. \tag{10.4a}$$

To represent advance we must have

$$\omega_k(t) \ge t \qquad \text{for } t_I \le t \le t_F. \tag{10.4b}$$

These restrictions permit a zero delay, e.g., $\omega_k(t) = t$, but preclude a reverse in direction from delay to advance, or vice versa.

A key aspect of the success of direct transcription methods is to exploit sparsity. For this reason the current version of our approach does not allow for the phase duration (t_I or t_F) to vary or for a delay to depend on either the state or the control or an optimization parameter. In these situations, while on a given grid the nonlinear programming (NLP) Jacobian and Hessian sparsity pattern can change from iteration to iteration. To apply the transcription method as given to these scenarios would require either treating the NLP as dense and incur the $O(n^3)$ linear algebra cost or recomputing the sparsity after each NLP iteration. Other alternatives are under investigation.

Let us denote the *delayed state vector* by $\mathbf{z}(\omega_k)$. Thus the vector $\mathbf{z}(\omega_k)$ of length n_z ($n_z \le n_y$) denotes the subset of the n_y state variables that are functions of the delay argument ω_k. The goal of this chapter is to extend the DAE model (10.1) to systems of the form

$$\dot{\mathbf{y}} = \mathbf{f}(\mathbf{y}(t), \mathbf{u}(t), \mathbf{z}(\omega_1), \ldots, \mathbf{z}(\omega_\nu), \mathbf{p}, t), \tag{10.5a}$$

$$\mathbf{0} = \mathbf{g}(\mathbf{y}(t), \mathbf{u}(t), \mathbf{z}(\omega_1), \ldots, \mathbf{z}(\omega_\nu), \mathbf{p}, t). \tag{10.5b}$$

Delay equations introduce an issue not encountered in a standard DAE problem. For a system defined by the DAE model (10.1a)–(10.1b) a complete description usually requires boundary conditions at the initial time t_I and/or the final time t_F. The boundary conditions can involve specified values for the states or nonlinear conditions $\psi(\mathbf{y}, \mathbf{u}, t) = 0$ and are enforced at specific points, namely at t_I and/or t_F. In contrast, for a delay equation such as (10.5a)–(10.5b), there must be a start-up region. Specifically, for delay problems as in (10.4a), we must evaluate $\mathbf{z}[\omega_k(t)]$. Thus we require

$$\mathbf{z}(\omega_k(t)) = \boldsymbol{\alpha}(\mathbf{p}, \omega_k(t)) \quad \text{for} \quad \omega_k(t) \le t_I, \tag{10.6}$$

where the *prehistory* $\boldsymbol{\alpha}(\mathbf{p}, \tau)$ must be a function defined on the start-up region $\omega_k(t) \le t_I$. A similar function may be required at the final time for problems with advances.

10.3 Direct Transcription Method

When solving an optimal control problem without delays as given by (10.1), the *direct transcription method* has been very effective [4]. Our goal is to extend the method to a new class of problems that include delay dynamics as given by (10.5). In general, the method has three basic steps:

1. **Direct Transcription** Transcribe the optimal control problem into an NLP problem by discretization.

2. **Sparse Nonlinear Program** Solve the sparse NLP.

3. **Mesh Refinement** Assess the accuracy of the approximation (i.e., the finite-dimensional problem), and if necessary refine the discretization, and then repeat the optimization steps.

As such the approach can be considered a *sequential nonlinear programming* or (SNLP) method. In our software implementation the sparse NLP can be solved using either a sequential quadratic programming (SQP) or primal-dual barrier (interior point) method.

Let us address a DDAE system of the form (10.5), and to simplify the presentation temporarily assume there are no parameters and a single delay function ω. Define new algebraic variables $\mathbf{v}(t)$, and require they be consistent with the delayed variables. The original DDAE system can now be written as

$$\dot{\mathbf{y}} = \mathbf{f}(\mathbf{y}(t), \mathbf{u}(t), \mathbf{v}(t), t), \tag{10.7a}$$

$$\mathbf{0} = \mathbf{g}(\mathbf{y}(t), \mathbf{u}(t), \mathbf{v}(t), t), \tag{10.7b}$$

$$\mathbf{0} = \mathbf{v}(t) - \mathbf{z}(\omega(t)). \tag{10.7c}$$

The basic idea is to first discretize the problem thereby creating a finite-dimensional approximation. Large-scale optimization methods can then be used to adjust the variables that define the discretization. The direct transcription approach introduces a discretization of the problem by subdividing the time domain into n_s segments or intervals

$$t_I = t_1 < t_2 < \cdots < t_M = t_F, \tag{10.8}$$

where the points are referred to as node, mesh, or grid points. We use $\mathbf{y}_k \equiv \mathbf{y}(t_k)$ to indicate the value of the state variable at a grid point, with a similar notation for the algebraic variables. Thus one treats

$$\mathbf{x}^{\mathsf{T}} = (\mathbf{y}_1, \mathbf{u}_1, \mathbf{v}_1, \ldots, \mathbf{y}_M, \mathbf{u}_M, \mathbf{v}_M) \tag{10.9}$$

as optimization variables in an NLP. We then approximate the differential equation using a nonlinear algebraic constraint. When the discretization is based on an implicit Runge–Kutta (IRK) scheme, the control problem is *transcribed* into a finite-dimensional nonlinear program. The software being developed uses the trapezoidal (TR) and Hermite–Simpson (HS) IRK methods.

For the TR method, we approximate the differential equations (10.7a) and algebraic constraints (10.7b) by a set of algebraic constraints given by

$$\mathbf{0} = \mathbf{y}_{k+1} - \mathbf{y}_k - \frac{h_k}{2}(\mathbf{f}_k + \mathbf{f}_{k+1}) = \boldsymbol{\zeta}_k, \qquad k = 1, \ldots, M-1, \tag{10.10a}$$

$$\mathbf{0} = \mathbf{g}_k, \qquad k = 1, \ldots, M, \tag{10.10b}$$

where (10.10a) are referred to as the *defect* constraints $\boldsymbol{\zeta}_k = \mathbf{0}$. This discretization is implicit because the optimization variables $(\mathbf{y}_k, \mathbf{u}_k, \mathbf{v}_k)$ appear as arguments in the nonlinear functions $\mathbf{f}_k \equiv \mathbf{f}(\mathbf{y}_k, \mathbf{u}_k, \mathbf{v}_k)$ and $\mathbf{g}_k \equiv \mathbf{g}(\mathbf{y}_k, \mathbf{u}_k, \mathbf{v}_k)$.

Now in order to evaluate the right-hand side functions in (10.7a) and (10.7b) we must express the consistency relationship (10.7c) in terms of the NLP optimization variables. When the delay argument is exterior to the phase, that is, when $\omega_k < t_I$ (or $\omega_k > t_M$), the value of the delayed state is given by the user-defined function $\boldsymbol{\alpha}(\mathbf{p}, \omega_k)$. When the delay argument is interior to the phase, $t_I \le \omega_k \le t_F$, let us define the interval J such that

$$t_J \le \omega(t_k) \le t_{J+1} \tag{10.11}$$

as illustrated in Figure 10.1. The value of J depends on k, but we omit that from our notation. Then for $\delta_k = (\omega_k - t_J)/h_J = (\omega_k - t_J)/(t_{J+1} - t_J)$ the interpolated value for the delayed state is just

$$\mathbf{z}(\omega(t_k)) \doteq \mathbf{z}(\omega_k) = c_1 \mathbf{z}_J + c_2 \mathbf{z}_{J+1} + c_3 \dot{\mathbf{z}}_J + c_4 \dot{\mathbf{z}}_{J+1}, \tag{10.12}$$

where the Hermite interpolation coefficients are

$$c_1 = (1-\delta_k)^2(1+2\delta_k),\ c_2 = \delta_k^2(3-2\delta_k),\ c_3 = \delta_k(1-\delta_k)^2 h_J,\ c_4 = -\delta_k^2(1-\delta_k)h_J. \tag{10.13}$$

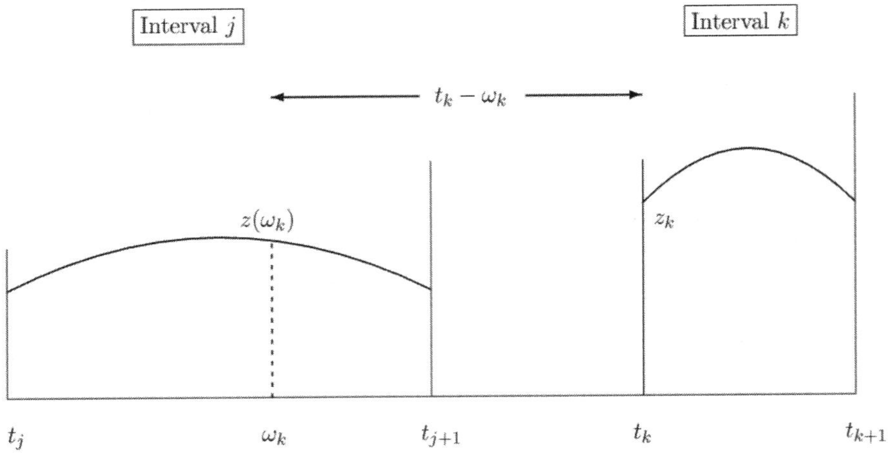

Figure 10.1. Consistency construction.

Thus to enforce consistency at the grid points we can impose the NLP constraints

$$\mathbf{v}(t_k) = \mathbf{z}(\omega(t_k)) = \begin{cases} \mathbf{z}(\omega_k) & \text{if } t_1 \le \omega_k \le t_M, \\ \alpha(\mathbf{p}, \omega_k) & \text{otherwise.} \end{cases} \tag{10.14}$$

Of course these consistency constraints are also implicit, since they involve the derivatives $\dot{\mathbf{z}}_J$ and $\dot{\mathbf{z}}_{J+1}$ which are given by the right-hand sides of (10.7a).

When using an HS discretization two things change. First, the NLP variables are

$$\mathbf{x}^{\mathsf{T}} = (\mathbf{y}_1, \mathbf{u}_1, \mathbf{v}_1, \mathbf{u}_{3/2}, \mathbf{v}_{3/2}, \ldots, \mathbf{y}_M, \mathbf{u}_M, \mathbf{v}_M), \tag{10.15}$$

where $\mathbf{u}_{k+\frac{1}{2}} = \mathbf{u}(\frac{1}{2}(t_j + t_{j+1}))$ is an additional NLP variable representing the control at the midpoint. Second, the discretization constraints (10.10) are replaced by

$$\mathbf{0} = \mathbf{y}_{k+1} - \mathbf{y}_k - \frac{h_k}{6}\left(\mathbf{f}_k + 4\mathbf{f}_{k+\frac{1}{2}} + \mathbf{f}_{k+1}\right) = \boldsymbol{\zeta}_k, \qquad k = 1, \ldots, M-1, \tag{10.16a}$$

$$\mathbf{0} = \mathbf{g}_k, \qquad k = 1, \ldots, M, \tag{10.16b}$$

$$\mathbf{0} = \mathbf{g}_{k+\frac{1}{2}}, \qquad k = 1, \ldots, M-1. \tag{10.16c}$$

To summarize, for the HS discretization the NLP variables (10.15) must be chosen to satisfy the NLP constraints (10.16). Furthermore, the NLP constraints (10.14) are imposed at both grid points and midpoints to ensure satisfying the consistency relationships (10.7c).

In order to efficiently solve the resulting large NLP it is imperative to exploit sparsity in the Jacobian and Hessian matrices. In particular, the Jacobian will involve derivatives of the constraints with respect to the variables $\mathbf{x}$ given by (10.9) or (10.15). These matrices will be sparse, because the subset

$$(\mathbf{y}_k, \mathbf{y}_j, \mathbf{y}_{j+1}, \mathbf{u}_k, \mathbf{u}_j, \mathbf{u}_{j+1}, \mathbf{v}_k, \mathbf{v}_j, \mathbf{v}_{j+1}, t_k)$$

contains the only variables that appear. In order to fully exploit sparsity it is important to exploit separability. This can be achieved by writing the constraint functions such that the linear terms are isolated from the nonlinear ones. Specifically using equation (4.114) in [4], we write

$$\begin{bmatrix} \mathbf{c}(\mathbf{x}) \\ F(\mathbf{x}) \end{bmatrix} = \mathbf{A}\mathbf{x} + \mathbf{B}\mathbf{q}(\mathbf{x}), \tag{10.17}$$

where the vector $\mathbf{q}$ consists of the nonlinear right-hand-side functions $\mathbf{f}_k$ and $\mathbf{g}_k$ evaluated at the grid points. Sparse finite difference estimates can be used to compute the Jacobian and Hessian matrices as described in [4]. Furthermore, because the functions $\mathbf{f}$ and $\mathbf{g}$ appear linearly in the transcribed formulation, sparsity in these right-hand-side functions can be exploited (cf. equation (4.115) in [4]).

There are two key ideas needed to extend the direct transcription method to accommodate models with delays. First, we introduce a new algebraic variable, i.e., the *pseudostate* $\mathbf{v}$, to represent the delayed state variable. We then introduce an algebraic (path) constraint, to ensure that $\mathbf{v}$ is consistent with the Hermite interpolant for the value of the real state at the delayed time. This leads to the NLP consistency equations (10.14), after transcribing the continuous problem into its finite-dimensional counterpart.

Although most of the computational effort is the same for problems with and without delay, there is one change required in the mesh refinement procedure. In particular, the refinement algorithm described in [4], which is based on work in [5], proceeds by systematically subdividing the interval containing the largest discretization error. In contrast, for delay problems, when the largest error occurs in a particular interval k, where $t_k \le t \le t_{k+1}$, we must also subdivide the delay interval J, where $t_J \le \omega(t_k) \le t_{J+1}$. Thus the "look back" effect, illustrated in Figure 10.1, also is implemented during the mesh refinement process. The calculations that follow were done with the direct transcription code SOCX which is available from Applied Mathematical Analysis, LLC. However, the discussion is relevant for any direct transcription solution of PDE systems with delays.

10.4 Delay PDE Example

The preceding section described how to extend the direct transcription method to ordinary differential-algebraic equations with delay. We now go one step farther and demonstrate the performance of the algorithm on a PDE with delay. First we describe the PDE problem, and then we discuss how it can be approximated by a system of ODEs.

Consider a modification to a reaction-diffusion model of a species. Define $Y(x,t)$ as the population density at time t and position x, where $0 \le t \le T$ and $0 \le x \le \pi$,

$$\frac{\partial Y(x,t)}{\partial t} = c_1 \frac{\partial^2 Y(x,t)}{\partial x^2} - c_2 Y(x,\omega(x,t))(1+Y(x,t)) + U(x,t) \tag{10.18a}$$

with boundary conditions

$$\frac{\partial Y(0,t)}{\partial x} = \frac{\partial Y(\pi,t)}{\partial x} = 0, \tag{10.18b}$$

$$Y(x,0) = \alpha(x) = 1 + \sin\left(2x - \frac{\pi}{2}\right). \tag{10.18c}$$

The (positive) diffusion parameter $c_1 = 1$, and the coefficient $c_2 = 0.5$ defines what the reaction consumes. The duration of the process is $T = 5$. U is a time and spatially dependent control which can represent, for example, stocking or harvesting. The goal is to minimize

$$F = \int_0^{\pi} \int_0^{T} c_3 U^2(x,t)\,dt\,dx + \int_0^{\pi} (Y(x,T) - h(x))^2\,dx, \tag{10.19}$$

where $h(x)$ is the desired final population profile and $c_3 = 0.1$ is a weight. Equations of the type (10.18a) have been widely studied [21]. Our particular version follows that of [16] with the addition of the source or control U.

For the sake of illustration we will consider problems with various delay functions $\omega(x,t)$. For examples with no spatial dependence we will consider the constant delay function

$$\omega(t) = t - r \tag{10.20a}$$

defined with delay time $r = 0.5$ as well as the variable delay functions

$$\omega(t) = t - r\cos\left(\frac{t\pi}{2T}\right), \tag{10.20b}$$

$$\omega(t) = t - r\left(1 - \frac{t}{T}\right), \tag{10.20c}$$

$$\omega(t) = t - \frac{rt}{T}, \tag{10.20d}$$

$$\omega(t) = t - 1 - \frac{1}{t+1}, \tag{10.20e}$$

$$\omega(t) = t - 1 - \log(t+1), \tag{10.20f}$$

$$\omega(t) = \frac{t}{2}. \tag{10.20g}$$

Delay function (10.20g) is a special case of (10.20d) which has been discussed several times in the literature [8], so we list it separately. For both (10.20b) and (10.20c) the delay starts at r and decreases to zero as time goes from zero to T. In contrast for (10.20d) the delay begins at zero and then increases to r at the final time T. The variable delay functions (10.20e)–(10.20g) are suggested in [20]. For (10.20g) the delay increases from zero to $T/2$. No prehistory is required for delay functions (10.20d) and (10.20g). For examples that require a prehistory when $\omega(x,t) < 0$ we set

$$Y((x,\omega(x,t)) = \alpha(x) \quad \text{for } 0 \le x \le \pi. \tag{10.21}$$

10.5 Method of Lines Formulation

Let us first transform the delay PDE (10.18a) into a system of ordinary DDEs using the method of lines. To do so introduce a spatial discretization, i.e.,

$$x_k = k\delta = k\frac{\pi}{n}, \quad k = 0,\ldots,n. \tag{10.22}$$

If we denote $Y(x_k,t) = y_k(t)$, $U(x_k,t) = u_k(t)$, and $\omega(x_k,t) = \omega_k(t)$, then we can introduce a second order difference approximation for $\frac{\partial^2 Y}{\partial x^2}$. Using this approximation, the PDE system

(10.18a) is replaced by the system of DDEs

$$\dot{y}_0 = \frac{c_1}{\delta^2}(y_1 - 2y_0 + y_{-1}) - c_2 y_0(\omega_0)(1 + y_0) + u_0, \tag{10.23a}$$

$$\dot{y}_k = \frac{c_1}{\delta^2}(y_{k+1} - 2y_k + y_{k-1}) - c_2 y_k(\omega_k)(1 + y_k) + u_k, \quad k = 1,\dots,n-1, \tag{10.23b}$$

$$\dot{y}_n = \frac{c_1}{\delta^2}(y_{n+1} - 2y_n + y_{n-1}) - c_2 y_n(\omega_n)(1 + y_n) + u_n. \tag{10.23c}$$

Using a second order approximation to the first derivative the boundary conditions (10.18b) lead to the algebraic equations

$$\frac{\partial Y(0,t)}{\partial x} \approx \frac{1}{2\delta}(y_1 - y_{-1}) = 0, \tag{10.24}$$

$$\frac{\partial Y(\pi,t)}{\partial x} \approx \frac{1}{2\delta}(y_{n+1} - y_{n-1}) = 0. \tag{10.25}$$

From these equations it follows that

$$y_{-1} = y_1, \tag{10.26}$$

$$y_{n+1} = y_{n-1}. \tag{10.27}$$

When the expressions for y_{-1} from (10.26) and y_{n+1} from (10.27) are substituted into the system (10.23) we obtain

$$\dot{y}_0 = \frac{2c_1}{\delta^2}(y_1 - y_0) - c_2 y_0(\omega_0)(1 + y_0) + u_0, \tag{10.28a}$$

$$\dot{y}_k = \frac{c_1}{\delta^2}(y_{k+1} - 2y_k + y_{k-1}) - c_2 y_k(\omega_k)(1 + y_k) + u_k, \quad k = 1,\dots,n-1, \tag{10.28b}$$

$$\dot{y}_n = \frac{2c_1}{\delta^2}(y_{n-1} - y_n) - c_2 y_n(\omega_n)(1 + y_n) + u_n. \tag{10.28c}$$

After introducing the method of lines approximation and denoting $\alpha(x_k) = \alpha_k$ the initial condition (10.18c) leads to the initial conditions

$$y_k(0) = \alpha_k, \qquad k = 0,\dots,n, \tag{10.29}$$

and the prehistory defined when $\omega(x,t) < 0$

$$y_k(t) = \alpha_k, \qquad k = 0,\dots,n. \tag{10.30}$$

The spatial integration in the objective can be approximated using a trapezoidal quadrature rule so that

$$F = \frac{1}{2}\delta f_0 + \delta \sum_{k=1}^{n-1} f_k + \frac{1}{2}\delta f_n, \tag{10.31a}$$

where

$$f_k = \int_0^T c_3 u_k^2(t)dt + (y_k(T) - h(x_k))^2. \tag{10.31b}$$

10.6 Delay PDE with Single Control

To simplify the development here as well as in Section 10.6.1, temporarily let us assume that the control depends only on time, that is, $U(x,t) = u(t)$, and the target profile is just $h(x) = 5$. For this example the dynamics that determine the states $(y_0,\ldots,y_n)$ and control u are just

$$\dot{y}_0 = \frac{2c_1}{\delta^2}(y_1 - y_0) - c_2 y_0(\omega)(1+y_0) + u, \tag{10.32a}$$

$$\dot{y}_k = \frac{c_1}{\delta^2}(y_{k+1} - 2y_k + y_{k-1}) - c_2 y_k(\omega)(1+y_k) + u, \quad k = 1,\ldots,n-1, \tag{10.32b}$$

$$\dot{y}_n = \frac{2c_1}{\delta^2}(y_{n-1} - y_n) - c_2 y_n(\omega)(1+y_n) + u. \tag{10.32c}$$

In this case the objective function given by (10.31) can be simplified to

$$F = \int_0^T c_3 u^2(t)\,dt + \frac{1}{2}\delta f_0 + \delta \sum_{k=1}^{n-1} f_k + \frac{1}{2}\delta f_n, \tag{10.33a}$$

where

$$f_k = \left[y_k(T) - h(x_k)\right]^2. \tag{10.33b}$$

For problems having a constant delay function (10.20a) it is possible to construct an equivalent system of ODEs with *no delay*. This technique, called the *method of steps* [1, 13], is described in Section 10.6.1 and can be used to verify results computed by the new delay algorithm.

10.6.1 Method of Steps Formulation

When the delay is constant as in (10.20a) the delay system (10.32) can be transformed into a system of ODEs with no delay using the method of steps. Specifically we break up the time domain into steps of length r and then define

$$\mathbf{s}_j(\rho) = \mathbf{y}(t), \quad (j-1)r \le t \le jr \tag{10.34}$$

for $j = 0,\ldots,N$, where the number of steps $N = T/r = 10$ and $0 \le \rho \le r$. Denote the vector $\mathbf{s}_j(\rho)$ of length $(n+1)$ representing the state on delay step j by

$$\mathbf{s}_j(\rho) = \begin{bmatrix} \mathbf{y}_0[\rho + (j-1)r] \\ \vdots \\ \mathbf{y}_n[\rho + (j-1)r] \end{bmatrix} \tag{10.35}$$

for $j = 0,\ldots,N$, and $0 \le \rho \le r$. For notational clarity it is convenient to introduce a matrix with $(n+1)$ rows and $(N+1)$ columns:

$$\begin{aligned} \mathbf{S}(\rho) &= \begin{bmatrix} \mathbf{s}_0(\rho) & \mathbf{s}_1(\rho) & \cdots & \mathbf{s}_N(\rho) \end{bmatrix} \\ &= \begin{bmatrix} \mathbf{y}_0(\rho - r) & \mathbf{y}_0(\rho - r + r) & & \mathbf{y}_0(\rho - r + Nr) \\ \vdots & \vdots & \cdots & \vdots \\ \mathbf{y}_n(\rho - r) & \mathbf{y}_n(\rho - r + r) & & \mathbf{y}_n(\rho - r + Nr) \end{bmatrix}. \end{aligned} \tag{10.36}$$

Note that each column of $\mathbf{S}$ corresponds to the state vector $\mathbf{y}$ on a particular step. Thus element $S_{k,j}$ denotes state vector k at delay step j.

A similar notational convention must be introduced for the control, namely for $j = 1,\ldots,N$:

$$u_j(\rho) = u(\rho + (j-1)r). \tag{10.37}$$

For steps $j = 1,\ldots,N$ the dynamics in the region $0 \le \rho \le r$ become

$$\dot{S}_{0,j} = \frac{2c_1}{\delta^2}\left(S_{1,j} - S_{0,j}\right) - c_2 S_{0,j-1}\left(1 + S_{0,j}\right) + u_j, \tag{10.38a}$$

$$\dot{S}_{k,j} = \frac{c_1}{\delta^2}\left(S_{k+1,j} - 2S_{k,j} + S_{k-1,j}\right) - c_2 S_{k,j-1}\left(1 + S_{k,j}\right) + u_j, \quad k = 1,\ldots,n-1, \tag{10.38b}$$

$$\dot{S}_{n,j} = \frac{2c_1}{\delta^2}\left(S_{n-1,j} - S_{n,j}\right) - c_2 S_{n,j-1}\left(1 + S_{n,j}\right) + u_j. \tag{10.38c}$$

To ensure continuity across the step boundaries, we must impose the boundary conditions

$$S_{k,j}(0) = S_{k,j-1}(r), \tag{10.39}$$

$$u_j(0) = u_{j-1}(r) \tag{10.40}$$

for the interior steps $j = 2,\ldots,N$ and all states $k = 0,\ldots,n$. The initial condition (10.29) leads to

$$S_{k,1}(0) = \alpha_k, \qquad k = 0,\ldots,n, \tag{10.41}$$

and the prehistory requires

$$S_{k,0}(\rho) = \alpha_k, \qquad k = 0,\ldots,n. \tag{10.42}$$

Now the objective function is given by

$$F = \sum_{j=1}^{N} \int_0^r c_3 u_j^2(\rho)d\rho + \frac{1}{2}\delta f_0 + \delta \sum_{k=1}^{n-1} f_k + \frac{1}{2}\delta f_n, \tag{10.43}$$

with

$$f_k = \left(S_{k,N}(r) - h(x_k)\right)^2.$$

10.6.2 Numerical Results

The partial differential delay problem defined by (10.32) has been solved for all of the cases listed in (10.20) using a spatial discretization with $n = 15$. The mesh in the time domain was refined until the discretization error $\epsilon \le 10^{-7}$ to give a solution with approximately eight significant figures of accuracy. The method of lines formulation results in a system with 16 state variables and 17 algebraic variables. The results are summarized in Table 10.1. Notice that the optimal objective function value obtained using the method of steps compares closely with the constant delay case in the second line of Table 10.1 as it should. In contrast, there is a significant difference in both the number of grid points and the solution time. Since the time domain is broken into 10 regions when using the method of steps, the "real" number of mesh points is $10 \times 38 = 380$, which is considerably more than the 170 points needed using the new delay approach. The method of steps also leads to an ODE system with 160 states and 10 control variables which also explains why the solution time for the new approach is nearly twice as fast as the method of steps.

Table 10.1. Performance summary; time dependent control. M = number of grid points, time is CPU in sec., F^* = optimal objective.

Description	M	Time	F^*
Method of steps	38	19.1291	3.64669763
Constant delay, cf. (10.20a)	170	11.0973	3.64661936
Variable delay, cf. (10.20b)	183	16.9664	8.37257003
Variable delay, cf. (10.20c)	178	11.4313	8.47992369
Variable delay, cf. (10.20d)	149	9.40557	3.76411998
Variable delay, cf. (10.20e)	194	12.0362	1.12890127
Variable delay, cf. (10.20f)	158	4.89126	0.492930043
Variable delay, cf. (10.20g)	125	7.62984	0.415342751

10.7 Delay PDE with Spatial Control

Let us now consider the original problem treating the control as a function of both space and time. The target profile is $h(x) = 6 - 3\cos(3x)$, and as before we choose $c_3 = 0.1$. In this case the system (10.28a)–(10.28c) is replaced by the system

$$\dot{y}_0 = \frac{2c_1}{\delta^2}(y_1 - y_0) - c_2 y_0(\omega)(1 + y_0) + u_0, \tag{10.44a}$$

$$\dot{y}_k = \frac{c_1}{\delta^2}(y_{k+1} - 2y_k + y_{k-1}) - c_2 y_k(\omega)(1 + y_k) + u_k, \quad k = 1, \dots, n-1, \tag{10.44b}$$

$$\dot{y}_n = \frac{2c_1}{\delta^2}(y_{n-1} - y_n) - c_2 y_n(\omega)(1 + y_n) + u_n. \tag{10.44c}$$

Thus to summarize the optimal control problem;

- determine the control variables $(u_0(t), \dots, u_n(t))$
- and state variables $(y_0(t), \dots, y_n(t))$
- to satisfy the differential equations (10.44)
- the initial conditions (10.29)
- the prehistory conditions (10.30)
- to minimize the objective function (10.31a).

10.7.1 Numerical Results

This problem has been solved for all of the delay functions (10.20) using a spatial discretization with $n = 15$, and Table 10.2 summarizes the computational performance. This problem formulation leads to a delay system with 16 states and 32 algebraic variables (16 "real" controls, and 16 "pseudostates").

Table 10.2. Performance summary; time and space dependent control. M = grid points, Ref = number of refinements, n = NLP variables, NDOF = number of degrees of freedom, time = CPU sec., and F^* = optimal objective.

Description	M	Ref	n	NDOF	Time	F^*
Constant delay, cf. (10.20a)	191	8	15248	6096	20.9858	23.9264815
Variable delay, cf. (10.20b)	207	8	16528	6608	16.3575	40.3828187
Variable delay, cf. (10.20c)	191	8	15248	6096	11.8072	40.6791184
Variable delay, cf. (10.20d)	199	9	15888	6352	22.4736	24.3653961
Variable delay, cf. (10.20e)	269	9	21488	8592	36.3875	13.3087632
Variable delay, cf. (10.20f)	254	11	20288	8112	19.9790	10.4068832
Variable delay, cf. (10.20g)	233	8	18608	7440	27.2819	10.0826024

10.7.2 Bounded Control

For the sake of illustration let us revisit the case with $\omega(t) = \frac{t}{2}$ and add lower bounds on the control variable, that is,

$$U(x,t) \geq 0. \tag{10.45}$$

As expected when the control bounds are imposed the optimal objective function value increases from $F^* = 10.0826024$ to $F^* = 14.6730062$. The solution is illustrated in Figures 10.2 and 10.3. It is clear from Figure 10.3 that the control bound prevents any negative values for $U(x,t)$. In particular, at the optimal solution some of the discretized control variables lie on the lower bound; that is, the *active set* is

$$\mathcal{A} = \left\{(k,j) \quad \text{such that} \quad |U^*(x_k,t_j)| \leq \epsilon\right\}$$

for some tolerance ϵ. These values are considered strictly active by the NLP algorithm. Figure 10.4 illustrates the control variable active set.

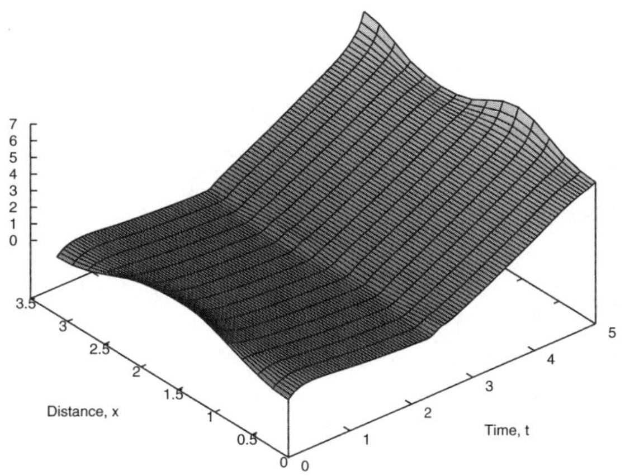

Figure 10.2. Optimal state $Y^*(x,t)$, $\omega(t) = \frac{t}{2}$, with spatial control with control bound (10.45).

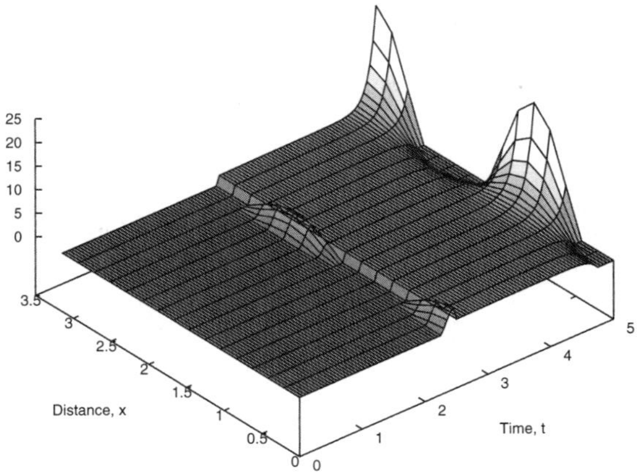

Figure 10.3. Optimal control $U^*(x,t)$, $\omega(t) = \frac{t}{2}$, with spatial control with control bound (10.45).

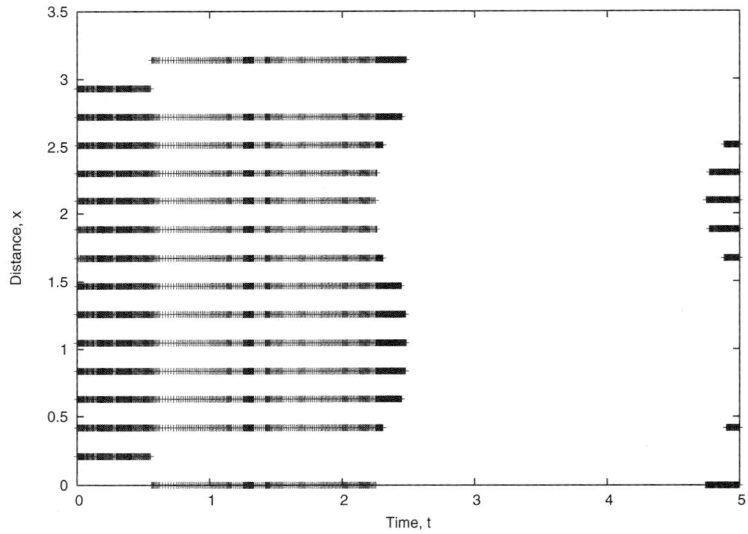

Figure 10.4. Optimal control active set, $U^*(x_k,t_j) = 0$, $\omega(t) = \frac{t}{2}$, with spatial control with control bound (10.45).

The active set is determined by the nonlinear program, and the computational cost is reflected in the overall problem solution times. Table 10.3 compares the performance of the sparse SQP and barrier algorithms for this example. Observe that the SQP algorithm which is an *active set* method is nearly four times as expensive as the primal-dual barrier algorithm. It is also interesting to note that the barrier algorithm required an additional mesh refinement iteration in comparison to the SQP algorithm. The mesh refinement algorithm corrects the small differences between the intermediate NLP solutions using a slightly different grid distribution.

Table 10.3. Performance comparison; SQP versus barrier NLP. Ref = grid refinement iteration, M = number of grid points, time is CPU in sec., ϵ is discretization error.

	SQP method			Barrier method		
Ref	M	ϵ	Time	M	ϵ	Time
1	11	1.1497×10^{-1}	.21797	11	1.1498×10^{-1}	.14298
2	21	3.1822×10^{-2}	.41694	21	3.1823×10^{-2}	.35395
3	21	9.6717×10^{-3}	2.9076	21	9.6716×10^{-3}	2.2317
4	41	6.8865×10^{-4}	9.3836	41	6.8819×10^{-4}	5.1452
5	77	3.9014×10^{-5}	39.045	76	3.8952×10^{-5}	13.165
6	95	7.2941×10^{-6}	58.623	94	7.2938×10^{-6}	20.785
7	189	1.5648×10^{-6}	284.59	187	1.5493×10^{-6}	54.485
8	206	3.1929×10^{-7}	338.99	204	3.1939×10^{-7}	61.543
9	243	1.3817×10^{-7}	440.23	243	1.3481×10^{-7}	82.291
10	248	9.8565×10^{-8}	469.76	248	1.7388×10^{-7}	70.759
11				253	9.8263×10^{-8}	75.775
Total			1644.2			386.68

10.8 Delay PDE with Spatial Control and Spatial Delay

Let us now consider an example in which the delay function is also spatially dependent. Instead of delay functions given by (10.20) let us consider a spatially dependent delay function similar to (10.20d):

$$\omega(x,t) = t - \frac{rt}{T}\left[\frac{4x(\pi - x)}{\pi^2}\right]. \tag{10.46}$$

Again using the method of lines formulation we must solve the system (10.28) with $n+1$ spatially dependent delay functions

$$\omega_j(t) = t - \frac{rt}{T}\left[\frac{4x_j(\pi - x_j)}{\pi^2}\right], \qquad x_j = j\delta, \qquad j = 0,\dots,n. \tag{10.47}$$

However, we can also exploit the spatial properties of this particular delay function to further simplify the formulation. First, observe that for the boundary lines at $x_0 = 0$ and $x_n = \pi$, there is *no* delay since $\omega_0(t) = \omega_n(t) = t$. Consequently, for this example $y_0(\omega_0) = y_0$ and $y_n(\omega_n) = y_n$. Furthermore, the spatial dependence is symmetric about the point $x = \pi/2$. Thus we have

$$\omega_j(t) = \omega_{n-j}(t), \qquad j < n/2. \tag{10.48}$$

When this spatial symmetry is reflected in the system (10.28) we obtain

$$\dot{y}_0 = \frac{2c_1}{\delta^2}(y_1 - y_0) - c_2 y_0 (1 + y_0) + u_0, \tag{10.49a}$$

$$\dot{y}_k = \frac{c_1}{\delta^2}(y_{k+1} - 2y_k + y_{k-1}) - c_2 y_k(\omega_k)(1 + y_k) + u_k, \quad k = 1, \dots, \nu, \tag{10.49b}$$

$$\dot{y}_k = \frac{c_1}{\delta^2}(y_{k+1} - 2y_k + y_{k-1}) - c_2 y_k(\omega_{n-k})(1 + y_k) + u_k, \quad k = \nu + 1, \dots, n-1, \tag{10.49c}$$

$$\dot{y}_n = \frac{2c_1}{\delta^2}(y_{n-1} - y_n) - c_2 y_n (1 + y_n) + u_n, \tag{10.49d}$$

where the quantity $n/2$ is treated as an integer, and $\nu \le n/2$.

10.8.1 Numerical Results

This example is characterized by delay and control functions that depend on both time and space. Figure 10.5 illustrates the behavior of the delay surface, and the optimal solution is illustrated in Figures 10.6 and 10.7. Notice that when the method of lines is applied to this problem, there is no state delay for the boundary lines at $x_0 = 0$ and $x_n = \pi$, leading to the ODEs (10.49a) and (10.49d). Consequently, the final problem has 16 states and 16 real controls, but only 14 pseudostates v_k needed to enforce consistency for the $\nu = 7$ distinct delay functions. A discretization error tolerance of 10^{-5} was used to refine the mesh and achieve approximately six significant figures of accuracy in the solution.

10.9 Summary and Conclusions

A new technique for computing the optimal control of delay-differential-algebraic dynamic systems has been introduced. The approach is illustrated by converting a problem described by a PDE with delays into a system of ODEs with delays. Results are presented for problems with constant delay, time varying delay, and delay that has both time and spatial dependence. Although we have focused on problems with delayed state variables, we expect that many of the ideas can be extended to control delays.

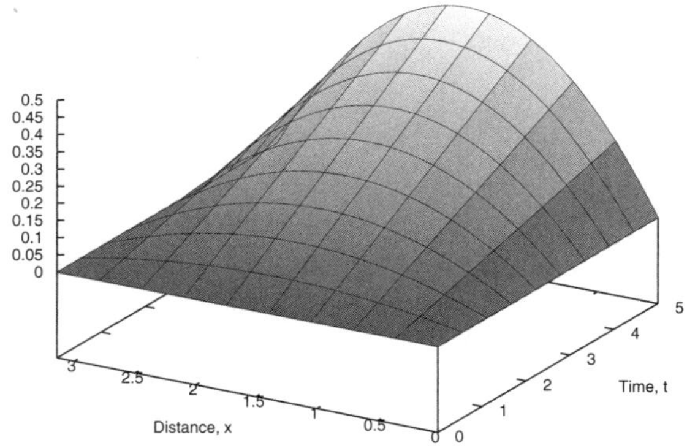

Figure 10.5. Delay function, $t - \omega(x,t) = \frac{rt}{T}\left(\frac{4x(\pi - x)}{\pi^2}\right)$.

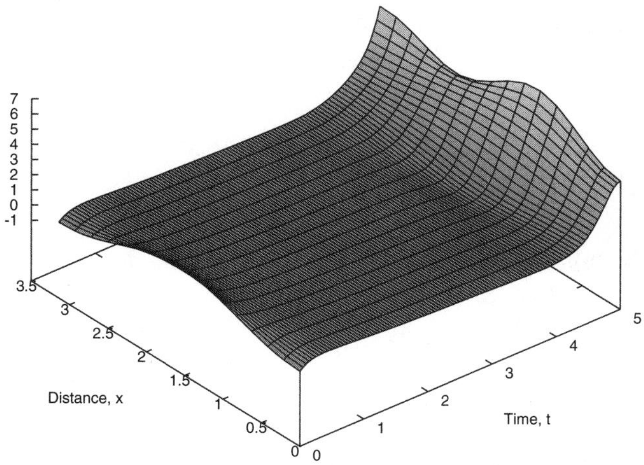

Figure 10.6. Optimal state $Y^*(x,t)$, $\omega(t) = t - \frac{rt}{T}\left(\frac{4x(\pi - x)}{\pi^2}\right)$, and spatial control.

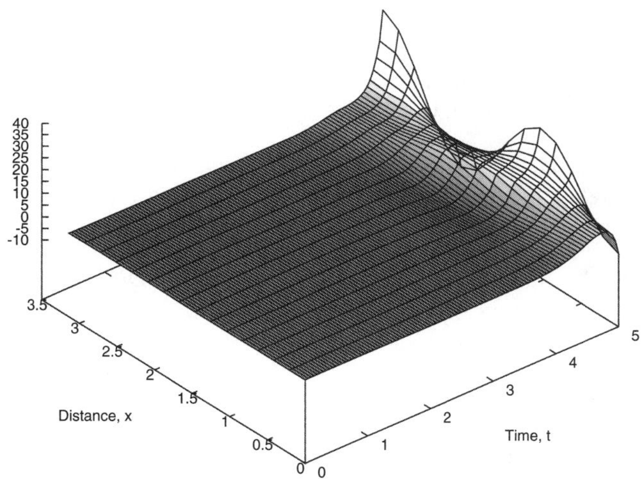

Figure 10.7. Optimal control $U^*(x,t)$, $\omega(t) = t - \frac{rt}{T}\left(\frac{4x(\pi - x)}{\pi^2}\right)$, and spatial control.

Acknowledgments

This research was supported in part by NSF grant DMS-0907832.

Bibliography

[1] U. M. ASCHER, R. M. M. MATTHEIJ, AND R. D. RUSSELL, *Numerical Solution of Boundary Value Problems for Ordinary Differential Equations*, Prentice–Hall, Englewood Cliffs, NJ, 1988.

[2] U. M. ASCHER AND L. R. PETZOLD, *The numerical solution of delay-differential-algebraic equations of retarded and neutral type*, SIAM J. Numer. Anal., 32 (1995), pp. 1635–1657.

[3] J. J. BATZEL, S. TIMISCHL-TESCHL, AND F. KAPPEL, *A cardiovascular-respiratory control system model including state delay with application to congestive heart failure in humans*, J. Math. Biol., 82 (2004), pp. 519–541.

[4] J. T. BETTS, *Practical Methods for Optimal Control and Estimation Using Nonlinear Programming*, 2nd ed., SIAM, Philadelphia, 2010.

[5] J. T. BETTS, N. BIEHN, S. L. CAMPBELL, AND W. P. HUFFMAN, *Compensating for order variation in mesh refinement for direct transcription methods*, J. Comput. Appl. Math., 125 (2000), pp. 147–158.

[6] C. BÜSKENS, L. GÖLLMANN, AND H. MAURER, *Optimal control of a stirred tank reactor with time delay*, in European Consortium of Mathematics in Industry, 1994.

[7] S. L. CAMPBELL AND R. MÄRZ, *Direct transcription solution of high index optimal control problems and regular Euler-Lagrange equations*, J. Comput. Appl. Math., 202 (2007), pp. 186–202.

[8] J. ČERMÁK AND J. JÁNKSKÝ, *On the asymptotics of the trapezoidal rule for the pantograph equation*, Math. Comp., 78 (2009), pp. 2107–2126.

[9] V. DESHMUKH, H. MA, AND E. BUTCHER, *Optimal control of parametrically excited linear delay differential systems via Chebyshev polynomials*, in Proceedings of the American Control Conference, Denver, CO, 2003.

[10] A. ENGELSONE, S. L. CAMPBELL, AND J. T. BETTS, *Direct transcription solution of higher-index optimal control problems and the virtual index*, Appl. Numer. Math., 57 (2007), pp. 281–296.

[11] L. GÖLLMANN, D. KERN, AND H. MAURER, *Optimal control problems with delays in state and control variables subject to mixed control-state constraints*, Optimal Control Appl. Methods, 30 (2009), pp. 341–365.

[12] V. GRETSCHKO, *Theorie und Numerik optimaler Steuerprozesse mit Retardierung und Steuer- und Zustandsbeshränkung*, Ph.D. thesis, Universität Münster, Münster, Germany, 2007.

[13] E. HAIRER, S. P. NORSETT, AND G. WANNER, *Solving Ordinary Differential Equations* I: *Nonstiff Problems*, Springer-Verlag, New York, 1993.

[14] M. LANDRY, S. A. CAMPBELL, K. MORRIS, AND C. O. AGUILAR, *Dynamics of an inverted pendulum with delayed feedback control*, SIAM J. Appl. Dyn. Syst., 4 (2005), pp. 333–351.

[15] H. MAURER, *Theory and applications of optimal control problems with control and state delays*, in Proceedings of the 53rd Australian Mathematical Conference, Adelaide, 2009.

[16] M. C. MEMORY, *Bifurcation and asymptotic behavior of solutions of a delay-differential equation with diffusion*, SIAM J. Math. Anal., 20 (1989), pp. 533–546.

[17] C. A. PAUL, *Designing Efficient Software for Solving Delay Differential Equations*, Tech. Rep. Numerical Analysis Report 368, Manchester Centre for Computational Mathematics, Department of Mathematics, University of Manchester, Manchester, England, 2000.

[18] L. F. SHAMPINE AND P. GAHINET, *Delay-differential-algebraic equations in control theory*, Appl. Numer. Anal., 56 (2005), pp. 574–588.

[19] R. F. STENGEL, R. GHIGLIAZZA, N. KULKARNI, AND O. LAPLACE, *Optimal control of innate immune response*, Optimal Control Appl. Methods, 23 (2002), pp. 91–104.

[20] Z.-Q. WANG AND L.-L. WANG, *A Legendre-Gauss collocation method for nonlinear delay-differential equations*, Discrete Contin. Dyn. Syst. Ser. B, 13 (2010), pp. 685–708.

[21] X.-P. YAN, *Stability and Hopf bifurcation for a delayed predator-prey system with diffusion effects*, Appl. Math. Comput., 192 (2007), pp. 552–566.

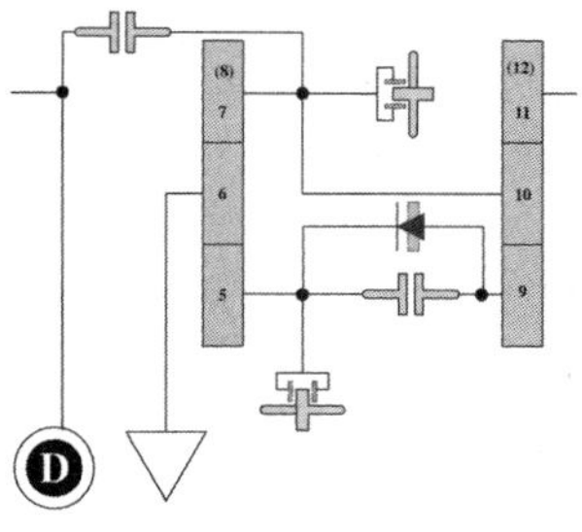

Chapter 11

Direct Transcription with Moving Finite Elements

Weifeng Chen, Kexin Wang, Zhijiang Shao, and Lorenz T. Biegler

A direct transcription formulation based on moving finite elements is developed for the solution of a class of optimal control problems. Novel features of the algorithm include the direct location of breakpoints for control profiles and a termination criterion based on a constant Hamiltonian profile. The algorithm is demonstrated on six optimal control problems drawn from the literature and from chemical process engineering.

11.1 Introduction

Direct transcription methods have seen considerable development over the last two decades. Here, both the discretized DAE model and the optimal control problem are formulated as a single NLP problem. Typically, the DAE is discretized to a set of algebraic equations over finite elements in time, and IRK methods are popular choices for this discretization. Moreover, with the development of large-scale NLP solvers, this approach can now be applied to challenging dynamic optimization problems, and a number of advantages can be realized. A key feature of direct transcription is that the embedded solution of the DAE model is no longer required. Moreover, exact first and second derivative information for the NLP is readily obtained from the algebraic formulation resulting from the discretization. Consequently, time-consuming, and often noisy, sensitivity calculations tied to the DAE solver are avoided. Also, failures of the embedded DAE solver are bypassed; this issue is especially important for DAE systems that have unstable forward modes. Finally, the large-scale NLP formulation is often sparse and structured, and this allows flexible decomposition strategies for efficient solution of the NLP.

On the other hand, an important issue in the development of direct transcription methods is the choice of the finite element mesh. This choice is guided by two objectives. First, the finite element mesh is selected to ensure accuracy of the DAE model. Given a suitably accurate IRK discretization, accurate solutions to a (semi-explicit index-one) DAE can be obtained with a sufficiently fine mesh. On the other hand, a mesh must also be selected that determines the optimal location of the control variable breakpoints. In [4], mesh selection and refinement are performed through the solution of a sequence of NLPs, each with a fixed mesh. This approach leads to a well-defined set of optimization problems, but it can be slow in the determination of breakpoints. Moreover, it is likely that the mesh refinement may require many more finite elements than needed to determine accurate solution profiles.

This chapter explores an alternate direct transcription strategy where *moving finite elements* are embedded within the solution of the optimization problem. The approach also requires a mesh refinement stage, but it contains two important features: direct determination of the breakpoint as part of the finite element mesh, and a stopping criterion based on the profile of the Hamiltonian function. The incorporation of moving finite elements also introduces a number of additional difficulties. A variable mesh introduces additional nonlinearity into the equality constraints, and therefore (additional) nonconvexity into the NLP formulation; this may lead to slower and more difficult convergence behavior, as the solution of the NLP may be insensitive to many of the mesh variables. Consequently, a careful NLP formulation and initialization is required for a successful optimization strategy, along with a bootstrapping strategy with the solution of intermediate subproblems.

In the next section, we motivate the optimal determination of breakpoints through the formulation and analysis of multiperiod dynamic optimization problems, and describe the problem class addressed in this chapter. The third section considers the discretization of these dynamic optimization problems and translates these time periods to finite elements. In particular, we incorporate error constraints that ensure state profiles are determined accurately, and that the set of active constraints remains constant within a finite element. Moreover, we relate the direct transcription solution to the multiperiod formulation through location of breakpoints and element junctions. The fourth section proposes an overall strategy for an adaptive moving finite element method, along with details of the implementation. These include the initialization of feasible profiles, solution of the NLP with moving finite elements, assessment of Hamiltonian profiles, and mesh refinement. This strategy is demonstrated on six examples in the fifth section. Here we consider two nonsingular examples and four singular problems which are drawn from the literature and from chemical process engineering applications. The last section concludes the chapter and offers directions for future work.

11.2 Multiperiod Dynamic Optimization

We consider the solution of optimal control problems of the form

$$\begin{aligned}
\min \quad & \Phi(z(t_f)) && (11.1a)\\
\text{s.t.} \quad & \frac{dz}{dt} = f(z(t), y(t), u(t), p), \quad z(t_0) = z_0(p), && (11.1b)\\
& g(z(t), y(t), u(t), p) = 0, && (11.1c)\\
& u_L \le u(t) \le u_U, p_L \le p \le p_U, && (11.1d)\\
& \psi(z(t_f)) \le 0, && (11.1e)
\end{aligned}$$

where (11.1b)–(11.1c) is a semi-explicit index-one DAE system with $t \in [t_0, t_f]$ as time, $z(t) \in \mathbf{R}^{n_z}$ and $y(t) \in \mathbf{R}^{n_y}$ as differential and algebraic state variables, and $u(t) \in \mathbf{R}^{n_u}$ and

$p \in \mathbf{R}^{n_p}$ as control and time-independent optimization variables. We assume that the objective function (11.1a) and final time conditions (11.1e) depend only on the differential state at final time. Otherwise, inequality constraints are given as simple bounds (11.1d). Because they lead to high-index arcs at the optimal solution, we do not consider bounds on the state variables in this study. Optimality conditions for state-constrained problems require analytic differentiation of the path constraint and lead to a much more difficult problem. Consequently, we leave this extension for future research.

Also we assume that $f(\cdot)$, $g(\cdot)$, $\Phi(\cdot)$, and $\psi(\cdot)$ are smooth, bounded functions with respect to their arguments. At the solution of (11.1), control profiles are allowed to be nonsmooth (e.g., discontinuous or nondifferentiable) at only a finite number of points (t_l), termed breakpoints; between breakpoints they are analytic functions of time. With these restrictions, we represent problem (11.1) as the following multiperiod problem over a number of time periods, $l = 1, \dots, N$, $t \in [t_{l-1}, t_l]$, with $t_N = t_f$, where t_l also serve as decision variables in problem (11.2):

$$\min \quad \Phi(z^N(t_f)) \tag{11.2a}$$

$$\text{s.t.} \quad \frac{dz^1}{dt} = f(z^1(t), y^1(t), u^1(t), p), \quad z^1(t_0) = z_0(p), \tag{11.2b}$$

$$\frac{dz^l}{dt} = f(z^l(t), y^l(t), u^l(t), p), \quad z^l(t_{l-1}) = z^{l-1}(t_{l-1}),\ l = 2, \dots, N, \tag{11.2c}$$

$$g(z^l(t), y^l(t), u^l(t), p) = 0, \quad l = 1, \dots, N, \tag{11.2d}$$

$$u_L \le u^l(t) \le u_U, p_L \le p \le p_U, \quad t \in (t_{l-1}, t_l],\ l = 1, \dots, N, \tag{11.2e}$$

$$\psi(z^N(t_f)) \le 0. \tag{11.2f}$$

We also assume that only differential state variables are assumed to be continuous across period boundaries, and that $z^l(t)$ and $y^l(t)$ are otherwise smooth, bounded functions of time. Finally, to further define the problem class for our moving finite element strategy, we assume that the solution of problem (11.2) has control profiles with $u^l_j(t) = u_{L,j}(t)$, $u^l_j(t) = u_{U,j}(t)$, or $u_{L,j}(t) < u^l_j(t) < u_{U,j}(t)$, $t \in (t_{l-1}, t_l)$, $j = 1, \dots, n_u$. In particular, we assume that all touchpoints (at the bounds of $u(t)$) are located only at t_l.

This allows us to define the solutions in each period of problem (11.2) by the following optimality conditions:

$$\frac{dz^1}{dt} = f(z^1(t), y^1(t), u^1(t), p), \quad z^1(t_0) = z_0(p), \tag{11.3a}$$

$$\frac{dz^l}{dt} = f(z^l(t), y^l(t), u^l(t), p), \quad z^l(t_{l-1}) = z^{l-1}(t_{l-1}),\ l = 2, \dots, N, \tag{11.3b}$$

$$g(z^l(t), y^l(t), u^l(t), p) = 0, \tag{11.3c}$$

$$\frac{d\lambda^l}{dt} = -\left[\frac{\partial f}{\partial z}\lambda^l + \frac{\partial g}{\partial z}\nu^l\right], \quad \lambda^l(t_l) = \lambda^{l+1}(t_l), \tag{11.3d}$$

$$\frac{\partial f}{\partial y^l}\lambda^l + \frac{\partial g}{\partial y^l}\nu^l = 0, \tag{11.3e}$$

$$u^l_j(t) \in (u_{j,L}, u_{j,U}), \quad \eta^l_L(t) = \eta^l_U(t) = 0, \tag{11.3f}$$

$$\vee\, u_j^l(t) = u_{j,L}, \quad \eta_L^l(t) \geq 0, \quad \eta_U^l(t) = 0,$$

$$\vee\, u_j^l(t) = u_{j,U}, \quad \eta_U^l(t) \geq 0, \quad \eta_L^l(t) = 0,\ t \in (t_{l-1}, t_l), \tag{11.3g}$$

$$\frac{\partial f}{\partial u^l}\lambda^l + \frac{\partial g}{\partial u^l}\nu^l + \eta_U^l - \eta_L^l = 0, \quad l = 1, \ldots, N, \tag{11.3h}$$

$$\frac{\partial z_0}{\partial p}\lambda(0) + \sum_{l=1}^{N} \int_{t_{l-1}}^{t_l} \frac{\partial f}{\partial p}\lambda^l + \frac{\partial g}{\partial p}\nu^l dt = 0, \tag{11.3i}$$

$$\lambda(t_f) = \frac{\partial \Phi(z_f)}{\partial z} + \frac{\partial \psi(z_f)}{\partial z}\eta_f, \tag{11.3j}$$

$$0 \leq \eta_f \perp \psi(z(t_f)) \leq 0, \tag{11.3k}$$

where $a \perp b$ indicates that the vector elements have either $a_i = 0$ or $b_i = 0$, or both. While problem (11.2) contains some restrictive assumptions, the multiperiod formulation and the optimization variables t_l allow the representation of many dynamic optimization problems, along with the direct optimization of their breakpoints. Also, as is well known [5, 7, 14], the Hamiltonian function for problem (11.1)

$$H = f(z, y, u, p)^T \lambda(t) + g(z, y, u, p)^T \nu(t) + (u_L - u)^T \eta_L(t) + (u - u_U)^T \eta_U(t) \tag{11.4}$$

is constant over time at the optimal solution. Using the conditions in (11.3) we have [7]

$$\begin{aligned}\frac{dH}{dt} &= \frac{\partial H}{\partial z}\frac{dz}{dt} + \frac{\partial H}{\partial \lambda}\frac{d\lambda}{dt} + \frac{\partial H}{\partial y}\frac{dy}{dt} + \frac{\partial H}{\partial \nu}\frac{d\nu}{dt} + \frac{\partial H}{\partial \eta_L}\frac{d\eta_L}{dt} + \frac{\partial H}{\partial \eta_U}\frac{d\eta_U}{dt} \\ &+ \frac{\partial H}{\partial u}\frac{du}{dt} + \frac{\partial H}{\partial p}\frac{dp}{dt} = 0, \quad t \in (t_{l-1}, t_l),\ l = 1, \ldots, N.\end{aligned} \tag{11.5}$$

We will see that this condition is especially useful for the detection of breakpoint locations.

In the next section, we approximate the solution of (11.2) with a direct transcription formulation, using the conditions in (11.3) and assumptions mentioned above.

11.3 NLP Formulations and Solution

To provide a precise definition of our direct transcription formulation, we introduce an orthogonal collocation formulation based on Lagrange polynomials using either Gauss or Radau points on finite elements. Here we choose $K+1$ interpolation points in element i and represent the state in a given element i as

$$z^K(t) = \sum_{j=0}^{K} \ell_j(\tau) z_{ij}, \quad \ell_j(\tau) = \prod_{k=0, \neq j}^{K} \frac{(\tau - \tau_k)}{(\tau_j - \tau_k)}, \quad t \in [t_{i-1}, t_i], \tau \in [0,1], \tag{11.6}$$

where $t = t_{i-1} + h_i \tau$, $\tau_0 = 0$, $0 < \tau_j \leq 1$, $j = 1, \ldots, K$, are shifted Gauss or Radau points and h_i is the length of element i. This polynomial representation has the desirable property that $z^K(t_{ij}) = z_{ij}$, where $t_{ij} = t_{i-1} + \tau_j h_i$. For $N > 1$, we also enforce continuity of the differential state profiles across element boundaries, written as

$$z_{i+1,0} = \sum_{j=0}^{K} \ell_j(1) z_{ij}, \quad i = 1,\dots,N-1, \tag{11.7a}$$

$$z_f = \sum_{j=0}^{K} \ell_j(1) z_{Nj}, \quad z_{1,0} = z_0. \tag{11.7b}$$

In addition, the control variables and algebraic states can also be represented by Lagrange interpolation profiles at K points:

$$u^K(t) = \sum_{j=1}^{K} \bar{\ell}_j(\tau) u_{ij}, \quad y^K(t) = \sum_{j=1}^{K} \bar{\ell}_j(\tau) y_{ij}, \bar{\ell}_j(\tau) = \prod_{k=1,\neq j}^{K} \frac{(\tau - \tau_k)}{(\tau_j - \tau_k)}. \tag{11.8}$$

Substituting (11.6) and (11.8) into (11.1b) and (11.1c) leads to the collocation equations written as

$$\sum_{j=0}^{K} \dot{\ell}_j(\tau_k) z_{ij} - h_i f(z_{ik}, y_{ik}, u_{ik}, p) = 0, \tag{11.9a}$$

$$g(z_{ik}, y_{ik}, u_{ik}, p) = 0, \quad i = 1,\dots,N,\ k = 1,\dots,K, \tag{11.9b}$$

where $\dot{\ell}_j(\tau) = \frac{d\ell_j(\tau)}{d\tau}$. The NLP formulation corresponding to the optimal control problem (11.1) can be written as

$$\min\ \Phi(z_f) \tag{11.10a}$$

$$\text{s.t.} \sum_{j=0}^{K} \dot{\ell}_j(\tau_k) z_{ij} - h_i f(z_{ik}, y_{ik}, u_{ik}, p) = 0, \tag{11.10b}$$

$$g(z_{ik}, y_{ik}, u_{ik}, p) = 0, \tag{11.10c}$$

$$u_L \le u_{ik} \le u_U, \quad p_L \le p \le p_U, \quad k = 1,\dots,K,\ i = 1,\dots,N, \tag{11.10d}$$

$$z_{i+1,0} = \sum_{j=0}^{K} \ell_j(1) z_{ij}, \quad i = 1,\dots,N-1, \tag{11.10e}$$

$$z_f = \sum_{j=0}^{K} \ell_j(1) z_{Nj}, \quad z_{1,0} = z(t_0), \tag{11.10f}$$

$$\psi(z_f) \le 0. \tag{11.10g}$$

Problem (11.10) allows for a number of formulation options, especially when extended to multiperiod problems. For the multielement formulation, it is important to note from (11.8) that the algebraic and control profiles are *not* specified at $\tau = 0$, and continuity is not enforced at the element boundary t_i for these profiles. For algebraic profiles defined by index-one constraints (11.9b), continuity of these profiles can be determined directly from the continuous differential variables at t_i. On the other hand, control profiles are allowed to be discontinuous at element boundaries.

In this way, we can directly relate the periods indexed by l in (11.2), each containing DAEs, and the finite elements indexed by i in (11.10), with collocation equations in each element, along with continuity of the differential variables and breakpoint locations for the controls. We note that (11.10) provides an approximation to (11.2) if we can maintain h_i sufficiently small, so that the collocation approximation is a sufficiently accurate approximation to the DAE system (11.2c)–(11.2d). This approximation can be enforced with the addition of error constraints as discussed next.

11.3.1 Treatment of Finite Elements

An important concern in the direct transcription approach is the appropriate selection of element lengths $h_i = t_i - t_{i-1}$. Clearly, if h_i is fixed in the NLP, then the resulting problem is less nonlinear and often easier to solve. In fact, for a linear DAE system, fixing h_i leads to a linearly constrained NLP. On the other hand, treating h_i as variables in (11.10), where they can "move" during the solution of the NLP, provides a number of advantages. Provided that values of h_i remain suitably small, variable finite elements can locate breakpoints in the optimal control profile as well as profile segments with active bounds. Moreover, as Runge–Kutta methods require state profiles to remain smooth *only within a finite element*, problem (11.10) can deal with variable finite elements using the following guidelines.

First, the number of finite elements N must be chosen sufficiently large. This is often estimated by trial and error through comparison with numerically integrated state profiles evaluated at different values of the decision variables p and $u(t)$. In the next section, this forms the first phase of our moving element strategy.

Moving finite element formulations are described extensively in [2, 16] for the solution of BVPs. Moreover, assuming that the state profiles are smooth over the domain of interest, the *global error* from the polynomial approximation, $e(t) = z(t) - z^K(t)$, can be estimated from

$$\max_{t\in[0,t_f]} \|e(t)\| \leq C_1 \max_{i\in\{1,\dots,N\}} \left(\max_{t\in[t_{i-1},t_i]} \|T_i(t)\| \right) + O(h_i^{K+2}), \tag{11.11}$$

where C_1 is a computable constant and $T_i(t)$ can be computed from the polynomial solution. Choices for $T_i(t)$ are reviewed in Russell and Christiansen [16]. In particular, one can compute residual DAEs at noncollocation points, t_{nc}, by choosing

$$T_i(t_{nc}) = \begin{bmatrix} \frac{dz^K(t_{i,nc})}{d\tau} - h_i f(z^K(t_{i,nc}), y^K(z(t_{i,nc})), u^K(t_{i,nc}), p) \\ g(z^K(t_{i,nc}), y^K(z(t_{i,nc})), u^K(t_{i,nc}), p) \end{bmatrix}.$$

As shown in [16], with $t_{i,nc} = t_{i-1} + h_i \tau_{nc}$, $\tau_{nc} \in [0,1]$, this leads to $\|e_i(t)\| \leq \bar{C}\ \|T_i(t_{i,nc})\|$, with the constant $\bar{C}$ given by

$$\bar{C} = \frac{1}{A} \int_0^{\tau_{nc}} \prod_{j=1}^{K} (s - \tau_j)\, ds, \quad A = \prod_{j=1}^{K} (\tau_{nc} - \tau_j).$$

With this estimate, N sufficiently large and a user-specified error tolerance ϵ, appropriate values of h_i can be determined by adding the constraints (11.12) to (11.10):

$$\sum_{i=1}^{N} h_i = t_f, \quad h_i \geq 0, \tag{11.12a}$$

$$\bar{C}\ \|T_i(t_{i,nc})\| \leq \epsilon. \tag{11.12b}$$

Adding (11.12) to (11.10) leads to more highly constrained NLPs, and successful implementation requires careful solution strategies, as considered in [18, 3].

In addition, we need to consider the restrictions on the control problems given by (11.3f)–(11.3g). For the polynomial approximations given by (11.8), we need to guard against two difficulties. First, it is clear that the approximate control profile may violate its bounds between collocation points. Second, the set of active bounds may change within a given element, and this may lead to nonsmoothness in the state profiles within that element, and poor approximations using (11.6). These problems are avoided if we represent the control profile as piecewise constant or piecewise linear approximations within each element, and the restrictions imposed by (11.3f)–(11.3g) can be obtained directly. For piecewise constant control we define

$$u_{ik} = u_i, \quad u_L \leq u_i \leq u_U, \tag{11.13}$$

while the piecewise linear case is given as

$$u_{ik} = u_{i,a}(1-\tau_k) + u_{i,b}\,\tau_k, \quad u_L \leq u_{i,a}.u_{i,b} \leq u_U. \tag{11.14}$$

The combined NLP formulation can therefore be completed by adding the constraints (11.12) and either (11.13) or (11.14) to (11.10).

11.3.2 Background Properties

The above ingredients for the moving finite element formulation are supported by the following properties:

- As shown in [12, 6], the KKT multipliers from the solution of (11.10) can be related to discretized approximations of adjoint profiles, $\lambda(t)$ and $\nu(t)$. In particular, the quantities $\omega_k \lambda_{ik}$ and $\omega_k h_i \nu_{ik}$ are the multipliers for (11.10b) and (11.10c), respectively, where $\omega_k = \int_0^1 \bar{\ell}_k(\tau)d\tau$ is the quadrature weight.

- For *nonsingular* control problems, it can be shown [12] for the *unconstrained* periods, i.e., (11.3f) holds (and with fixed state and adjoint boundary values), that $\|z^K(t) - z(t)\|$, $\|u^K(t) - u(t)\|$, $\|y^K(t) - y(t)\|$, $\|\lambda^K(t) - \lambda(t)\|$, and $\|\nu^K(t) - \nu(t)\|$ have convergence rates of $O(h^q)$, with $h = \max_i h_i$, and $q = K$ or $K+1$ for Radau and Gauss collocation, respectively. (Since $u(t)$ is allowed only to be piecewise constant (11.13) or linear (11.14) here, we would expect q to be reduced to 1 or 2, respectively.)

- For the constrained periods with (11.3g), control variables are fixed, and therefore state and adjoint DAEs follow the same $O(h^q)$ convergence rates as well.

- To ensure that the breakpoints are located correctly, we use these adjoint approximations to approximate the Hamiltonian and apply a condition based on (11.5). This is also related to an approach described in [17].

11.4 Moving Finite Element Optimization Strategy

We now describe a systematic approach for initializing and solving the direct transcription NLP with mesh refinement. In particular, careful NLP formulation and initialization are required for a successful optimization strategy, along with a bootstrapping strategy with the solution of intermediate subproblems. While this approach shares some concepts with the mesh refinement strategy described in [4], key differences are the use of breakpoints for control profiles and the Hamiltonian termination criterion based on (11.5).

A straightforward way to generate an initial mesh is to fix $u(t)$ and p to initial (guessed) values $\bar{u}(t)$ and $\bar{p}$ and solve a sequence of NLPs indexed by $i = 1,2,3,\ldots$ and given by

$$\bar{h}_i = \arg\left\{ \max h_i \right. \tag{11.15a}$$

$$\text{s.t.} \sum_{j=0}^{K} \dot{\ell}_j(\tau_k) z_{ij} - h_i f(z_{ik}, y_{ik}, \bar{u}_{ik}, \bar{p}) = 0, \tag{11.15b}$$

$$g(z_{ik}, y_{ik}, \bar{u}_{ik}, \bar{p}) = 0, \quad k = 1, K, \tag{11.15c}$$

$$-\epsilon \le \bar{C}\; T_i(t_{i,nc}) \le \epsilon, \tag{11.15d}$$

$$\left. 0 \le h_i \le t_f - \sum_{i'=1}^{i-1} \bar{h}_{i'} \right\}, \tag{11.15e}$$

where $z_{1,0} = z_0$, and for $i \ge 2$, $z_{i,0} = \sum_{j=0}^{K} \ell_j(1) z_{i-1,j}$, solved in the previous interval. This approach is essentially an initial value solver for the DAE system using a fully implicit finite element method with time steps $\bar{h}_i$; it is particularly suitable for DAEs with stable forward modes. This approach leads to a feasible initialization for the following NLP with the N moving finite elements determined from (11.15):

$$\min \Phi(z_f) \tag{11.16a}$$

$$\text{s.t.} \sum_{j=0}^{K} \dot{\ell}_j(\tau_k) z_{ij} - h_i f(z_{ik}, y_{ik}, u_{ik}, p) = 0, \tag{11.16b}$$

$$g(z_{ik}, y_{ik}, u_{ik}, p) = 0, \tag{11.16c}$$

$$u_L \le u_{ik} \le u_U, \quad p_L \le p \le p_U, \tag{11.16d}$$

$$k \in \{1,\ldots,K\},\ i \in \{1,\ldots,N\},$$

$$z_{i+1,0} = \sum_{j=0}^{K} \ell_j(1) z_{ij}, \quad i = 1,\ldots,N-1, \tag{11.16e}$$

$$z_f = \sum_{j=0}^{K} \ell_j(1) z_{Nj}, \quad z_{1,0} = z(t_0), \tag{11.16f}$$

$$\psi(z_f) \le 0, \tag{11.16g}$$

$$-\epsilon \leq \bar{C}\, T_i(t_{i,nc}) \leq \epsilon, \tag{11.16h}$$

$$u_{ik} = u_i, \quad u_L \leq u_i \leq u_U, \quad \text{or} \tag{11.16i}$$

$$u_{ik} = u_{i,a}(1-\tau_{ik}) + u_{i,b}\tau_{ik}, \quad u_L \leq u_{i,a}, u_{i,b} \leq u_U, \tag{11.16j}$$

$$0 \leq h_i \leq t_f, \quad \sum_{i=1}^{N} h_i = t_f. \tag{11.16k}$$

Here the finite element lengths are initialized to nominal values generated in (11.15). We find that the error constraints (11.16h) are quite useful to prevent large changes of the breakpoints. On the other hand, since the optimal solution is unknown a priori, solution of (11.16) may lead to a suboptimal profiles if there are insufficient elements to represent the optimal solution. Consequently, a provision is needed for element addition.

To address this issue we develop a Hamiltonian-based termination criterion. Once this criterion is solved within tolerance, we declare a sufficiently accurate solution to (11.1). As shown above, KKT multipliers of the collocation equations provide an approximation on the adjoint variables $\lambda(t)$ and $\nu(t)$. Consequently, we can approximate the Hamiltonian profile at the collocation points as follows:

$$H_{(i-1)K+k} = \lambda_{ik} f(z_{ik}, y_{ik}, u_{ik}, p) + \nu_{ik} g(z_{ik}, y_{ik}, u_{ik}, p) \tag{11.17}$$

and assess whether it is constant over time. We could also use this information at the solution of (11.16) to ensure that the adjoint equations (11.3d), (11.3e), and (11.3h) are satisfied to a suitable error tolerance. However, because deriving error constraints based on (11.3d), (11.3e), and (11.3h) may be cumbersome, we leave this step as an option in our solution strategy and present numerical results without this step. In our tests so far, we have found that error constraints on (11.3d)–(11.3e), (11.3h) were not needed as long as the termination criterion on (11.17) was satisfied.

The resulting mesh refinement strategy is given as follows:

1. (Optional) Divide the time domain equally with a small number of fixed finite elements and solve (11.10) and (11.13) or (11.14) to provide an initial guess for $u(t)$.

2. Having an initial guess for $u(t)$ and p, solve a sequence of NLPs (11.15) for $i = 1,2,3\ldots$ while $t_f - \sum_{i'=1}^{i-1} \bar{h}_{i'} > 0$ to determine an initial finite element mesh $\bar{h}_i$.

3. Combine adjacent finite elements if $u(t)$ can be expressed with the same interpolation function (within a small tolerance). Solve (11.16) and check the length of finite elements. Delete all elements with $h_i \leq \epsilon_h\, t_f$. Resolve (11.16) and, if necessary, reapply element deletion until no small elements remain.

4. Calculate $H_{(i-1)K+k}$ from (11.17) and check

   $$|H_j - H_{j-1}| \leq \max(\gamma_1 |H_j|, \gamma_2), \quad j = 2,\ldots,KN,$$

   where γ_1 and γ_2 are small positive constants. If the Hamiltonian criterion (and, optionally, error conditions for (11.3d), (11.3e), and (11.3h)) are satisfied, Stop. Otherwise, if these conditions are not satisfied for element i, divide this element into two equally spaced elements and return to step 3.

11.5 Examples

We consider six examples to illustrate the mesh refinement strategy. The first two examples are small, nonsingular optimal control problems. The first has a nondifferential control profile with an active segment at the upper bound. The mesh refinement strategy is used to locate this segment. The second example has no active bounds and demonstrates how mesh refinement works for this simpler case.

We also consider four singular control problems.These problems are characterized by breakpoints in their control problems, and care is required in the selection of finite elements. The first two singular problems are small examples that have analytical solutions, while the last two examples are based on industrial-scale process problems.

The mesh refinement strategy was implemented without checking whether error conditions for (11.3d), (11.3e), and (11.3h) were satisfied. The approach was programmed in the AMPL [8] modeling environment and all of these examples are solved with IPOPT 3.8.3 [19] running on an Intel(R) CPU T2300, 1.66 GHz processor with 1.5 GB RAM. For these examples, we use the following constants: $K = 3$, $\epsilon = 10^{-4}$, $\epsilon_h = 10^{-4}$, $\gamma_1 = 10^{-3}$, and $\gamma_2 = 10^{-6}$. The largest example was solved in less than 60 CPUs, and the others required considerably less computation than this.

11.5.1 Nonsingular Control Problems

Each of the following two examples has only two states and a single control. For both examples, the piecewise linear form (11.14) is used to represent the control profile.

Batch Temperature Profile with Parallel Reactions

Consider a nonisothermal batch reactor with first order parallel reactions $A \to B, A \to C$, where the goal is again to find a (transformed) temperature profile that maximizes the final amount of product B after one hour ($t_f = 1$). The optimal control problem can be stated as

$$\min \quad -b(1) \tag{11.18a}$$

$$\text{s.t.} \quad \frac{da}{dt} = -a(t)(u(t) + u(t)^2/2), \tag{11.18b}$$

$$\frac{db}{dt} = a(t)u(t), \tag{11.18c}$$

$$a(0) = 1, \quad b(0) = 0, \quad u(t) \in [0,5]. \tag{11.18d}$$

For the first case, we solve only problems (11.15) and (11.16). This leads to 11 finite elements and allows some movement of the finite elements, but it does not ensure a constant Hamiltonian. For the second case, the complete moving finite element strategy is performed. The number of finite elements now increases to 17. Here, the NLP (11.16) is solved 3 times and the objective is −0.57354398, with the kink observed at $t_s = 0.9513$ where the active control segment begins. The numerical results for both cases are shown

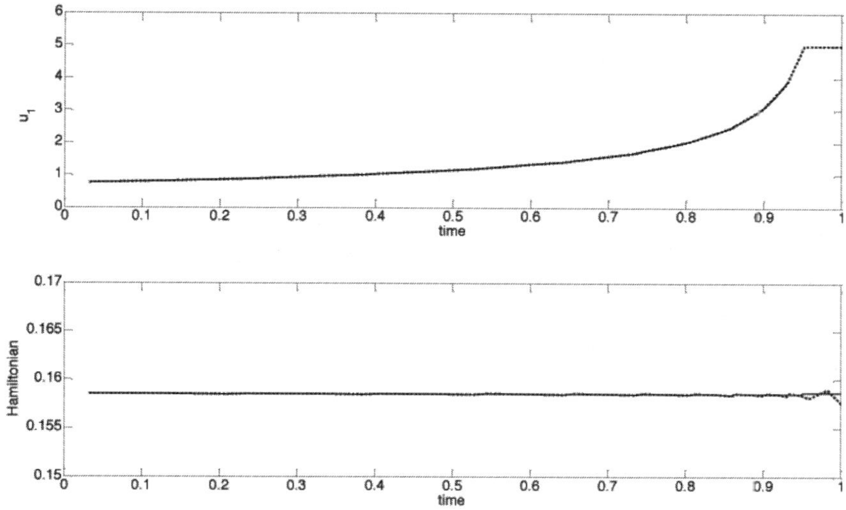

Figure 11.1. Control and Hamiltonian profiles for batch reactor in Case 1 (without element addition, dotted line) and Case 2 (with element addition, solid line).

in Figure 11.1 for the control profile and the Hamiltonian. Through the placement of additional finite elements, we observe a significant improvement in the Hamiltonian profile.

Rayleigh Problem

In this example we find $u(t)$ that solves

$$\min \quad \int_0^{t_f=2.5} x_1^2 + u^2 dt \tag{11.19a}$$

$$\text{s.t.} \quad \frac{dx_1}{dt} = x_2, \tag{11.19b}$$

$$\frac{dx_2}{dt} = -x_1 + (1.4 - 0.14x_2^2)x_2 + 4u, \tag{11.19c}$$

$$x_1(0) = -5, \quad x_2(0) = -5. \tag{11.19d}$$

For the first case, we solve only problems (11.15) and (11.16). This leads to 21 finite elements movement of the finite elements, but again does not ensure a constant Hamiltonian. Also, note that a spike appears at about $t = 0.1$ in the control profile, which is due to a very small ($h = 10^{-7}$) finite element.

For the second case, the complete moving finite element strategy is performed. The number of finite elements now increases to 52 and the NLP (11.16) is solved 5 times and the objective is 29.3760878. The numerical results in both steps are shown as Figure 11.2 for the control and Hamiltonian profiles. Here, the spike disappears and smooth profiles are found.

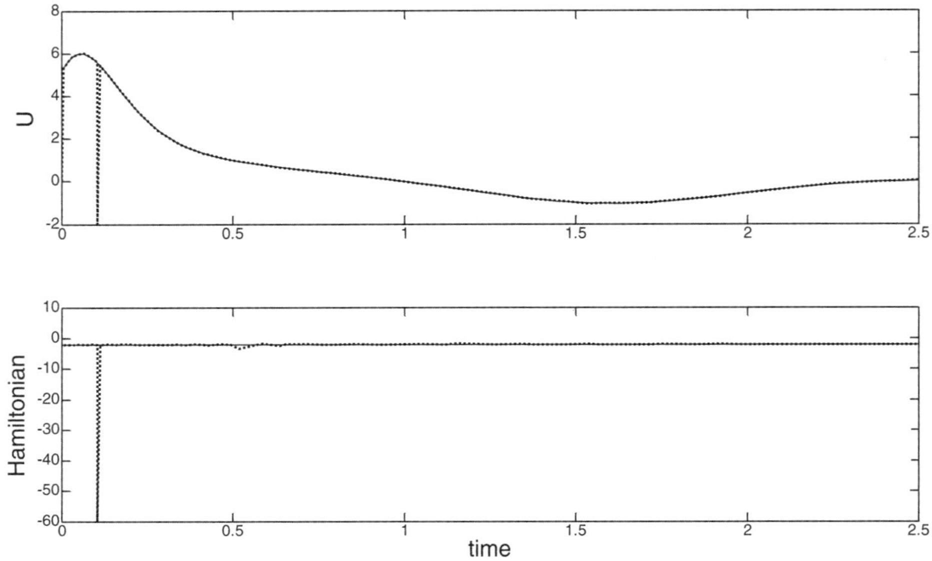

Figure 11.2. Control and Hamiltonian profiles for Rayleigh problem in Case 1 (without element addition, dotted line) and Case 2 (with element addition, solid line).

11.5.2 Singular Control Problems

Optimal control problems with singular arcs lack coercivity conditions (see, e.g., [4, 9, 12, 13]) that are required for a convergent direct transcription strategy. Consequently, in the absence of problem-specific information, they can lead to ill-conditioned discretizations and widely inaccurate discretizations as the mesh is refined. Here we show that moving finite elements with Hamiltonian-based termination can avoid ill-conditioning and lead to an accurate solution by limiting the refinement of the mesh.

Aly–Chan Singular Control Problem

Consider the singular optimal control problem [1] given by

$$\min_{u} z_3\left(\frac{\pi}{2}\right) \tag{11.20a}$$

$$\text{s.t. } \frac{dz_1}{dt} = z_2; \quad z_1(0) = 0; \tag{11.20b}$$

$$\frac{dz_2}{dt} = u; \quad z_2(0) = 1; \tag{11.20c}$$

$$\frac{dz_3}{dt} = \frac{1}{2}z_2^2 - \frac{1}{2}z_1^2; \quad z_3(0) = 0; \tag{11.20d}$$

$$-1 \le u \le 1. \tag{11.20e}$$

The analytic solution of this problem has a singular arc ($u(t) = -\sin t$) over the entire time domain, $t \in [0, \pi/2]$. On the other hand, Figure 11.3 presents the results of solving

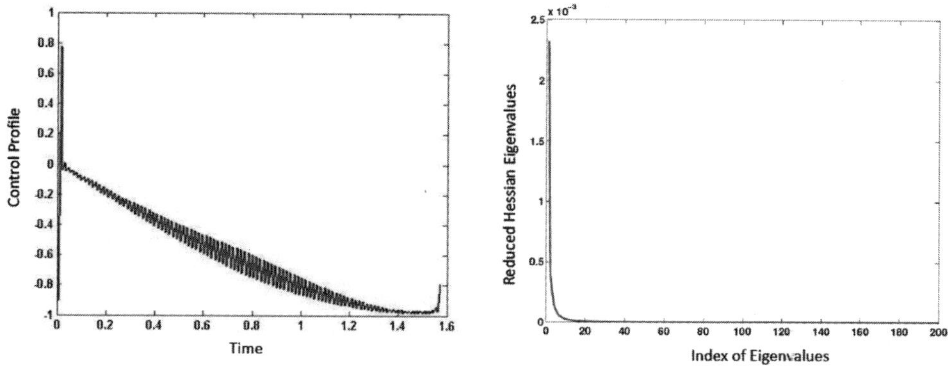

Figure 11.3. Control profile for Aly–Chan problem obtained with $N = 100$, $K = 2$ (left). Eigenvalues for the reduced Hessian in decreasing size (right) [5].

the problem (11.20) with $K = 2$ and $N = 100$ and equally spaced elements. We see that for large N, the control profile is highly oscillatory with spikes at the initial and final times. Moreover, the spectrum of eigenvalues for the reduced Hessian of the NLP quickly approaches zero and for $N = 100$ its condition number is 1.5×10^9.

To solve this singular problem, we apply the moving finite element strategy with the control profile represented by piecewise linear elements (11.14) and $K = 3$, leading to a solution with 10 finite elements. In the second step, the entire moving finite element strategy is performed. The number of finite elements now increases to 16, the NLP (11.16) is solved 3 times, and the objective is $-7.5026486 \times 10^{-9}$. The numerical results in both steps are shown in Figure 11.4 for the control and Hamiltonian profiles. Compared to the analytical control profile, we observe slight errors in the first, very small element (where $h_1 = 0.0075$). Beyond this, the average error in the control profile is 0.003.

Catalyst Mixing Problem

We next consider the catalyst mixing problem analyzed by Jackson [11]. The reactions $A \Longleftrightarrow B \rightarrow C$ take place in a tubular reactor at constant temperature. The first reaction is reversible and is catalyzed by Catalyst I while the second irreversible reaction is catalyzed by Catalyst II. The goal of this problem is to determine the optimal mixture of catalysts along the length t of the reactor in order to maximize the amount of product C. This problem has an analytic solution, and the control problem is given by bang-singular-bang characteristics.

The optimal catalyst mixing problem can be stated as

$$\min \quad a(t_f) + b(t_f) - a_0 \tag{11.21a}$$

$$\text{s.t.} \quad \frac{da(t)}{dt} = -u(k_1 a(t) - k_2 b(t)), \tag{11.21b}$$

$$\frac{db(t)}{dt} = u(k_1 a(t) - k_2 b(t)) - (1-u)k_3 b(t), \tag{11.21c}$$

$$a(0) = 1, \quad b(0) = 0, \quad u(t) \in [0,1], \tag{11.21d}$$

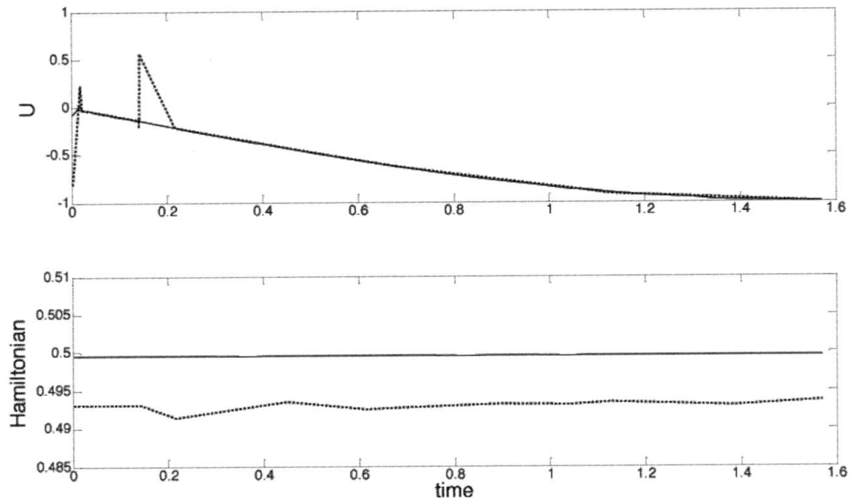

Figure 11.4. Control and Hamiltonian profiles for Aly–Chan problem in Case 1 (without element addition, dotted line) and Case 2 (with element addition, solid line).

with $t_f = 4$, $k_1 = k_3 = 1$, and $k_2 = 10$. As derived in [11, 5], the analytical solution is $u(t) = 1$ for $t \leq t_1 = 0.136299$, $u(t) = 0.227142$ for $t_1 \leq t \leq t_2 = 3.725230$, and $u(t) = 0$ for $t_2 \leq t \leq t_f$.

To solve this problem, we first consider the solution of problems (11.15) and (11.16) without the Hamiltonian termination criterion. This leads to profiles with 11 finite elements. We then compare this solution with the complete mesh refinement strategy. Here the number of finite elements is 15, the NLP (11.16) is solved twice, leading to an objective value of -0.19181435, and the optimal control profile is slightly shifted. The control and Hamiltonian profiles are compared in Figure 11.5. Our solution matches the analytic control profile with an error less than 2×10^{-4}, and the analytic switching points have an error less than 10^{-5}.

Lee–Ramirez Bioreactor

We next consider the bioreactor control problem by Lee and Ramirez [15]. Here two control profiles (glucose and inducer feeding rate) are determined to maximize the volume of product at $t_f = 10$ and the reactor model is an index-one DAE. This problem can be stated as

$$\begin{aligned} \min \quad & -x_4(t_f)x_1(t_f) - a_0 \\ \text{s.t.} \quad & \frac{dx_1}{dt} = u_1 + u_2, \; x_1(0) = 1, \\ & \frac{dx_2}{dt} = g_1 x_2 - (u_1 + u_2)x_2/x_1, \; x_2(0) = 0.1, \\ & \frac{dx_3}{dt} = u_1 c_1/x_1 - (u_1 + u_2)x_3/x_1 - g_1 x_2/c_2, \; x_3(0) = 40, \end{aligned}$$

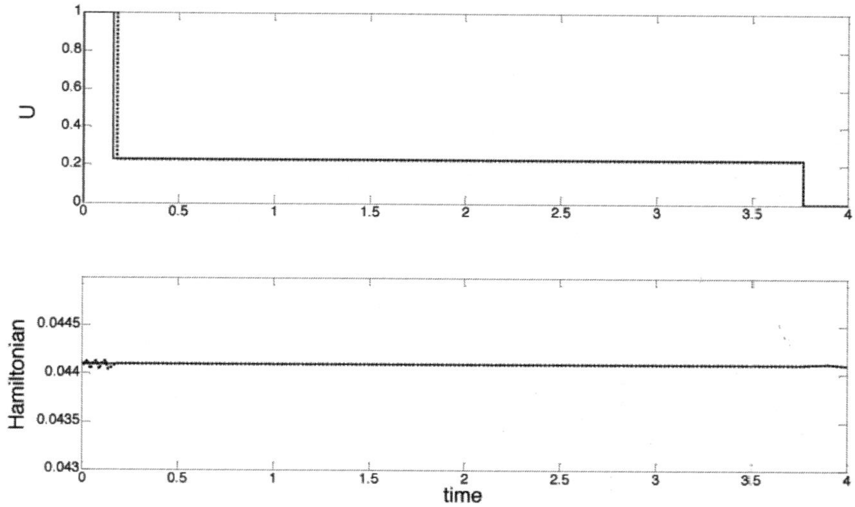

Figure 11.5. Control and Hamiltonian profiles for catalyst mixing problem in Case 1 (without element addition, dotted line) and Case 2 (with element addition, solid line).

$$\frac{dx_4}{dt} = g_2 x_2 - (u_1 + u_2)x_4/x_1,\ x_4(0) = 0,$$
$$\frac{dx_5}{dt} = u_2 c_3/x_1 - (u_1 + u_2)x_5/x_1,\ x_5(0) = 0,$$
$$\frac{dx_6}{dt} = g_3 x_6,\ x_6(0) = 1,$$
$$\frac{dx_7}{dt} = g_3(1 - x_7),\ x_7(0) = 0,$$
$$d_1 = 14.35 + x_3 + (x_3^2/111.5),\ d_2 = 0.22 + x_5,\ d_3 = 0.22x_7/d_2,$$
$$g_1 = x_3(x_6 + 0.22x_7/d_2)/d_1,$$
$$g_2 = 0.233x_3(0.0005 + x_5)/(0.022 + x_5)/d_1,$$
$$g_3 = 0.09x_5/(0.034 + x_5),$$
$$0 \le u(t) \le 1,$$

where $c_1 = 100$, $c_2 = 0.51$, and $c_3 = 4$.

We consider two problem formulations. We first consider piecewise linear profiles (11.14) with the case where problems (11.15) and (11.16) are solved without the Hamiltonian termination criterion. This leads to profiles with 26 finite elements. We then compare this with the complete mesh refinement strategy. Here the number of finite elements increases to 34 and the NLP (11.16) is solved 4 times, leading to an objective value of 6.151600711. The Hamiltonian and piecewise linear control profiles are compared in Figure 11.6. With mesh refinement we see an improvement in the optimal control profiles, but some small oscillations still remain in u_1. This is consistent with the nearly flat response surface for singular control problems.

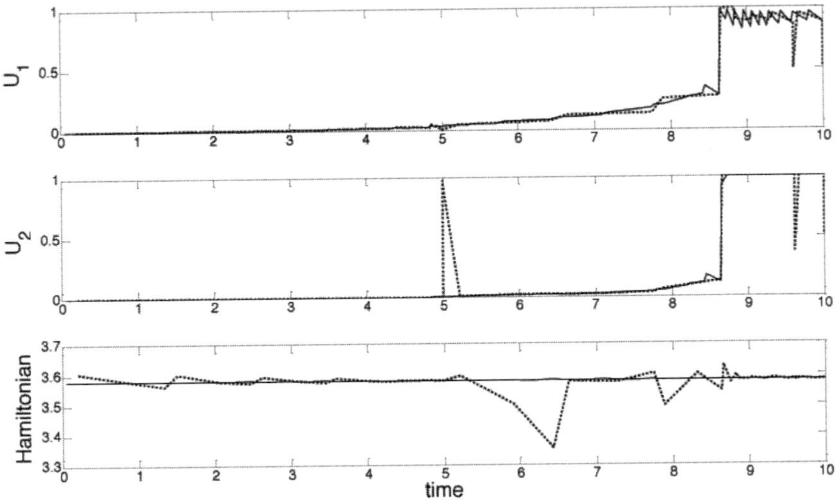

Figure 11.6. Piecewise linear control (11.14) and Hamiltonian profiles for Lee–Ramirez problem in Case 1 (without element addition, dotted line) and Case 2 (with element addition, solid line).

To improve this result, we next consider the regularizing influence of piecewise constant profiles (11.13). Again, we first consider the solution of problems (11.15) and (11.16) without the Hamiltonian termination criterion. This leads to profiles with 18 finite elements and several spikes. With the complete mesh refinement strategy, the number of finite elements increases to 34 and the NLP (11.16) is solved 4 times, leading to an objective value of 6.151596365. The Hamiltonian and piecewise constant control profiles are compared in Figure 11.7.

Distillation Column Optimization

Finally, we determine an optimal product profile in order to achieve desired product purity. The binary distillation column separates components A and B. The dynamic model is based on an equimolar overflow assumption (i.e., vapor V and liquid L flow rates do not vary with tray index j), with vapor-liquid equilibrium described by the Wilson equation. The index-one DAE model has 32 differential state variables (due to 30 trays, a reboiler, and condenser), 32 algebraic variables, and the distillate flow as the control variable. This problem is originally described in [10] and is paraphrased below:

$$\min \quad \int_0^{t_f=240} (x_1 - 0.995)^2 dt$$

$$\text{s.t.} \quad \frac{dx_1}{dt} = \frac{1}{M_{cond}} V(y_2 - x_1),\ x_1(0) = x_{1,0},$$

$$\frac{dx_j}{dt} = \frac{1}{M_{tray}} L(x_{j-1} - x_j) - V(y_j - y_{j+1}),\ x_j(0) = x_{j,0}, j = 2, \ldots, 16,$$

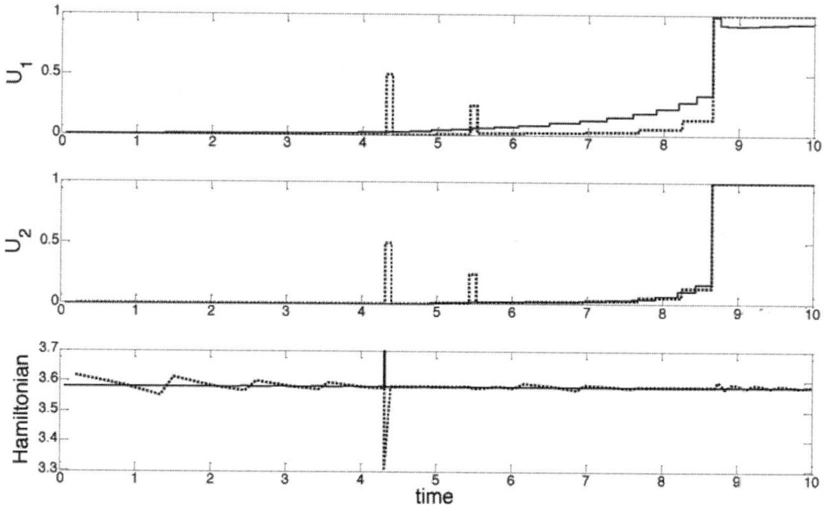

Figure 11.7. Piecewise constant control (11.13) and Hamiltonian profiles for Lee–Ramirez problem in Case 1 (without element addition, dotted line) and Case 2 (with element addition, solid line).

$$\frac{dx_{17}}{dt} = \frac{1}{M_{tray}}(Fx_f + Lx_{16} - (F+L)x_{17} - V(y_{17} - y_{18}), \; x_{17}(0) = x_{17,0},$$

$$\frac{dx_j}{dt} = \frac{1}{M_{tray}}(F+L)(x_{j-1} - x_j) - V(y_j - y_{j+1}), \; x_j(0) = x_{j,0}, j = 18,\ldots,31,$$

$$\frac{dx_{32}}{dt} = \frac{1}{M_{reb}}(F+L)x_{31} - (F - u_1(t))x_{32} - Vy_{32}, \; x_{32}(0) = x_{32,0},$$

$$P = x_j \gamma_{A,j} P^{sat}_{A,j} + (1 - x_j)\gamma_{B,j} P^{sat}_{B,j},$$

$$y_j P = x_j \gamma_{A,j} P^{sat}_{A,j},$$

$$P^{sat}_{A,j} = P^{sat}_A(T_j), \quad P^{sat}_{B,i} = P^{sat}_B(T_j),$$

$$\gamma_{A,j} = \gamma_A(x_j, T_j), \quad \gamma_{B,j} = \gamma_B(x_j, T_j),$$

where (for tray j) M is molar holdup, V and L are vapor and liquid flow rates, respectively, y and x are vapor and liquid mole fractions, respectively, F is the feed flow rate, γ refers to activity coefficients defined by the Wilson equation, P^{sat} is the vapor pressure, and P and T are pressure and temperature, respectively. Additional problem data and expressions for γ_A, γ_B, P^{sat}_A, and P^{sat}_B can be found at http://www.hedengren.net/research/models.htm.

To solve this problem, we first consider the solution of problems (11.15) and (11.16) without the Hamiltonian termination criterion, thus leading to 9 finite elements. On the other hand, with the complete mesh refinement strategy, the number of finite elements is decreased to 6 and the NLP (11.16) is solved 6 times, leading to an objective value of

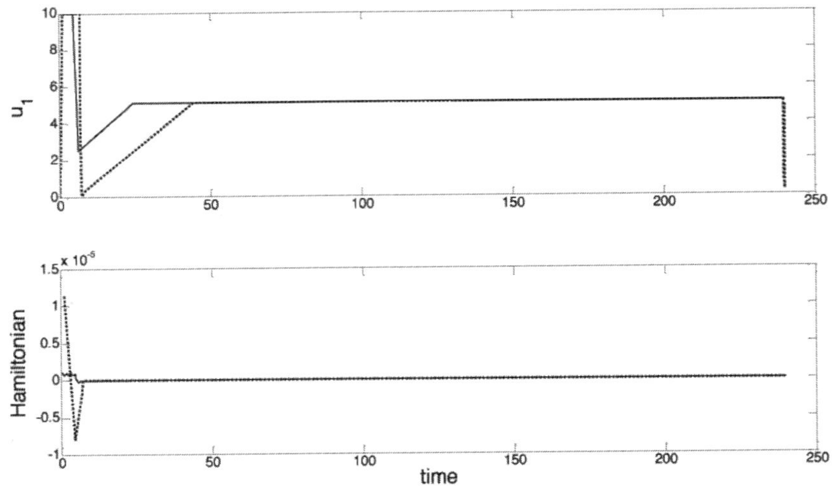

Figure 11.8. Control and Hamiltonian profiles for distillation problem in Case 1 (without element addition, dotted line) and Case 2 (with element addition, solid line).

$6.45851470 \times 10^{-4}$. Note how the finite elements have been consolidated in step 3 of the algorithm. The control and Hamiltonian profiles for both cases are compared in Figure 11.8.

11.6 Conclusions and Future Work

In this study we develop a mesh refinement strategy for a class of dynamic optimization problems based on direct transcription. This approach is based on moving finite elements embedded within an NLP formulation. Novel features of the algorithm include the direct location of breakpoints for control profiles and a termination criterion based on a constant Hamiltonian profile. The algorithm is demonstrated on six optimal control problems drawn from the literature and from chemical process engineering. All of these show that improved solutions can be obtained by enforcing a constant Hamiltonian profile. Moreover, based on a user-defined tolerance, this criterion provides a bound on a sufficient number of finite elements.

Motivated by concepts from classical optimal control theory, the mesh refinement strategy is based on a number of heuristics. Initialization of the states mimics the solution of IVPs with an IRK method. The NLP addressed in the second step includes degrees of freedom that allow for variable breakpoint locations in the control profiles. Finally, as with boundary value solvers, additional finite elements need to be considered, and their selection is based on approximation error as well as a constant Hamiltonian profile.

While the mesh refinement strategy is grounded on properties of the direct transcription formulation, the convergence analysis is still open and will be addressed in future work. In addition, we plan to deal with more general classes of optimal control problems that include state variable constraints. Finally, a number of features will be considered to improve the mesh refinement strategy within an integrated optimization platform. These include efficient warm start strategies for mesh refinement and restarting the NLP, as well as an automated direct transcription and refinement strategy for dynamic process optimization.

Acknowledgments

The work of Chen, Wang, and Shao was partially supported by the 973 Program of China (2012CB720500) and the NSF of Zhejiang Province (Y1110243).

Bibliography

[1] G. M. ALY AND W. C. CHAN, *Application of a modified quasilinearization technique to totally singular optimal problems*, Internat. J. Control, 17 (1973), pp. 809–815.

[2] U. ASCHER, R. MATTHEIJ, AND R. RUSSELL, *Numerical Solution of Boundary Value Problems for Ordinary Differential Equations*, SIAM, Philadelphia, 1995.

[3] M. BARTL, P. LI, AND L. T. BIEGLER, *Improvement of state profile accuracy in nonlinear dynamic optimization with the quasi-sequential approach*, AIChE J., 57 (2011), pp. 2185–2197.

[4] J. BETTS, *Practical Methods for Optimal Control and Estimation Using Nonlinear Programming*, Adv. Des. Control 19, SIAM, Philadelphia, 2010.

[5] L. T. BIEGLER, *Nonlinear Programming: Concepts, Algorithms and Applications to Chemical Processes*, SIAM, Philadelphia, 2010.

[6] L. T. BIEGLER, A. M. CERVANTES, AND A. WÄCHTER, *Advances in simultaneous strategies for dynamic process optimization*, Chem. Eng. Sci., 57 (2002), pp. 575–593.

[7] A. E. BRYSON AND Y. C. HO, *Applied Optimal Control*, Hemisphere, Mineola, NY, 1975.

[8] R. FOURER, D. M. GAY, AND B. W. KERNIGHAN, *AMPL: A Modeling Language for Mathematical Programming*, Duxbury/Brooks/Cole, Pacific Grove, CA, 2002.

[9] W. W. HAGER, *Runge-Kutta methods in optimal control and the transformed adjoint system*, Numer. Math., 87 (2000), pp. 247–282.

[10] J. HAHN AND T. F. EDGAR, *An improved method for nonlinear model reduction using balancing of empirical gramians*, Comput. Chem. Eng., 26 (2002), pp. 1379–1397.

[11] R. JACKSON, *Optimal use of mixed catalysts for two successive chemical reactions*, J. Optim. Theory Appl., 2 (1968), pp. 27–39.

[12] S. KAMESWARAN AND L. T. BIEGLER, *Convergence rates for direct transcription of optimal control problems using collocation at Radau points*, Comput. Optim. Appl., 41 (2008), pp. 81–126.

[13] S. KAMESWARAN AND L. T. BIEGLER, *Simultaneous dynamic optimization strategies: Recent advances and challenges*, Comput. Chem. Eng., 30 (2006), pp. 1560–1575.

[14] E. B. Lee and L. Markus, *Foundations of Optimal Control Theory*, John Wiley and Sons, New York, 1967.

[15] J. Lee and W. F. Ramirez, *Optimal fed-batch control of induced foreign protein production by recombinant bacteria*, AIChE. J., 40 (1994), pp. 899–907.

[16] R. D. Russell and J. Christiansen, *Adaptive mesh selection strategies for solving boundary value problems*, SIAM J. Numer. Anal., 15 (1978), pp. 59–80.

[17] P. Tanartkit and L.T. Biegler, *A nested simultaneous approach for dynamic optimization problems* II: *The outer problem*, Comput. Chem. Eng., 21 (1997), pp. 1365–1388.

[18] S. Vasantharajan and L.T. Biegler, *Simultaneous strategies for optimization of differential-algebraic systems with enforcement of error criteria*, Comput. Chem. Eng., 14 (1990), pp. 1083–1100.

[19] A. Wächter and L. T. Biegler, *On the implementation of an interior-point filter line-search algorithm for large-scale nonlinear programming*, Math. Program., 106 (2006), pp. 25–57 (https://projects.coin-or.org/Ipopt).

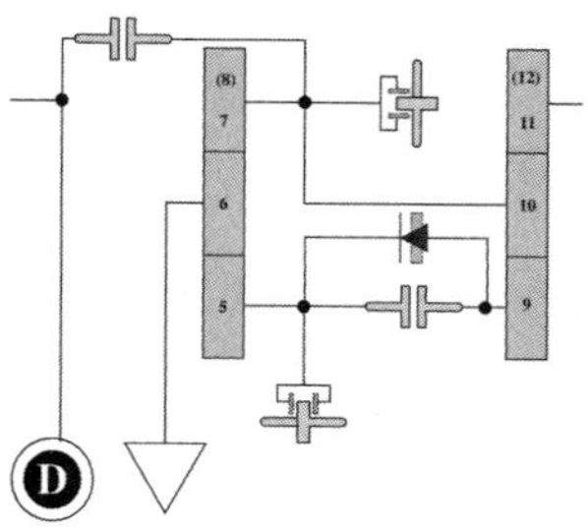

Chapter 12
Solving Parameter Estimation Problems with SOCX

Jason J. Boisvert, Matthew W. Donaldson, and Raymond J. Spiteri

Mathematical models often involve unknown parameters that must be fit to experimental data. SOCX is a software library for the solution of optimal control problems, of which parameter estimation problems form a subclass. Although it is powerful software, its user interface is challenging and there is little published work regarding broad measures of its performance.

EASY-FIT is software for solving parameter estimation problems that has a large database of problems and uses a high-level language called PCOMP that allows for easy problem specification. In this chapter we evaluate the performance of SOCX on a large set of problems from the EASY-FIT database. In order to facilitate this evaluation, we develop a source-to-source translator called EF2SOCX that automatically converts problems from the PCOMP language to input suitable for SOCX, thus significantly improving the SOCX interface as well. We find that EF2SOCX is able to correctly translate all 700 *out of the* 828 *problems in the EASY-FIT database for which it was designed. Of these,* 682 *(i.e., over* 97%*) of the problems are successfully solved by SOCX. Furthermore, based on the problems that were not solved, we are able to make some suggestions for future development to SOCX, including an increased size of problem class, the addition of more stable collocation methods, and methods for global optimization.*

12.1 Introduction

Optimal control problems involve the optimization of some performance index over time subject to a given dynamical system. Optimal control is applied in many fields ranging from industrial engineering and military applications to economics, medicine, and biology [3, 5, 14].

Analytical solutions of optimal control problems generally do not exist. Consequently, the solutions of optimal control problems must be approximated numerically. Several numerical software packages exist for the numerical solution of optimal control problems, including SOCX [4], PROPT [21], COOPT [20], RIOTS_95 [19], DIRCOL [22], and MISER3 [13].

Mathematical models often involve unknown parameters that must be fit to experimental data. These *parameter estimation problems* form a subclass of the more general optimal control problem. More precisely, parameter estimation problems involve fitting a mathematical model with unknown parameters to data such that a measure of the misfit is minimized.

EASY-FIT is an interactive software system used to identify model parameters in explicit model functions and dynamical systems of equations [9]. There is a large database of parameter estimation problems associated with EASY-FIT. The problems range from simple textbook examples to research problems (perhaps recast as parameter estimation problems) taken from scientific literature. They are specified in a Fortran-like high-level language called PCOMP. This makes the interface to EASY-FIT quite user friendly.

The software package SOCX (Sparse Optimal Control Extended) software library is a collection of FORTRAN90 subroutines developed by Applied Mathematical Analysis, LLC, to solve optimal control problems. Despite the potential wide applicability of SOCX, there seem to be no broad, systematic studies of its performance. Besides the issue of finding a large and diverse collection of relevant problems to use as a test set, the lack of such studies may in part be due to the challenging interface of SOCX. For example, a basic problem specification for SOCX typically requires about 500 lines of Fortran code, thus posing a significant barrier to manually conducting a large, systematic study.

In this chapter we study the performance of SOCX on the large set of parameter estimation problems associated with EASY-FIT. In order to facilitate this study, we developed a source-to-source translator called EF2SOCX that automatically converts problems specified in PCOMP to FORTRAN input suitable for SOCX, thus significantly improving the SOCX interface. The chapter is organized as follows. In Section 12.2, we formulate the parameter estimation problems considered; they involve explicit model functions, ODEs, and DAEs. In Section 12.3, we describe the functionality of EF2SOCX. In Section 12.4, we present the results of running SOCX on the 700 problems from the 828 problems in the EASY-FIT database that can be translated by EF2SOCX. Finally, in Section 12.5 we summarize our conclusions and provide suggestions for future development with SOCX.

12.2 Parameter Estimation Problems and Software

In solving optimal control problems, we seek a time-varying control function that yields a trajectory of a dynamical system that simultaneously satisfies equations of motion, boundary conditions, and equality and/or inequality constraints, while optimizing a performance index (or cost function). Parameter estimation problems form a subclass of optimal control problems in which the parameters represent an integer number of constant (time-independent) controls.

We now give a brief mathematical formulation of parameter estimation problems and the three categories of parameter estimation problems considered in this chapter. Details on the mathematical formulation of optimal control problems can be found in Section 12.6. We also provide details on the numerical methods upon which the SOCX and EASY-FIT software are based.

12.2.1 Parameter Estimation Problems

The underlying idea in solving parameter estimation problems is to estimate an integer number of constant unknown parameters $\mathbf{p} = (p_1, \ldots, p_{n_u})^T$ of a mathematical model by minimizing the norm of the difference between given data from theoretically predicted values of a model function at certain time values, e.g., a least squares fit [18]. Important aspects of parameter estimation problems include the following:

1. The optimization problem is *finite dimensional.*

2. The parameters are *constant.*

3. The performance index *quantifies a misfit to data.*

For the purposes of this study, we assume the form of the performance index to be least squares. Hence formally, we want to solve a least squares problem of the form

$$\begin{aligned}
&\min_{\mathbf{p}} \sum_{k=1}^{m_{exp}} \sum_{i=1}^{m_t} \sum_{j=1}^{m_c} [w_{ij}^k (h_k(\mathbf{p}; \mathbf{y}(\mathbf{p}; t_i, c_j), \mathbf{z}(\mathbf{p}; t_i, c_j), t_i, c_j) - \hat{h}_{ij}^k)]^2, \\
&\mathbf{p}_L \leq \mathbf{p} \leq \mathbf{p}_U \ , \ \mathbf{p} \in \mathbf{R}^{n_u}, \qquad (12.1) \\
&\mathbf{g}_=(\mathbf{p}; \mathbf{y}(\mathbf{p}; t, \mathbf{c}), \mathbf{z}(\mathbf{p}; t, \mathbf{c}), t, \mathbf{c}) = \mathbf{0}, \\
&\mathbf{g}_\geq(\mathbf{p}; \mathbf{y}(\mathbf{p}; t, \mathbf{c}), \mathbf{z}(\mathbf{p}; t, \mathbf{c}), t, \mathbf{c}) \geq \mathbf{0},
\end{aligned}$$

where $\mathbf{h}(\mathbf{p}; \mathbf{y}(\mathbf{p}; t, \mathbf{c}), \mathbf{z}(\mathbf{p}; t, \mathbf{c}), t, \mathbf{c}), t, \mathbf{c})$ is a vector-valued fitting function depending on the unknown parameter vector $\mathbf{p}$, the solution $\mathbf{y}(\mathbf{p}; t, \mathbf{c})$, $\mathbf{z}(\mathbf{p}; t, \mathbf{c})$ of a differential-algebraic dynamical system, the time t, and the *concentration variables* $\mathbf{c}$. Following the nomenclature of [18], $\mathbf{c}$ represents the *known* parameters in a given experiment, as opposed to the ones that are fit to the data. There are m_{exp} experimental data sets with m_t time values for m_c concentration values, corresponding to $M = m_t m_c m_{exp}$ measurements. The experimental values at time t_i, concentration c_j, and experimental data set h_k are denoted by $\hat{h}_{ij}^k$. We assume that there are weight factors $w_{ij}^k \geq 0$ given by the user that reflect the individual influence of a measurement value on the fit. Weights may be set to zero if, e.g., there are no measurements available for a particular time value. We may also have nonlinear restrictions in the form of equality or inequality constraints $\mathbf{g}_= : \mathbf{R}^{n_u} \times \mathbf{R}^{n_y} \times \mathbf{R}^{n_z} \times \mathbf{R} \times \mathbf{R}^{m_c} \rightarrow \mathbf{R}^{n_=}$ and $\mathbf{g}_\geq : \mathbf{R}^{n_u} \times \mathbf{R}^{n_y} \times \mathbf{R}^{n_z} \times \mathbf{R} \times \mathbf{R}^{m_c} \rightarrow \mathbf{R}^{n_\geq}$. We assume that all constraint functions are continuously differentiable with respect to $\mathbf{p}$. All inequalities are understood to hold componentwise. A general dynamical system is given by DAEs that describe a time-dependent process and that depend on $\mathbf{p}$.

The norm of the difference between measured experimental data and theoretically predicted values of a model function at certain time and concentration values is called the *residual*, which is a vector with components denoted by r_{ij}^k, $i = 1, 2, \ldots, m_t$, $j = 1, 2, \ldots, m_c$, and $k = 1, 2, \ldots, m_{exp}$. We note that for each of the three categories of parameter estimation problems described in this section it is possible in EASY-FIT to have a global scaling strategy in addition to the individual weight factors for each measurement value. The four options are

- 0: no additional scaling,
- 1: division of residuals by square root of sum of squares of measurement values for each data set, i.e.,
$$\frac{r_{ij}^k}{\sqrt{\sum_{i=1}^{m_t}\sum_{j=1}^{m_c}(\hat{h}_{ij}^k)^2}},$$
- -1: division of each single residual by corresponding absolute measurement value, i.e.,
$$\frac{r_{ij}^k}{|\hat{h}_{ij}^k|},$$
- -2: division of each single residual by corresponding squared measurement value, i.e.,
$$\frac{r_{ij}^k}{[\hat{h}_{ij}^k]^2}.$$

This global scaling strategy assumes measurements $\hat{h}_{ij}^k$ do not vanish, as appropriate, so that the scalings are always defined. Otherwise, residual scaling in the presence of zero measurements can be achieved through appropriate assignment of the weight factors w_{ij}^k.

There are six categories of parameter estimation problems in the EASY-FIT database, those that are defined in terms of explicit model functions, Laplace transformations, systems of ODEs, systems of DAEs, systems of time-dependent PDEs in one spatial dimension, and systems of partial differential-algebraic equations (PDAEs) in one spatial dimension. Of these six categories, EF2SOCX is presently designed to translate three of them: explicit model functions, systems of ODEs, and systems of DAEs. We now take a more in-depth look at these three categories.

Explicit Model Functions

Problems for which the vector-valued model function is available in explicit form, i.e., $\mathbf{h} = \mathbf{h}(\mathbf{p}; t, \mathbf{c})$ in (12.1) does not depend on the solution $\mathbf{y}(t)$ to a set of ODEs or DAEs, are said to belong to the class of parameter estimation problems with *explicit model functions*.

ODE-Dependent Model Functions

Problems for which the model function $\mathbf{h} = \mathbf{h}(\mathbf{p}; \mathbf{y}(\mathbf{p}; t, \mathbf{c}), t, \mathbf{c})$ in (12.1) depends on the solution $\mathbf{y}(\mathbf{p}; t, \mathbf{c})$ of a system of n_y IVPs

$$\dot{\mathbf{y}} = \mathbf{f}(\mathbf{p}; \mathbf{y}, t, \mathbf{c}), \quad \mathbf{y}(t_I) = \mathbf{y}^I(\mathbf{p}; \mathbf{c}), \tag{12.2}$$

are said to belong to the class of parameter estimation problems with *ODE-dependent model functions*. The initial values $\mathbf{y}^I(\mathbf{p}; \mathbf{c})$ may depend on one or more of the system parameters to be estimated and on the concentration parameter $\mathbf{c}$.

We note that dynamical constraints should be defined only in the form of inequalities $\mathbf{g}_{\geq}(\mathbf{p};\mathbf{y}(\mathbf{p};t_i,\mathbf{c}),t,\mathbf{c}) \geq \mathbf{0}$; equality constraints are to be treated as algebraic equations that are part of the DAE-dependent model functions; see below.

DAE-Dependent Model Functions

Problems for which the model function $\mathbf{h}(\mathbf{p};\mathbf{y}(\mathbf{p};t,\mathbf{c}),\mathbf{z}(\mathbf{p};t,\mathbf{c}),t,\mathbf{c})$ in (12.1) depends on the solution of a DAE consisting of n_y differential variables $\mathbf{y}(\mathbf{p};t,\mathbf{c})$ and n_z algebraic variables $\mathbf{z}(\mathbf{p};t,\mathbf{c})$

$$\begin{aligned} \dot{\mathbf{y}} &= \mathbf{f}(\mathbf{p};\mathbf{y},\mathbf{z},t,\mathbf{c}), \quad & \mathbf{y}(t_I) = \mathbf{y}^I(\mathbf{p};\mathbf{c}), \\ \mathbf{0} &= \mathbf{g}_{=}(\mathbf{p};\mathbf{y},\mathbf{z},t,\mathbf{c}), \quad & \mathbf{z}(t_I) = \mathbf{z}^I(\mathbf{p};\mathbf{c}), \end{aligned} \tag{12.3}$$

are said to belong to the class of parameter estimation problems with *DAE-dependent model functions*. The initial values of the differential variables $\mathbf{y}^I(\mathbf{p};\mathbf{c})$ and algebraic variables $\mathbf{z}^I(\mathbf{p};\mathbf{c})$ may depend on the system parameters to be estimated and on the concentration parameter $\mathbf{c}$.

The system is called an index-one DAE if the algebraic equations can be solved (at least in principle) for $\mathbf{z}$ for all t. This is possible if and only if the Jacobian matrix

$$\frac{\partial \mathbf{g}}{\partial \mathbf{z}}(\mathbf{p};\mathbf{y}(\mathbf{p};t,\mathbf{c}),\mathbf{z}(\mathbf{p};t,\mathbf{c}),t,\mathbf{c})$$

has full rank [1].

In EASY-FIT, there are implicit solvers that are able to solve index-two and index-three DAEs by transforming these high-index problems to index one by successive differentiation of the algebraic equations [12]. However, it is generally not trivial to perform this reduction in a numerically stable fashion, and this option is presently unavailable in $\mathbb{SOCX}$; hence we do not discuss the treatment of high-index DAEs further and accordingly assume that the problems considered are such that index reduction is not necessary for all parameter realizations.

12.2.2 Software

Software for the numerical solution of optimal control problems can be classified as being based on either *direct methods*, which solve an NLP problem based on a problem such as (12.1), and *indirect methods*, which solve the necessary (and sometimes sufficient) conditions for an optimum of (12.1). Both EASY-FIT and $\mathbb{SOCX}$ are based on direct methods; so further discussion of indirect methods is omitted. More information on indirect methods can be found in [3, 5].

Direct methods for optimal control problems lead to infinite-dimensional optimization problems for (piecewise) continuous variables $\mathbf{y}(t)$, $\mathbf{z}(t)$, and controls $\mathbf{u}(t)$. Although a parameter estimation problem involves a finite number of constant parameters, it may still be subject to ODEs (12.2) or DAEs (12.3) and hence theoretically remains an infinite-dimensional optimization problem. The *transcription* method is then used to reduce the infinite-dimensional optimization problem to a typically large, sparse NLP problem, which by definition has a *finite* set of variables and constraints, through discretization of the differential equations and constraints. Accordingly, the main aspects of software for the numerical solution of parameter estimation problems based on direct methods are the methods for the (1) transcription, (2) NLP solution, and (3) solution refinement. The iterative nature of this process leads to it sometimes being referred to as *sequential nonlinear programming*.

EASY-FIT has the choice of using two different NLP codes. The first is the code DFNLP by Schittkowski [17] that introduces additional variables and equality constraints and solves the resulting constrained NLP problem by the SQP algorithm NLPQL of Schittkowski [16]. The second is the subroutine DN2GB developed by Dennis, Gay, and Welsch [7] that is a variation of Newton's method, capable of handling upper and lower bounds for variables. In certain situations the algorithm reduces to a Gauss–Newton or Levenberg–Marquardt algorithm. The method is stabilized by a trust-region technique along with an adaptive choice of a model Hessian to achieve convergence. SOCX also has the choice of two different NLP codes, one based on SQP and the other on a primal-dual (or interior-point) method. It is noteworthy that the NLP methods in both EASY-FIT and SOCX are based on *local* optimization and hence give solutions that are dependent on starting values. Methods based on *global* optimization techniques are generally considered to be too expensive for use on the large, sparse NLP problems that arise from optimal control problems.

When applicable, any ODEs (12.2) or DAEs (12.3) present in the problem formulation must be transcribed into the NLP paradigm. EASY-FIT utilizes the initial-value nature of (12.2) and (12.3) by solving them using Runge–Kutta methods for IVPs. Accordingly, solution refinement in the sense of error estimation and control and mesh refinement for the IVPs is performed by the integrator. On the other hand, SOCX takes a boundary-value approach to transcribing (12.2) and (12.3) into the NLP paradigm. Specifically (12.2) and (12.3) are discretized by the TR or HS methods (the first two members of the Lobatto IIIA family of Runge–Kutta methods), and the ensuing collocation conditions are imposed as constraints in the NLP problem. Solution refinement via error estimation and control and mesh refinement is performed in a manner similar to that used by BVP solvers; further details are available in [3].

12.3 Functionality of the EF2SOCX Translator

The SOCX software package remains relatively untested in terms of published studies on solving a large number of problems and evaluating the results. This may be caused in part by the fairly steep learning curve involved in mastering SOCX combined with the significant programming effort required to specify and solve problems due to its challenging interface.

Parameter estimation problems are common in industry and in the scientific literature. The parameter estimation software package EASY-FIT comes with a large test set of 828 parameter estimation problems involving explicit, ODE-dependent, and DAE-dependent model functions, as described in Section 12.2. These problems are programmed in a high-level language called PCOMP, allowing for concise code for problem specification and a syntax similar to FORTRAN77. In this chapter we report on a comprehensive study of SOCX on a large set of parameter estimation problems. This study is made possible through the development of a source-to-source translator called EF2SOCX that can convert problems specified in PCOMP to FORTRAN input suitable for SOCX. EF2SOCX has been designed to follow the same rules as PCOMP. A list of all declarations supported by PCOMP is given in [18]. For a more general and complete description of PCOMP, see [8, 18].

The existence of EF2SOCX is essential when considering the amount of effort required to use SOCX. A typical SOCX code used to solve a basic parameter estimation problem can easily reach 500 lines of FORTRAN90 code. It is therefore unrealistic to program a large number of problems manually for solution by SOCX. EF2SOCX provides an automated process to read input from EASY-FIT and write an output file that is directly executable by SOCX, thus making it feasible to solve a large number of problems as well as significantly reducing the number of potential low-level coding errors on the part of the user.

The effectiveness of EF2SOCX is evaluated in terms of how well it can translate problems from the EASY-FIT database for which it was designed. This can be determined by evaluating the performance of SOCX on these problems by analyzing the SOCX results in relation to those obtained by EASY-FIT. This provides us with an out-of-the-box measure as to how well SOCX performs with its default settings. Clearly the knowledge of an experienced user can have a great influence on the performance of a software package such as SOCX. However, the focus of this study is not on comparing the performance of SOCX with that of EASY-FIT; indeed there are two fundamental reasons why we think such a comparison is not possible. First, there are many parameters that are specific to each piece of software; in other words, a certain parameter in one piece of software may not have an equivalent in the other, thus eliminating the possibility of a completely fair comparison. Second, it can be expected that there is some amount of performance tuning of EASY-FIT to the problems in its database. Consequently, any performance comparison with EASY-FIT using its own database would likely favor EASY-FIT.

As already mentioned, the performance of SOCX is greatly influenced by its inputs, but in particular, the initial guess supplied for the solution is arguably the most important and yet the most subjective. In this chapter, we limit our discussion of the input requirements for SOCX to include only methods for constructing an initial guess for the dynamic variables $\mathbf{d}(t)$; see (12.5). We refer the reader to [4] for a complete description of the input requirements for SOCX.

Several options are available in SOCX for constructing an initial guess for the dynamic variables. We note that whenever initial or final values for the dynamic variables are missing, guesses for these values are provided by SOCX. For example, the guesses for the initial conditions of the control variables are set to 0; i.e., $\mathbf{u}(t_I) = \mathbf{0}$. Also, the guesses for the final conditions of the dynamic variables are set to the initial conditions; i.e., $\mathbf{d}(t_F) = \mathbf{d}(t_I)$. The initial guess method is defined in SOCX using the array INIT. Popular choices include the following.

INIT(1) = 1: Construct a linear initial guess between $\mathbf{d}(t_I)$ and $\mathbf{d}(t_F)$; i.e., linearly interpolate the boundary conditions. If N_{GRID} is the number of grid points used in the discretization, then for grid point j the independent variable is

$$t_j = (j-1)\left[\frac{t_F - t_I}{N_{\text{GRID}} - 1}\right] + t_I,$$

and the dynamic variables are

$$\mathbf{d}(t_j) = (j-1)\left[\frac{\mathbf{d}(t_F) - \mathbf{d}(t_I)}{N_{\text{GRID}} - 1}\right] + \mathbf{d}(t_I),$$

where $1 \le j \le N_{\text{GRID}}$.

INIT(1) = 6: Construct an initial guess by solving an IVP with a linear control approximation. The control variables $\mathbf{u}(t)$ are approximated by

$$\mathbf{u}(t) = \frac{t_F - t}{t_F - t_I}\mathbf{u}(t_I) + \frac{t - t_I}{t_F - t_I}\mathbf{u}(t_F).$$

In the case of parameter estimation problems, the controls are constant. By default (INIT(2)=1), the state variables $\mathbf{y}(t)$ are approximated by the solution of the ODEs

$$\dot{\mathbf{y}} = \mathbf{f}[\mathbf{y}(t), \mathbf{u}(t), t]\,,$$

solved as an IVP with a variable step size Runge–Kutta–Taylor integrator. With INIT(2) = 3, this option uses a variable order, variable-stepsize backward differentiation formula (BDF) suitable for stiff problems [11].

INIT(1) = 9: Construct an initial guess using linear interpolation/extrapolation of the M measurement data points with grid points coinciding with the time values of the measurements.

12.4 Numerical Experiments

To test the performance of EF2SOCX as a translator and SOCX as a solver on a large number of parameter estimation problems, we use EF2SOCX to translate 828 problems involving explicit model functions, ODEs, and DAEs from the EASY-FIT database. By definition, these are problems that EASY-FIT has solved. However, EF2SOCX has not been designed to translate all of these problems. We describe below the subsets of the problems that cannot be translated and give the corresponding reasons. Excluding the problems that cannot be translated, we are left with a test set of size 700. We then describe a three-stage initial-guess strategy that gives SOCX more information about the EASY-FIT solution in an effort to increase the odds of obtaining a converged solution. A summary of the test results is given in the form of tables that we discuss in detail.

12.4.1 Untranslatable Problems

At present, EF2SOCX has not been designed to translate the following five subsets of problems in the EASY-FIT database.

Type I: *Models with discontinuities in their ODE or DAE right-hand side functions.* These discontinuities may cause slow convergence, small stepsize selection, or even divergence in the numerical methods used by SOCX. Although it is possible for SOCX to solve problems that involve discontinuities, it is generally best to avoid such formulations. Several suggestions as to how to do so are given in [3].

Type II: *Models with multiple phases and constraints applied at interior, i.e., nonboundary, points of a phase.* These models are supported by SOCX; however, because the treatment of these constraints in SOCX typically requires analysis on a problem-by-problem basis, it does not seem possible to automate the translation process in a generic fashion. For example, the required SOCX code could have constraints that depend on quantities such as concentration variables, breakpoints, and initial values involving unknown parameters. These problems can be translated by EF2SOCX if the user defines the models in PCOMP so that the constraints are applied at boundary points of a phase. This is done by defining a breakpoint to coincide with the point at which the constraint is applied so that the constraint is applied at the end of the phase. The state variables should be continuous across these breakpoints because they were continuous in the original model. In particular, single-phase models with constraints applied at interior points of a phase can be translated by EF2SOCX by inserting a breakpoint at the point at which the constraint is applied and forming an equivalent multiphase model but with the constraint applied at the end of the phase.

Type III: *Models with discontinuous solutions across breakpoints with initial values that depend on an unknown parameter or state variable.* These models are supported by SOCX; however, as with Type II, the treatment of the required constraints in SOCX typically requires analysis on a problem-by-problem basis because the constraints could depend on quantities such as concentration variables, breakpoints, and initial values involving unknown parameters. Accordingly, the translation process cannot be generically automated.

Type IV: *Models with variable breakpoints, i.e., breakpoints that are treated as parameters to be optimized.* These models are not directly supported by SOCX and therefore cannot be translated by EF2SOCX.

Type V: *High-index DAEs, i.e., DAEs with index higher than* 1. These models are not directly supported by SOCX and therefore cannot be automatically translated by EF2SOCX.

The process of using EF2SOCX to translate a parameter estimation problem specified in the format of the EASY-FIT problem database for solution by SOCX involves running EF2SOCX with two files. The first file contains the model functions defined in PCOMP, and the second contains the problem-specific constants. EF2SOCX performs some simple checks to ensure a problem does not fall into Types II–V listed above and generates a warning if so. This automatically generated warning prevents EF2SOCX output from compiling because it is printed at the beginning of the SOCX input file containing the generated FORTRAN90 code. Despite this, users may attempt to compile EF2SOCX output by simply removing any translator warnings from this file. Problems of Type I and specifications of problems of Type III that are not straightforward must be identified manually on a problem-by-problem basis. A breakdown of the identification of the total number of untranslatable problems for each of three categories of models is given in Table 12.1.

Table 12.1. Number of problems from EASY-FIT database identified as untranslatable.

Model type	EF2SOCX	Manual	Total
Explicit	0	0	0
ODE-dependent	69	38	107
DAE-dependent	18	3	21

If EF2SOCX output can be compiled, the executable is then run, and the output is checked to see whether SOCX terminated successfully. The results are summarized in

Table 12.2. We see that none of the problems produced a FORTRAN90 compiling error; i.e., they were translated to compilable FORTRAN90 code by EF2SOCX. Moreover, out of the 700 problems translated, SOCX was able to terminate successfully on 98.7% explicit model problems, 95.2% ODE-dependent problems, and 77.8% DAE-dependent problems by directly using EF2SOCX output, i.e., with the use of many SOCX default settings. These results demonstrate that the use of SOCX default settings is relatively robust, but they also suggest that the larger the problem class, the more likely increased user intervention beyond the use of default settings is required to obtain convergence to a solution.

Table 12.2. Testing results obtained from EF2SOCX and SOCX using only default settings.

Model type	Problems	F90 errors	Successful termination
Explicit	232	0.0%	(229) 98.7%
ODE-dependent	441	0.0%	(420) 95.2%
DAE-dependent	27	0.0%	(21) 77.8%

12.4.2 Initial Guess Strategies

In the previous subsection, we have established that EF2SOCX reliably generates compilable FORTRAN90 code that can successfully terminate on a large number of parameter estimation problems using only default settings. Often one of the most influential yet subjective aspects to solving a parameter estimation problem in practice is the choice of initial guess. With the hope of increasing the probability of successful terminations beyond the use of default settings, various initial guess strategies were applied to the dynamic variables and the unknown parameters to progressively give more information about the solution obtained by EASY-FIT. These values are a solution to the parameter estimation problem in EASY-FIT, and because we are trying to solve the same problems with SOCX, they should also be close to a solution that can be found by SOCX.

The initial guess for a parameter estimation problem consists of several parts. For example, an initial guess must be provided for the unknown parameters $\mathbf{p}$. In the case of parameter estimation problems that involve ODEs or DAEs, an initial guess for the dynamic variables $\mathbf{d}(t)$ from (12.5) must also be provided. A good initial guess for the discretization nodes in time when solving an ODE or DAE is also a nonnegligible issue in general; however, it is not treated in this study. The initial guess strategies applied are grouped into the following three stages.

Stage 1 consists of setting two related initial guesses, namely the EASY-FIT initial guess for the unknown parameters and the parameter bounds. In this stage, EF2SOCX creates code that passes the initial guess supplied in the EASY-FIT specification to SOCX. The parameter bounds $\mathbf{p}_L$ and $\mathbf{p}_U$ are set according to the following strategy. If a parameter bound given in the EASY-FIT specification does not exceed 10 in absolute value, this value is passed to SOCX; otherwise the parameter bound is left unspecified. In general, for the problems in the EASY-FIT database, it seems reasonable to assume that an upper bound no greater than 10 (or a lower bound no less than -10) on an unknown parameter is literal. This choice to modify the bounds was made because it became apparent that an upper bound of say 10^6, which is commonly used in EASY-FIT, perhaps to treat infinite or unknown bounds, causes poor scaling within the numerical algorithms of $\mathbb{SOCX}$ when the magnitude of the optimal parameter values are not this large. If the parameter bounds are not specified, then $\mathbb{SOCX}$ computes bounds for the unknown parameters internally.

In $\mathbb{SOCX}$, the variables and constraints are scaled in an attempt to improve the conditioning and efficiency of the underlying NLP problem. Variable scaling is performed by estimating the smallest and largest variable values from the user-supplied variable bounds. If bounds are not supplied, they are computed internally in an attempt to produce a well-scaled Jacobian, e.g., by normalizing the rows and columns of the Jacobian to have the same magnitude, along with several other heuristics; see [4]. Effectively, if there are no definite bounds on the unknown parameters, our experience indicates that it is better to not specify them when using $\mathbb{SOCX}$.

Stage 1 also uses the four different methods described in Section 12.3 for constructing an initial guess for the dynamic variables $\mathbf{d}(t)$. The method is defined using the array INIT. We now describe in detail the corresponding strategies for obtaining an initial guess for $\mathbf{d}(t)$ at each stage.

We do not have access to the EASY-FIT initial guess for $\mathbf{d}(t)$, but we do have access to the initial guess and bounds used for the unknown parameters. Therefore our first attempt at using the same initial guess as EASY-FIT uses the same initial guess for the unknown parameters and the same unknown parameter bounds as used by EASY-FIT. We also use the default initial guess method for the dynamic variables in $\mathbb{SOCX}$, i.e., linear interpolation of the boundary conditions (INIT(1)=1). If this does not lead to successful termination by $\mathbb{SOCX}$, the first modification is to modify the parameter bounds by removing nonliteral bounds used by EASY-FIT, i.e., leaving them unspecified in $\mathbb{SOCX}$, and continuing to use the default initial guess method for $\mathbf{d}(t)$ in $\mathbb{SOCX}$.

We augment this strategy with different initial guess methods for $\mathbf{d}(t)$ available in $\mathbb{SOCX}$. We use linear interpolation of measurement data (INIT(1)=9), followed by linear control approximation (INIT(1)=6) and the Runge–Kutta–Taylor integrator (INIT(2)=1), and finally linear control approximation (INIT(1)=6) and the BDF integrator (INIT(2)=3). For each of these three methods, we first use the same unknown parameter bounds as EASY-FIT, and then the bounds are unspecified. This combination of four initial guess methods for the dynamic variables, and either specifying the parameter bounds as per the EASY-FIT data or leaving them unspecified, forms Stage 1.

Stage 2 consists of attempting to obtain a successful termination by $\mathbb{SOCX}$ by using the optimal EASY-FIT parameter values $\mathbf{p}^*$ as the initial guess for the unknown parameters in $\mathbb{SOCX}$. Again, we begin by using the same parameter bounds as EASY-FIT, followed by unrestricting the parameter bounds. The four different initial guess methods for the dynamic variables used in Stage 1 are used in the same order.

At this point, if $\mathbb{SOCX}$ has failed to converge to a solution, we proceed to Stage 3, the final stage. This stage consists of using the optimal EASY-FIT parameter values $\mathbf{p}^*$ as the initial guess for the unknown parameters in $\mathbb{SOCX}$ along with tight bounds. We note that we cannot set the bounds to be exactly the EASY-FIT optimal values because this removes the degrees of freedom associated with the unknown parameters and may lead to an ill-posed problem. In other words, if an unknown parameter declared in $\mathbb{SOCX}$ is fixed to a specific value, i.e., the lower bound is equal to the upper bound, then the degree of freedom previously associated with the parameter is lost from the NLP problem. Whenever possible we set the bounds to ±0.1% of the EASY-FIT optimal values. If the difference between the upper and lower parameter bounds turns out to be smaller than the allowable SOCX constraint tolerance CONTOL $= 1.49 \times 10^{-8}$, we set the bounds to the EASY-FIT optimal values ±CONTOL*/2, where CONTOL* $= 1.5 \times 10^{-8} \gtrsim$ CONTOL is used to ensure that the difference in the bounds is larger than CONTOL in double-precision floating-point arithmetic. This strategy for tightly setting the bounds is summarized in equations (12.4). Specifically, for $i = 1, 2, \ldots, n_u$, we set

$$p_{L,i} = p_i^* - \max(0.001 \cdot |p_i^*|, \text{CONTOL}^*/2), \tag{12.4a}$$

$$p_{U,i} = p_i^* + \max(0.001 \cdot |p_i^*|, \text{CONTOL}^*/2). \tag{12.4b}$$

The entire range of initial guess strategies is summarized in Table 12.3 along with the results obtained from their application to the translatable problems from the EASY-FIT database.

Table 12.3. Percentages of successful terminations for various initial guess strategies.

d guess	**p** guess and bounds	Explicit	ODEs	DAEs
Stage 1				
INIT(1)=1	$\mathbf{p}_0$ and $\mathbf{p}_L, \mathbf{p}_U$	(92) 39.7%	(111) 25.2%	(7) 25.9%
	$\mathbf{p}_0$ and unspecified bounds	(103) 44.4%	(151) 34.2%	(4) 14.8%
INIT(1)=9	$\mathbf{p}_0$ and $\mathbf{p}_L, \mathbf{p}_U$	(3) 1.3%	(19) 4.3%	(1) 3.7%
	$\mathbf{p}_0$ and unspecified bounds	(8) 3.4%	(35) 7.9%	(2) 7.4%
INIT(1)=6	$\mathbf{p}_0$ and $\mathbf{p}_L, \mathbf{p}_U$	—	(13) 2.9%	(1) 3.7%
INIT(2)=1	$\mathbf{p}_0$ and unspecified bounds	—	(23) 5.2%	(1) 3.7%
INIT(1)=6,	$\mathbf{p}_0$ and $\mathbf{p}_L, \mathbf{p}_U$	—	(6) 1.4%	0
INIT(2)=3	$\mathbf{p}_0$ and unspecified bounds	—	(12) 2.7%	0
Stage 2				
INIT(1)=1	$\mathbf{p}^*$ and $\mathbf{p}_L, \mathbf{p}_U$	(4) 1.7%	(2) 0.5%	(1) 3.7%
	$\mathbf{p}^*$ and unspecified bounds	(4) 1.7%	(7) 1.6%	(1) 3.7%
INIT(1)=9	$\mathbf{p}^*$ and $\mathbf{p}_L, \mathbf{p}_U$	(2) 0.9%	(1) 0.2%	0
	$\mathbf{p}^*$ and unspecified bounds	(1) 0.4%	(4) 0.9%	0
INIT(1)=6	$\mathbf{p}^*$ and $\mathbf{p}_L, \mathbf{p}_U$	—	(3) 0.7%	0
INIT(2)=1	$\mathbf{p}^*$ and unspecified bounds	—	(9) 2.0%	(1) 3.7%
INIT(1)=6,	$\mathbf{p}^*$ and $\mathbf{p}_L, \mathbf{p}_U$	—	(1) 0.2%	0
INIT(2)=3	$\mathbf{p}^*$ and unspecified bounds	—	(5) 1.1%	0
Stage 3				
INIT(1)=1	$\mathbf{p}^*$ and $\mathbf{p}_L$, $\mathbf{p}_U$ from (12.4)	(8) 3.4%	(9) 2.0%	(1) 3.7%
INIT(1)=9	$\mathbf{p}^*$ and $\mathbf{p}_L$, $\mathbf{p}_U$ from (12.4)	(4) 1.7%	(2) 0.5%	(1) 3.7%
INIT(1)=6 INIT(2)=1	$\mathbf{p}^*$ and $\mathbf{p}_L$, $\mathbf{p}_U$ from (12.4)	—	(5) 1.1%	0
INIT(1)=6 INIT(2)=3	$\mathbf{p}^*$ and $\mathbf{p}_L$, $\mathbf{p}_U$ from (12.4)	—	(2) 0.5%	0

12.4.3 Results and Discussion

The following experiments were performed using EASY-FIT Version 4.0 on an Intel Celeron 533 MHz processor under Windows XP Professional and SOCX version 2011.09 on a 2×2.26 GHz Quad-Core Intel Xeon Mac Pro, compiled with GNU gfortran 4.6.1 under Mac OS X 10.7. Unless otherwise noted, default values for settings were used whenever possible. Specific nondefault settings were as follows. The maximum number of mesh refinements was set to 10 (the SOCX default is 5), the maximum number of objective function evaluations was set to 50000 (the SOCX default is 10000), and the maximum number of iterations was set to 200 (the SOCX default is 100). By default, SOCX dynamically adapts its use of specific routines used to solve a given problem in an attempt to minimize the time required. It turns out that this adaptivity can affect the success of the overall algorithm. Accordingly, for reproducibility of the experiments reported below, this adaptivity of SOCX was disabled by having EF2SOCX include the commands

```
CALL INSOCX( 'ITSWCH=2' )
CALL INSOCX( 'MTSWCH=3' )
CALL INSOCX( 'NSSWCH=1' )
```

at the beginning of the SOCX driver file generated for every problem.

We consider a SOCX *solution* to a problem to be the converged numerical estimates of the optimal set of unknown parameters, state variables, predicted model values, and objective function value as produced by SOCX.

Using the above methods in the order described, we get the results given in Table 12.3. We note that only the strategy with a default initial guess for the dynamic variables and the EASY-FIT initial unknown parameter guess and bounds is run on all 700 translatable problems. All subsequent strategies are run only on the remaining unsolved problems. Also, the percentages reported are all relative to the total number of translatable problems for the respective category of models.

Using linear interpolation of the boundary conditions (INIT(1)=1) and the initial guess and bounds for the parameters from EASY-FIT, 39.7% of the translatable explicit model problems, 25.2% of the translatable ODE-dependent problems, and 25.9% of the translatable DAE-dependent problems were solved. By also removing the parameter bounds, a total of 84.1% of the translatable explicit model problems, 59.4% of the translatable ODE-dependent problems, and 40.7% of the translatable DAE-dependent problems were solved. By the end of Stage 1, 88.8% of the translatable explicit model problems, 83.9% of the translatable ODE-dependent problems, and 59.3% of the translatable DAE-dependent problems were solved. Across all model types, Stage 1 strategies solved almost 85% ($592/700 \approx 84.6\%$) of the total number of translatable problems. It is clear that removing the nonliteral parameter bounds allows SOCX to solve many more problems than it does otherwise. In fact, across all models in Stage 1, 339 *more* problems were solved with unspecified bounds than with the original bounds from the EASY-FIT database.

After combining Stages 1 and 2, 93.5% of the translatable explicit model problems, 91.2% of the translatable ODE-dependent problems, and 70.4% of the translatable DAE-dependent problems were solved. Combining all three stages, 98.7% of the translatable explicit model problems, 95.2% of the translatable ODE-dependent problems, and 77.8% of the translatable DAE-dependent problems were solved. A "—" in Table 12.3 indicates that the solution method is not applicable to the model category.

In total, 670 (or 95.7%) of the 700 translatable problems were solved with code generated by EF2SOCX using the three-stage initial-guess strategy described above. There were 3 translatable explicit model problems, 21 translatable ODE-dependent problems, and 6 translatable DAE-dependent problems that had further modifications of their initial guess applied to them in order to obtain solutions. The 3 remaining explicit model problems were solved by treating the problems on a case-by-case basis, e.g., by removing redundant constraints or dummy fitting functions. These problems were eventually solved using INIT(1)=9. A total of 6 ODE-dependent problems and 3 DAE-dependent problems were solved by treating them on a case-by-case basis, e.g., by modifying the initial guess from EASY-FIT to be $\mathcal{O}(1)$. Two of the 6 ODE problems were solved by re-enabling the default dynamic run-time adaptation feature in SOCX. However, in general several attempts using this feature may still be necessary before SOCX is able to converge to a solution. SOCX was unable to solve the remaining 15 translatable ODE-dependent and 3 DAE-dependent problems. A discussion of the possible reasons behind the failure of SOCX to solve these problems is now given.

The translatable ODE- and DAE-dependent problems that did not result in successful termination in SOCX were generally deemed to be too ill-conditioned for solution by SOCX with the numerical methods available in it. A common observation in these problems is that SOCX is unable to obtain a solution to the ODEs or DAEs even when the optimization step is removed. In other words, no feasible solution for the discrete constraints could be computed. Specifically, we set up SOCX to solve the BVP only by fixing the unknown parameters to specific values and removing the discrete data. This reduced problem may not be solvable for one or more of the following reasons: (1) the discrete nonlinear constraints are ill-conditioned; (2) the initial guess is not sufficiently good; or (3) the gradient calculations are unreliable. It is unclear at present why these problems are too difficult for even the BDF integrator to solve. Solution of the underlying ODE/DAE via shooting may ameliorate these issues, but in principle such an approach only shifts the ill-conditioning to the gradient of the objective function, making the optimization in SOCX more difficult. It may be useful to try another solver that uses collocation, e.g., COLDAE [2], to see if it can solve the ODEs/DAEs, but as such this is beyond the scope of this study.

There was one problem, called BCBPLUS, that was reported by SOCX itself to be ill-conditioned. To compute the sparsity pattern of the right-hand side of the system of ODEs, SOCX perturbs the variables $\mathbf{p}$, $\mathbf{y}$ and then tries to evaluate the right-hand side functions. During this perturbation evaluation a “function error” is detected. This message is generated by SOCX when the magnitude of the right-hand side functions exceeds 10^{16}. This leads to two negative effects. First, SOCX cannot detect right-hand side sparsity and therefore assumes the right-hand side is dense, making the computation of the finite difference derivatives much more expensive. Second, the numerical values of the Jacobian of the right-hand side are extremely large, resulting in the failure of the NLP algorithms.

Another problem, called FEDBATCH, could be solved by SOCX, but with a more sophisticated initial guess technique not currently supported by EF2SOCX. FEDBATCH is a four-phase problem, and it was eventually solved using the method of *continuation* [1]. In other words, it was too difficult to obtain a sufficiently accurate initial guess to the problem in a heuristic manner. Instead we solved an easier problem consisting of only one phase using the optimal parameter values of EASY-FIT and fixing tight bounds on these values. The solution at the end of phase 1 is then used as the guess for the dynamic variables at the start of phase 2. This procedure is then repeated until the time interval was expanded to include all four phases. The concept of continuation is a powerful method for obtaining solutions to many problems where standard initial guess methods fail to provide a sufficiently accurate initial guess [6].

As a final performance measure, we categorize the objective function values of the successful terminations produced by $\mathbb{SOCX}$ in terms of those by EASY-FIT in Table 12.4. The objective function value is described as either smaller than, comparable to, or greater than the objective function value of EASY-FIT. We say the objective function values are comparable if they differ by less than 10%. We deem this to also be an acceptable amount in absolute terms because the magnitude of most EASY-FIT objective function values is in the range 10^{-4} to 10^{-2}, thus leading to at least two matching decimal places between the two objective function values. The results are based on using the Stage 1 and Stage 2 initial-guess strategies in Table 12.3, performed in the same order. Stage 3 strategies are not included because they essentially force $\mathbb{SOCX}$ to find the same solution as EASY-FIT, implying the solutions may not be the best that $\mathbb{SOCX}$ can achieve. For example, if $\mathbb{SOCX}$ converges to a solution using the default method for the dynamic variables, the initial unknown parameter values of EASY-FIT, and the parameter bounds of EASY-FIT, it is possible that using a more sophisticated initial guess method for the dynamic variables could produce a solution with a smaller objective function value. The same may be said of EASY-FIT; however it is suspected, and in some cases confirmed, that EASY-FIT is tuned to solve certain problems by using nondefault values. For example, many problems use different tolerance values. Also, integration methods used to solve the system of differential equations vary from problem to problem.

Table 12.4. Comparison of $\mathbb{SOCX}$ and EASY-FIT objective function values (Stages 1 and 2 only).

Model type	Smaller objective	Comparable objective	Larger objective
Explicit	(48) 22.1%	(98) 45.2%	(71) 32.7%
ODEs	(41) 10.2%	(213) 53.0%	(148) 36.8%
DAEs	(2) 10.5%	(7) 36.8%	(10) 52.6%

For all three model types, approximately 64% of the problems produced a solution with an objective function value not larger than the one given by EASY-FIT. The percentages are 67.3%, 63.2%, and 47.4% for the explicit models, ODE-dependent, and DAE-dependent problems, respectively. Table 12.4 indicates that $\mathbb{SOCX}$ and EASY-FIT found different solutions for almost half of the problems solved using only Stages 1 and 2. Therefore, it would also be informative to use global optimization software to determine whether the best solutions found are indeed global optima or to locate them. We note that approximately 10 explicit models were manually coded for solution with the global optimization software package GlobSol [15]. Most of these problems were run for several days using a 1.33 GHz PowerPC G4 processor with 768 MB of RAM and failed to converge over that time. Others were run for 2 weeks and also failed to converge over that time, reinforcing the claim made in Section 12.2 that current computational methods for solving global optimization problems generally require large amounts of computation time.

Overall we find the following:

- EF2SOCX successfully translated and either solved or provided a solid basis upon which to solve over 97% of the problems for which it was designed;
- leaving nonliteral bounds on the unknown parameters unspecified greatly increases the ability of SOCX to converge to a solution;
- with a sufficiently good initial guess for the dynamic variables, over 95% of the total number of problems could be solved;
- just over 72% of the total number of problems could be solved by using the default initial guess method for the dynamic variables, showing that the default method is indeed effective;
- when considering the problems that could be translated but not solved, the most common difficulty was that solving the ODEs and DAEs produced ill-conditioned constraints.

12.5 Conclusions and Future Work

EASY-FIT and SOCX are two software packages that can solve parameter estimation problems. To facilitate the specification of solving parameter estimation problems in SOCX and at the same time test SOCX on a large test set of such problems, we developed and used a source-to-source translator called EF2SOCX to translate the input for EASY-FIT into the corresponding input for SOCX. We measured the effectiveness of EF2SOCX in terms of its ability to correctly translate all the problems for which it was designed. Consequently we were able to measure the out-of-the-box effectiveness of SOCX in terms of the solutions it produced on the test set with its default values. For the problems that could be translated, we found that 100% of the explicit models, 96.6% of the ODEs, and 88.9% of the DAEs could be solved by either using the problem file generated by EF2SOCX or by making slight modifications to the EF2SOCX generated problem file. These percentages account for over 97% (682/700) of the 700 translatable problems from the EASY-FIT database.

In the process of attempting to solve the 700 translatable problems from the EASY-FIT database, we found that if there are no physically motivated bounds on the unknown parameters, it is better to leave them unspecified when using SOCX than to use arbitrary values.

We also found a large fraction of the parameter estimation problems tested have locally optimal solutions. From Table 12.4, we see about half (50.2%) of the total problems solved using only Stages 1 and 2 have multiple optima, motivating the appeal of global optimization functionality.

Finally, we suggest some possible improvements for SOCX. First, it is likely the usability of SOCX would improve greatly from a user interface such as EF2SOCX or the one included with EASY-FIT. Second, because the solutions found by SOCX are rather sensitive to the input parameters, users would benefit from additional advice on how to choose them. Third, the applicability of SOCX would be significantly broadened if it could seamlessly handle all the problem classes that EASY-FIT can, particularly problems of Types IV and V. Finally, it may be possible to improve the performance of $\mathbb{SOCX}$ on stiff or ill-conditioned problems by means of a collocation method with greater stability.

Part of our future work includes attempting to solve the ODEs and DAEs that were too difficult to solve in $\mathbb{SOCX}$ with another BVP solver that uses a Gauss–Legendre-based collocation method, such as COLDAE [2], as well as exploring the use of global optimization methods.

12.6 Appendix on Optimal Control Problems

Optimal control problems, also referred to as *dynamic optimization problems*, involve the optimization of a *performance index* over time, subject to a given dynamical system or other constraints. The decision variables of the optimization aspect of the problem are known as the *controls*, and the solution of the dynamical system for a prescribed control is called the *state*. Different types of constraints may be imposed on both the control and the state [5].

As described by [3], an optimal control problem can be formulated as a collection of N phases, where the independent variable t for phase K is defined in $t_I^{(K)} \le t \le t_F^{(K)}$. A *phase* is defined to be a portion of a trajectory in which the dynamics of the system maintain the same form. Within phase K, the dynamics of the system are described by $n_d^{(K)}$ *dynamic variables* $\mathbf{d}^{(K)}(t)$ made up of $n_y^{(K)}$ state variables $\mathbf{y}^{(K)}(t)$ and $n_u^{(K)}$ control variables $\mathbf{u}^{(K)}(t)$:

$$\mathbf{d}^{(K)}(t) = \begin{bmatrix} \mathbf{y}^{(K)}(t) \\ \mathbf{u}^{(K)}(t) \end{bmatrix}, \tag{12.5}$$

where $\mathbf{y}^{(K)} \in \mathbf{R}^{n_y^{(K)}}$, $\mathbf{u}^{(K)} \in \mathbf{R}^{n_u^{(K)}}$, and $n_d^{(K)} = n_y^{(K)} + n_u^{(K)}$.

Typically, the dynamics of the system are defined by a set of ODEs,

$$\dot{\mathbf{y}}^{(K)} = \mathbf{f}^{(K)}[\mathbf{y}^{(K)}(t), \mathbf{u}^{(K)}(t), t], \tag{12.6a}$$

where $\mathbf{f}^{(K)} : \mathbf{R}^{n_y^{(K)}} \times \mathbf{R}^{n_u^{(K)}} \times \mathbf{R} \to \mathbf{R}^{n_y^{(K)}}$. The solution must satisfy algebraic constraints of the form

$$\mathbf{g}_L^{(K)} \le \mathbf{g}^{(K)}[\mathbf{y}^{(K)}(t), \mathbf{u}^{(K)}(t), t] \le \mathbf{g}_U^{(K)}, \tag{12.6b}$$

where $\mathbf{g} : \mathbf{R}^{n_y^{(K)}} \times \mathbf{R}^{n_u^{(K)}} \times \mathbf{R} \to \mathbf{R}^{n_g^{(K)}}$ is composed of $n_=^{(K)}$ equality constraints and $n_\geq^{(K)}$ inequality constraints, including simple bounds on the state variables

$$\mathbf{y}_L^{(K)} \le \mathbf{y}^{(K)}(t) \le \mathbf{y}_U^{(K)} \tag{12.6c}$$

and control variables

$$\mathbf{u}_L^{(K)} \le \mathbf{u}^{(K)}(t) \le \mathbf{u}_U^{(K)}. \tag{12.6d}$$

The phases are linked by boundary conditions of the form

$$\begin{aligned} \boldsymbol{\psi}_L \le \boldsymbol{\psi} \Big[& \mathbf{y}^{(1)}(t_I^{(1)}), \mathbf{u}^{(1)}(t_I^{(1)}), t_I^{(1)}, \\ & \mathbf{y}^{(1)}(t_F^{(1)}), \mathbf{u}^{(1)}(t_F^{(1)}), t_F^{(1)}, \ldots, \\ & \mathbf{y}^{(N)}(t_I^{(N)}), \mathbf{u}^{(N)}(t_I^{(N)}), t_I^{(N)}, \\ & \mathbf{y}^{(N)}(t_F^{(N)}), \mathbf{u}^{(N)}(t_F^{(N)}), t_F^{(N)} \Big] \le \boldsymbol{\psi}_U, \end{aligned} \tag{12.6e}$$

where $\boldsymbol{\psi}_L$ and $\boldsymbol{\psi}_U$ are vectors of appropriate size. These boundary conditions allow the values of the dynamic variables at the beginning and/or end of any phase to be related to those of any other phase.

The optimal control problem is to determine the n_u-dimensional time-varying control vector $\mathbf{u}(t)$, consisting of the controls $\mathbf{u}^{(K)}(t)$, $K = 1,2,\ldots,N$, on each phase, that minimizes the performance index,

$$\begin{aligned} J = \phi\Big[&\mathbf{y}^{(1)}(t_I^{(1)}), \mathbf{u}^{(1)}(t_I^{(1)}), t_I^{(1)}, \\ &\mathbf{y}^{(1)}(t_F^{(1)}), \mathbf{u}^{(1)}(t_F^{(1)}), t_F^{(1)}, \ldots, \\ &\mathbf{y}^{(N)}(t_I^{(N)}), \mathbf{u}^{(N)}(t_I^{(N)}), t_I^{(N)}, \\ &\mathbf{y}^{(N)}(t_F^{(N)}), \mathbf{u}^{(N)}(t_F^{(N)}), t_F^{(N)}\Big] \\ &+ \sum_{K=1}^{N} \left\{ \int_{t_I^{(K)}}^{t_F^{(K)}} \mathbf{q}^{(K)}\left[\mathbf{y}^{(K)}(t), \mathbf{u}^{(K)}(t), t\right] dt \right\}, \end{aligned} \tag{12.6f}$$

and that may involve *quadrature functions* $\mathbf{q}^{(K)}$. Equations (12.6) specify a general optimal control problem with multiple phases, multiple states, multipoint boundary conditions, and a general form for the performance index.

Solving optimal control problems numerically can be challenging. Often a wide range of allowable parameter values can result in solutions that are only locally optimal. In such cases it may be desirable to use global optimization methods. However, if rigorous proofs of global optimality are also desired, the added computational expense associated with these methods generally makes them impractical except on small problems [10].

Breakpoints

In practice it is common for the form of the dynamics (12.3) of a system to change during the trajectory. Accordingly, conditions on the solution at these *breakpoints* must be enforced, e.g., continuity of the solution. A typical example is a pharmacokinetic application with an initial infusion of a drug followed by subsequent infusions of drug doses by injection. Each infusion of a drug dose represents a breakpoint in the model because the dynamics depend on the amount of drug dose injected at each infusion. As is the case in this example, it is possible that the solution (i.e., the concentration of drug in the blood stream) is discontinuous at a breakpoint. *Variable breakpoints* are breakpoints that are treated as parameters to be optimized, e.g., in a pharmacokinetic problem, the times at which an infusion is to be administered.

For simplicity we assume an ODE model with initial values as given in (12.2); similar definitions are easily made for the DAE model (12.3). Formally we describe the model by

$$\dot{\mathbf{y}}^{(1)} = \mathbf{f}^{(1)}(\mathbf{p};\mathbf{y}^{(1)},t,\mathbf{c}), \quad \mathbf{y}^{(1)}(t_I) = \bar{\mathbf{y}}^{(1)}(\mathbf{p};\mathbf{c})$$

for $t_I \le t \le \tau_1$, and

$$\dot{\mathbf{y}}^{(K)} = \mathbf{f}^{(K)}(\mathbf{p};\mathbf{y}^{(K)},t,\mathbf{c}), \quad \mathbf{y}^{(K)}(\tau_K) = \bar{\mathbf{y}}^{(K)}(\mathbf{p};\mathbf{c},\mathbf{y}^{(K-1)}(\mathbf{p};\tau_{K-1},\mathbf{c}))$$

for $\tau_{K-1} \le t \le \tau_K$, $K = 2,\ldots,N = m_b + 1$, where m_b is the number of breakpoints τ_{K-1}, and $\tau_0 = t_I < \tau_1 < \cdots < \tau_{m_b} < t_F$. The initial values of each subsystem are given by

$\bar{\mathbf{y}}^{(K)}(\mathbf{p};\mathbf{c},\mathbf{y}^{(K-1)}(\mathbf{p};\tau_{K-1},\mathbf{c}))$ and may depend on the parameter vector $\mathbf{p}$ to be estimated, the concentration $\mathbf{c}$, and the solution of the previous phase with breakpoint τ_{K-1}. In the numerical algorithm for the solution to such problems, the integration of the differential equations must be restarted at each breakpoint.

Acknowledgments

This work was supported by grants from NSERC Canada and MITACS. We would like to thank J.T. Betts for many helpful discussions.

Bibliography

[1] U. M. ASCHER AND L. R. PETZOLD, *Computer Methods for Ordinary Differential Equations and Differential-Algebraic Equations*, SIAM, Philadelphia, 1998.

[2] U. M. ASCHER AND R. J. SPITERI, *Collocation software for boundary value differential-algebraic equations*, SIAM J. Sci. Comput., 15 (1994), pp. 938–952.

[3] J. BETTS, *Practical Methods for Optimal Control Using Nonlinear Programming*, 2nd ed., SIAM, Philadelphia, 2010.

[4] J. BETTS AND W. HUFFMAN, *Sparse Optimal Control Software* SOCS, Tech. Rep. Mathematics and Engineering Analysis MEA-LR-085, Boeing Information and Support Services, The Boeing Company, Seattle, WA, 1997.

[5] A. BRYSON AND Y.-C. HO, *Applied Optimal Control*, Blaisdell, New York, 1969.

[6] H. DAVIS, *Introduction to Nonlinear Differential and Integral Equations*, Dover, New York, 1962.

[7] J. DENNIS, JR., D. GAY, AND R. WELSCH, *Algorithm* 573: *Nl2sol—an adaptive nonlinear least-squares algorithm*, ACM Trans. Math. Software, 7 (1981), pp. 369–383.

[8] M. DOBMANN, M. LIEPELT, K. SCHITTKOWSKI, AND C. TRASSL, *PCOMP: A Fortran Code for Automatic Differentiation, Language Description and Users Guide*, Tech. Rep., Dept. of Mathematics, University of Bayreuth, Bayreuth, Germany, 1995.

[9] EASY-FIT, *Software for Parameter Estimation*, http://www.ai7.uni-bayreuth.de/easy_fit.htm, 2009.

[10] W. ESPOSITO AND C. FLOUDAS, *Deterministic global optimization in nonlinear optimal control problems*, J. Global Optim., 17 (2000), pp. 97–126.

[11] C. GEAR, *The automatic integration of stiff ordinary differential equations*, in Information Processing 68, Vol. 1, North–Holland, Amsterdam, 1969, pp. 187–193.

[12] E. HAIRER AND G. WANNER, *Solving Ordinary Differential Equations* II: *Stiff and Differential-Algebraic Problems*, Springer Ser. Comput. Math. 14, Springer-Verlag, Berlin, 1991.

[13] L. JENNINGS, M. FISHER, K. TEO, AND C. GOH, *MISER3: Software for Solving Optimal Control Problems*, http://www.cado.uwa.edu.au/miser/, 2005.

[14] H. JOSHI, *Optimal control of an HIV immunology model*, Optimal Control Appl. Methods, 23 (2002), pp. 199–213.

[15] R. KEARFOTT, *Rigorous Global Search: Continuous Problems*, Kluwer Academic Publishers, Dordrecht, The Netherlands, 1996.

[16] K. SCHITTKOWSKI, *NLPQL: A FORTRAN subroutine solving constrained nonlinear programming problems*, Ann. Oper. Res., 5 (1985), pp. 485–500.

[17] K. SCHITTKOWSKI, *Solving Nonlinear Least Squares Problems by a General Purpose SQP-Method*, in Trends in Mathematical Optimization, Internat. Schriftenreihe Numer. Math. 84, Birkhäuser, Basel, 1988, pp. 295–309.

[18] K. SCHITTKOWSKI, *Numerical Data Fitting in Dynamical Systems: A Practical Introduction with Applications and Software*, Kluwer Academic Publishers, Norwell, MA, 2002.

[19] A. SCHWARTZ, E. POLAK, AND Y. CHEN, *RIOTS: A MATLAB Toolbox for Solving Optimal Control Problems*, http://www.schwartz-home.com/~adam/RIOTS/, 1997.

[20] R. SERBAN AND L. PETZOLD, *COOPT – a software package for optimal control of large-scale differential-algebraic equation systems*, Math. Comput. Simulation, 56 (2001), pp. 187–203.

[21] TOMLAB, *Matlab Optimal Control Software*, http://tomdyn.com/, 2009.

[22] O. VON STRYK, *DIRCOL: A Direct Collocation Method for the Numerical Solution of Optimal Control Problems*, http://www.sim.tu-darmstadt.de/sw/dircol/dircol_app.html, 1999.

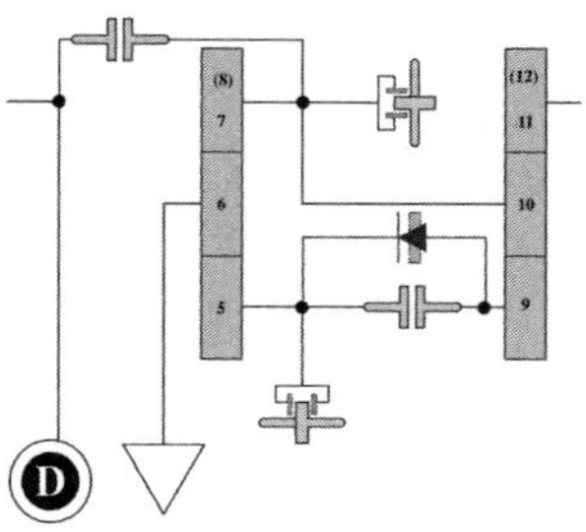

Chapter 13

Control of Integrated Chemical Process Systems Using Underlying DAE Models

Michael Baldea and Prodromos Daoutidis

Modern chemical plants adopt integrated design configurations, featuring material recycling and energy recovery. This chapter explores the dynamic impact of process integration. We define a generic class of integrated processes and establish a connection between steady-state design features and the transient behavior of the process. Using singular perturbation tools, we derive a low-dimensional DAE representation of the core process dynamics, followed by postulating a hierarchical control framework that exploits our dynamics results in the design of well-conditioned controllers with desirable closed-loop properties.

13.1 Introduction

Modern chemical plants are designed in "lean" configurations, featuring material recycling and energy recovery, a reduced number of physical units and diminished material inventory. Along with the elimination of intermediate storage (buffer) tanks, these traits intuitively result in significant dynamic interactions between the process units and lead to an intricate dynamic behavior at the process level. Consequently, the design and implementation of advanced, model-based control systems aimed at improving plant performance is a difficult matter, with the complexity, large dimension and ill-conditioning of the process models being major hindrances [5, 16, 18, 19, 20].

In this chapter, we explore the dynamic impact of process integration. We begin by defining a generic class of integrated process systems, in which an extensive quantity that obeys conservation laws (e.g., material, energy) is recovered from the product and recycled to the process feed. We establish a connection between this steady-state design feature and the transient behavior of the process. Using singular perturbation tools, we derive a low-dimensional DAE representation of the core process dynamics, followed by postulating a hierarchical control framework that exploits our dynamics results in the design of well-conditioned controllers with desirable closed-loop properties.

13.2 Preliminaries

13.2.1 Singularly Perturbed ODE Systems

The analysis of ODE systems with small parameters ε, $0 < \varepsilon \ll 1$, is generally referred to as perturbation analysis or perturbation theory. We introduce a few central definitions below and refer the reader to standard texts (e.g., those by Kokotovic, Khalil, and O'Reilly [11] and Khalil [10]) for more details.

Definition 13.1. *Consider the dynamical system*

$$\dot{\mathbf{x}} = f(\mathbf{x},\mathbf{y},t,\varepsilon), \quad \mathbf{x}(0) = \mathbf{x}^0, \tag{13.1}$$

$$\varepsilon\dot{\mathbf{y}} = \mathbf{g}(\mathbf{x},\mathbf{y},t,\varepsilon), \quad \mathbf{y}(0) = \mathbf{y}^0, \tag{13.2}$$

with $\mathbf{x} \in \mathbf{R}^n$, $\mathbf{y} \in \mathbf{R}^m$ *and* $0 < \varepsilon \ll 1$. *The system* (13.1)–(13.2) *is referred to as a singular perturbation problem.*

As $\varepsilon \ll 1$ is assumed to be quite small, one can intuitively consider the limit case $\varepsilon \to 0$. This limit leads to a problem that differs significantly from the original one: when $\varepsilon = 0$, the dimension of the state space of (13.1)–(13.2) collapses from $n+m$ to n, because the differential equation (13.2) degenerates into an algebraic equation and the original system collapses to a system of DAEs, where the bar is used to indicate that the variables belong to a system with $\varepsilon = 0$:

$$\dot{\bar{\mathbf{x}}} = \mathbf{f}(\bar{\mathbf{x}},\bar{\mathbf{y}},t,0), \tag{13.3}$$

$$\mathbf{0} = \mathbf{g}(\bar{\mathbf{x}},\bar{\mathbf{y}},t,0). \tag{13.4}$$

Definition 13.2. *The model of* (13.1)–(13.2) *is said to be in a* standard *singularly perturbed form if, in a domain of interest,* (13.4) *has* $k \geq 1$ *distinct (isolated) roots* $\bar{\mathbf{y}} = \bar{\Phi}_i(\bar{\mathbf{x}},t)$, $i = 1,\ldots,k$. *Whenever this condition is violated, the system* (13.1)–(13.2) *is said to be in a nonstandard singularly perturbed form* [13].

In our developments, we will also make use of the Lie derivative and the Lie bracket, and of the concept of involutivity, defined, respectively, as follows.

Definition 13.3. *Let* $\mathbf{x} \in \mathbf{R}^n$ *be a vector,* $h(\mathbf{x}) : \mathbf{R}^n \to \mathbf{R}$ *a scalar function, and* $\mathbf{g}(\mathbf{x}) : \mathbf{R}^n \to \mathbf{R}^n$ *a vector function. The Lie derivative of function h with respect to function* $\mathbf{g}$ *(or directional derivative of h along* $\mathbf{g}$*) is defined as* $L_{\mathbf{g}}h(\mathbf{x}) = \frac{\partial h}{\partial \mathbf{x}}(\mathbf{x})\mathbf{g}(\mathbf{x})$.

Let $\mathbf{f}(\mathbf{x}) : \mathbf{R}^n \to \mathbf{R}^n$ *be a vector function. The* Lie bracket *of the vector functions* $\mathbf{g}(\mathbf{x})$ *and* $\mathbf{f}(\mathbf{x})$ *is defined as* $[\mathbf{f}(\mathbf{x}),\mathbf{g}(\mathbf{x})] = \frac{\partial \mathbf{g}}{\partial \mathbf{x}}(\mathbf{x})\mathbf{f}(\mathbf{x}) - \frac{\partial \mathbf{f}}{\partial \mathbf{x}}(\mathbf{x})\mathbf{g}(\mathbf{x})$.

Definition 13.4. *An m-dimensional distribution* $\mathbf{G}(\mathbf{x}) = \mathbf{span}\{\mathbf{g}_1(\mathbf{x}),\ldots,\mathbf{g}_m(\mathbf{x})\}$ *is* involutive *if the Lie bracket* $[\mathbf{g}_i(\mathbf{x}),\mathbf{g}_j(\mathbf{x})]$, *for all* $i,j = 1,\ldots,m$, *of any pair of vector fields belonging to* $\mathbf{G}(\mathbf{x})$, *is a vector field which belongs to* $\mathbf{G}(\mathbf{x})$.

13.2.2 Process Systems

The analysis of process systems [8] focuses on finite-dimensional systems $\mathcal{S}$ whose dynamics can be represented by equations of the type

$$\dot{\mathbf{v}} = \phi(\mathbf{v}) + \gamma(\mathbf{v},\mathbf{u}), \tag{13.5}$$

where $\mathbf{v} \in V \subset \mathbf{R}^{v}$ are the state variables and $\mathbf{u} \in U \subset \mathbf{R}^{u}$ are inputs.

An inventory [8] for the system $\mathcal{S}$ is an additive, continuous ($\mathcal{C}^1$) function $x : V \to \mathbf{R}^+$, so that, if $\mathcal{S}$ consists of subsystems $\mathcal{S}_1$ and $\mathcal{S}_2$, and v_1 and v_2 are, respectively, the states of $\mathcal{S}_1$ and $\mathcal{S}_2$, we can write

$$x(\mathbf{v}) = x(v_1) + x(v_2); \tag{13.6}$$

that is, the inventory of a system is the sum of the inventories of its subsystems.

We assume that there exists a set of inventories, $\mathbf{x} \in X \subset \mathbf{R}^{v}$, such that the nonlinear mapping $\mathbf{x} = \mathcal{X}(\mathbf{v})$ is invertible; that is, the state of the system $\mathcal{S}$ can be completely characterized in terms of inventories. Using (13.5) and the continuity of the inventory function, for a vector of inventories $\mathbf{x} = [x_1, \ldots, x_v]^T$ that fulfills the assumption above, we can write

$$\dot{\mathbf{x}} = \mathbf{f}(\mathbf{x}) + \mathbf{g}(\mathbf{x},\mathbf{u}), \tag{13.7}$$

where $\mathbf{f}(\mathbf{x}) = L_{\phi}\mathbf{x}(\mathbf{v}) + \mathbf{f}^*$ and $\mathbf{g}(\mathbf{x},\mathbf{u}) = L_{\gamma}\mathbf{x}(\mathbf{v}) - \mathbf{f}^*$. $\mathbf{f}$ is called the rate of production and $\mathbf{f}^*$ is the vector of steady-state production rates. $\mathbf{g}$, the rate of supply, admits the decomposition

$$\mathbf{g}(\mathbf{x},\mathbf{u}) = \sum_{i=1}^{m} \mathbf{g}_i F_i, \tag{13.8}$$

where F_i, $i = 1, \ldots, m$, are boundary flows, describing the rate of addition/depletion of inventory (energy, material) through the m ports of the system, and $\mathbf{g}_i$, $i = 1, \ldots, m$, are vectors of appropriate dimensions.

13.3 Generic Model of Integrated Process Systems

We will consider a generic class of integrated chemical process systems, such as the one in Figure 13.1, consisting of N units in series. We will use the terms “integrated” and “integration” to capture the presence of a recycle connection R, intended to transfer inventory from the last unit to the first. Using (13.7), and assuming that the individual units can be modeled as lumped parameter systems, the mathematical model describing the evolution of an inventory (e.g., material, energy) of this system can be written as

$$\dot{\mathbf{x}} = \mathbf{f}(\mathbf{x}) + \sum_{j=0,N} \mathbf{g}_j(\mathbf{x}) u_j + \mathbf{Rc} \sum_{j=1}^{N-1} k_j \mathbf{g}_j(\mathbf{x}) u_j + \mathbf{Rc}\mathbf{g}_R(\mathbf{x}) u_r, \tag{13.9}$$

where $u_j = (F_j / F_{j,s})$ represent (possibly manipulated) dimensionless variables that correspond to the inventory flows, $k_j = F_{j,s}/F_{R,s}$, $j = 1 \ldots N$, and $\mathbf{g}_j(\mathbf{x})$ and $\mathbf{g}_R(\mathbf{x})$ are vector functions of appropriate dimensions. The subscript s denotes steady-state values. Note that

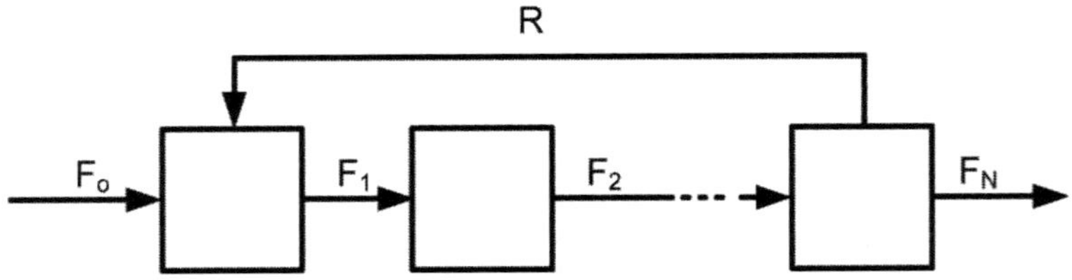

Figure 13.1. Generic integrated process system, featuring N units and an inventory recycle connection.

the model explicitly identifies the terms that involve the process port flows ($j = 0, N$), the internal inventory flows ($j = 1, \ldots, N-1$), and the recycle flow ($j = R$).

In order to investigate the impact on the presence and magnitude of inventory recycling on the process dynamics, (13.9) also makes use of the recycle number **Rc**, which we define below.

Definition 13.5. *The recycle number of an integrated process is a* process dimensionless number, *expressed as the ratio of the (steady-state) rate at which inventory is recycled to the rate at which inventory is introduced in the process through the inlet port:*

$$\mathbf{Rc} = \frac{R_s}{F_{os}}. \tag{13.10}$$

Evidently, for a given inventory inlet rate (set by F_o), the value of **Rc** varies with the rate at which inventory is recycled. Two limit cases are of interest:

Case I: $\mathbf{Rc} \ll 1$. In this case, the rate at which inventory is recycled is much smaller than the feed rate, the internal inventory flows are comparable to the feed rate, and $k_j \gg 1$, $\mathbf{Rc}k_j = \tilde{k}_j = \mathcal{O}(1), j = 1, \ldots, N-1$. In the limit as the recycle rate tends to zero, ($\mathbf{Rc} \to 0$), the last term in (13.9) vanishes, and the model reduces to the model of a series system:

$$\dot{\mathbf{x}} = \mathbf{f}(\mathbf{x}) + \sum_{j=0,N} \mathbf{g}_j(\mathbf{x})u_j + \sum_{j=1}^{N-1} \tilde{k}_j \mathbf{g}_j(\mathbf{x})u_j. \tag{13.11}$$

Intuitively, we can expect that the presence of a (small) inventory recycle flow will not have a significant impact on the dynamics of the process system.

Case II: $\mathbf{Rc} \gg 1$. In this case the inventory recycle rate is much larger than the feed rate, the internal flows are equally large, and $k_j = \mathcal{O}(1), j = 1, \ldots N-1$. In this context, the model of (13.9) can be rewritten as

$$\dot{\mathbf{x}} = \mathbf{f}(\mathbf{x}) + \sum_{j=0,N} \mathbf{g}_j(\mathbf{x})u_j + \mathbf{Rc}\left[\sum_{j=1}^{N-1} k_j \mathbf{g}_j(\mathbf{x})u_j + \mathbf{g}_R(\mathbf{x})u_r\right]. \tag{13.12}$$

As **Rc** is a large number, the contribution of the third term in (13.12) to the evolution of $\mathbf{x}$ is significant. Consequently, we can expect the dynamic behavior of systems with large recycle to deviate from the dynamics of the nominal system with N units in series.

From the point of view of chemical process systems, Case II is of elevated interest given the current trend towards ever tighter integration of chemical plants through material recycling and energy recovery. While significant inventory recycling is entirely a *steady-state design* feature of a chemical process, this feature has a profound impact on process *dynamics and control*, and the remainder of the chapter is dedicated to investigating this connection.

To this end, let us rewrite the model in (13.12) in a more general form:

$$\dot{\mathbf{x}} = \mathbf{f}(\mathbf{x}) + \mathbf{G}^s(\mathbf{x})\mathbf{u}^s + \frac{1}{\varepsilon}\mathbf{G}^l(\mathbf{x})\mathbf{u}^l, \tag{13.13}$$

where, as above, $\mathbf{x}$ is the vector of unit inventories, $\mathbf{u}^s \in \mathbf{R}^{m^s}$ is the vector of scaled input variables that correspond to the *small* input/output of inventory from the process, $\mathbf{u}^l \in \mathbf{R}^{m^l}$ is the vector of scaled input variables that correspond to the *large* internal inventory flows (including inventory recycling), $\varepsilon = 1/\mathbf{Rc}$, and $\mathbf{G}^s(\mathbf{x})$ and $\mathbf{G}^l(\mathbf{x})$ are matrices of appropriate dimensions.

Equation (13.13) is a nonstandard singularly perturbed system of equations. The rational approach for addressing the control of such systems involves the properly coordinated synthesis of separate fast and slow controllers [11, 13], so that overall stability, output tracking, and disturbance rejection performance can be achieved. The design of such controllers and the closed-loop analysis must be performed on the basis of separate reduced order models that describe the dynamics in the fast and slow time scales, issues that are considered in the following sections.

13.4 Reduced Order Modeling

13.4.1 Fast Dynamics

We define a fast, "stretched" time scale $\tau = t/\varepsilon$. Rewriting (13.13) in this time scale and considering the limit case $\varepsilon \to 0$ (which physically corresponds to an infinitely high recycle number or, equivalently, an infinitely high inventory recycle flow rate), we obtain a description of the fast dynamics of the process:

$$\frac{d\mathbf{x}}{d\tau} = \mathbf{G}^l(\mathbf{x})\mathbf{u}^l. \tag{13.14}$$

Note that the above model of the fast dynamics involves only the (large) flow rates $\mathbf{u}^l$ of the inventory recycle and internal inventory streams and does not involve the (smaller) flow rates $\mathbf{u}^s$ of input and output of inventory to and from, respectively, the process. Examining (13.9), it is intuitive that the flow rates of the internal streams do not affect the *total inventory* in the process, and that the total inventory is affected only by the flow rates $\mathbf{u}^s$ of the input/output streams. In other words, (13.14) effectively describes the dynamics of the individual units in the recycle loop—note that, by definition, τ is in the order of magnitude of the time constant of an individual process unit—and does not capture the overall (process-level) changes in inventory. We can use this observation to further infer the following:

- The differential equation (13.14) is not linearly independent. By consequence, the steady-state condition
$$\mathbf{0} = \mathbf{G}^l(\mathbf{x})\mathbf{u}^l \tag{13.15}$$
for the fast dynamics (13.14) does not specify a set of isolated equilibrium points, but rather a low-dimensional equilibrium subspace (manifold), in which a slow component of the system dynamics evolves.

- The slow component of the process dynamics is associated with the evolution of the total inventory of the process.

- Based on physical considerations, at most one equation is required to completely capture the above overall, process-level inventory balance. Thus, we can expect that the dimension of the system of equations describing the slow dynamics of the process system to be at most one, and the equilibrium manifold (13.15) of the fast dynamics to be at most one-dimensional.

Remark 13.6. *In general, $\mathcal{C}$ chemical components are present in a process system, and the inventory of each component must be traced individually. In this case, $\mathcal{C}$ equations are required to completely capture the process-level material balance. Extending the arguments above, the dimension of the system of equations describing the slow dynamics will thus be at most $\mathcal{C}$, and the equilibrium manifold* (13.15) *will be at most $\mathcal{C}$-dimensional. When material and energy balances are considered simultaneously, the dimension of the slow system must by further augmented by one, to $\mathcal{C}+1$, to account for the total enthalpy of the process. All subsequent arguments in this chapter consider this general case.*

Remark 13.7. *The algebraic constraints of* (13.15) *incorporate a set of (yet unknown) manipulated inputs, $\mathbf{u}^l$. The equilibrium manifold described by* (13.15) *is thus* control dependent [13].

13.4.2 Slow Dynamics

In order to obtain the description of the slow dynamics, we make the following (typically true [2, 4, 6, 9, 14]) assumption.

Assumption 13.8. *The matrix $\mathbf{G}^l(\mathbf{x})$ can be decomposed as $\mathbf{G}^l(\mathbf{x}) = \mathbf{B}(\mathbf{x})\bar{\mathbf{G}}^l(\mathbf{x})$, with $\mathbf{B}(\mathbf{x}) \in \mathbf{R}^{n\times(n-(\mathcal{C}+1))}$ being a full column rank matrix and the matrix $\bar{\mathbf{G}}^l(\mathbf{x}) \in \mathbf{R}^{(n-(\mathcal{C}+1))\times m^l}$ having linearly independent rows.*

We also assume the following.

Assumption 13.9. *The distribution* $\mathrm{span}\{B_1(\mathbf{x}),\ldots,B_{n-(C+1)}(\mathbf{x})\}$*, where $B_i \in \mathrm{col}(\mathbf{B})$ is involutive in X.*

Multiplying (13.13) by ε and considering the limit of an infinitely high recycle flow rate (i.e., $\varepsilon \to 0$) in the original time scale t, we obtain the algebraic constraints in (13.15) or, equivalently, the linearly independent constraints
$$\bar{\mathbf{G}}^l(\mathbf{x})\mathbf{u}^l = \mathbf{0} \tag{13.16}$$

which correspond to the quasi-steady state of the fast dynamics and must be satisfied in the slow time scale.

Also in the limit $\varepsilon \to 0$, the terms $\frac{\bar{\mathbf{G}}^l(\mathbf{x})\mathbf{u}^l}{\varepsilon}$, which correspond to the *differences* of large flow rates present in the inventory equations of every process unit, become indeterminate. Defining $\mathbf{z} = \lim_{\varepsilon \to 0} \frac{\bar{\mathbf{G}}^l(\mathbf{x})\mathbf{u}^l}{\varepsilon}$ as the vector of these finite, but unknown terms, the system in (13.13) becomes

$$\begin{aligned} \dot{\mathbf{x}} &= \mathbf{f}(\mathbf{x}) + \mathbf{G}^s(\mathbf{x})\mathbf{u}^s + \mathbf{B}(\mathbf{x})\mathbf{z}, \\ \mathbf{0} &= \bar{\mathbf{G}}^l(\mathbf{x})\mathbf{u}^l, \end{aligned} \tag{13.17}$$

which represents a DAE model of the slow dynamics of the process, induced by the presence of significant inventory recycling.

As anticipated in Remark 13.7, the constraints in (13.16) depend on $\mathbf{u}^l$. In other words, the slow dynamics of the process cannot be completely characterized (e.g., by obtaining a reduced order ODE representation or state-space realization) prior to defining $\mathbf{u}^l$ as a function of the process state variables (or measured outputs) via an appropriate control law. These issues are addressed in the following section.

13.5 Control of Integrated Process Systems

13.5.1 Hierarchical Controller Design

The two-time scale behavior of the inventory of integrated processes suggests the use of a hierarchical control structure with two tiers of control action:

- distributed control, addressing control objectives for individual process units in the fast time scale;
- supervisory control, addressing control objectives for the overall process in the slow time scale.

To this end, let us complete the description of (13.13) with a vector of output variables $\mathbf{y}$:

$$\begin{aligned} \dot{\mathbf{x}} &= \mathbf{f}(\mathbf{x}) + \mathbf{G}^s(\mathbf{x})\mathbf{u}^s + \frac{1}{\varepsilon}\mathbf{G}^l(\mathbf{x})\mathbf{u}^l, \\ \mathbf{y} &= \mathbf{h}(\mathbf{x}), \end{aligned} \tag{13.18}$$

and let us further define $\mathbf{y} = \begin{bmatrix} \mathbf{y}^l \\ \mathbf{y}^s \end{bmatrix}$, where $\mathbf{y}^l$ denote the subset of the output variables that are associated with control objectives for the individual process units (typically involving the control of local inventories, i.e., the stabilization of liquid holdups or gas pressures) and $\mathbf{y}^s$ those that are associated with control objectives for the overall process, e.g. production rate, total inventory, and product quality.

The above time scale decomposition provides a transparent framework for the selection of manipulated inputs that can be used for control in the two time scales. Specifically, it establishes that the output variables $\mathbf{y}^l$ need to be controlled in the fast time scale, using the large flow rates $\mathbf{u}^l$, while the control of the variables $\mathbf{y}^s$ is to be considered in the slow time scale, using the variables $\mathbf{u}^s$. Moreover, the reduced order approximate models for the fast (see (13.14)) and slow (the state-space realization of (13.17)) dynamics can serve as a basis for the synthesis of well-conditioned nonlinear controllers in each time scale.

13.5.2 Control of the Fast Dynamics

Due to the dependence of the state space where the slow dynamics evolves on the large flow rates (Remark 13.7), the fast controller design must precede the slow controller design.

The design of the fast distributed controllers for the individual units can, in general, be addressed as a collection of individual control problems, where the strictness of the operational requirements for each unit dictates the complexity of the corresponding controller; typical applications rely on the use of simple linear controllers e.g., proportional (P), proportional-integral (PI), or proportional-integral-derivative (PID).

13.5.3 Control in the Slow Time Scale

The design and implementation of the fast controllers allows for the derivation of a minimal order realization of the DAE model of the slow dynamics in (13.17). In order to proceed with our arguments, we make the following assumption.

Assumption 13.10. *There exists a linear state feedback controller*

$$\mathbf{u}^l = \mathbf{K}^l\mathbf{x} \tag{13.19}$$

such that the matrix $L_{\mathbf{B}}[\tilde{\mathbf{G}}^l(\mathbf{x})\mathbf{K}^l\mathbf{x}]$*, where (with some abuse of notation)* $L_{\mathbf{B}}[\tilde{\mathbf{G}}^l(\mathbf{x})\mathbf{K}^l\mathbf{x}]$ *denotes the Lie derivative of* $\tilde{\mathbf{G}}^l(\mathbf{x})\mathbf{K}^l\mathbf{x}$ *along the columns of* $\mathbf{B}$*, is uniformly Hurwitz in* X*.*

The implementation of the linear feedback (13.19) allows us to obtain an equivalent description of the original model (13.13) that is in a standard singularly perturbed form.

Proposition 13.11. *Consider a process system of the form* (13.13) *under Assumptions* 13.8 *and* 13.9 *and the control law of Assumption* 13.10*. Then, there exists an* ε*-independent coordinate transformation invertible in* $\mathbf{x}$

$$\begin{bmatrix} \zeta \\ \eta \end{bmatrix} = \mathbf{T}(\mathbf{x}) = \begin{bmatrix} \phi(\mathbf{x}) \\ \tilde{\mathbf{G}}^l(\mathbf{x})K^l\mathbf{x} \end{bmatrix}, \tag{13.20}$$

with $\zeta \subset \mathbf{R}^{\mathfrak{C}+1}$*,* $\eta \subset \mathbf{R}^{n-(\mathfrak{C}+1)}$*, under which the system* (13.13) *takes the standard singularly perturbed form*

$$\dot{\zeta} = \tilde{\mathbf{f}}(\zeta,\eta) + \tilde{\mathbf{G}}^s(\zeta,\eta)\mathbf{u}^s, \tag{13.21}$$

$$\varepsilon\dot{\eta} = \varepsilon\bar{\mathbf{f}}(\zeta,\eta) + \varepsilon\bar{\mathbf{G}}^s(\zeta,\eta)\mathbf{u}^s + Q(\zeta,\eta)\eta, \tag{13.22}$$

with $\tilde{\mathbf{f}}(\zeta,\eta) = L_{\mathbf{f}(\mathbf{x})}\phi(\mathbf{x})$*,* $\tilde{\mathbf{G}}^s(\zeta,\eta) = L_{\mathbf{G}^s(\mathbf{x})}\phi(\mathbf{x})$*,* $\bar{\mathbf{f}}(\zeta,\eta) = L_{\mathbf{f}(\mathbf{x})}[\tilde{\mathbf{G}}^l(\mathbf{x})K^l\mathbf{x}]$*,* $\bar{\mathbf{G}}^s(\zeta,\eta) = L_{\mathbf{G}^s(\mathbf{x})}[\tilde{\mathbf{G}}^l(\mathbf{x})K^l\mathbf{x}]$*,* $Q(\zeta,\eta) = L_{\mathbf{B}(\mathbf{x})}[\tilde{\mathbf{G}}^l(\mathbf{x})\mathbf{K}^l\mathbf{x}]$*, evaluated at* $\mathbf{x} = T^{-1}(\zeta,\eta)$*.*

Proof. Substituting the expression for the linear feedback (13.19) in the original model (13.13), the resulting expression satisfies the conditions of Theorem 6.1 of [13], and therefore the ε-independent coordinate change exists. □

Using this result, we can derive an expression for the slow dynamics of (13.13).

Proposition 13.12. *Consider the process system* (13.13) *and suppose Assumptions* 13.8–13.10 *hold. Then,*

$$\dot{\zeta} = \tilde{\mathbf{f}}(\zeta, \mathbf{0}) + \tilde{\mathbf{G}}^s(\zeta, \mathbf{0})\mathbf{u}^s \tag{13.23}$$

represents a minimal order state-space realization of the slow dynamics of the system.

Proof. Let us rederive an expression for the fast dynamics in the transformed coordinates. Using the same definition of the fast time scale τ and considering the limit case $\varepsilon \to 0$, we obtain

$$\frac{d\eta}{d\tau} = Q(\zeta, \eta)\boldsymbol{\eta}. \tag{13.24}$$

Let us now consider the same limiting case $\varepsilon \to 0$ in the original time scale t, in which case we obtain the following DAE representation of the slow dynamics:

$$\dot{\zeta} = \tilde{\mathbf{f}}(\zeta, \eta) + \tilde{\mathbf{G}}^s(\zeta, \eta)\mathbf{u}^s, \tag{13.25}$$

$$\mathbf{0} = Q(\zeta, \eta)\boldsymbol{\eta}. \tag{13.26}$$

Following Assumption 13.10, the matrix $Q(\zeta, \eta)$ is nonsingular and hence the algebraic constraints of (13.26) are equivalent to $\mathbf{0} \equiv \boldsymbol{\eta}$. Equation (13.23) thus represents a ($\mathcal{C}$+1)-dimensional state-space realization of the slow dynamics of the system (13.13). Further, Assumption 13.10 fulfills the conditions of Proposition 3.3 of [13] and, consequently, the ODE representation (13.23) is of minimal order. □

Corollary 13.13. *The fast component of the system dynamics (see* (13.24)*) is exponentially stable in X.*

Proof. According to Assumption 13.10, the matrix Q is uniformly Hurwitz in X and the fast component of the dynamics is therefore exponentially stable over this domain. □

13.6 Closed-Loop Stability

From a practical perspective, the model in (13.25) should be used to design a (multivariable) controller that manipulates the inputs $\mathbf{u}^s$ to fulfill the process level inventory control objectives $\mathbf{y}^s$. It is important to note that the availability of a low order ODE model of the process-level inventory dynamics affords significant flexibility in designing the supervisory control system, as any of the available (inversion- or optimization-based [12, 15, 21]) control methods for nonlinear ODE systems can be applied. The stability of the composite control comprising of the linear feedback (13.19) and such a supervisory controller is characterized below.

Theorem 13.14. *Consider an integrated process system of the type* (13.13) *under the composite control consisting of*

- *the linear feedback* (13.19) *and*
- *any supervisory controller that exponentially stabilizes the slow dynamics.*

Then, there exists $\varepsilon^* > 0$ such that the origin ($\mathbf{x} = 0$) of the full order system is asymptotically stable in X for all $0 < \varepsilon < \varepsilon^*$.

Proof. We begin by observing that the exponential stability (under the feedback (13.19)) of the fast component of the dynamics in X allows for the supervisory controller for the entire system to be synthesized based on the slow model.

Subsequently, the same arguments presented in Theorem 11.4 in [10] can be used to show that there exists an $\varepsilon^* > 0$ such that the origin of (13.13) under this composite control inherits the asymptotic stability property of the closed-loop slow system for all $\varepsilon < \varepsilon^*$. □

Remark 13.15. *The composite control approach in Theorem* 13.14 *is also beneficial from an implementation point of view: the reduced dimensions and improved conditioning (reduced stiffness) of the supervisory controller (compared to a controller based on the original model* (13.13)) *will result in reduced online calculation times and less sensitivity to noise and disturbances. The proposed control system would have to be supplemented with an appropriate logic-based controller to account for inherently transient situations such as plant startup or shutdown.*

Remark 13.16. *The framework developed above can, in principle, be applied when Assumption* 13.9 *is not satisfied: when the distribution* $\text{span}\{B_1(\mathbf{x})\ldots B_{n-(C+1)}(\mathbf{x})\}$ *is not involutive, a standard singularly perturbed form of the type* (13.21) *can still be obtained via* ε*-dependent coordinate change* [13].

Remark 13.17. *Special instances of the general framework described above have been successfully used for analyzing the dynamics of and designing control systems for numerous integrated process systems of elevated practical interest. The reader is referred to* [1, 2, 4, 9, 14] *and to Section* 13.7 *for details.*

13.7 Case Study: Control of an Energy Integrated Process

We consider an energy integrated process designed to convert a feed stream rich in compound A (of molar concentration c_{Ao}) into compound B in a high temperature, mildly exothermic, first-order reaction carried out in an adiabatic reactor (Figure 13.2). The process feed stream is at ambient temperature and is heated to the elevated operating temperature of the reactor by recovering a large amount of heat from the product stream in a

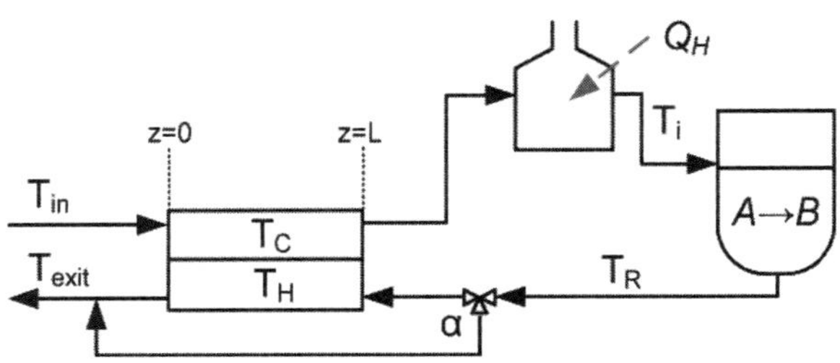

Figure 13.2. Reactor-FEHE process with fired furnace and exchanger bypass stream. *Used with permission* [3].

feed-effluent heat exchanger (FEHE). For improved operability, the plant features a heater that is used at full capacity in startup mode and as a trim heater during operation, as well as a bypass stream which is used to regulate heat recovery in the FEHE. The operational objectives for this process are the stabilization of the reactor holdup and operating temperature, as well as controlling the product stream temperature and maintaining its purity within the specifications of a downstream separation system. The reactor effluent flow rate, the bypass flow rate, and the heater duty are available as manipulated inputs.

Assuming that the temperature and composition dependence of the heat capacity and density of the process streams is not significant, that there is no phase change in the heat exchanger, and that the material holdup in both legs of the heat exchanger and in the heater is constant, the model of the process can be written as

$$\begin{aligned}
\frac{dV}{dt} &= F_{in} - F,\\
\frac{dc_A}{dt} &= \frac{F_{in}}{V}(c_{Ao} - c_A) - k_o e^{(-E/RT_R)} c_A,\\
\frac{dT_R}{dt} &= \frac{F_{in}}{V}(T_i - T_R) - \frac{1}{\rho C_p} k_o e^{(-E/RT_R)} c_A \Delta H,\\
\frac{\partial T_H}{\partial t} &= -v_H \frac{\partial T_H}{\partial z} - \frac{UA}{\rho C_p}\frac{T_H - T_C}{V_H},\\
\frac{\partial T_C}{\partial t} &= -v_C \frac{\partial T_C}{\partial z} + \frac{UA}{\rho C_p}\frac{T_H - T_C}{V_C}\\
\frac{dT_i}{dt} &= \frac{F_{in}}{V_f}(T_{C_{z=L}} - T_i) + \frac{Q_H}{\rho C_p V_f}
\end{aligned} \tag{13.27}$$

with the boundary conditions $T_{H_{z=L}} = T_R$, $T_{C_{z=0}} = T_{in}$. In the model above, F_{in} is the volumetric flow rate of the input stream to the system and F is the reactor outlet flow rate; V, V_H, V_C, and V_f are the holdups of the reactor, the FEHE hot and cold legs, and the furnace, respectively. T_{in} is the cold leg inlet temperature to the FEHE. T_H and T_C are the hot and cold leg temperatures for the FEHE. T_i and T_R are the reactor inlet and outlet temperatures and T_{exit} is the temperature of the product stream exiting the process. Q_H is the furnace heat duty, v_H and v_C are the velocities of the fluid in the hot and cold compartments of FEHE, L is the length of the exchanger, and z is the spatial coordinate. The nominal values of the process variables and parameters are presented in Table 13.1.

The FEHE bypass ratio, α, is defined as the ratio of the flow rate of the bypass stream and the flow rate of the feed stream, F. Using this definition and the assumption that the heat capacity and density of the streams do not change, we can calculate the temperature of the stream exiting the process as

$$T_{exit} = \alpha T_R + (1-\alpha) T_{H(z=0)}. \tag{13.28}$$

13.7.1 System Analysis

In order to carry out numerical simulations, the spatial derivatives in (13.27) were discretized using a backwards finite difference scheme on a grid of 1001 nodes. It was verified that increasing the number of nodes does not result in significant changes in the simulation results and that the energy balance at the process level is closed.

Table 13.1. Nominal values of process parameters for the reactor-FEHE process.

k_o	1.2667×10^7	s^{-1}	c_A	55.19	mol/l
E	142870.0	J/mol	T_R	922.0	K
ΔH	-54.828	kJ/mol	T_i	909.62	K
ρC_p	4.184×10^6	$J/m^3/K$	T_{exit}	363.53	K
UA	83680.0	W/K	$T_{C,z=0}$	860.53	K
α	0.1		Q_H	118.42	kW
V	0.1	m^3	T_{in}	300.0	K
V_H	0.1	m^3	F_{in}	5.7667×10^{-4}	m^3/s
V_C	0.09	m^3	F	5.7667×10^{-4}	m^3/s
V_f	0.01	m^3	c_{Ao}	1000.0	mol/m^3
L	5.0	m			

The temperature profiles of the heat exchanger (Figures 13.3(a) and 13.3(b)) show a pinch at the cold end (the temperature of the hot stream leaving the exchanger $T_{H,z=0} = 301.48\ K$ is very close to the inlet temperature). This indicates that the process is tightly integrated, with a high degree of energy recovery.

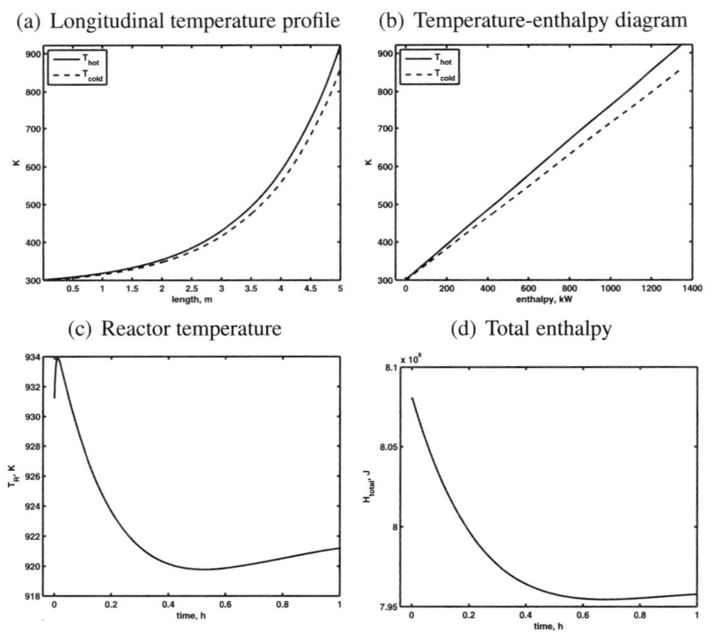

Figure 13.3. Steady-state temperature profiles in the FEHE (a), (b) and dynamic response of the system (c), (d). *Used with permission* [3].

A dynamic simulation of the 2006-dimensional ODE system from initial conditions slightly perturbed from the values in Table 13.1 confirmed the theoretical analysis developed earlier in this chapter. Typical of processes with tight energy integration, the temperatures of the individual units exhibit a two-time scale behavior: the corresponding plots (Figure 13.3(c)) show a fast transient, followed by a slow approach to steady state. Conversely, as anticipated in Section 13.4, the variation of the total enthalpy (Figure 13.3(d)) is slow.

13.7.2 Reduced Order Modeling

The results of the analysis above indicate that the process under consideration belongs to the general category of integrated processes. Specifically, it features a high-energy recycle. The rational development of a control framework would therefore entail a time scale decomposition and model reduction, as discussed earlier in this chapter. However, the presence of the partial differential, advection terms in (13.27) prevents us from applying the methods developed in Section 13.4 directly. In order to circumvent this difficulty, let us first develop an alternative modeling construct based entirely on ODEs. Figure 13.4 provides a block-diagram representation of the energy flows in the process. Postponing the derivation of an expression of the heat exchanger duty H_{rec} (originally described by the partial differential terms) until later in this section, we can use this representation to develop a lumped parameter model of the process:

$$
\begin{aligned}
\dot{V} &= F_{in} - F, \\
\dot{c}_A &= \frac{F_{in}}{V}(c_{Ao} - c_A) - k_o e^{(-E/RT_R)} c_A, \\
\dot{T}_R &= \frac{F_{in}}{V}(T_i - T_R) - \frac{1}{\rho C_p} k_o e^{(-E/RT_R)} c_A \Delta H, \\
\dot{T}_H &= \frac{F_{in}(1-\alpha)}{V_H}(T_R - T_H) - \frac{H_{rec}}{\rho C_p V_H}, \\
\dot{T}_C &= \frac{F_{in}}{V_C}(T_{in} - T_C) + \frac{H_{rec}}{\rho C_p V_C}, \\
\dot{T}_i &= \frac{F_{in}}{V_f}(T_C - T_i) + \frac{Q_H}{\rho C_p V_f}.
\end{aligned}
\tag{13.29}
$$

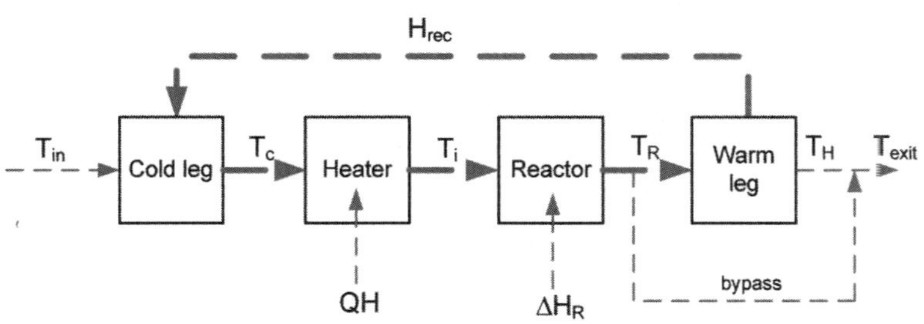

Figure 13.4. Block diagram of the reactor-FEHE process. H_{rec} represents the duty of the heat exchanger. *Used with permission* [3].

Based on this representation, we can define the recycle number **Rc** as an *energy recovery number*:

$$\mathbf{Rc} = \frac{H_{rec,s}}{[F_{in}\rho C_p(T_{in} - T_{ref})]_s}, \tag{13.30}$$

with T_{ref} being a reference temperature. In order to accomplish a high degree of thermal integration, **Rc** should be significant; conversely, we can expect that $\varepsilon = 1/\mathbf{Rc}$ will be small, i.e., $\varepsilon \ll 1$. Using arguments similar to those developed in Section 13.4, we also define the following $\mathcal{O}(1)$ quantities:

$$k_R = \frac{[F_{in}\rho C_p(T_R - T_{ref})]_s}{H_{rec,s}}, \qquad u_R = \frac{F_{in}\rho C_p(T_R - T_{ref})}{[F_{in}\rho C_p(T_R - T_{ref})]_s}, \tag{13.31}$$

$$k_i = \frac{[F_{in}\rho C_p(T_i - T_{ref})]_s}{H_{rec,s}}, \qquad u_i = \frac{F_{in}\rho C_p(T_i - T_{ref})}{[F_{in}\rho C_p(T_i - T_{ref})]_s}, \tag{13.32}$$

$$k_C = \frac{[F_{in}\rho C_p(T_c - T_{ref})]_s}{H_{rec,s}}, \qquad u_C = \frac{F_{in}\rho C_p(T_C - T_{ref})}{[F_{in}\rho C_p(T_C - T_{ref})]_s} \tag{13.33}$$

and $u_{rec} = H_{rec}/H_{rec,s}$. With this notation, we can rewrite the model in (13.29) (note that $\alpha/\varepsilon = \mathcal{O}(1)$; i.e., the energy flow associated with the bypass stream is much smaller than the amount of energy recycled to the process) as

$$\begin{aligned}
\dot{V} &= F_{in} - F, \\
\dot{c}_A &= \frac{F_{in}}{V}(c_{Ao} - c_A) - k_o e^{(-E/RT_R)} c_A, \\
\dot{T}_R &= -\frac{1}{\rho C_p} k_o e^{(-E/RT_R)} c_A \Delta H + \frac{1}{\varepsilon}\frac{[F_{in}(T_{in} - T_{ref})]_s}{V_f} V(k_i u_i - k_R u_R), \\
\dot{T}_H &= -\frac{F(1-\alpha)}{V_H} T_H - \frac{1}{V_H}\alpha F T_R + \frac{1}{\varepsilon}\frac{[F_{in}(T_{in} - T_{ref})]_s}{V_H}(k_R u_R - u_{rec}), \\
\dot{T}_C &= \frac{F_{in}}{V_C} T_{in} + \frac{1}{\varepsilon}\frac{[F_{in}(T_{in} - T_{ref})]_s}{V_C}(-k_C u_C + u_{rec}), \\
\dot{T}_i &= \frac{Q_H}{\rho C_p V_f} + \frac{1}{\varepsilon}\frac{[F_{in}(T_{in} - T_{ref})]_s}{V_f}(k_C u_C - k_i u_i),
\end{aligned} \tag{13.34}$$

which is in the form of (13.13). We can now proceed with the analysis as described in Section 13.4. Let us define the stretched, fast time variable $\tau = t/\varepsilon$. Rewriting the model in terms of τ and considering the limit $\varepsilon \to 0$, we obtain a description of the fast dynamics of the process:

$$
\begin{aligned}
\frac{dV}{d\tau} &= 0,\\
\frac{dc_A}{d\tau} &= 0,\\
\frac{dT_R}{d\tau} &= \frac{[F_{in}(T_{in}-T_{ref})]_s}{V}(k_i u_i - k_R u_R),\\
\frac{dT_H}{d\tau} &= \frac{[F_{in}(T_{in}-T_{ref})]_s}{V_H}(k_R u_R - u_{rec}),\\
\frac{dT_C}{d\tau} &= \frac{[F_{in}(T_{in}-T_{ref})]_s}{V_C}(-k_C u_C + u_{rec}),\\
\frac{dT_i}{d\tau} &= \frac{[F_{in}(T_{in}-T_{ref})]_s}{V_f}(k_C u_C - k_i u_i)
\end{aligned} \tag{13.35}
$$

with the corresponding (linearly independent) quasi-steady-state constraints:

$$
0 = k_R u_R - u_{rec};\;\; 0 = k_i u_i - k_R u_R;\;\; 0 = k_C u_C - k_i u_i. \tag{13.36}
$$

Subsequently, the slow component of the process dynamics takes the form of (13.17):

$$
\begin{aligned}
\dot{V} &= F_{in} - F,\\
\dot{c}_A &= \frac{F_{in}}{V}(c_{Ao} - c_A) - k_o e^{(-E/RT_R)} c_A,\\
\dot{T}_R &= -\frac{1}{\rho C_p} k_o e^{(-E/RT_R)} c_A \Delta H + \frac{[F_{in}(T_{in}-T_{ref})]_s}{V_f} V z_2,\\
\dot{T}_H &= -\frac{F(1-\alpha)}{V_H} T_H - \frac{1}{V_H}\alpha F T_R + \frac{[F_{in}(T_{in}-T_{ref})]_s}{V_H} z_1,\\
\dot{T}_C &= \frac{F_{in}}{V_C} T_{in} - \frac{[F_{in}(T_{in}-T_{ref})]_s}{V_C}(z_1 + z_2 + z_3),\\
\dot{T}_i &= \frac{Q_H}{\rho C_p V_f} + \frac{[F_{in}(T_{in}-T_{ref})]_s}{V_f} z_3,\\
0 &= F_{in}\rho C_p T_R - H_{rec};\quad 0 = T_i - T_R;\quad 0 = T_C - T_i,
\end{aligned} \tag{13.37}
$$

where $z_1 = \lim_{\varepsilon\to 0}\frac{1}{\varepsilon}(k_R u_R - u_{rec})$, $z_2 = \lim_{\varepsilon\to 0}\frac{1}{\varepsilon}(k_i u_i - k_R u_R)$, $z_3 = \lim_{\varepsilon\to 0}\frac{1}{\varepsilon}(k_C u_C - k_i u_i)$. The expressions of the algebraic constraints in (13.37) have been obtained by substituting the definitions in (13.31) and the definition of u_{rec} in (13.36). In order to determine the algebraic variables z_i, we must differentiate these algebraic constraints and, to this end, we must provide an expression for H_{rec}, the duty of the FEHE.

The process conditions in the FEHE (counter-current flow with no phase change and constant physical properties) lend themselves naturally to calculating the duty based on the log mean temperature difference (LMTD), $H_{rec} = UA$ LMTD, with

$$
\text{LMTD} = \frac{(T_R - T_C) - (T_H - T_{in})}{\ln[(T_R - T_C)/(T_H - T_{in})]} \simeq \left[\frac{(T_R - T_C)^{1/3} + (T_H - T_{in})^{1/3}}{2}\right]^3. \tag{13.38}
$$

The Underwood approximation [17] in (13.38) is used to circumvent the issues that arise from the fact that the original expression of LMTD is not well defined in certain physical situations (e.g., when the temperature gradient in the exchanger is constant, $T_R = T_C$ and $T_H = T_{in}$). Substituting the definition of H_{rec} in (13.37), after differentiating the constraints once to obtain the algebraic variables and applying a coordinate change of the type in (13.20), namely

$$\begin{aligned}
\delta &= T_R,\\
\eta_1 &= F_{in}\rho C_p T_R - 0.5\,UA\left[(T_R - T_C)^{1/3} + (T_H - T_{in})^{1/3}\right]^3,\\
\eta_2 &= T_i - T_R,\\
\eta_3 &= T_C - T_i,
\end{aligned} \tag{13.39}$$

we obtained the following minimal order state space realization of the slow dynamics of the type in (13.23):

$$\begin{aligned}
\dot{V} &= F_{in} - F,\\
\dot{c_A} &= \frac{F_{in}}{V}(c_{Ao} - c_A) - k_o e^{(-E/R\delta)} c_A,\\
\dot{\delta} &= \big[-kc_A \Delta H\, V\mathrm{UA} + \rho C_p T_{in}\,\mathrm{UA}(F_{in} - F)\\
&\quad + 8(\rho \mathrm{C_p})^2 F\mathrm{F_{in}}\,\delta(\alpha - 1) + \rho \mathrm{C_p}\, F\mathrm{T_{in}}\alpha\,\mathrm{UA}\\
&\quad - \alpha\,\rho \mathrm{C_p}\,\delta\, F\,\mathrm{UA} + \mathrm{QHUA}\big]/\mathrm{DEN},\\
\mathrm{DEN} &= \rho C_p\left[\mathrm{UA}(\mathrm{V_C} + \mathrm{V_f} + V) + 8\,\mathrm{V_h F_{in}}\,\rho \mathrm{C_p}(1 - \alpha)\right].
\end{aligned} \tag{13.40}$$

13.7.3 Controller Design

According to the theoretical framework developed in this chapter, the control system for this process should be carried out in a hierarchical fashion, starting with the stabilization of the inventories in each unit, which, in this case, is equivalent to stabilizing the unit temperatures. Examining the fast submodel in (13.35), it is easy to observe that it does not contain any of the manipulated inputs of the process. At first glance, this could constitute a matter of concern, as there are seemingly no means to control the fast dynamics. However, (13.35) contains no energy source or energy sink terms, describing, in effect, the flow of a fluid through the path comprised of the cold leg of the heat exchanger (where it is heated), the heater (where it undergoes no temperature change), the reactor (where, again, it undergoes no temperature change), and, finally, the warm leg of the heat exchanger, where it is cooled. Based on physical considerations (even under the ideal assumption that no energy is lost to the environment), the fast component of the process dynamics is clearly stable and requires no feedback control action.

Turning to the slow dynamics, we are interested in stabilizing the reactor holdup, controlling the process outlet temperature, and controlling the temperature in the reactor to manage the conversion. The first two objectives can be addressed using simple linear controllers. In order to stabilize the reactor holdup, we implemented a PI controller using the reactor outlet flow rate, F, as a manipulated input:

$$F = F_s + K_v\left[(V_{set} - V) + \frac{1}{\tau_v}\int_0^t (V_{set} - V)\,d\tau\right], \tag{13.41}$$

with $F_s = 5.7667 \cdot 10^{-4}\ m^3/s$, $K_v = 5\ s^{-1}$, and $\tau_v = 18\,\text{min}$. The control of T_{exit} was addressed using α as a manipulated input and the PI control law

$$\alpha = \alpha_o + K_c \left[(T_{exit,set} - T_{exit}) + \frac{1}{\tau_I} \int_0^t (T_{exit,set} - T_{exit})\, d\tau \right], \qquad (13.42)$$

with $\alpha_o = 0.1$, $K_c = 0.0018\ K^{-1}$, and $\tau_I = 10\ s$. The low reset time used for this controller is justified by the fast nature of the mixing process.[3]

Implementing the reactor temperature controller merits some discussion. While T_R is not a true slow variable (it has a two-time scale behavior, as illustrated in Figure 13.3(c)), as we argued above, the fast transient of the process (and, inherently, of T_R) is stable. We are thus interested in controlling the *slow* component of the reactor temperature, which governs in effect the behavior of the entire process. To this end, we conveniently chose the coordinate transformation (13.39) so that the energy balance in (13.40) is written in terms of the reactor temperature T_R, rather than in terms of the total enthalpy of the process. The nonstiff, low-dimensional model (13.40) is ideally suited for control purposes. We used it to design an input-output linearizing controller with integral action [7] that manipulates QH and enforces a first order response in the T_R dynamics, i.e., $\beta dT_R/dt + T_R = T_{R,set}$, with $\beta = 15\,\text{min}$.

13.7.4 Simulation Results and Discussion

The aforementioned controllers were implemented on the full order 2006-dimensional discretization of the original, distributed-parameter model, and their performance was tested through simulation. The simulation scenario considers the response of the closed-loop system to a 10% increase in the production rate, imposed at $t = 1\,\text{h}$ by increasing the feed flow rate. An unmeasured $10K$ increase in the feed temperature occurs at $t = 1\,\text{h}$ as well. The setpoint of the reactor temperature controller $T_{R,set}$ is increased by $2K$ at $t = 1\,\text{h}$ in order to maintain reactor conversion at the higher production rate.

The results in Figure 13.5 show that the proposed controller has excellent performance. These results further confirm the wide applicability of the theoretical framework proposed in this chapter for developing an understanding of the dynamics of integrated processes, and for designing effective control and energy management strategies.

13.8 Conclusions

This chapter introduced and defined the concept of an integrated process system. We established that the dynamic behavior of process systems with significant inventory recycling exhibits a time scale separation. In the fast (unit) time scale, the dynamic coupling between the units is weak and can be ignored. The interactions between units do, however, become significant over long periods of time, and we showed that integrated processes exhibit a slow, system-level dynamics. Using singular perturbation tools, we developed a method for deriving a DAE model (and the corresponding minimum order ODE realization) of this core dynamics. We then proposed a framework for exploiting the two-time scale behavior

[3] Equation (13.28) assumes instant and perfect mixing at the exit of the warm leg of the heat exchanger. In order to facilitate numerical simulations, this output was filtered with a first order filter with a time constant $\tau_{filter} = 2\,\text{s}$ prior to being used in the feedback controller.

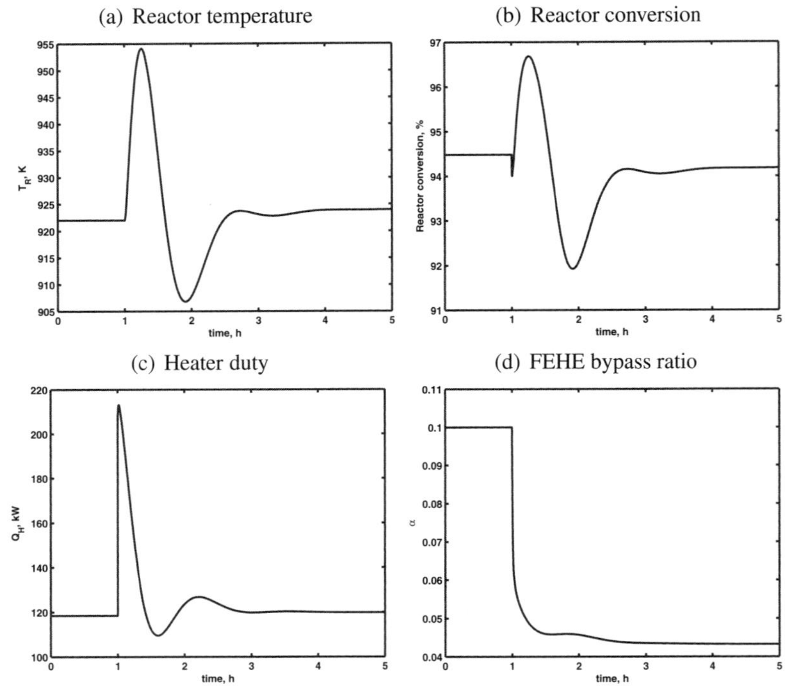

Figure 13.5. Evolution of the process variables for a 10% increase in the production rate, in the presence of unmeasured feed temperature disturbances. *Used with permission* [3].

in a hierarchical controller design with well-defined stability properties. Finally, we studied the application of the proposed analysis and control framework on an energy integrated process, demonstrating excellent closed-loop performance.

Bibliography

[1] M. Baldea and P. Daoutidis, *Model reduction and control of reactor–heat exchanger networks*, J. Process Control, 16 (2006), pp. 265–274.

[2] M. Baldea and P. Daoutidis, *Control of integrated process networks–a multi-time scale perspective*, Comput. Chem. Eng., 31 (2007), pp. 426–444.

[3] M. Baldea and P. Daoutidis, *Dynamics and Nonlinear Control of Integrated Process Systems*, Cambridge University Press, Cambridge, UK, 2012.

[4] M. Baldea, P. Daoutidis, and A. Kumar, *Dynamics and control of integrated networks with purge streams*, AIChE J., 52 (2006), pp. 1460–1472.

[5] P. Christofides, J. Davis, N. El-Farra, D. Clark, K. Harris, and J. Gipson, *Smart plant operations: Vision, progress and challenges*, AIChE J., 53 (2007), pp. 2734–2741.

[6] M. CONTOU-CARRÈRE, M. BALDEA, AND P. DAOUTIDIS, *Dynamic precompensation and output feedback control of integrated process networks*, Ind. Eng. Chem. Res., 43 (2004), pp. 3528–3538.

[7] P. DAOUTIDIS AND C. KRAVARIS, *Dynamic output feedback control of minimum-phase nonlinear processes*, Chem. Eng. Sci., 47 (1992), pp. 837–849.

[8] C. A. FARSCHMAN, K. P. VISWANATH, AND B. E. YDSTIE, *Process systems and inventory control*, AIChE J., 44 (1998), pp. 1841–1857.

[9] S. JOGWAR, M. BALDEA, AND P. DAOUTIDIS, *Dynamics and control of process networks with large energy recycle*, Ind. Eng. Chem. Res., 48 (2009), pp. 6087–6097.

[10] H. K. KHALIL, *Nonlinear Systems*, 3rd ed., Prentice–Hall, Englewood Cliffs, NJ, 2002.

[11] P. V. KOKOTOVIC, H. K. KHALIL, AND J. O'REILLY, *Singular Perturbations in Control: Analysis and Design*, Academic Press, London, 1986.

[12] C. KRAVARIS AND J. KANTOR, *Geometric methods for nonlinear process control*, Parts 1-2, Ind. Eng. Chem. Res., 29 (1990), pp. 2295–2323.

[13] A. KUMAR AND P. DAOUTIDIS, *Control of Nonlinear Differential Equation Systems*, Res. Notes Math. 397, Chapman & Hall/CRC, Boca Raton, FL, 1999.

[14] A. KUMAR AND P. DAOUTIDIS, *Dynamics and control of process networks with recycle*, J. Process Control, 12 (2002), pp. 475–484.

[15] D. MAYNE, J. RAWLINGS, C. RAO, AND P. SCOKAERT, *Constrained model predictive control: Stability and optimality*, Automatica, 36 (2000), pp. 789–814.

[16] R. SCATTOLINI, *Architectures for distributed and hierarchical model predictive control—A review*, J. Process Control, 19 (2009), pp. 723–731.

[17] A. UNDERWOOD, *Simple formula to calculate mean temperature difference*, Chem. Eng., 77 (1970), p. 192.

[18] S. VASUDEVAN AND G. RANGAIAH, *Criteria for performance assessment of plantwide control systems*, Ind. Eng. Chem. Res., 49 (2010), pp. 9209–9221.

[19] J. WASSICK, *Enterprise-wide optimization in an integrated chemical complex*, Comput. Chem. Eng., 33 (2009), pp. 1950–1963.

[20] S. XU AND J. BAO, *Distributed control of plantwide chemical processes*, J. Process Control, 19 (2009), pp. 1671–1687.

[21] V. ZAVALA AND L. BIEGLER, *The advanced-step NMPC controller: Optimality, stability and robustness*, Automatica, 45 (2009), pp. 86–93.

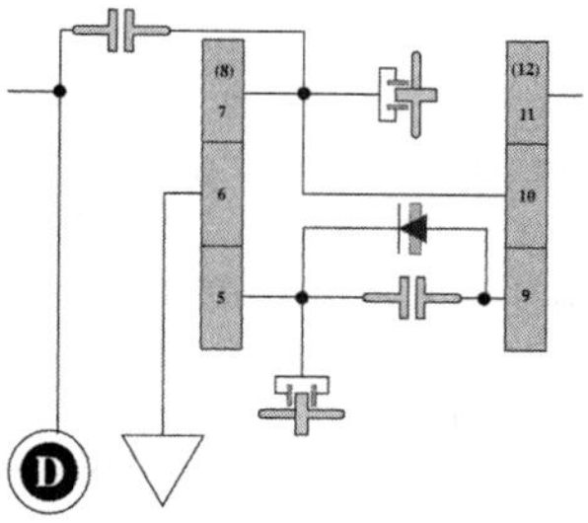

Chapter 14
DMPC for Building Temperature Regulation

Yudong Ma, Stefan Richter, and Francesco Borrelli

We study the problem of temperature regulation in a network of building thermal zones. The control objective is to keep zone temperatures within a comfort range while consuming the least energy using predictive knowledge of weather and occupancy. First, we present a simplified thermal zone dynamic model and energy consumption models for system components. Model identification and validation based on historical measured data are presented. Second, a distributed model-based predictive control (DMPC) is designed for optimal heating and cooling. The DMPC is implemented using sequential quadratic programming (SQP) dual decomposition and the fast gradient method. Simulation results show good performance and computational tractability of the resulting scheme.

14.1 Introduction

The building sector consumes about 40% of the energy used in the United States and is responsible for nearly 40% of greenhouse gas emissions. It is therefore economically, socially, and environmentally significant to reduce the energy consumption of buildings.

This work focuses on the modeling and predictive control of networks of thermal zones. The system considered in this chapter consists of an air handling unit (AHU) and a set of variable air volume (VAV) boxes which serves a network of thermal zones. The AHU is equipped with a cooling coil, a damper, and a fan. The damper mixes return air and outside air. The cooling coil cools down the mixed air, and the fan drives the air to the VAV boxes. Each VAV box has a damper controlling the mass flow rate of air supplied to thermal zones. A heating coil in each VAV box can reheat the supply air when necessary.

The chapter is divided in two parts. The objective of the first part is to develop low order models suitable for real-time predictive optimization. This part is extracted from [14]. In this work we model the system as a network of two-mass nonlinear systems. We present identification and validation results based on historical data collected from Bancroft Library at the University of California, Berkeley. The results are promising and show that the models well capture thermal zone dynamics when the external load (due to occupancy, weather, and equipment) is minor. Historical data are then used to compute the envelope of the external load by comparing nominal models and measured data when external disturbances are not negligible.

In the second part, a distributed model-based predictive control (DMPC) is designed for regulating heating and cooling in order to minimize energy consumption while satisfying comfort constraints. The main idea of predictive control is to use the model of the plant to predict the future evolution of the system [5, 13, 17]. At each sampling time, starting at the current state, an open-loop optimal control problem is solved over a finite horizon. The optimal command signal is applied to the process only during the following sampling interval. At the next time step a new optimal control problem based on new measurements of the state is solved over a shifted horizon. For complex constrained multivariable control problems, model predictive control (MPC) has become the accepted standard in the process industries [2]: its success is largely due to its almost unique ability to simply and effectively handle hard constraints on control inputs and states.

The size of the centralized predictive control problem rapidly grows when a realistic number of rooms together with a meaningful control horizon are considered. Therefore the real-time implementation of an MPC scheme is a challenge for the low-cost embedded platforms currently used for HVAC control algorithms. The techniques presented in this chapter enable the implementation of an MPC algorithm by distributing the computational load on a set of VAV box embedded controllers coordinated by the embedded controllers on the AHU system. Compared to existing DMPC schemes [3, 4, 22, 25], the proposed method is tailored to the specific class of problems considered in this work. In particular, it makes use of sequential quadratic programming (SQP) [7, 19], proximal minimization [1], and dual decomposition [12] to handle the system nonlinearities and the decentralization, respectively. The SQP and proximal minimization methods are used to derive a strictly convex quadratic program (QP) from the original nonlinear optimization problem. The dual decomposition scheme takes advantage of the separability of the dual Lagrangian QP problem. By doing so, the dual QP is solved iteratively by updating dual and primal variables in a distributed fashion. In this chapter we show that if the centralized MPC problem is properly formulated, the resulting primal and dual update laws can be easily derived. Simulation results show good performance and computational tractability of the resulting scheme.

We remark that the evaluation of optimal controllers for building climate regulation has been studied in the past by several authors (see [8, 9, 10, 20] and the references therein). Compared to existing literature, this chapter focuses on distributing the computational load of the predictive controller on multiple, low-cost, embedded platforms.

The chapter is organized as follows. Section 14.2 introduces the system and the simplified thermal zone models to be used for the predictions in MPC schemes. In Section 14.3 the distributed MPC control algorithm is outlined. A numerical example is presented in Section 14.4. Finally, conclusions are drawn in Section 14.5.

14.2 System Model

The objective of this section is to introduce a simplified HVAC system architecture and develop a control-oriented model for it. We consider an AHU and a fan serving multiple

VAV boxes controlling air temperature and flows in a network of thermal zones (next called "rooms" for brevity). Figure 14.1 depicts the system architecture: the AHU uses a mixture of outside air and return air to generate cool air by using a cooling coil (usually driven by chilled water; see [16] for optimal generation of chilled water). The cool air then is distributed by a fan to VAV boxes connected with each room. The damper position in the VAV box controls the mass flow rate of air entering a room. In addition, a heating coil in the VAV box is used to warm up the supply air if needed.

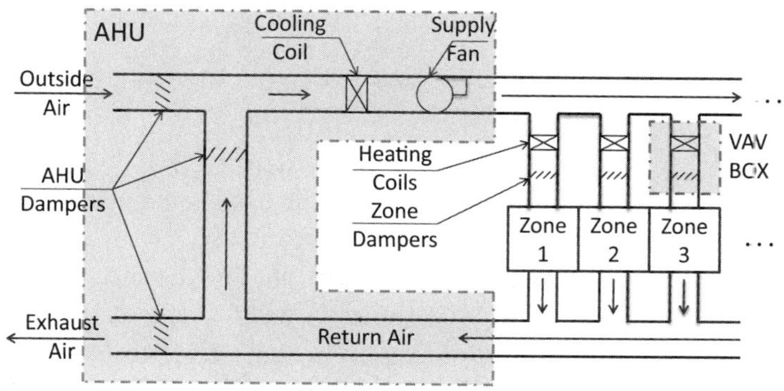

Figure 14.1. HVAC system scheme.

In order to develop a simplified yet descriptive model, the following assumptions are introduced:

(A1) The system pressure dynamics are not considered.

(A2) The dynamics of each component (AHU, VAV boxes, and fan) are neglected. This implies that the supply air temperature and flow set points are tracked perfectly.

(A3) Air temperature is constant through the ducts.

(A4) The amount of air exiting the rooms is the same as the amount of air entering the rooms. This neglects the infiltration and exhalation of air through the building envelope.

14.2.1 Simplified System Model

We use an undirected graph structure to represent the rooms and their dynamic couplings in the following way. We associate the ith room with the ith node of a graph, and if an edge (i, j) connecting the ith and jth node is present, the rooms i and j are subject to direct heat transfer. The graph $\mathcal{G}$ will be defined as

$$\mathcal{G} = (\mathcal{V}, \mathcal{A}), \tag{14.1}$$

where $\mathcal{V}$ is the set of nodes (or vertices) $\mathcal{V} = \{1, \ldots, N_v\}$ and $\mathcal{A} \subseteq \mathcal{V} \times \mathcal{V}$ the set of edges (i, j), with $i \in \mathcal{V},\ j \in \mathcal{V}$. We denote by $\mathcal{N}^i$ the set of neighboring nodes of i, i.e., $j \in \mathcal{N}^i$ if and only if $(i, j) \in \mathcal{A}$.

Now consider a single room $j \in \mathcal{V}$. The air enters the room j with a mass flow rate $\dot{m}_s^j$. It is assumed that in the AHU, the outside air fully mixes with the return air without delay, and the mixing proportion δ between the return air and outside air is controlled by the damper configurations in the AHU system to obtain

$$T_m = \delta T_r + (1-\delta)T_{oa}, \tag{14.2}$$

where T_{oa} is the outside air temperature, and T_m is the temperature of the mixed air. T_r is the return air temperature calculated as weighted average temperature of return air from each room:

$$T_r = \sum_{i\in\mathcal{V}} \dot{m}_s^i T^i / \sum_{i\in\mathcal{V}} \dot{m}_s^i. \tag{14.3}$$

The return air is not recirculated when $\delta = 0$, and no outside fresh air is used when $\delta = 1$. δ can be used to save energy through recirculation, but it has to be strictly less than one to guarantee a minimal outdoor fresh air delivered to the rooms.

We model the room as a two-mass system. C_1^j is the fast-dynamic mass that has lower thermal capacitance (e.g., air around VAV diffusers), and C_2^j represents the slow-dynamic mass that has higher thermal capacitance (e.g., solid parts which include floor, walls, and furniture). We remark that the phenomenon of fast and slow dynamics has been observed in [11]. The thermal dynamic model of a room is

$$\begin{aligned}
C_1^j \frac{dT_1^j}{dt} &= \dot{m}_s^j c_p (T_m - \Delta T_c + \Delta T_h^j - T_1^j) + (T_2^j - T_1^j)/R^j \\
&\quad + (T_{oa} - T_1^j)/R_{oa}^j + \sum_{i\in\mathcal{N}^j} (T_1^i - T_1^j)/R_{ij} + P_d^j, && (14.4a)\\
C_2^j \frac{dT_2^j}{dt} &= (T_1^j - T_2^j)/R^j \quad \forall j \in \mathcal{V}, && (14.4b)\\
T_m &= \delta T_r + (1-\delta)T_{oa}, && (14.4c)\\
\sum_{i\in\mathcal{V}} \dot{m}_s^i T_r &= \sum_{i\in\mathcal{V}} \dot{m}_s^i T^i, && (14.4d)\\
T^j &= T_1^j \quad \forall j \in \mathcal{V}, && (14.4e)
\end{aligned}$$

where T_1^j and T_2^j are system states representing the temperature of the lumped masses C_1^j and C_2^j, respectively. The supply air temperature difference across the cooling coil in the AHU is denoted as ΔT_c, and ΔT_h^j is the air temperature difference across the heating coil in the jth VAV box. T^j is the perceived temperature of room j, which is assumed to be equal to the temperature of the fast-dynamic mass C_1^j. R_{oa}^j is the thermal resistance between room j and outside air, and c_p is the specific heat capacity of room air. R^j models the heat resistance between C_1^j and C_2^j, $R_{ij} = R_{ji}$ models thermal resistances between room i and the adjacent room j, and P_d^j is an unmeasured load induced by external factors such as occupancy, equipment, and solar radiation.

The model (14.4) is tested to capture the temperature dynamics of a thermal zone in the Bancroft Library located on the campus of UC Berkeley. By using historical data we have identified the model parameters for each thermal zone and validated the resulting model. The dimension of the conference room is $5 \times 4 \times 3$ m, and it has one door and no windows. As a result, the effect of solar radiation is negligible. The major source of load derives from occupants and electronic equipment. The conference room has one neighboring office room ($\mathcal{N}^1 = \{2\}$).

The model parameters ($[C_1^1, C_2^1, R^1, R_{12}, R_{oa}^1]$) are identified by using a nonlinear regression algorithm using measured data collected on July 4th, 2010, over 24 hours. This corresponds to a Sunday when the conference room has no occupants ($P_d^1 = 0$). Measurements of room temperature (T^1), supply air temperature ($T_s^1 = T_m - \Delta T_c + \Delta T_h^1$), mass flow rate of the supply air ($\dot{m}_s^1$), the neighboring room temperature (T^2), and outside air temperature T_{oa} are used for the identification. The identified parameters values are reported in Table 14.1.

Table 14.1. Identification results for conference room model on July 4th, 2010.

Parameter	Value	Parameter	Value
C_1^1	9.163×10^3 kJ/K	R_{12}	2.000 K/kW
C_2^1	1.694×10^5 kJ/K	R_{oa}^1	57 K/kW
R^1	1.700 K/kW		

The identification results plotted in Figure 14.2 show that the proposed model successfully captures the thermal dynamics of the conference room without occupants. In Figure 14.2 the solid line depicts the measured room temperature trend and the dashed line is the room temperature predicted by model (14.4) when driven by the measured inputs.

The proposed model (14.4) with the identified parameters in Table 14.1 is validated against measurements during other weekends. Figure 14.3 plots the validation results for July 11th, 2010. One can observe that the predictions match well the experimental data.

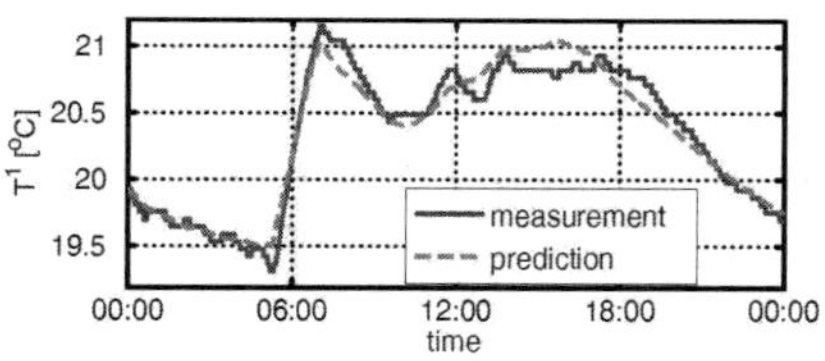

Figure 14.2. Identification results of the thermal zone model (14.4) [14].

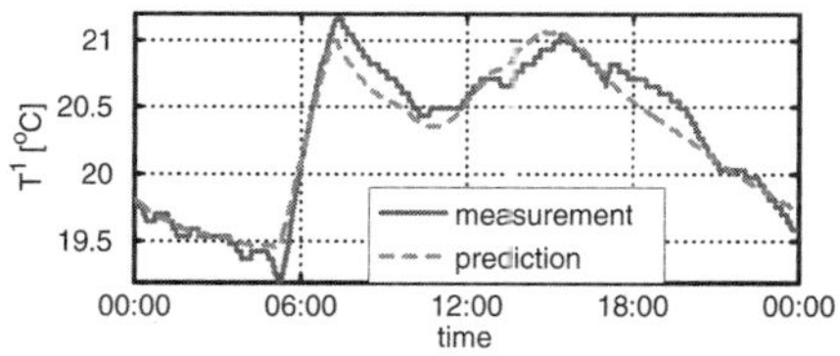

Figure 14.3. Validation results of simplified room model (14.4) [14].

The load prediction $P_d^1(t)$ is important for designing predictive feedback controllers and assessing potential energy savings. The disturbance load envelopes can be learned from historical data, shared calendars, and weather predictions. For instance, the conference room discussed earlier has two regularly scheduled group meetings around 10:00 and 14:00 every Wednesday. By using historical data we can observe this from the data. Figure 14.4 depicts the envelope-bounded load during all Wednesdays in July, 2010. The envelope is computed as pointwise minimum and maximum difference between the measured data and the nominal model (i.e, model (14.4) with the identified parameters in Table 14.1 and $P_d^1 = 0$). Two peaks can be observed in the disturbance load envelope in Figure 14.4,

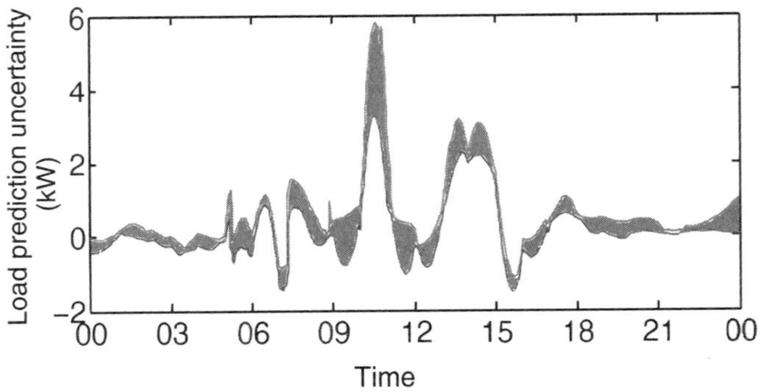

Figure 14.4. Envelope bounds of disturbance load profile (kw) calculated using model (14.4) for Wednesdays in July, 2010 [14].

which correspond to the two regularly scheduled group meetings. In the remainder of this paper, a nominal MPC is designed based on the load prediction profile with the highest probability. Ongoing research is focusing on stochastic MPC, where the load envelope with corresponding probability distribution function can be used at the controller design stage.

14.2.2 Simplified Energy Consumption Model

The components at the lower level of the architecture that use energy include dampers, supply fans, heating coils, and cooling coils as shown in Figure 14.1. The supply fan needs electrical power to drive the system, while the heating and cooling coils consume the energy of the chilled and hot water. It is assumed that the power to drive the dampers is negligible. A simple energy consumption model for each component is presented next.

Fan Power

The fan power can be approximated as a second order polynomial function of the total supply air mass flow rate[4] $\dot{m}_{\text{fan}}$ driven by the fan:

$$P_f = c_0 + c_1 \dot{m}_{\text{fan}} + c_2 \dot{m}_{\text{fan}}^2, \tag{14.5}$$

where c_0, c_1, c_2 are parameters to be identified by fitting recorded data. The simplified fan model (14.5) is tested on the recorded data from the UC Berkeley Bancroft Library from October 1st to October 10th, 2010. The identification results plotted in Figure 14.5 suggest that the polynomial function successfully predicts the electricity consumption of the fan.

Cooling and Heating Coils

Cooling coils and heating coils are air-water heat exchangers. There has been extensive studies to develop simplified yet descriptive models of coil units [6, 24, 26]. The authors

[4]The supply air mass flow rate by the fan is equal to the summation of air flow to each room $\dot{m}_{\text{fan}} = \sum_{j \in \mathcal{V}} \dot{m}_s^j$.

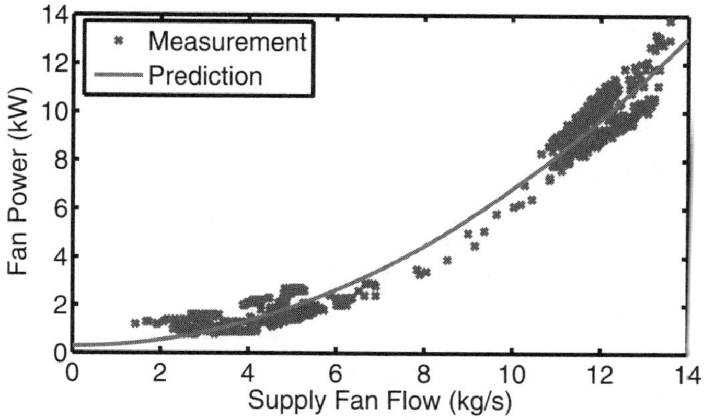

Figure 14.5. Identification results for the simplified fan power consumption model (14.5) [15].

in [24] developed simple empirical equations with four parameters by using a finite difference method to capture the transient response of counterflow heat exchangers. In [6], the authors presented an improved simulation model based on ASHRAE Secondary HVAC Toolkit, and in [26] a simplified control-oriented cooling coil unit model is presented based on energy and mass conservation laws.

In this work, we use a simple coil model with constant efficiency (η_c for the cooling coils and η_h for the heating coils). With this simplification the energy consumption model is a static function of the load on the air side

$$P_c = \frac{\sum_{j \in \mathcal{V}} \dot{m}_s^j c_p \Delta T_c}{\eta_c \, \mathrm{COP}_c}, \quad P_h^j = \frac{c_p \dot{m}_s^j \Delta T_h^j}{\eta_h \, \mathrm{COP}_h}, \tag{14.6}$$

where P_c is the electrical power consumption related to the generation of chilled water consumed by the cooling coils in AHU and P_h^j is the power used to generate the hot water consumed by the heating coils in the VAV box connected with room j. ΔT_c is the temperature difference through the cooling coils, ΔT_h^j is the temperature difference through the heating coils in the VAV box j, COP_c is the chilling coefficient of performance, and COP_h is the heating coefficient of performance. The coefficient of performance (COP) is defined as

$$\mathrm{COP} = \frac{E_{\text{thermal}}}{E_{\text{input}}}. \tag{14.7}$$

COP captures the efficiency of the exchange system, i.e., the amount of thermal energy E_{thermal} (J) generated by the system with one Joule of energy consumed. The input energy E_{input} can be from different resources such as electricity, fuel, and gas for different systems. Model (14.6) is oversimplified as compared to the aforementioned literature. However, the model is adequate to capture the energy consumption if coils are operating in a narrow performance range.

14.2.3 Constraints

The states and control inputs of system (14.4) are subject to the following constraints (for all $j \in \mathcal{V}$):

$T^j \in [\underline{T},\ \overline{T}]$. Comfort range.

$\dot{m}_s^j \in [\underline{\dot{m}},\ \overline{\dot{m}}]$. The maximum mass flow rate of air supplied to a room is limited by the size of VAV boxes. The minimum mass flow rate is imposed to guarantee a minimal ventilation level.

$\Delta T_c \in [\underline{\Delta T_c},\ \overline{\Delta T_c}]$. The temperature decrement of the supply air (cooled by the cooling coil) is constrained by the capacity of the AHU.

$\Delta T_h^j \in [\underline{\Delta T_h},\ \overline{\Delta T_h}]$. The temperature increment of the supply air (heated by the heating coil) is constrained by the capacity of the VAV boxes.

$\delta \in [\underline{\delta}, \overline{\delta}]$. The AHU damper position is positive and less than $\overline{\delta} < 1$ to make sure that there is always fresh outside air supplied to office rooms.

14.2.4 Model Summary

The thermal dynamic model (14.4) is compactly rewritten as

$$\frac{dx^j}{dt} = f_c(x^j, u^j, u^a, T_m, w^j) + \sum_{i \in \mathcal{N}^j} E_i^j x^i \quad \forall j \in \mathcal{V}, \tag{14.8a}$$

$$g_c(x^{j\in\mathcal{V}}, u^{j\in\mathcal{V}}, u^a, w^{j\in\mathcal{V}}, T_r, T_m) = 0, \tag{14.8b}$$

where $x^j = (T_1^j,\ T_2^j)$ is the state of the jth room, $u^j = (\dot{m}_s^j,\ \Delta T_h^j)$ are the control inputs to the jth VAV box, and $u^a = (\delta,\ \Delta T_c)$ collects the AHU control inputs. The vector $w^j = (P_d^j,\ T_{oa})$ is the disturbance assumed to be perfectly known. Equation (14.8b) lumps up the algebraic equations (14.4c) and (14.4d) that describe the static model for return air temperature and mixed air temperature, respectively. The set $x^{j\in\mathcal{V}}$ is defined as $\{x^1, x^2, \ldots, x^{\mathcal{N}_v}\}$. Note that the room dynamics in the network are coupled through states (the second term in (14.8a)) and inputs (δ and ΔT_c are common to all rooms).

The continuous-time dynamic system (14.8) is discretized. We consider the following control law with a uniform sampling time Δt:

$$u^j(t) = u_k^j \quad \forall j \in \mathcal{V}, \tag{14.9a}$$

$$u^a(t) = u_k^a, \tag{14.9b}$$

$$t \in [k\Delta t,\ (k+1)\Delta t).$$

With the control law defined by (14.9), the system equations (14.8) are discretized over $t \in [k\Delta t,\ (k+1)\Delta t)$ using the trapezoidal method to obtain

$$\frac{x_{k+1}^j - x_k^j}{\Delta t} = \frac{1}{2} f_c(x_k^j, u_k^j, u_k^a, T_{mk}, w_k^j) + \frac{1}{2} f_c(x_{k+1}^j, u_k^j, u_k^a, T_{mk}^+, w_{k+1}^j)$$
$$+ \sum_{i \in \mathcal{N}^j} E_i^j \frac{x_k^i + x_{k+1}^i}{2} \quad \forall j \in \mathcal{V}, \tag{14.10a}$$
$$g_c(x_k^{j\in\mathcal{V}}, u_k^{j\in\mathcal{V}}, u_k^a, w_k^{j\in\mathcal{V}}, T_{rk}, T_{mk}) = 0, \tag{14.10b}$$
$$g_c(x_{k+1}^{j\in\mathcal{V}}, u_k^{j\in\mathcal{V}}, u_k^a, w_{k+1}^{j\in\mathcal{V}}, T_{rk}^+, T_{mk}^+) = 0, \tag{14.10c}$$

where T_{rk}^+ and T_{mk}^+ are the temperature of return air and mixed air when time approaches $(k+1)\Delta t$, respectively. Note that T_{rk}^+ (T_{mk}^+) is generally different from T_{rk+1} (T_{mk+1}) due to the discontinuity introduced by the control law (14.9).

If $u_k^c = (u_k^a, T_{rk}, T_{mk}, T_{rk}^+, T_{mk}^+)$, then the discretized room model (14.10) can be compactly rewritten as

$$f(x_{k+1}^j, x_k^j, u_k^j, u_k^c, w_k^j, w_{k+1}^j) + \sum_{i \in \mathcal{N}^j} E_i^j \frac{x_k^i + x_{k+1}^i}{2} = 0 \quad \forall j \in \mathcal{V}, \tag{14.11a}$$
$$g(x_{k+1}^{j\in\mathcal{V}}, x_k^{j\in\mathcal{V}}, u_k^{j\in\mathcal{V}}, u_k^c, w_k^{j\in\mathcal{V}}, w_{k+1}^{j\in\mathcal{V}}) = 0, \tag{14.11b}$$
$$u_k^c \in \mathcal{U}^c; \quad x_k^j \in \mathcal{X}, \ u_k^j \in \mathcal{U}^j \quad \forall j \in \mathcal{V}. \tag{14.11c}$$

The constraints (14.11c) are defined in Section 14.2.3.

In the next section we will also use the following linearized version of model (14.11) around the trajectory of states and control inputs $(\bar{x}_k^1, \dots, \bar{x}_k^{N_v}, \bar{u}_k^1, \dots, \bar{u}_k^{N_v}, \bar{u}_k^c)$, $k = 0, 1, \dots, N-1$:

$$T_k^j dx_{k+1}^j + A_k^j dx_k^j + B_k^j du_k^j + B_k^c du_k^c + f_k^{ej} + \sum_{i \in \mathcal{N}^j} E_i^j \frac{dx_{k+1}^i + dx_k^i}{2} = 0 \tag{14.12a}$$
$$\forall j \in \mathcal{V}, \tag{14.12b}$$
$$G_k^c du_k^c + \sum_{j \in \mathcal{V}} G_{j,k}^+ dx_{k+1}^j + \sum_{j \in \mathcal{V}} G_{j,k}^x dx_k^j + \sum_{j \in \mathcal{V}} G_{j,k}^u du_k^j + g_k^e = 0, \tag{14.12c}$$
$$\bar{u}_k^c + du_k^c \in \mathcal{U}^c, \ \bar{x}_k^j + dx_k^j \in \mathcal{X}, \ \bar{u}_k^j + du_k^j \in \mathcal{U}^j \quad \forall j \in \mathcal{V}, \tag{14.12d}$$
$$T_k^j = \left.\frac{\partial f}{\partial x_{k+1}^j}\right|_{\bar{x}_{k+1}^j}, \ A_k^j = \left.\frac{\partial f}{\partial x_k^j}\right|_{\bar{x}_k^j}, \ B_k^j = \left.\frac{\partial f}{\partial u_k^j}\right|_{\bar{u}_k^j}, \ B_k^c = \left.\frac{\partial f}{\partial u_k^c}\right|_{\bar{u}_k^c}, \tag{14.12e}$$
$$G_{j,k}^+ = \left.\frac{\partial g}{\partial x_{k+1}^j}\right|_{\bar{x}_{k+1}^j}, \ G_{j,k}^x = \left.\frac{\partial g}{\partial x_k^j}\right|_{\bar{x}_k^j}, \ G_{j,k}^u = \left.\frac{\partial g}{\partial u_k^j}\right|_{\bar{u}_k^j}, \ G_k^c = \left.\frac{\partial g}{\partial u_k^c}\right|_{\bar{u}_k^c}, \tag{14.12f}$$
$$f_k^{ej} = f(\bar{x}_{k+1}^j, \bar{x}_k^j, \bar{u}_k^j, \bar{u}_k^c, w_k^j, w_{k+1}^j) + \sum_{i \in \mathcal{N}^j} E_i^j \frac{\bar{x}_{k+1}^i + \bar{x}_k^i}{2} \quad \forall j \in \mathcal{V}, \tag{14.12g}$$
$$g_k^e = g(\bar{x}_{k+1}^{j\in\mathcal{V}}, \bar{x}_k^{j\in\mathcal{V}}, \bar{u}_k^{j\in\mathcal{V}}, \bar{u}_k^c, w_k^{j\in\mathcal{V}}, w_{k+1}^{j\in\mathcal{V}}), \tag{14.12h}$$

where dx_k^j, du_k^j, and du_k^c are the deviations of states and control inputs around the trajectory and f_k^e and g_k^e are residuals of nonlinear equality constraints (14.11a) and (14.11b).

14.3 DMPC

In this section we formalize the MPC control problem [5] and provide details on the distributed MPC (DMPC) design. We are interested in solving at each time step t the following optimization problem:

$$\min_{\mathbf{U},\mathbf{X}} J(\mathbf{U},\mathbf{X}) = \sum_{k=0}^{N-1} \left\{ P_c + P_{\text{fan}} + \sum_{j\in\mathcal{V}} P_h^j \right\} \Delta t \tag{14.13a}$$

s.t.

$$f(x^j_{k+1|t}, x^j_{k|t}, u^j_{k|t}, u^c_{k|t}, w^j_{k|t}, w^j_{k+1|t}) + \sum_{i\in\mathcal{N}^j} E^j_i (x^i_{k+1|t} + x^i_{k|t})/2 = 0$$

$$\forall j \in \mathcal{V},\ k = 0,1,\ldots,N-1, \tag{14.13b}$$

$$g(x^{j\in\mathcal{V}}_{k+1|t}, x^{j\in\mathcal{V}}_{k|t}, u^{j\in\mathcal{V}}_{k|t}, u^c_{k|t}, w^{j\in\mathcal{V}}_{k|t}, w^{j\in\mathcal{V}}_{k+1|t}) = 0,$$

$$k = 0,1,\ldots,N-1, \tag{14.13c}$$

$$x^j_{k|t} \in \mathcal{X}^j \quad \forall j \in \mathcal{V},\ k = 1,\ldots,N, \tag{14.13d}$$

$$u^j_{k|t} \in \mathcal{U}^j,\ u^c_{k|t} \in \mathcal{U}^c \quad \forall j \in \mathcal{V},\ k = 0,\ldots,N-1, \tag{14.13e}$$

$$x^j_{0|t} = x^j(t) \quad \forall j \in \mathcal{V}, \tag{14.13f}$$

where $\mathbf{U} = (u^1_{0|t},\ldots,u^1_{N-1|t},\ldots,u^{N_v}_{0|t},\ldots,u^{N_v}_{N-1|t},u^c_{0|t},\ldots,u^c_{N-1|t})$ lumps up u^c and $u^{j\in\mathcal{V}}$ over the prediction horizon, and $\mathbf{X} = (x^1_{0|t},\ldots,x^1_{N|t},\ldots x^{N_v}_{0|t},\ldots,x^{N_v}_{N|t})$ is the vector of system state predictions over the prediction horizon. Let $\mathcal{T} = (\mathbf{X},\mathbf{U})$ be the vector collecting all optimization variables. The cost function in (14.13) is the total energy consumed by all VAV boxes and the AHU system over the prediction horizon.

In (14.13), $x^j_{k|t}$ denotes the state of room j at time $t + k\Delta t$ predicted at time t starting from the current state $x^j_{0|t} = x^j(t)$ for all $j \in \mathcal{V}$.

Let $(\mathbf{U}^\star,\mathbf{X}^\star)$ be the optimal solution of problem (14.13). Then, only the first element of every control sequence is implemented to the system, i.e., $u^j(t) = u^{j\star}_{0|t}$, $u^a(t) = u^{a\star}_{0|t}$.

The optimization (14.13) is repeated at time $t + \Delta t$, with the updated state $x_{0|t+\Delta t} = x(t+\Delta t)$, yielding a *moving or receding horizon control* strategy.

The MPC problem (14.13) has a nonconvex cost (14.13a) which includes bilinear terms for the energy consumption of cooling and heating coils, bilinear equality constraints (14.13b) and (14.13c), and box constraints on system states and control inputs. The size of the nonlinear optimization problem rapidly grows when a realistic number of rooms and a meaningful horizon length N are considered. In order to solve the MPC problem (14.13) in a distributed fashion, we apply SQP, proximal minimization, and dual decomposition. Next we show the main idea of these techniques and implementation details for the specific class of problems considered in this chapter.

The SQP procedure is an efficient method to solve NLP problems [7, 19]. The basic idea is to linearize the nonlinear constraints around a candidate solution and replace the objective with a quadratic function around this guess. The solution to the resulting QP is then used to update the candidate solution. The iterations are repeated until convergence is achieved [7]. Note that the Hessian of the cost (14.13a) is, in general, indefinite since the cost is nonconvex.

The proximal minimization algorithm is adopted to solve the indefinite QP problem obtained from the SQP procedure by iteratively solving a set of subproblems whose cost functions have extra convex quadratic terms [1]. By doing so, the primal cost function of the subproblem is strictly convex, which implies differentiability of the dual cost function. This enables faster convergence in the procedure of dual decomposition [1].

The concept of dual decomposition traces back to the 1970's [12], and it has been extensively studied since then [1, 21]. The QP derived from the proximal minimization algorithm is a separable convex optimization problems and therefore the gradient of the dual problem can be calculated in a distributed fashion, and the dual variables can be optimized separately by using gradient or subgradient based approaches. The primal optimal solution then can be reconstructed from the dual variables.

To summarize, the optimal solution to problem (14.13) is obtained through three nested iterative algorithms. The outer iteration solves the original nonlinear optimization problem (14.13) by solving a sequence of QPs, and the second and third levels of iteration solve the QP in a distributed fashion by using proximal minimization and dual decomposition, respectively.

The details of the proposed algorithms are described next. Problem (14.13) is time-variant because of the disturbance load profile w. With abuse of notation and for the sake of simplicity, in the rest of the chapter we will remove the term "$|t$" from the lower indices.

14.3.1 Level 1: Modified SQP

At the SQP iteration n_s, problem (14.13) is linearized by replacing the nonlinear system dynamics (14.13b) with the linearized ones (14.12) at a candidate solution $\mathcal{T}^{n_s} = (\mathbf{U}^{n_s}, \mathbf{X}^{n_s})$. The cost function (14.13a) is approximated by a quadratic function around $\mathcal{T}^{n_s}$ while neglecting the off-diagonal terms. The resultant optimization problem (14.14) is convex:

$$\min_{\mathbf{dU},\mathbf{dX}} dJ(\mathbf{dU},\mathbf{dX}) = \sum_{k=0}^{N-1} \left\{ \sum_{j\in\mathcal{V}} \left(\frac{1}{2} du_k^{j\,T} Q_k^j du_k^j + c_k^{j\,T} du_k^j \right) + \frac{1}{2} du_k^{c\,T} Q_k^c du_k^c + c_k^{c\,T} du_k^c \right\} \tag{14.14a}$$

s.t.

$$T_k^j dx_{k+1}^j + A_k^j dx_k^j + B_k^j du_k^j + B_k^c du_k^c + f_k^{e\,j} + \sum_{i\in\mathcal{N}^j} E_i^j \frac{dx_{k+1}^i + dx_k^i}{2} = 0 \quad \forall j\in\mathcal{V},\ k=0,1,\dots,N-1, \tag{14.14b}$$

$$G_k^c du_k^c + \sum_{j\in\mathcal{V}} G_{j,k}^{+} dx_{k+1}^j + \sum_{j\in\mathcal{V}} G_{j,k}^{x} dx_k^j + \sum_{j\in\mathcal{V}} G_{j,k}^{u} du_k^j + g_k^e = 0, \quad k=0,1,\dots,N-1, \tag{14.14c}$$

$$dx_k^j \in d\mathcal{X}^j \quad \forall j\in\mathcal{V},\ k=1,\dots,N, \tag{14.14d}$$

$$du_k^j \in d\mathcal{U}^j,\ du_k^c \in d\mathcal{U}^c \quad \forall j\in\mathcal{V},\ k=0,\dots,N-1, \tag{14.14e}$$

where Q_k^c and Q_k^j are diagonal matrices obtained by removing the off-diagonal terms of the Hessian of (14.13a) at candidate solution $\mathcal{T}^{n_s}$. The vector $\mathbf{dU} = (du_0^1, \dots, du_{N-1}^1,$

$\ldots,du_0^{N_v},\ldots,du_{N-1}^{N_v},du_0^c,\ldots,du_{N-1}^c)$ collects control input difference from the candidate solution $\mathbf{U}^{n_s}$, and the vector $\mathbf{dX}=(dx_1^1,\ldots,dx_N^1,\ldots,dx_1^{N_v},\ldots,dx_N^{N_v})$ collects system state deviations from the trajectory $\mathbf{X}^{n_s}$. The constraint sets $d\mathcal{X}^j=\mathcal{X}^j-x_k^{j\,n_s}$, $d\mathcal{U}^j=\mathcal{U}^j-u_k^{j\,n_s}$, and $d\mathcal{U}^c=\mathcal{U}^c-u_k^{c\,n_s}$ define the feasible state variations $\mathbf{dX}$ and control input variations $\mathbf{dU}$, respectively.

The optimal solution $d\mathcal{T}^\star=\left(\mathbf{dX}^\star,\ \mathbf{dU}^\star\right)$ to problem (14.14) is computed by the iterative algorithms at levels 2 and 3 described next. The vector $d\mathcal{T}^\star$ is used to update the candidate solution as

$$\mathcal{T}^{n_s+1}=\mathcal{T}^{n_s}+\alpha\cdot d\mathcal{T}^\star.$$

In this work, a constant step length α is applied. At the SQP iteration n_s+1, the process of linearizing problem (14.13) and solving problem (14.14) is repeated. The SQP algorithm is terminated if

$$\|d\mathcal{T}^\star\|\le\kappa, \tag{14.15}$$

where κ is a predefined convergence tolerance.

Remark 14.1. *The convergence of the proposed SQP algorithm for general nonconvex programs is not guaranteed. Extensive numerical tests have failed in finding an instance of the problem considered in this chapter where the proposed algorithm would not converge. It is noted that the convergence rate of the SQP procedure can be improved when an adaptive step length α is selected using an advanced line search algorithm* [19].

14.3.2 Level 2: Proximal Minimization

The iterative algorithm at the second level solves problem (14.14) by using the proximal minimization algorithm proposed in [1]. The quadratic objective in problem (14.14) is positive semidefinite. Its solution is obtained by optimizing a sequence of subproblems obtained by adding a quadratic term to the original cost. At the proximal minimization iteration n_p, we consider the subproblem

$$\begin{aligned}\min_{\widehat{\mathbf{dU}},\widehat{\mathbf{dX}}}\ d\widehat{J}^{n_p}(\widehat{\mathbf{dU}},\widehat{\mathbf{dX}})=&\sum_{k=0}^{N-1}\left\{\sum_{j\in\mathcal{V}}\left(\frac{1}{2}d\widehat{u}_k^{jT}Q_k^j d\widehat{u}_k^j+c_k^{jT}d\widehat{u}_k^j\right)+\frac{1}{2}d\widehat{u}_k^{cT}Q_k^c d\widehat{u}_k^c+c_k^{cT}d\widehat{u}_k^c\right\}\\&+\frac{\rho}{2}\left(\|\widehat{\mathbf{dU}}-\widehat{\mathbf{dU}}^{n_p-1}\|_2^2+\|\widehat{\mathbf{dX}}-\widehat{\mathbf{dX}}^{n_p-1}\|_2^2\right)\end{aligned} \tag{14.16a}$$

s.t.

$$T_k^j d\widehat{x}_{k+1}^j+A_k^j d\widehat{x}_k^j+B_k^j d\widehat{u}_k^j+B_k^c d\widehat{u}_k^c+f_k^{e\,j}+\sum_{i\in\mathcal{N}^j}E_i^j\frac{d\widehat{x}_{k+1}^i+d\widehat{x}_k^i}{2}=0\quad\forall j\in\mathcal{V},\ k=0,1,\ldots,N-1, \tag{14.16b}$$

$$G_k^c d\widehat{u}_k^c+\sum_{j\in\mathcal{V}}G_{j,k}^{+}d\widehat{x}_{k+1}^j+\sum_{j\in\mathcal{V}}G_{j,k}^{x}d\widehat{x}_k^j+\sum_{j\in\mathcal{V}}G_{j,k}^{u}d\widehat{u}_k^j+g_k^e=0,\quad k=0,1,\ldots,N-1, \tag{14.16c}$$

$$d\widehat{x}_k^j\in d\mathcal{X}^j\quad\forall j\in\mathcal{V},\ k=1,\ldots,N, \tag{14.16d}$$

$$d\widehat{u}_k^j\in d\mathcal{U}^j,\ d\widehat{u}_k^c\in d\mathcal{U}^c\quad\forall j\in\mathcal{V},\ k=0,\ldots,N-1, \tag{14.16e}$$

where $\rho > 0$ is a strictly positive constant predetermined such that the resulting cost (14.16a) is positive definite. $\left(\mathbf{d}\widehat{\mathbf{X}}^{n_p-1}, \mathbf{d}\widehat{\mathbf{U}}^{n_p-1}\right)$ is the optimal solution to problem (14.16) at iteration $n_p - 1$. When $n_p = 1$, let $\mathbf{d}\widehat{\mathbf{X}}^0 = \mathbf{0}$, $\mathbf{d}\widehat{\mathbf{U}}^0 = \mathbf{0}$.

The optimal solution $(\mathbf{d}\widehat{\mathbf{X}}^\star, \mathbf{d}\widehat{\mathbf{U}}^\star)$ to problem (14.16) is computed by the algorithm at level 3. We set $\mathbf{d}\widehat{\mathbf{X}}^{n_p} = \mathbf{d}\widehat{\mathbf{X}}^\star$ and $\mathbf{d}\widehat{\mathbf{U}}^{n_p} = \mathbf{d}\widehat{\mathbf{U}}^\star$ and terminate the iterative algorithm if

$$\|d\widehat{\mathbf{U}}^{n_p} - d\widehat{\mathbf{U}}^{n_p-1}\|_2 \le \kappa, \ \|d\widehat{\mathbf{X}}^{n_p} - d\widehat{\mathbf{X}}^{n_p-1}\|_2 \le \kappa. \tag{14.17}$$

It has been proved in [1] that if the optimization problem (14.14) is convex, then the vector $(\mathbf{d}\widehat{\mathbf{X}}^{n_p}, \mathbf{d}\widehat{\mathbf{U}}^{n_p})$ converges to an optimum $(\mathbf{dX}^\star, \mathbf{dU}^\star)$ of problem (14.14).

14.3.3 Level 3: Dual Decomposition and Fast Gradient Method

The dual decomposition algorithm is used to solve problem (14.16) by solving its dual problem. The dual problem of the QP (14.16) is formulated by assigning dual variables λ_k^j and μ_k to the constraints (14.16b) and (14.16c), respectively. The dual problem can be formulated as follows:

$$\max_{\lambda,\mu \text{ free}} \min_{\mathbf{d}\widehat{\mathbf{U}},\mathbf{d}\widehat{\mathbf{X}}} d\widehat{J}^{n_p} + L^c + L^f \tag{14.18a}$$

$$\text{s.t.}$$

$$d\widehat{x}_k^j \in d\mathcal{X}^j \quad \forall j \in \mathcal{V},\ k = 1,\dots,N, \tag{14.18b}$$

$$d\widehat{u}_k^j \in d\mathcal{U}^j \quad \forall j \in \mathcal{V},\ k = 0,\dots,N-1, \tag{14.18c}$$

$$d\widehat{u}_k^c \in d\mathcal{U}^c \quad \forall k = 0,\dots,N-1, \tag{14.18d}$$

where $\widehat{dJ}^{n_p}$ is the cost defined in (14.16a). The term

$$L^c = \sum_{k=1}^{N} \mu_k^T \left(G_k^c d\widehat{u}_k^c + \sum_{j\in\mathcal{V}} G_{j,k}^{+} d\widehat{x}_{k+1}^j + \sum_{j\in\mathcal{V}} G_{j,k}^{x} d\widehat{x}_k^j + \sum_{j\in\mathcal{V}} G_{j,k}^{u} d\widehat{u}_k^j + g_k^e \right)$$

is the dual term corresponding to constraint (14.16c). The term

$$L^f = \sum_{j\in\mathcal{V}} \sum_{k=1}^{N} \lambda_k^{j\,T} \left(T_k^j d\widehat{x}_{k+1}^j + A_k^j d\widehat{x}_k^j + B_k^j d\widehat{u}_k^j + B_k^c d\widehat{u}_k^c + f_k^{e\,j} + \sum_{i\in\mathcal{N}^j} E_i^j \frac{d\widehat{x}_{k+1}^i + d\widehat{x}_k^i}{2} \right)$$

is the dual term for constraint (14.16b).

In our previous work [14], the dual problem is solved by a projected subgradient method with a constant step size, which suffers from relatively slow convergence. The algorithm convergence speed can be improved by applying the fast gradient method since the dual function becomes Lipschitz-smooth when a proximal term is added [18, 23].

We note that the cost function (14.18a) and the constraints of the inner minimization problem (14.18) are separable. This special structure allows us to solve problem (14.18) using the fast gradient method in a distributed way as described next.

Three sets of variables are updated in the fast gradient method [23], namely the dual variables λ and μ, the auxiliary variables $\bar{\lambda}$ and $\bar{\mu}$ of the same dimension as the dual variables, and a parameter γ controlling the step size β. They are initialized as $\lambda^0 = \lambda^{\star n_p-1}$,

$\mu^0 = \mu^{\star n_p - 1}$, $\bar{\lambda}^0 = \lambda^0$, $\bar{\mu}^0 = \mu^0$, and $\gamma^0 = \frac{\sqrt{5}-1}{2}$. Note that for $n_p = 1$, we set $\lambda^{\star n_p - 1} = \mathbf{0}$ and $\mu^{\star n_p - 1} = \mathbf{0}$.

The optimal solution $\left(\lambda^{\star n_p - 1},\ \mu^{\star n_p - 1}\right)$ to the dual problem at the second level iteration $n_p - 1$ is used as a warm start for faster convergence.

At the third level iteration n_d, the dual and auxiliary variables are updated by using the fast gradient method as follows [23]:

$$\lambda_k^{j\,n_d} = \bar{\lambda}_k^{j\,n_d - 1} + \frac{1}{L} h_{\lambda_k^j}(\bar{\lambda}_k^{j\,n_d - 1}) \quad \forall j \in \mathcal{V} \quad \forall k = 1,2,\ldots,N, \tag{14.19a}$$

$$\mu_k{}^{n_d} = \bar{\mu}_k{}^{n_d - 1} + \frac{1}{L} h_{\mu_k}(\bar{\mu}_k^{n_d - 1}) \quad \forall k = 0,1,\ldots,N-1, \tag{14.19b}$$

$$\gamma^{n_d} = \frac{\gamma^{n_d - 1}}{2}\left(\sqrt{\left(\gamma^{n_d-1}\right)^2 + 4} - \gamma^{n_d - 1}\right), \tag{14.19c}$$

$$\beta = \frac{\gamma^{n_d - 1}(1 - \gamma^{n_d - 1})}{\left(\gamma^{n_d - 1}\right)^2 + \gamma^{n_d}}, \tag{14.19d}$$

$$\bar{\lambda}_k^{j\,n_d} = \lambda_k^{j\,n_d} + \beta(\lambda_k^{j\,n_d} - \lambda_k^{j\,n_d - 1}) \quad \forall j \in \mathcal{V},\ \forall k = 1,2,\ldots,N, \tag{14.19e}$$

$$\bar{\mu}_k{}^{n_d} = \mu_k{}^{n_d} + \beta(\mu_k{}^{n_d} - \mu_k{}^{n_d - 1}) \quad \forall k = 0,1,\ldots,N-1, \tag{14.19f}$$

where L is the Lipschitz constant of the dual gradient calculated as in [23]. $h_{\lambda_k^j}(\bar{\lambda}_k^{j\,n_d - 1})$ and $h_{\mu_k}(\bar{\mu}_k^{n_d - 1})$ are the gradients of the dual cost function (14.18a) at $\bar{\lambda}_k^{j\,n_d - 1}$ and $\bar{\mu}_k{}^{n_d - 1}$, respectively. The fast gradient method algorithm is terminated if

$$\|h_{\lambda_k^j}(\bar{\lambda}_k^{j\,n_d})\|_2 \le \kappa \quad \forall j \in \mathcal{V},\ k = 1,2,\ldots,N, \tag{14.20a}$$

$$\|h_{\mu_k}(\bar{\mu}_k^{n_d})\|_2 \le \kappa \quad k = 1,2,\ldots,N. \tag{14.20b}$$

The computation of the gradients $h_{\lambda_k^j}(\bar{\lambda}_k^{j\,n_d})$ and $h_{\mu_k}(\bar{\mu}_k^{n_d})$ is the most time-consuming step of the algorithm. The proposed algorithm uses the following approach.

First, we compute the primal variables in (14.18) associated with the set of dual variables $\left(\bar{\lambda}^{n_d}, \bar{\mu}^{n_d}\right)$:

$$d\widehat{x}_k^{\star j\,n_d} = \Pi_{d\mathcal{X}^j}\left(d\widehat{x}_k^{j\,n_p - 1} + \frac{1}{\rho}\left(- G_{j,k}^{+\ T} \bar{\mu}_{k-1}^{n_d} - G_{j,k}^{x\ T} \bar{\mu}_k{}^{n_d} - A_k^{j\,T} \bar{\lambda}_k^{j\,n_d} - T_k^{j\,T} \bar{\lambda}_{k-1}^{j\,n_d} \right.\right.$$
$$\left.\left. - \sum_{i \in \mathcal{N}^j} E_j^{i\,T} \frac{\bar{\lambda}_{k-1}^{i\,n_d} + \bar{\lambda}_k^{i\,n_d}}{2} \right)\right) \quad \forall j \in \mathcal{V},\ \forall k = 1,2,\ldots,N, \tag{14.21a}$$

$$d\widehat{u}_k^{\star j\,n_d} = \Pi_{d\mathcal{U}^j}\left((\rho I + Q_k^j)^{-1} (\rho d\widehat{u}_k^{j\,n_p - 1} - c_k^j - G_{j,k}^{u\ T} \bar{\mu}_k{}^{n_d} - B_k^{j\,T} \bar{\lambda}_k^{j\,n_d}) \right)$$
$$\forall j \in \mathcal{V},\ \forall k = 0,1,\ldots,N-1, \tag{14.21b}$$

$$d\widehat{u}_k^{\star c\,n_d} = \Pi_{d\mathcal{U}^c}\left((\rho I + Q_k^c)^{-1} \left(\rho d\widehat{u}_k^{c\,n_p - 1} - c_k^c - G_{c,k}^{u\ T} \bar{\mu}_k{}^{n_d} - \sum_{j \in \mathcal{V}} B_k^{c\,T} \bar{\lambda}_k^{j\,n_d} \right)\right)$$
$$\forall k = 0,1,\ldots,N-1, \tag{14.21c}$$

where $\Pi_{\mathcal{S}}(\star)$ is the operation of projecting $\star$ onto the convex set $\mathcal{S}$.[5] Then, the gradients of the dual cost function (14.18a) at $(\bar{\mu}^{n_d}, \bar{\lambda}^{n_d})$ can be computed as follows [23]:

$$h_{\lambda_k^j}(\bar{\lambda}_k^{j\,n_d}) = T_k^j \, d\widehat{x}_{k+1}^{\star j\; n_d} + A_k^j d\widehat{x}_k^{\star j n_d} + B_k^j d\widehat{u}_k^{\star j n_d} + B_k^c d\widehat{u}_k^{\star c n_d} + f_k^{e\,j}$$
$$+ \sum_{i \in \mathcal{N}^j} E_i^j \frac{d\widehat{x}_{k+1}^{\star i\; n_d} + d\widehat{x}_k^{\star i n_d}}{2} \tag{14.22a}$$
$$\forall j \in \mathcal{V},\ \forall k = 0, 1, \ldots, N-1,$$
$$h_{\mu_k}(\bar{\mu}_k^{n_d}) = G_k^c d\widehat{u}_k^{\star c n_d} + \sum_{j \in \mathcal{V}} G_{jk}^{+} d\widehat{x}_{k+1}^{\star j n_d} + \sum_{j \in \mathcal{V}} G_{jk}^{x} d\widehat{x}_k^{\star j n_d}$$
$$+ \sum_{j \in \mathcal{V}} G_{j,k}^{u} d\widehat{u}_k^{\star j n_d} + g_k^e \tag{14.22b}$$
$$\forall k = 0, 1, \ldots, N-1.$$

In summary, the proposed MPC problem is solved locally by three nested levels of iterations: the outer iteration solves the original nonlinear optimization problem (14.13) by solving a sequence of QPs (14.14). The second and third levels of iterations solve the QP (14.14) in a distributed fashion by using proximal minimization and dual decomposition. Algorithm 14.1 summarizes the main steps of the proposed DMPC scheme.

ALGORITHM 14.1. DMPC for building control.

Initial:	Let $\mathbf{U}^1$, $\mathbf{X}^1$ be initial guesses for primal variables for problem (14.13). Set the SQP iteration index to $n_s = 1$.
Step 1:	Let n_s be the current SQP iteration index. Linearize the system model (14.11) at $(\mathbf{U}^{n_s}, \mathbf{X}^{n_s})$ to obtain the coefficients in (14.14). Let $n_p = 1$, and set $\widehat{\mathbf{dU}}^0 = \mathbf{0}$ and $\widehat{\mathbf{dX}}^0 = \mathbf{0}$ in (14.16). Also let $\lambda^{\star 0} = \mathbf{0}$ and $\mu^{\star 0} = \mathbf{0}$.
Step 2:	Let n_p be the current proximal minimization iteration index, and set $n_d = 1$. Initialize $\widehat{\lambda}^1 = \lambda^1 = \lambda^{\star n_p - 1}$ and $\widehat{\mu}^1 = \mu^1 = \mu^{\star n_p - 1}$
Step 3:	Let n_d be the current dual decomposition iteration index. Update primal variables $(\widehat{d\mathbf{U}}^{\star}, \widehat{d\mathbf{X}}^{\star})$ as in the distributed algorithm (14.21).
Step 4:	Exchange the updated primal variables and calculate the gradients (14.22) for the dual function.
Step 5:	Update the dual sequences λ^{n_d+1}, μ^{n_d+1}, $\widehat{\lambda}^{n_d+1}$, $\bar{\lambda}^{n_d+1}$, and γ^{n_d+1} as in the distributed algorithm (14.19).
Step 6:	If condition (14.20) is satisfied, go to the next step. Otherwise, set $n_d = n_d + 1$ and go to Step 3.
Step 7:	Update $\widehat{d\mathbf{U}}^{n_p} = \widehat{d\mathbf{U}}^{\star n_d}$ and $\widehat{d\mathbf{X}}^{n_p} = \widehat{d\mathbf{X}}^{\star n_d}$. If condition (14.17) is satisfied, go to the next step. Otherwise, set $n_p = n_p + 1$ and go to Step 2.
Step 8	Set $d\mathbf{U}^{\star} = \widehat{d\mathbf{U}}^{n_p}$ and $d\mathbf{X}^{\star} = \widehat{d\mathbf{X}}^{n_p}$. If condition (14.15) is satisfied, terminate and the optimal control sequence is $\mathbf{U}^{n_s}$. Otherwise, update $\mathbf{U}^{n_s+1} = \mathbf{U}^{n_s} + \alpha \cdot d\mathbf{U}^{\star}$ and $\mathbf{X}^{n_s+1} = \mathbf{X}^{n_s} + \alpha \cdot d\mathbf{X}^{\star}$, set $n_s = n_s + 1$, and go to Step 1.

[5]Note that as Q_k^j and Q_k^c are diagonal matrices, the matrix inversion in (14.21b) and (14.21c) can be easily evaluated. The projection is easy to calculate as $d\mathcal{X}^j$, $d\mathcal{U}^j$, and $d\mathcal{U}^c$ are box constraints.

14.4 Simulation Results

This section presents a numerical example to show the effectiveness of the proposed controller design methodology.

We compare the proposed control methodology with a baseline control logic (BC), which is a simplified version of a production control logic. The BC works as follows. When all the room temperatures are within the comfort range, the mass flow rate of the supply air ($\dot{m}_s^j$) is set to its minimum and the valves of cooling and heating coils are closed. When a room temperature hits the lower bound, the air mass flow rate to the room is maintained at its minimum, and the supply air temperature will be adjusted by the heating coil in the corresponding VAV box so that the room temperature stays at the lower bound value. When a room temperature violates the upper constraints, the AHU supply air temperature is set to its minimum, and the mass flow rate of the supply air is controlled so that the room temperature is within the comfort range.

The numerical example considers a network of 10 rooms. All the rooms have the same model parameters as in Table 14.1 identified for the conference room in the Bancroft Library in Section 14.2. The undirected graph describing the topology of the room network is $\mathcal{G} = (\mathcal{V}, \mathcal{A})$, where $\mathcal{V} = \{1,2,\ldots,10\}$ and $\mathcal{A} = \{(1,2),(2,3),\ldots,(9,10)\}$. The weather information is downloaded from July 2nd to July 3rd, 2009, at UC Berkeley, and the temperature profile is plotted in Figure 14.6(a). Because of the warm weather, only cooling is critical in the considered scenario. We assume that during 15:00 and 8:00 the next day, the rooms are empty without occupancy, leaving a minimum internal load of 0.01 kW due to lighting or other electrical devices. We use an internal load profile different for each room. In particular, we compute the internal load for room j as

$$P_d{}^j = (1+0.2j)P_{dn}, \quad j = 1,2,\ldots,10.$$

In our simulations, the parameters for controllers in Section 14.3 are listed in Table 14.2. The sampling time Δt is chosen to be 30 minutes, and the prediction horizon is one day ($N = 48$). The comfort constraints are defined as [21, 25]°C from 6:00 to 18:00 when there is occupancy in the buildings, and the comfort set is relaxed to [19, 27]°C when offices are unoccupied.

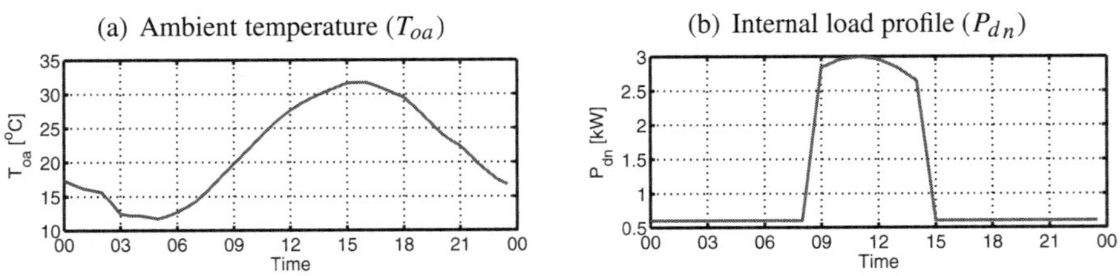

Figure 14.6. Simulation setups.

Figure 14.7 shows the simulation results for the network of rooms controlled by the baseline controller. Figure 14.7(a) shows that all the room temperatures are within the comfort range defined by the dotted lines. During early morning till 04:00, all zone temperatures are within the comfort range. As a result, the supply fan maintains only the

Table 14.2. Parameters for the numerical example.

Param.	Value	Param.	Value	Param.	Value	Param.	Value
$\underline{\dot{m}}$	0.005 kg/s	$\overline{\dot{m}}$	5 kg/s	η_c	0.7	η_h	0.8
$\underline{\Delta T_c}$	0 K	$\overline{\Delta T_c}$	8 K	P_f	0.08	α	0.25
$\underline{\Delta T_h}$	0 K	$\overline{\Delta T_h}$	8 K	$\underline{\delta}$	0	$\overline{\delta}$	0.8
COP_c	5	COP_h	0.9	N	48	κ	1×10^{-3}

minimum required air mass flow rate to each zone, and the valve of the cooling coils in the AHU is fully closed. The occupancy load at noon results in a peak total air mass flow rate of 7.2 kg/s. The cooling coils are operating at maximum capacity as soon as one of the zone temperatures hits the upper constraints so that the thermal comfort can be guaranteed. The return air damper position is fully closed to take advantage of free cooling when the ambient temperature is lower than the zone temperature.

The performance of the proposed DMPC controller is reported in Figure 14.8. It cools down the room temperature to the lower bounds of the comfort range during the early morning (Figure 14.8(a)) while the baseline controller remains inactive until the room temperature hits the upper bounds around 4:00 (Figure 14.7(a)). This *precooling* saves energy, since during the early morning the lower ambient temperature enables free cooling. The free cooling is illustrated in Figure 14.8(d). The MPC algorithm decides to open the cooling coil valve from 8:00, which is 3 hours later than the schedule proposed by baseline control logics in Figure 14.7(d).

Moreover, it is noted that instead of cooling all zones simultaneously, MPC cools down zones consecutively as Figure 14.8(a) illustrates. This feature significantly reduces the peak total air flow rate from 7.2 kg/s of BC to 5.8 kg/s (Figure 14.8(b)) and thus saves fan energy consumption (note that we use a quadratic penalty of total supply air mass flow rate in (14.13a)). The simulation results suggested that the precooling and consecutive cooling strategies induced by DMPC enable a 43.5% energy saving compared to the baseline.

DMPC Algorithm Complexity

The proposed DMPC algorithm (Algorithm 14.1) can be implemented in a network of embedded processors with low computational capacity since Steps 3 and 4 of Algorithm 14.1 require only a few algebraic operations and simple projections. Figure 14.9 shows that a large number of iterations is required. This imposes a requirement for high network communication speed.

The DMPC algorithm (Algorithm 14.1) was coded in MATLAB and runs on a single PC with Intel Core Duo CPU 3.00 GHz. The runtime of the DMPC algorithm is estimated based on the assumption that the computation of Steps 3 and 4 in Algorithm 14.1 are executed in parallel on $N_v + 1$ units including the controller units equipped on each VAV box and the AHU unit, and that the communication time is neglected. The results are reported in Figure 14.10 for different numbers of thermal zones considered. The dashed line shows the runtime of DMPC algorithm *when implemented on $N_v + 1$ CPUs in parallel*, and the solid line depicts the time required to solve problem (14.13) by the Interior Point OPTimizer (IPOPT) interfaced via AMPL *on one CPU*. Both the DMPC algorithm and the IPOPT start from the same initial guess of optimal solutions. One can notice that when

(a) Dashed lines are room temperatures ($T^{j\in\mathcal{V}}$), and solid lines define the comfort sets.

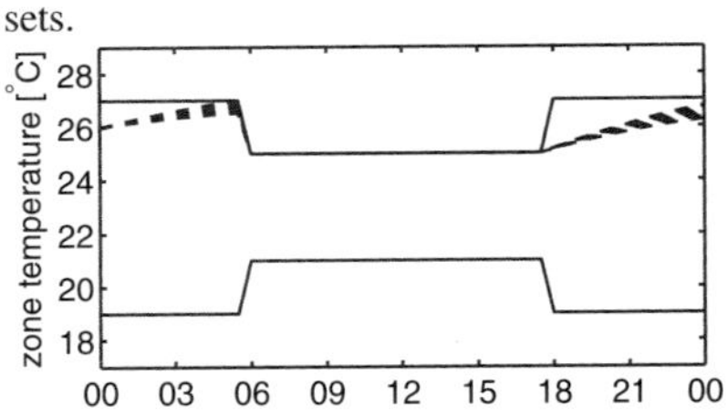

(b) Dashed lines are mass flow rate of the supply air to each zone ($\dot{m}_s^{j\in\mathcal{V}}$), and the solid line is the total mass flow rate $\dot{m}_{\text{fan}}$.

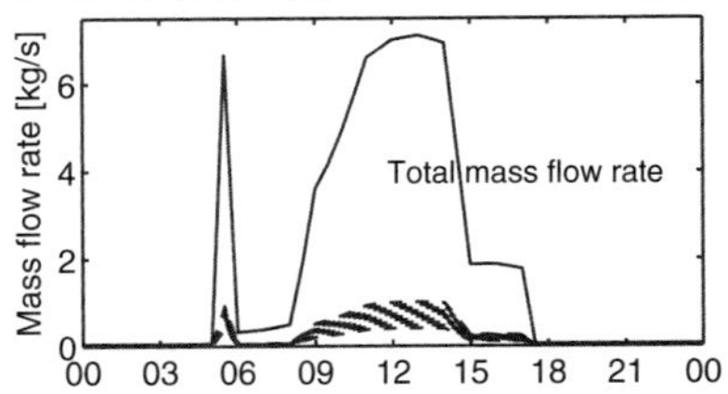

(c) Return air damper position (δ)

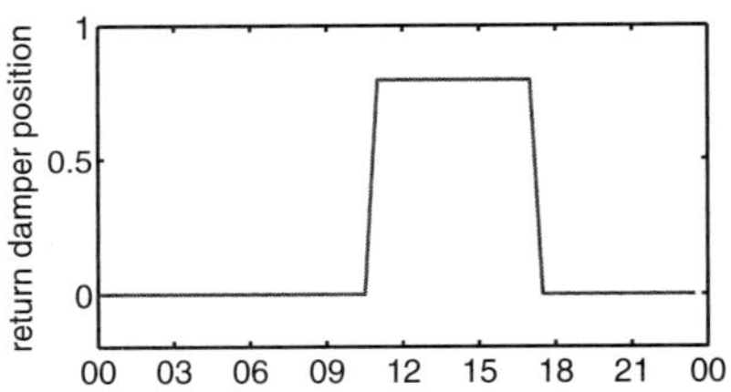

(d) Temperature difference across cooling coil (ΔT_h)

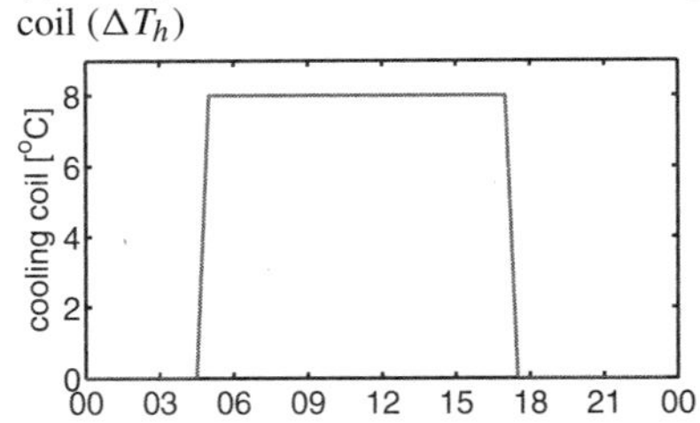

Figure 14.7. System behavior for simplified baseline control logics.

(a) Dashed lines are room temperatures ($T^{j\in\mathcal{V}}$), and solid lines define the comfort sets.

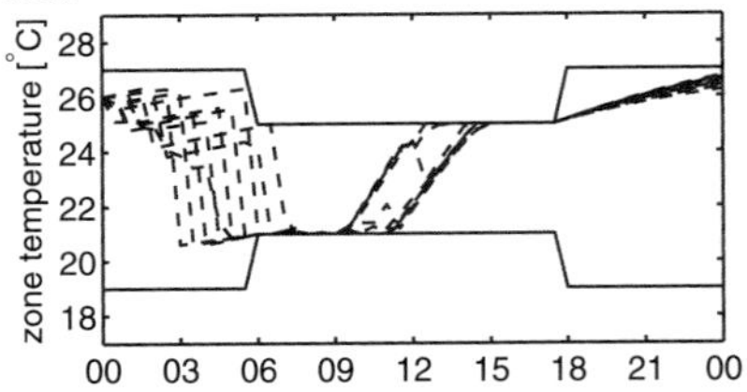

(b) Dashed lines are mass flow rate of the supply air to each zone ($\dot{m}_s^{j\in\mathcal{V}}$), and the solid line is the total mass flow rate $\dot{m}_{\text{fan}}$.

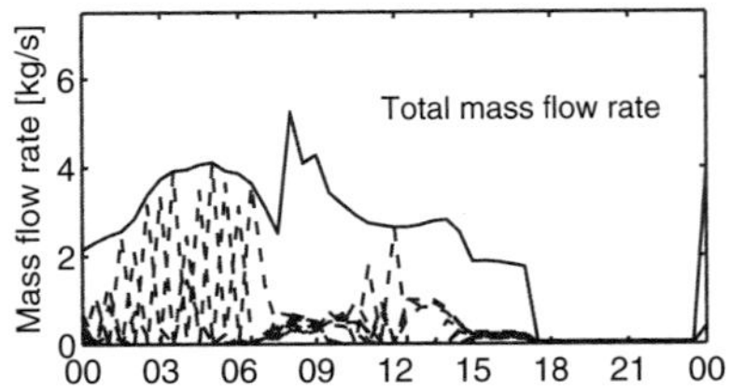

(c) Return air damper position (δ)

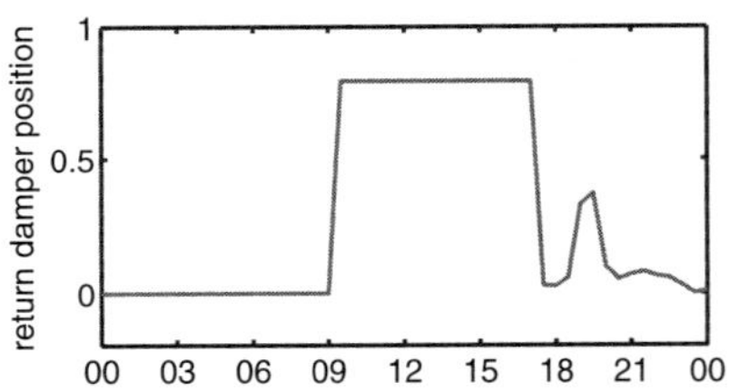

(d) Temperature difference across cooling coil (ΔT_h)

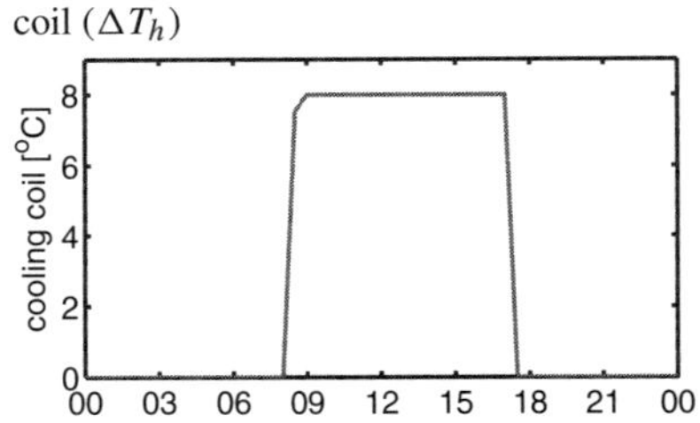

Figure 14.8. System behavior for distributed model predictive control.

the number of zones is less than three, IPOPT is faster than DMPC on a single PC. As the number of zones and the size of problem (14.13) increase, one can notice that DMPC could be implemented with a faster control sampling rate than IPOPT.

Figure 14.11 plots the variations of the optimal cost $J^{\star}_{DMPC}$ obtained by Algorithm 14.1 relative to the optimal cost $J^{\star}_{IPOPT}$ by IPOPT when solving MPC problem (14.13)

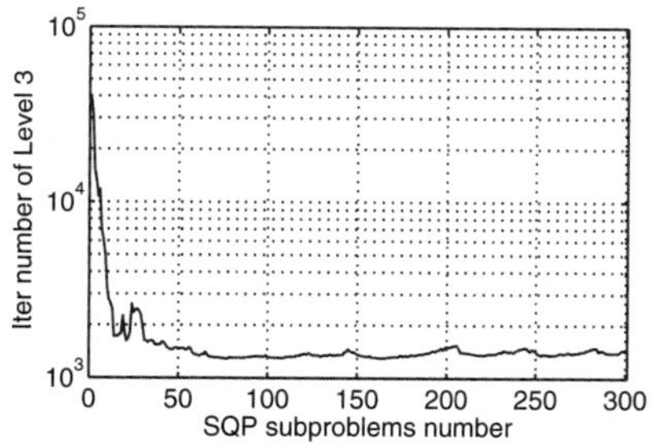

Figure 14.9. Number of dual decomposition iterations at time $t = 0$.

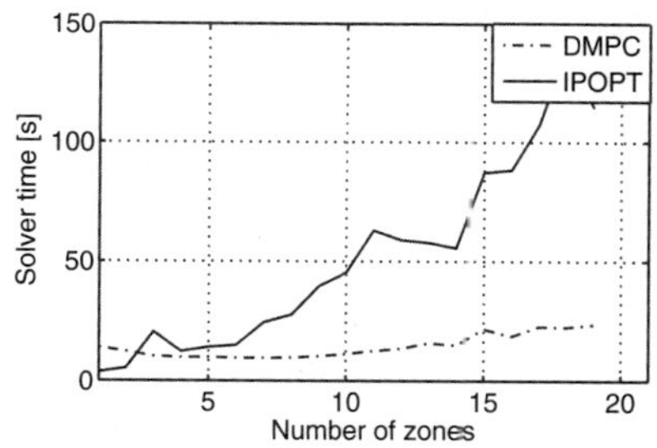

Figure 14.10. Comparison between CPU time for DMPC and IPOPT.

with different numbers of zones N_v. It is noted that DMPC results in slightly higher optimal costs than IPOPT. The reasons for this include the selection of convergence tolerance κ in Table 14.2.

Fast Gradient Method Improvement

The advantage of applying the fast gradient method instead of the classical gradient method in level 3 of the DMPC scheme is illustrated in Figure 14.12. We focus on the MPC problem (14.13) at $t = 0$. The modified SQP algorithm converged in 301 iterations. The dashed line in Figure 14.12 depicts the total number of iterations $\sum_{i=1}^{n_v} n_d^i$ required to solve the subproblems generated from the SQP algorithm when the fast gradient method in Section 14.3.3 is applied. The solid line depicts the number of iterations for the projected classical gradient method with constant step size.

In this work, n_d^i is the number of iterations to solve the ith subproblem (14.16) obtained in the proximal minimization level. If the fast gradient method is used, the number of iterations required to satisfy the stopping criterion is, on average, about four times less than for the classical gradient.

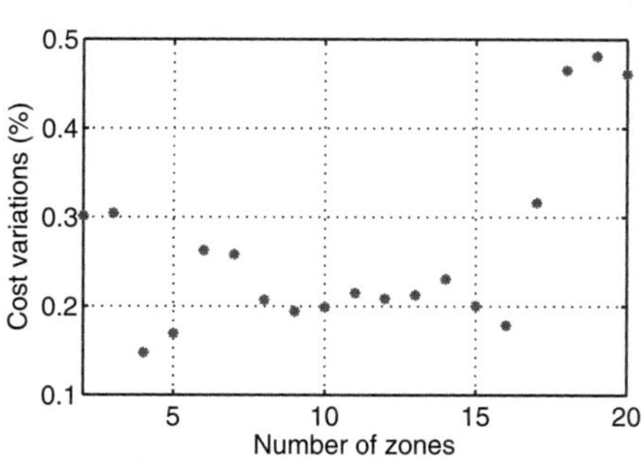

Figure 14.11. Cost variations $(J^\star_{DMPC} - J^\star_{IPOPT})/J^\star_{IPOPT}$.

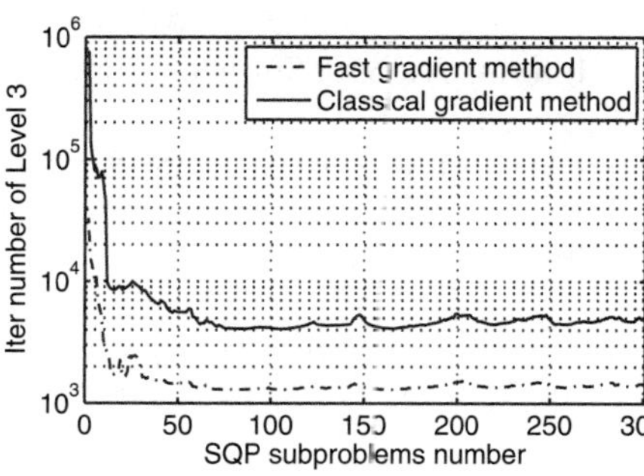

Figure 14.12. Number of dual decomposition iterations at time $t = 0$.

14.5 Conclusions

In this study a simplified two-mass room model is presented. Validation results show that the model captures the thermal dynamics of a thermal zone with negligible external load. Based on this model and predictive information of thermal loads and weather, a distributed model predictive control is designed to regulate thermal comfort while minimizing energy consumption. We have presented a three-level iterative algorithm for the solution of the nonlinear MPC problem. The key advance of this approach is the ability to solve the nonlinear MPC problem in a distributed manner and also in parallel. The resulting scheme is suitable for being implemented on a set of distributed low-cost processors. Simulation results show interesting behavior and short computation time. Future research will focus on analyzing stability issues of the resultant nonlinear MPC controller and improving the convergence rate of the proposed algorithm using adaptive step length schemes and parameters.

Bibliography

[1] D. P. Bertsekas and J. N. Tsitsiklis, *Parallel and Distributed Computation*, Prentice–Hall, Englewood Cliffs, NJ, 1989.

[2] F. Borrelli, *Constrained Optimal Control of Linear and Hybrid Systems*, Lecture Notes in Control and Inform. 290, Springer-Verlag, Berlin, 2003.

[3] F. Borrelli and T. Keviczky, *Distributed LQR design for identical dynamically decoupled systems*, IEEE Trans. Automat. Control, 53 (2008), pp. 1901–1912.

[4] F. Borrelli, T. Keviczky, and G. Stewart, *Decentralized constrained optimal coantrol approach to distributed paper machine control*, in Proceedings of the 44th IEEE Conference on Decision and Control and European Control Conference. CDC-ECC '05, 2005, pp. 3037–3042.

[5] E. Camacho and C. Bordons, *Nonlinear model predictive control: An introductory review*, in Assessment and Future Directions of Nonlinear Model Predictive Control, R. Findeisen, F. Allgöwer, and L. Biegler, eds., Lecture Notes in Control and Inform. Sci. 358, Springer, Berlin, Heidelberg, 2007, pp. 1–16.

[6] R. Chillar and R. Liesen, *Improvement of the ASHRAE secondary HVAC toolkit simple cooling coil model for simulation*, in Proceedings of the SimBuild 2004 Conference, Boulder, CO, 2004.

[7] S. Han, *A globally convergent method for nonlinear programming*, J. Optim. Theory Appl., 22 (1977), pp. 297–309.

[8] G. Henze, C. Felsmann, and G. Knabe, *Evaluation of optimal control for active and passive building thermal storage*, Int. J. Therm. Sci., 43 (2004), pp. 173–183.

[9] G. Henze, M. Krarti, and M. Brandemuehl, *Guidelines for improved performance of ice storage systems*, Energy and Buildings, 35 (2003), pp. 111–127.

[10] G. Henze, J. Pfafferott, S. Herkel, and C. Felsmann, *Impact of adaptive comfort criteria and heat waves on optimal building thermal mass control*, Energy and Buildings, 39 (2007), pp. 221–235.

[11] J. Jang, *System Design and Dynamic Signature Identification for Intelligent Energy Management in Residential Buildings*, Ph.D. thesis, University of California at Berkeley, Berkeley, CA, 2008.

[12] L. Lasdon, *Duality and decomposition in mathematical programming*, IEEE Trans. Systems Sci. Cybern., 4 (1968), pp. 86–100.

[13] J. Lee, M. Morari, and C. Garcia, *State-space interpretation of model predictive control*, Automatica J. IFAC, 30 (1994), pp. 707–717.

[14] Y. Ma, G. Anderson, and F. Borrelli, *A distributed predictive control approach to building temperature regulation*, in Proceedings of the 2011 American Control Conference, 2011.

[15] Y. Ma, F. Borrelli, A. Daly, and A. Kelman, *Predictive control for energy efficient buildings with thermal storage: Modeling, simulation, and experiments*, IEEE Control Syst. Mag., 32 (2012), pp. 44–64.

[16] Y. Ma, F. Borrelli, B. Hencey, B. Coffey, S. Bengea, and P. Haves, *Model predictive control for the operation of building cooling systems*, in Proceedings of the 2010 American Control Conference, 2010, pp. 5106 –5111.

[17] D. Mayne, J. Rawlings, C. Rao, and P. Scokaert, *Constrained model predictive control: Stability and optimality*, Automatica J. IFAC, 36 (2000), pp. 789–814.

[18] Y. Nesterov, *Introductory Lectures on Convex Optimization: A Basic Course*, Kluwer Academic Publishers, Dordrecht, The Netherlands, 2004.

[19] J. Nocedal and S. J. Wright, *Numerical Optimization*, Springer-Verlag, New York, 1999.

[20] F. Oldewurtel, A. Parisio, C. Jones, M. Morari, D. Gyalistras, M. Gwerder, V. Stauch, B. Lehmann, and K. Wirth, *Energy efficient building climate control using stochastic model predictive control and weather predictions*, in Proceedings of the 2010 American Control Conference, 2010, pp. 5100–5105.

[21] A. Rantzer, *Dynamic dual decomposition for distributed control*, in Proceedings of the American Control Conference, 2009, pp. 884–888.

[22] J. Rawlings and B. Stewart, *Coordinating multiple optimization-based controllers: New opportunities and challenges*, J. Process Control, 18 (2008), pp. 839–845.

[23] S. Richter, M. Morari, and C. Jones, *Towards computational complexity certification for constrained MPC based on Lagrange relaxation and the fast gradient method*, in Proceedings of the 50th IEEE Conference on Decision and Control, 2011.

[24] F. ROMIE, *Transient response of the counterflow heat exchanger*, J. Heat Trans., 106 (1984), pp. 620–626.

[25] B. STEWART, A. VENKAT, J. RAWLINGS, S. WRIGHT, AND G. PANNOCCHIA, *Cooperative distributed model predictive control*, Systems Control Lett., 59 (2010), pp. 460–469.

[26] Y. WANG, W. CAI, Y. SOH, S. LI, L. LU, AND L. XIE, *A simplified modeling of cooling coils for control and optimization of HVAC systems*, Energy Conversion and Management, 45 (2004), pp. 2915–2930.

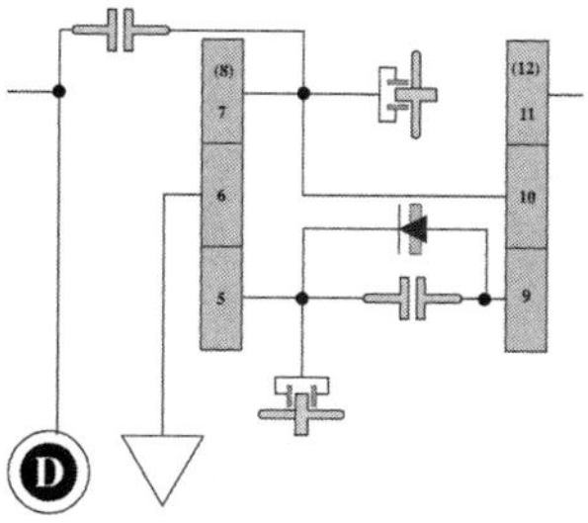

Chapter 15

Dynamic Regularization, Level Set Shape Optimization, and Computed Myography

Kees van den Doel and Uri M. Ascher

We consider inverse problems of surface recovery from noisy boundary data, where the forward problem involves the inversion of elliptic PDEs. Instances arise in electromagnetic data inversion, impedance tomography, potential problems, and computed myography. Two unrelated applications are described, both giving rise to a situation where the usual Tikhonov-type approach leads to numerical difficulty, reminiscent of fully implicit DAEs. A dynamical regularization scheme alleviates this difficulty.

15.1 Introduction

The material covered in this section is well known. We use it to introduce notation along with the problems under consideration leading to the more advanced specific issues addressed in the following two sections.

The goal in the problems considered in this chapter is to recover a distributed parameter function $u(\mathbf{x})$, $\mathbf{x} \in \mathcal{R}^d$, given a *forward operator* $F(u)$ and *data* b, such that

$$b = F(u) + \epsilon, \tag{15.1}$$

where ϵ is noise.

For any admissible function u, F predicts the data. Thus, we seek u such that the predicted data matches the observed data up to the noise level but not beyond: for instance, a solution u such that $F(u) = b$ would necessarily display undesirable effects of overfitting when a nonnegligible amount of noise is present. Let us assume for convenience that b and F are organized as vector arrays with m components each. Ideally, then, we wish the misfit to satisfy

$$\|F(u) - b\| \approx \|\epsilon\|,$$

say in the ℓ_2-norm.

Furthermore, $F(u)$ is obtained as the approximate solution of a system of differential equations with algebraic constraints, projected to the locations where the data is measured. Such inverse problems are common in practice, arising, for instance, in medical imaging [11, 16, 17, 31], geophysical exploration [9, 19, 25, 32, 34], and data assimilation [2, 8].

To be more specific, we concentrate on problems where the forward operator can be written as the matrix-vector product

$$F(u) = Qy, \tag{15.2a}$$

where the field y solves a finite element or finite difference discretization of an elliptic PDE

$$Ay = q, \tag{15.2b}$$

with A or q depending on u. Thus we imagine some mesh discretizing the domain on which the PDE is defined plus its boundary, and reshape the mesh functions for both y and u as vector arrays. The matrix Q is then $m \times n$, where n is the number of mesh points. In the applications we consider, typically $m \ll n$. There may be a positivity constraint imposed on $u(\mathbf{x})$, but we disallow this below for notational simplicity.

There is also a *multi-experiment* version for this problem, whereby

$$F^i(u) = Q^i y^i, \quad i = 1, \ldots, s, \tag{15.3a}$$

with the fields y^i solving the discretized elliptic PDEs

$$A(u)y^i = q^i \tag{15.3b}$$

for s sources q^i and data sets b^i. Thus, $F = (F^1, \ldots, F^s)$ and $b = (b^1, \ldots, b^s)$ are supervectors. Note that the same $A(u)$ appears for different experimental settings i. Let us also define the sensitivity matrices

$$J^i = \frac{\partial F^i}{\partial u}, \tag{15.3c}$$

dropping the superscript i when $s = 1$. Note that J has the dimensions of the matrix Q.

A basic approach for solving the problem (15.1)–(15.2a) is to consider the constrained nonlinear least squares minimization problem

$$\min_{u,y} \frac{1}{2} \|Qy - b\|^2 \tag{15.4a}$$

$$\text{s.t. } Ay = q. \tag{15.4b}$$

This formulation is obviously reminiscent of a typical optimal control problem [28, 29] and admits additional algebraic equality and inequality constraints. Assuming further that A is invertible we can write (15.4) in the unconstrained form

$$\min_{u} \frac{1}{2} \|F(u) - b\|^2, \tag{15.5}$$

where $F(u) = QA^{-1}q$.

Of course, we do not really intend to actually solve the *underdetermined* (15.5) to the (bitter) end in the presence of noise. Moreover, with our elliptic PDEs in the forward modeling this derivation defines a highly ill-posed problem that as such requires regularization. In a Tikhonov-type regularization [18, 35] one solves instead

$$\min_u \phi(u;\beta) \equiv \frac{1}{2}\|F(u)-b\|^2 + \beta R(u), \tag{15.6}$$

where R is a regularization operator such that for any parameter value $\beta > 0$ the problem (15.6) is stable and has a locally unique solution. The parameter value is ideally determined depending on the noise ϵ. Casting this in the constrained form (15.4) again reveals a special case of a usual formulation of an optimal control problem.

The necessary conditions for a minimum in (15.6) are

$$\nabla\phi = J^T(F(u)-b) + \beta R'(u) = 0. \tag{15.7}$$

The choice of R adds a priori information and corresponds to a prior in a Bayesian probability framework [27]. Different regularization operators yield different solutions in the presence of uncertainty such as noise. Let us further specify R as the same-mesh discretization of the operator

$$\frac{1}{2}\int |\nabla u|^2,$$

which penalizes nonsmoothness in the function $u(\mathbf{x})$. This is a frequent choice in practice. Then we can write

$$R'(u) = Zu, \tag{15.8}$$

where Z is a discretization matrix of a Laplacian under appropriate boundary conditions.

In (15.6) we then have a least squares problem. If F is linear, $F(u) = Ju$, then (15.7) are the usual normal equations

$$\left(J^T J + \beta Z\right)u = J^T b. \tag{15.9}$$

If F is nonlinear, then we consider a generalized Gauss–Newton (GN) iteration. Thus, for a current iterate u we evaluate J and F and update $u \leftarrow u + \delta u$, where δu solves the normal equations

$$\left(J^T J + \beta_0 Z\right)\delta u = -\left(J^T(F-b) + \beta R'(u)\right). \tag{15.10}$$

Setting $\beta_0 = \beta$ yields the usual GN iteration, but we make the distinction that β_0 serves the purpose of stabilizing the iteration as in a Levenberg–Marquardt or trust region approach [30], whereas β is the Tikhonov parameter that is involved with determining the solution and relates to the noise.

For both the linear case (15.9) and the nonlinear case (15.10) we have a potential numerical difficulty when β or β_0, respectively, are small, because the matrix $J^T J$ has a large nontrivial null space. Thus, the small term βZ or $\beta_0 Z$ that is negligible in the active space becomes dominant in the null space, and this shift can cause numerical sensitivity. A similar scenario arises in singular-singularly perturbed differential equations that become fully implicit DAEs in the limit [3, 6, 10, 37]. In addition, there is of course the general difficulty of solving a close to singular optimization problem (15.6) when β is small in the nonlinear case.

Just because there is a *potential* difficulty, however, does not automatically imply that it will always arise. In our recent work we were surprised to discover for two different applications that this difficulty was for the same problem *consistently* benign in some circumstances and rather significant, to the point of changing our algorithms, in others. The rest of this chapter is therefore devoted to descriptions of these two problems, their instances where the difficulty arose, and the remedies we found.

15.2 Computed Myography

This section is based on our papers [16] and, more importantly, [17].

The purpose of the following inverse problem is source localization in electromyography. We wish to reliably determine the level of electric activity of individual muscles in a human limb using surface EMG (sEMG) voltage measurements. Applications include prosthetic control and muscle function assessment.

To achieve this objective we have built the system depicted schematically in Figure 15.1.

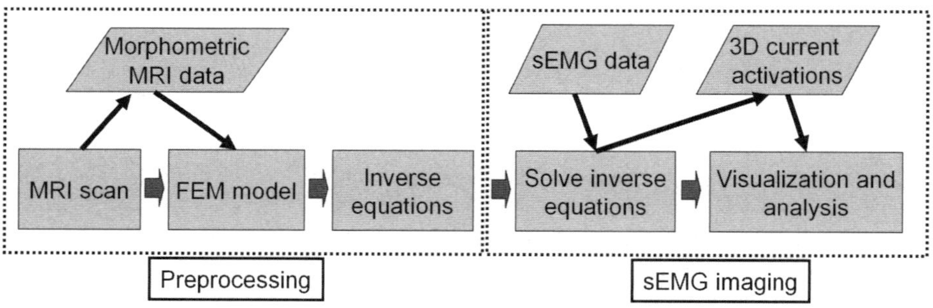

Figure 15.1. Diagram of the CMG system. The thick arrows indicate sequential processing steps, the black arrows indicate data produced or used, the rectangles represent processes, and the slanted rectangles represent data. The preprocessing is done once for a subject, and the resulting model can be used to monitor muscle function over a period of time as long as no significant change in morphology takes place.

The forward problem involves boundary solution values y of the PDE problem

$$-\nabla(\sigma \boldsymbol{\nabla} y) = q \quad \text{in } \Omega, \tag{15.11a}$$

$$\frac{\partial y}{\partial \nu} = 0 \quad \text{on } \partial\Omega_1, \tag{15.11b}$$

$$\frac{\partial y}{\partial \nu} = -\Sigma y \quad \text{on } \partial\Omega_2, \tag{15.11c}$$

with the domain Ω and its boundary depicted in Figure 15.2. MRI is used to approximately determine $\sigma(\mathbf{x})$ and Σ in a preprocessing step, and the objective of the inverse problem is to determine the source $q(\mathbf{x})$.

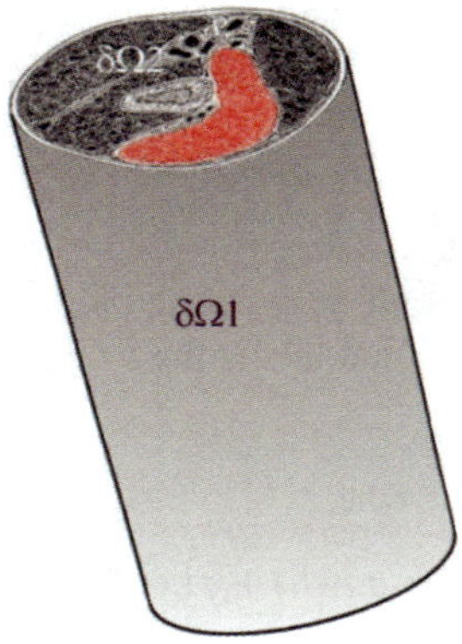

Figure 15.2. Schematic geometry of a limb.

The source q consists of polarization waves: these are intracellular action potentials (IAP) traveling over muscle fibers. In a muscle there are 100–1000 motor units (MU). Such a unit is a collection of muscle fibers excited synchronously. The wave shape of the IAP is known to be a smeared *tripole*. Tripoles are distributed randomly over active muscles: the more active the muscle the denser. On a larger scale the source may therefore be viewed as a density function.

We use the commercial system COMSOL for the finite element discretization of the forward PDE problem (15.11) leading to (15.2b).

In [16] we sought the function $u(\mathbf{x}) = q(\mathbf{x})$ based on measurements of y on the boundary $\partial\Omega_1$. A Tikhonov-type formulation essentially as described in the previous section was employed, resulting in a large but linear least squares problem (15.6) for u whose normal equations are written as (15.9). The parameter β was selected by the discrepancy principle (see, e.g., [18]) such that

$$\|Ju - b\| \approx 1.5\varepsilon\|b\|, \tag{15.12}$$

where the approximate noise level $\varepsilon \approx \|\epsilon\|$ is assumed known, an often realistic assumption.

The CGLS method [23], preconditioned with Z, was applied to solve (15.9). This is a commonly used method which is crucial here, because it allows us to avoid constructing the very large and dense matrix $J^T J$. Only matrix-vector multiplications of the form $w = Jv = Q(A^{-1}v)$ and $J^T w$ for various vectors v are required, and the corresponding PDE solutions $A^{-1}v$ and $w^T A^{-1}$ are approximately obtained using the conjugate gradient (CG) method, preconditioned with incomplete Cholesky decomposition that is carried out once. To obtain a highly efficient method for practical purposes, though, we must compute such PDE solutions to only a rough accuracy.

No noteworthy numerical difficulties were encountered using this method in [16].

However, the quality of the recovered solution was often not sufficient for medical imaging purposes. For an example see Section 5 in [17], especially Figures 10 and 12 there. Indeed, source recovery from a potential problem with data on the boundary is well known to be impossible without additional a priori information [24, 26].

Such additional information was introduced in [17]. Specifically, since we know that in our experimental setup $q(\mathbf{x})$ consists of tripoles we represent the sought source in terms of a tripole basis. This is written as

$$q \equiv \mathcal{D}u. \tag{15.13a}$$

The operator $\mathcal{D}$ is a discrete second derivative along the direction of the muscle fibers. Orienting the coordinate system so that this is the z direction we have

$$\begin{aligned} \mathcal{D} &= \partial_{zz}^{\tilde{h}}, \quad \text{where} \\ \partial_{zz}^{\tilde{h}} f(x,y,z) &= (f(x,y,z+\tilde{h}) + f(x,y,z-\tilde{h}) - 2f(x,y,z))/\tilde{h}^2. \end{aligned} \tag{15.13b}$$

Here $\tilde{h}$ is a small positive scalar that is part of the problem modeling. In our PDE discretizations we make sure that our step sizes (mesh widths) h are never larger than this $\tilde{h}$.

We now have $F = Ju$, with $J = QA^{-1}D$, where D is an $n \times n$ matrix corresponding to the discrete unidirectional second derivative operator $\mathcal{D}$. The obtained reconstruction results are much improved, and we again refer the reader to [17] for comparative examples in two and three space dimensions along with details of the corresponding experiments.

Returning to the theme of this chapter, however, carrying out the numerical method just outlined has also brought up unexpected numerical difficulties. Although we had thought that the CGLS iteration should converge faster with D than without it (because the search for a solution is in a better structured subspace), in effect we obtained results as depicted in Figure 15.3. Note that the number of iterations when using D has significantly increased.

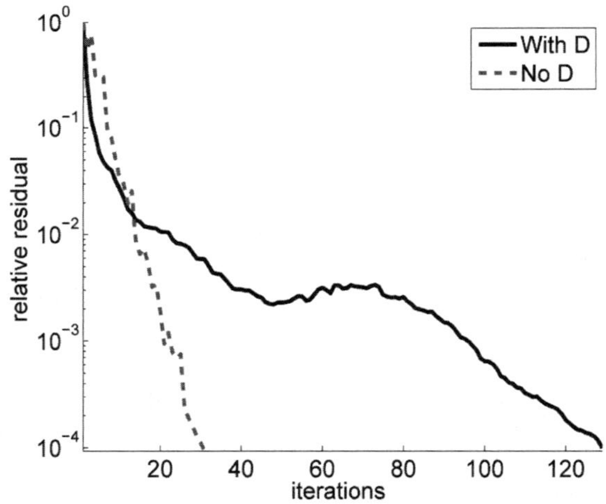

Figure 15.3. Iteration progress for the CMG problem using Tikhonov regularization on a 32^2 uniform grid with 5% noise in the data. Shown are relative residual plots with $D \sim \partial_{zz}^h$ and without D for $\beta = 10^{-3}$, evaluating J-products and preconditioner to tolerance level 10^{-3}. The vertical axis corresponds to the relative residual $\|(J^TJ + \beta Z)u - J^Tb\|/\|J^Tb\|$.

The differential operator in three space dimensions of which $A^{-1}D$ is a discretization (with σ a frozen constant) has the symbol [36]

$$\frac{-\theta_3^2}{(-\sigma)(\theta_1^2+\theta_2^2+\theta_3^2)} \tag{15.14}$$

for a Fourier mode $\exp\{\imath \mathbf{x}\cdot\boldsymbol{\theta}\}$ with frequency $\boldsymbol{\theta} = (\theta_1,\theta_2,\theta_3)$. This symbol is bounded for any $\boldsymbol{\theta}$, but it does not tend to 0 when θ_3 increases, unlike the symbol of A^{-1} without D. Thus, there is more complicated structure here, and it helps to exacerbate the potential trouble of having small $\beta > 0$ as mentioned in the previous section.

This has forced us to switch from Tikhonov to the method of *iterative regularization* [21, 22]. We set $\beta = 0$ and apply a limited number of CGLS iterations preconditioned by Z to the underdetermined problem (15.5), written here as

$$\min_u \|Ju - b\|.$$

The iteration process is terminated when $\|Ju - b\|$ stops decreasing or if the discrepancy principle holds.

The following *theorem* is proved in [22]: using iterative regularization, as the iteration number k increases,

- $\|Ju_k - b\|$ decreases monotonically;
- the penalty $R(u_k)$ increases monotonically.

Furthermore, note that the effect of the choice of R is manifested here through the preconditioner! The preconditioner Z is in fact a smoother.

Having gotten rid of small but positive β, it turns out that this method works efficiently even if matrix-vector products Jv are only roughly calculated. In fact, having D now works to reduce rather than increase the iteration count. The results depicted in Figure 15.4 are typical.

15.3 Level Sets and Shape Optimization

This section is based on our papers [12, 13, 14], and to a lesser extent also [15].

The famous electrical impedance tomography (EIT) problem in medical imaging and direct current (DC) resistivity problem in geophysical exploration both involve finding a positive conductivity function $\sigma(\mathbf{x})$ in a domain Ω (which represents part of the human body or the earth, respectively), based on sparse measurements of the solution $y(\mathbf{x})$ of the PDE

$$\nabla\big(\sigma \boldsymbol{\nabla} y\big) = q, \quad \mathbf{x} \in \Omega, \tag{15.15}$$

for given sources $q(\mathbf{x})$ subject to homogeneous Neumann boundary conditions [9, 11, 19, 31, 32]. It is well known that if the entire Dirichlet-to-Neumann map (equivalently, the boundary data $y|_{\partial\Omega}$ expressed in terms of the source q for any $q(\mathbf{x})$ restricted to $\partial\Omega$) is given, then $\sigma(\mathbf{x})$ can be uniquely recovered provided that it is at least differentiable on the domain Ω (see, e.g., [7, 31]).

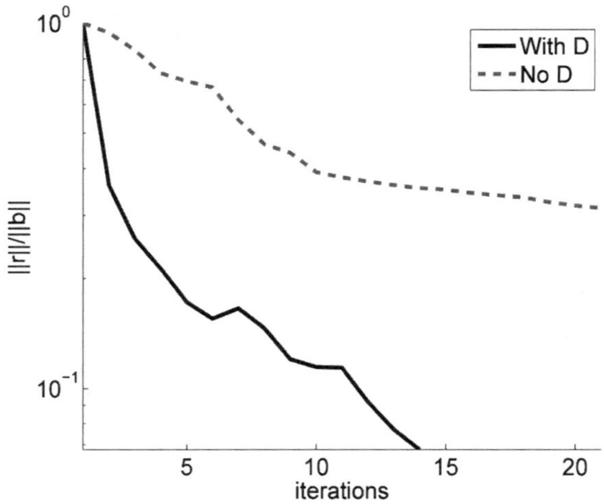

Figure 15.4. Iteration progress for the CMG problem using iterative regularization on a 32^2 uniform grid with 5% noise in the data. Results with evaluation of J-products and preconditioner to tolerance level 10^{-2} are displayed. The vertical axis corresponds to the relative residual $\|Ju - b\|/\|b\|$.

Our problem here differs from the CMG problem in Section 15.2 in that the inverse problem is nonlinear. Moreover, there is justification in hoping for improved reconstruction quality when many experiments are used, so measuring data for varied locations of sources and receivers is common practice. Thus, it is the notation (15.3) that we should keep in mind here.

In what follows until (15.16) we look for a function $u(\mathbf{x})$ satisfying $\sigma = e^u$. This smooth pointwise transformation automatically yields a positive σ. The reduced range of values in u can often be tolerated in practice and even be of help in some geophysical applications.

Using a staggered mesh discretization as described, e.g., in [4], we obtain a discrete inverse problem which can be cast in the notation of Section 15.1 with $A = A(u)$. A Tikhonov regularization as described there (in a possibly weighted version, as the depth direction is different from the horizontal directions in geophysics) is typically applied, and the resulting large, sparse optimization problem (15.6)–(15.7) is typically solved as described in the two previous sections.

When $\sigma(\mathbf{x})$, and hence $u(\mathbf{x})$, may be assumed to have bounded derivatives the procedure outlined above often works well in practice, even for small values of the Tikhonov parameter $\beta > 0$; see, e.g., [4, 20, 32].

If the sought conductivity function is only piecewise smooth, representing a highly heterogeneous medium, then the situation rapidly becomes far more difficult to handle, both theoretically and practically [5, 31]. Our fighting chances improve again, though, if we are allowed to assume that $\sigma(\mathbf{x})$ is piecewise constant such that at each $\mathbf{x}$ it may assume only one of two values, σ_I or σ_{II}. This then is a shape optimization problem with σ_I the conductivity of a uniform body set against a uniform background whose conductivity is σ_{II}. The inverse problem is much more stable under this further restriction [1].

Iterating with a discontinuous, even if piecewise constant, function remains perilous. In [12, 13] we considered a level set function representation for the present problem, where we define u by $\sigma = e^{P(u;0)} = e^{P(u)}$, with

$$P(\xi;h) = \frac{\ln\sigma_I - \ln\sigma_{II}}{2}\tanh(\xi/h) + \frac{\ln\sigma_I + \ln\sigma_{II}}{2}. \tag{15.16}$$

The function $P(u;h)$ depends on the resolution, or grid width h. It is a scaled and mollified version of the Heaviside step function, and its derivative magnitude is at most $O(\frac{|\ln\sigma_I - \ln\sigma_{II}|}{h})$. Thus, as $h \to 0$ the sought function $u(\mathbf{x})$ satisfying

$$\nabla\left(e^{P(u)}\boldsymbol{\nabla} y^i\right) = q^i, \quad i = 1,\ldots,s, \tag{15.17}$$
$$\frac{\partial y^i}{\partial \nu}\Big|_{\partial\Omega} = 0$$

has bounded first derivatives, whereas $\sigma(\mathbf{x})$ is generally discontinuous.

Applying the same Tikhonov-type method as described in Section 15.1 was then the next obvious step; see [12]. Note that the least squares regularization is applied to the smoother u, not to the discontinuous σ. However, as it turned out this method now performs poorly in the sense of yielding unpredictable reconstruction results (and not just in poorer iteration counts as in Section 15.2). Comparison with the good, smooth case shows that the difference is rooted in the fact that instead of having smooth Jacobian matrices J^i we now have elements in them that are either $O(h^{-1})$ (the mesh width h is of course small, like all h's) or nearly 0. This roughness makes computations with small $\beta > 0$ hazardous.

Our remedy in [12, 13] was to set $\beta = 0$, obtaining regularization by other means and calling the method *dynamical regularization*. For a linear inverse problem the method version in [13] (but not [12]) coincides with the much earlier method of iterative regularization described in Section 15.2.

Alternatively, it is possible more generally to entertain $\beta > 0$ for a while longer and turn our attention to β_0 in (15.10). Applying iterative regularization as described earlier involves setting $\beta_0 = 0$ and employing only a small number of preconditioned CG (inner) iterations towards "solving" the now singular linear system. This in turn corresponds to a stabilized Tikhonov iteration [14, 33]. (Actually, it is not quite pure Tikhonov, because once the last outer iteration has been carried out there is still residue of this additional inner regularization and also the outer iteration stopping criterion is slightly different; but let's not split hairs about these.)

Despite the above we have found it convenient in [14] to work with the dynamical regularization method, setting $\beta = 0$ and relying even more on the finite number of preconditioned CG steps to yield the necessary regularization. However, the difference from the now stabilized Tikhonov method is no longer as major as before. We refer the reader to [14] for further details and numerical experiments.

15.4 Summary

We have seen two practical situations in applications in which the quest for better reconstructions, i.e., solutions for the inverse problem at hand that not only fit the data well but also better fit our additional information and notion of what is correct, yielded additional numerical difficulties when employing Tikhonov-type regularization. These difficulties have been addressed by using dynamical or iterative regularization, setting the Tikhonov parameter to zero in a DAE-like manner, and by further stabilizing (regularizing) the linear iteration using only a few preconditioned CG iterations. Note in addition that using only a few such iterations also provides a big boost to efficiency.

We remark that a difficulty such as considered here can arise only when considering highly ill-posed problems. These can be much harder to solve in a minimally satisfactory way than mildly ill-posed ones (such as may arise in image processing, for instance), and they often require additional information in a nontrivial way that may lead to very different numerical solution techniques.

Bibliography

[1] A. Alessandrini and S. Vessella, *Lipschitz stability for the inverse conductivity problem*, Adv. in Appl. Math., 35 (2005), pp. 207–241.

[2] J. Anderson and S. Anderson, *A Monte Carlo implementation of the nonlinear filtering problem to produce ensemble assimilations and forecasts*, Mon. Weather Rev., 127 (1999), pp. 2741–2758.

[3] U. Ascher, *On some difference schemes for singular singularly-perturbed boundary value problems*, Numer. Math., 46 (1985), pp. 1–30.

[4] U. Ascher and E. Haber, *A multigrid method for distributed parameter estimation problems*, Electron. Trans. Numer. Anal., 18 (2003), pp. 1–18.

[5] U. M. Ascher, E. Haber, and H. Huang, *On effective methods for implicit piecewise smooth surface recovery*, SIAM J. Sci. Comput., 28 (2006), pp. 339–358.

[6] U. M. Ascher and L. R. Petzold, *Computer Methods for Ordinary Differential Equations and Differential-Algebraic Equations*, SIAM, Philadelphia, 1998.

[7] J. Barcelo, T. Barcelo, and A. Ruiz, *Stability of the inverse conductivity problem in the plane for less regular conductivities*, J. Differential Equations, 173 (2001), pp. 231–270.

[8] K. Bergemann and S. Reich, *A localization technique for ensemble Kalman filters*, Quart. J. Royal Meteorological Soc., 136 (2010), pp. 701–707.

[9] L. Borcea, *Nonlinear multigrid for imaging electrical conductivity and permittivity at low frequency*, Inverse Problems, 17 (2001), pp. 329–360.

[10] K. Brenan, S. Campbell, and L. Petzold, *Numerical Solution of Initial-Value Problems in Differential-Algebraic Equations*, North–Holland, New York, 1989; reprinted, SIAM, Philadelphia, 1996.

[11] M. Cheney, D. Isaacson, and J. C. Newell, *Electrical impedance tomography*, SIAM Rev., 41 (1999), pp. 85–101.

[12] K. v. d. Doel and U. Ascher, *On level set regularization for highly ill-posed distributed parameter estimation problems*, J. Comput. Phys., 216 (2006), pp. 707–723.

[13] K. v. d. Doel and U. Ascher, *Dynamic level set regularization for large distributed parameter estimation problems*, Inverse Problems, 23 (2007), pp. 1271–1288.

[14] K. v. d. Doel and U. Ascher, *Adaptive and stochastic algorithms for EIT and DC resistivity problems with piecewise constant solutions and many measurements*, submitted.

[15] K. v. d. Doel, U. Ascher, and A. Leitao, *Multiple level sets for piecewise constant surface reconstruction in highly ill-posed problems*, J. Sci. Comput., 43 (2010), pp. 44–66.

[16] K. v. d. Doel, U. Ascher, and D. Pai, *Computed myography: Three dimensional reconstruction of motor functions from surface EMG data*, Inverse Problems, 24 (2008), pp. 1–17.

[17] K. v. d. Doel, U. Ascher, and D. Pai, *Source localization in electromyography using the inverse potential problem*, Inverse Problems, 27 (2011), 025008.

[18] H. W. Engl, M. Hanke, and A. Neubauer, *Regularization of Inverse Problems*, Kluwer, Dordrecht, The Netherlands, 1996.

[19] E. Haber, U. Ascher, and D. Oldenburg, *Inversion of 3D electromagnetic data in frequency and time domain using an inexact all-at-once approach*, Geophysics, 69 (2004), pp. 1216–1228.

[20] E. Haber, M. Chung, and F. Herrmann, *An Effective Method for Parameter Estimation with PDE Constraints with Multiple Right Hand Sides*, manuscript, 2010.

[21] M. Hanke and P. C. Hansen, *Regularization methods for large scale problems*, Surveys Math. Indust., 3 (1993), pp. 253–315.

[22] P. C. Hansen, *Rank-Deficient and Discrete Ill-Posed Problems: Numerical Aspects of Linear Inversion*, SIAM, Philadelphia, 1998.

[23] P. C. Hansen, J. G. Nagy, and D. P. O'Leary, *Deblurring Images: Matrices, Spectra, and Filtering*, SIAM, Philadelphia, 2006.

[24] H. L. F. Helmholtz, *Ueber einige gesetze der verheilung elektrischer ströme in körperlicher leitern mit anwendung auf die thierisch-elektrischen versuche*, Ann. Physik und Chemie, 9 (1853), pp. 211–233.

[25] F. Herrmann, Y. Erlangga, and T. Lin, *Compressive simultaneous full-waveform simulation*, Geophysics, 74 (2009), p. A35.

[26] F. Hettlich and W. Rundell, *Iterative methods for the reconstruction of an inverse potential problem*, Inverse Problems, 12 (1996), pp. 251–266.

[27] J. Kaipo and E. Somersalo, *Statistical and Computational Inverse Problems*, Springer, Berlin, 2005.

[28] J. Lions, *Some Methods in the Mathematical Analysis of Systems and their Control*, Gordon & Breach, New York, 1981.

[29] T. MATHEW, M. SARKIS, AND C. SCHAERER, *Analysis of block matrix preconditioners for elliptic optimal control problems*, Numer. Linear Algebra Appl., 14 (2007), pp. 257–279.

[30] J. NOCEDAL AND S. WRIGHT, *Numerical Optimization*, Springer, New York, 1999.

[31] L. PAIVARINTA, A. PANCHENKO, AND G. UHLMANN, *Complex geometrical optics solutions for Lipschitz conductivities*, Rev. Mat. Iberoamericana, 19 (2003), pp. 57–72.

[32] A. PIDLISECKY, E. HABER, AND R. KNIGHT, *RESINVM3D: A MATLAB 3D resistivity inversion package*, Geophysics, 72 (2007), pp. H1–H10.

[33] A. RIEDER AND A. LECHLEITER, *Towards a general convergence theory for inexact newton regularizations*, Numer. Math., 114 (2010), pp. 521–548.

[34] J. ROHMBERG, R. NEELAMANI, C. KROHN, J. KREBS, M. DEFFENBAUGH, AND J. ANDERSON, *Efficient seismic forward modeling and acquisition using simultaneous random sources and sparsity*, Geophysics, 75 (2010), pp. WB15–WB27.

[35] A. N. TIKHONOV AND V. Y. ARSENIN, *Methods for Solving Ill-Posed Problems*, John Wiley and Sons, New York, 1977.

[36] U. TROTTENBERG, C. OOSTERLEE, AND A. SCHULLER, *Multigrid*, Academic Press, San Diego, CA, 2001.

[37] A. VASILE'EVA AND V. BUTUZOV, *Singularly Perturbed Equations in the Critical Case*, Math Research Center Monograph, 1980. Translated from the Russian, Nauka, 1973.

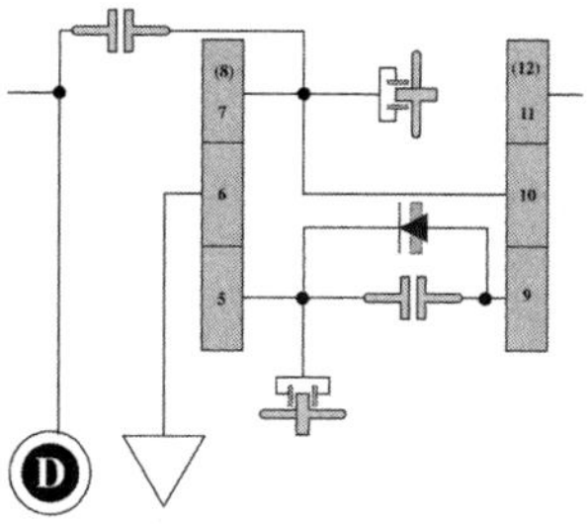

Chapter 16

The Application of Pontryagin's Minimum Principle for Endpoint Optimization of Batch Processes

Srinivas Palanki

In this chapter, a methodological framework is developed for endpoint optimization of batch processes. The traditional approach to solving this problem requires the numerical solution of a DAE system, which leads to computational problems especially when the DAE index is high. Pontryagin's Minimum Principle is used to formulate a solution for the optimal inputs in a feedback control structure rather than in an open-loop numerical solution. It is shown that this framework can account for changes in initial conditions as well as process disturbances. The proposed methodology is illustrated via a simulation example.

16.1 Introduction

The batch and semibatch mode of operation is widely used in the speciality chemicals and pharmaceutical industry. This mode is preferred in manufacturing situations where the production volumes are low, isolation is required for reasons of safety of sterility, the materials involved are hard to handle, the process is not well characterized, and when there is a requirement for the product to reach the market quickly. Strictly speaking, there is no addition of reactant or removal of product in a batch reactor once the operation starts, whereas in a semibatch reactor, one could add a reactant or remove a product during operation. However, from a process systems viewpoint, the key feature that differentiates batch and semibatch processes from continuous processes is that continuous processes have a well-defined steady state and, in theory, operate for infinite time, whereas batch and semibatch processes do not operate at a steady state and end in finite time. For this reason, the term "batch process" in this chapter will include semibatch processes as well.

Batch processes form an important class of industrial operations. In the face of increased competition, batch process optimization provides a unified framework for reducing production costs, improving product quality, and meeting environmental and safety regulations. From an industrial perspective, the main processing objective is of economic nature and it is necessary to operate the process optimally in the face of uncertainty. Furthermore, since most batch operations are repeated multiple times, there is a potential to optimize process operations from batch to batch. This provides the motivation to develop an optimization strategy that can be implemented in real time.

16.2 Problem Formulation and Mathematical Preliminaries

In batch chemical processes, typical manipulated inputs are temperature, pH, and flow rate of the limiting reactant. While flow rate appears as an affine function, temperature and pH appear as nonaffine functions in the state model. In addition, batch processes are subject to physical constraints (e.g., constraints on reactor volume and on flow rate) that appear as inequality constraints in the state model. The objective is to optimize a performance index, which is a measure of desired product specifications on quality and quantity, at a fixed final time. This endpoint optimization problem can be mathematically formulated as follows:

$$\min\ J = \phi(x(t_f)) \tag{16.1}$$

subject to

$$\begin{aligned} \dot{x} &= f(x,u) + g(x)v, \\ x(0) &= x_0, \\ S(x,u,v) &\leq 0, \end{aligned} \tag{16.2}$$

where u is a *vector* of m manipulated inputs that appear as nonaffine functions, while v is a *scalar* manipulated input that appears as an affine function, x is an n-vector of system states, t_f is the final time, f and g are smooth vector functions, S is a vector of p path constraints (which include both state and input constraints), and J is a scalar performance index to be minimized. Furthermore, the differential equations represented by (16.2) are linearly independent. It is assumed that the system dynamics, objective function, and the constraints are given by analytic expressions.

The objective is to find time profiles of the nonaffine inputs u and the affine input v such that the performance index J is minimized at the final time, t_f. If the dynamical system, its initial conditions, and the constraints defined by (16.2) are well defined and do not change, it is possible to utilize a dynamic programming perspective to numerically compute an optimal input vector profile that minimizes the objective function described by (16.1) and then implement this function in open-loop fashion. Conceptually, one is looking for numerical values of the input vector that satisfy the algebraic constraints $S(x,u,v) \leq 0$ and lead to state trajectories by solving $\dot{x} = f(x,u) + g(x)v$ that minimize the cost function J. When the algebraic constraints $S(x,u,v)$ are active, the optimal inputs are computed via the numerical solution of DAEs described by (16.2). This is a particularly difficult problem to solve when the DAE index is high.

Most batch process models are developed from scanty laboratory data using simplified assumptions, and for this reason, these dynamic models are subject to parametric uncertainty. Furthermore, initial conditions vary from batch to batch and unmodeled disturbances affect the system, especially when laboratory-scale batch processes are scaled up to industrial sized reactors. For these reasons, it is relevant to develop a methodological framework that solves for the optimal inputs in a feedback control structure rather than in an open-loop numerical solution. Such a framework can be used as the basis of a real-time optimization scheme that can lead to endpoint optimization in the face of changes in initial conditions, process disturbances, and modeling uncertainties. A feedback control-theoretic approach to solve the endpoint optimization problem posed by (16.1) and (16.2) requires tools from nonlinear systems theory and differential geometry. We review some mathematical preliminaries in this section.

Given $f(x,t)$, $g(x,t)$ C^∞ vector fields on $\mathcal{R}^n$, the Lie bracket $[f,g]$ is a vector field defined by

$$[f,g] = \frac{\partial g}{\partial x} f - \frac{\partial f}{\partial x} g, \tag{16.3}$$

where $\frac{\partial f}{\partial x}$ and $\frac{\partial g}{\partial x}$ are Jacobians and $[f,g]$ is also a C^∞-vector fields on $\mathcal{R}^n$. One can define iterated Lie brackets, and the following notation is standard:

$$\begin{aligned} ad_f^0 g &= g(x,t), \\ ad_f^1 g &= [f,g] + \frac{\partial}{\partial t} ad_f^0 g, \\ ad_f^2 g &= [f, ad_f^1 g] + \frac{\partial}{\partial t} ad_f^1 g, \\ &\vdots \\ ad_f^k g &= [f, ad_f^{k-1} g] + \frac{\partial}{\partial t} ad_f^{k-1} g. \end{aligned} \tag{16.4}$$

Consider the dynamical system represented by (16.2) with the *inequality* constraint replaced by an *equality* constraint vector $S(x,u,v) = 0$. The *relative degree*, r_i, of an output $S_i(x,u,v)$ with respect to an input u_i or v is defined as the number of times the output S_i has to be differentiated with respect to time before an input appears explicitly. Note that if one regards the algebraic constraints as outputs of an explicit nonlinear system and this explicit system can be decoupled via static state feedback, then the index is directly related to relative degree and the system may be reduced to an index-zero system by the application of the decoupling feedback law. However, when the system cannot be decoupled via static state feedback, one has to consider the use of dynamic extension algorithms for index reduction.

The mathematical concepts introduced above are illustrated via this simple example. Consider a batch chemical reactor where the following reaction is taking place:

$$\begin{aligned} &A \rightarrow B \rightarrow C, \\ &A \rightarrow D. \end{aligned} \tag{16.5}$$

The species A is being fed to the reactor at a feed concentration of s_F. All reactions follow first order kinetics with rate constants k_B (for $A \rightarrow B$), k_C (for $B \rightarrow C$), and k_D (for $A \rightarrow D$). The objective is to maximize the amount of B produced in one hour by manipulating the volumetric feed rate v. Furthermore, the maximum reactor volume is V_{max}. The reactor initially has only m_{A0} moles of species A, and the initial reactor volume is V_0. This endpoint optimization problem can be mathematically expressed as

$$\min J = -x_2 \tag{16.6}$$

subject to

$$\frac{d}{dt}\begin{bmatrix} x_1 \\ x_2 \\ x_3 \\ x_4 \end{bmatrix} = \begin{bmatrix} -k_B x_1 - k_D x_1 \\ k_B x_1 - k_C x_2 \\ k_D x_1 \\ 0 \end{bmatrix} + \begin{bmatrix} s_F \\ 0 \\ 0 \\ 1 \end{bmatrix} v \tag{16.7}$$

and the initial conditions

$$x(0) = [m_{A0}\; 0\; 0\; V_0]^T \tag{16.8}$$

and the constraint

$$x_4 - V_{max} \leq 0, \tag{16.9}$$

where x_1, x_2, and x_3 are the molar amounts of species A, B, and D, respectively, and x_4 is the reactor volume.

It can be seen that the problem is in the form described by (16.1) and (16.2). Lie brackets for the dynamical system represented by (16.7) are as follows:

$$\begin{aligned} ad_f^0 g &= \begin{bmatrix} s_F \\ 0 \\ 0 \\ 1 \end{bmatrix}, \\ ad_f^1 g &= \begin{bmatrix} (k_B + k_D)s_F \\ 0 \\ 0 \\ 0 \end{bmatrix}. \end{aligned} \tag{16.10}$$

If we replace the inequality constraint represented by (16.9) with an equality constraint, it can be shown that successive time derivatives lead to

$$\begin{aligned} C &= x_4 - V_{max}, \\ \frac{dC}{dt} &= \dot{x}_4 = v. \end{aligned} \tag{16.11}$$

Since v appears in the first time derivative of the constraint equation, the relative degree of this constraint with respect to the input is 1.

In the next section, we will utilize a control-theoretic framework to develop state feedback laws that can be used in a real-time optimization framework for endpoint optimization.

16.3 The Classical Optimal Control Perspective and Optimal State Feedback Synthesis

By Pontryagin's Minimum Principle (PMP), the minimization problem described by (16.1) and (16.2) is equivalent to minimizing the Hamiltonian

$$H = \lambda^T f + \lambda^T g v + \mu^T S, \tag{16.12}$$

where λ is an n-vector of Lagrange multipliers associated with the system states and μ is a p-vector of nonnegative Lagrange multipliers associated with the system constraints that satisfy

$$\begin{aligned}\dot{\lambda}^T &= -\lambda^T \frac{\partial f}{\partial x} - \lambda^T \frac{\partial g}{\partial x} v - \mu^T \frac{\partial S}{\partial x}, \\ \lambda^T(t_f) &= \frac{\partial \phi}{\partial x}|_{t_f}, \\ \mu^T S &= 0.\end{aligned} \tag{16.13}$$

The first order necessary conditions for optimality are given by

$$H_{u_i} = 0, \qquad i = 1,2,\ldots,m, \tag{16.14}$$

$$H_v = 0. \tag{16.15}$$

Substituting (16.12) in (16.14) and (16.15), we get

$$\lambda^T \frac{\partial f}{\partial u_i} + \mu^T \frac{\partial S}{\partial u_i} = 0, \qquad i = 1,2,\ldots,m, \tag{16.16}$$

$$\lambda^T g + \mu^T \frac{\partial S}{\partial v} = 0. \tag{16.17}$$

To solve for the optimal inputs u_i and v, from (16.16) and (16.17), one has to solve for both the states (x) and the Lagrange multipliers (λ, μ). However, the boundary conditions for x and λ are split; i.e., the initial conditions of x and the terminal conditions of λ are known. Thus, the classical PMP formulation leads to a two-point BVP, which is computationally expensive to solve. The presence of both differential and algebraic equations in this formulation leads to numerical difficulties particularly when the DAE index is high. Furthermore, in this formulation, the numerical problem has to be solved separately every time there is a change in initial conditions or model parameters. The optimal solution has to be implemented in open-loop fashion, and so process uncertainties and disturbances are not attenuated. This provides the motivation to develop a general framework for endpoint optimal state feedback synthesis that can be implemented in real time for both batch-to-batch as well as within-the-batch optimization.

The PMP formulation presented above shows that the optimal solution consists of several different intervals; an input can either be at its constraint value or in the interior of its constraint values. The first order necessary conditions for optimality represented by (16.16) and (16.17) have two parts, a system-dependent part ($\lambda^T \frac{\partial f}{\partial u_i}$ or $\lambda^T g$) and a constraints-dependent part ($\mu^T \frac{\partial S}{\partial u_i}$ or $\mu^T \frac{\partial S}{\partial v}$).

16.3.1 Active Path Constraint

When a path constraint is active, it is possible to compute the optimal input by differentiating the active path constraint r_i times until an input appears explicitly. This results in a standard regulation problem of relative degree r_i for each active constraint. When the input is computed from an active path constraint, this part of the optimal solution does not depend on the Lagrange multipliers.

16.3.2 Solution Inside Feasible Region

We first compute an optimal state feedback for a nonaffine input, u_i, inside the feasible region where no path constraints are active. From (16.14), since $H_{u_i} = 0$, all successive time derivatives are also equal to zero. Taking $n-1$ time derivatives of the Hamiltonian results in the following set of first order necessary conditions of optimality for a nonaffine input u_i:

$$\begin{aligned}
\lambda^T \left(I\frac{d}{dt} - \frac{\partial f}{\partial x} - \frac{\partial g}{\partial x}v \right)^0 f_{u_i} &= 0, \\
\lambda^T \left(I\frac{d}{dt} - \frac{\partial f}{\partial x} - \frac{\partial g}{\partial x}v \right)^1 f_{u_i} &= 0, \\
&\vdots \\
\lambda^T \left(I\frac{d}{dt} - \frac{\partial f}{\partial x} - \frac{\partial g}{\partial x}v \right)^{n-1} f_{u_i} &= 0.
\end{aligned} \tag{16.18}$$

The first equation in (16.18) is H_{u_i}. The second equation in (16.18) is obtained by taking the time derivative of H_{u_i} and using the first equation in (16.13) to eliminate $\dot{\lambda}^T$. This procedure is repeated for each successive time derivative. After each time derivative, the terms are arranged so that they follow the same pattern. It is observed that the Lagrange multiplier vector, λ, appears as an affine function in (16.18). Since λ is trivially not equal to zero (since the path constraints are not active), it is possible to eliminate the n Lagrange multipliers from the n linear equations (in λ_i) represented by (16.18), resulting in the following state feedback law:

$$\det\left[\left(I\frac{d}{dt} - \frac{\partial f}{\partial x} - \frac{\partial g}{\partial x}v \right)^0 f_{u_i} \mid \cdots \mid \left(I\frac{d}{dt} - \frac{\partial f}{\partial x} - \frac{\partial g}{\partial x}v \right)^{n-1} f_{u_i} \right] = 0. \tag{16.19}$$

There will be m such equations for the m nonaffine inputs. If the columns defined by $\left(I\frac{d}{dt} - \frac{\partial f}{\partial x} - \frac{\partial g}{\partial x}v\right)^k f_{u_i}$ are linearly *dependent*, it is necessary to find a reduced order state feedback. In other words, it is necessary to find a set of linearly independent rows and columns that make the determinant represented by (16.19) lose rank *structurally*. Note that the state feedback laws derived from (16.19) involve time derivatives of the inputs and are thus *dynamic* feedback laws.

Now, we compute the optimal state feedback for the affine input, v, inside the feasible region where no path constraints are active. Since v appears as an affine function in the state model, the optimal control problem with respect to input v is singular. The singular extremal is characterized by its degree of singularity, s, which is defined as the number of times the expression $H_v = 0$ has to be differentiated with respect to time before the input v appears explicitly. If the degree of singularity is infinite, then v will not appear in successive time derivatives of $H_v = 0$, even after $n-1$ time derivatives. Mathematically, this can be represented as

$$\begin{aligned}
\lambda^T g &= 0, \\
\lambda^T ad_f^1 g &= 0, \\
&\vdots \\
\lambda^T ad_f^{n-1} g &= 0,
\end{aligned} \tag{16.20}$$

where $ad_f^k g$ represents the kth-fold Lie bracket of the time-varying system represented by (16.2). The n Lagrange multipliers, λ_i, appear as affine functions in the n equations represented by (16.20) and since $\lambda \neq 0$, the Lagrange multipliers can be eliminated. Thus, for a system with an infinite degree of singularity with respect to v, the optimal solution involves regulating the system to a surface S defined by

$$S = \det\left[g \mid ad_f^1 g \mid \cdots \mid ad_f^{n-1} g\right] = 0. \tag{16.21}$$

In this situation, the endpoint optimization problem is reduced to a standard regulation problem of finding the necessary input to regulate the system represented by (16.2) to the surface represented by (16.21).

If the degree of singularity is $n-1$, then v will appear in the $(n-1)$th time derivative of $H_v = 0$ as shown below:

$$\begin{aligned} \lambda^T g &= 0, \\ \lambda^T ad_f^1 g &= 0, \\ &\vdots \\ \lambda^T ad_f^{n-2} g &= 0, \\ \lambda^T ad_f^{n-1} g + \lambda^T [g, ad_f^{n-2} g] v &= 0. \end{aligned} \tag{16.22}$$

The n Lagrange multipliers, λ_i, appear as affine functions in the n equations represented by (16.22) and since $\lambda \neq 0$, the Lagrange multipliers can be eliminated to yield the following state feedback law:

$$v = -\frac{\det\left[g \mid ad_f^1 g \mid \cdots \mid ad_f^{n-1} g\right]}{\det\left[g \mid ad_f^1 g \mid \cdots \mid ad_f^{n-2} g \mid [g, ad_f^{n-2} g]\right]} \tag{16.23}$$

This state feedback law provides a convenient expression that can be implemented in real time.

If the degree of singularity is less than $n-1$, then the $(n-1)$ time derivative of $H_v = 0$ to eliminate the Lagrange multipliers will result in expressions that involve time derivatives of the input v as shown below:

$$\begin{aligned} \lambda^T g &= 0, \\ \lambda^T ad_f^1 g &= 0, \\ &\vdots \\ \lambda^T ad_f^{s-1} g &= 0, \\ \lambda^T ad_f^{s} g + \lambda^T [g, ad_f^{s-1} g] v &= 0, \\ \lambda^T \left(I\frac{d}{dt} - \frac{\partial f}{\partial x} - \frac{\partial g}{\partial x} v\right)^{s+1} g &= 0, \\ &\vdots \\ \lambda^T \left(I\frac{d}{dt} - \frac{\partial f}{\partial x} - \frac{\partial g}{\partial x} v\right)^{n-1} g &= 0. \end{aligned} \tag{16.24}$$

The n Lagrange multipliers, λ_i, appear as affine functions in the n equations represented by (16.24) and since $\lambda \neq 0$, the Lagrange multipliers can be eliminated to yield the following dynamic state feedback law:

$$\det\left[g \mid ad_f^1 g \mid \cdots \mid ad_f^s g + [g, ad_f^{s-1} g]v \mid \cdots \mid \left(I\frac{d}{dt} - \frac{\partial f}{\partial x} - \frac{\partial g}{\partial x}v \right)^{n-1} g \right] = 0. \quad (16.25)$$

16.3.3 Piecing Together the Optimal Solution

In the previous subsection, it was shown that two types of intervals are possible between switching instants depending on whether the solution is determined by active path constraints or is in the interior of the constraints. Analytical expressions for the inputs can be obtained for each type of intervals. This procedure is not dependent on the computation of Lagrange multipliers and provides expressions for all possible types of intervals that might occur in the optimal solution. To piece together the optimal solution, it is necessary to iterate between (i) the sequence of various types of intervals and (ii) the switching times between various intervals. Thus, the analytical approach presented here can be used as a method of parsimonious parameterization of the optimal solution. Once the nominal solution has been obtained via iteration on the switching instants and interval sequences, it is possible to implement this scheme in a real-time setting using the analytical expressions for optimal state feedback in each interval.

An implicit assumption is made here that the switching instants are robust to plant perturbations and are computed off-line. If the batch operation is perturbed significantly from the set of conditions where the switching instants were initially determined, this can lead to suboptimal performance. Since the batch operation is repeated many times, it is possible to compute new switching instants at the end of each batch operation using data from the current batch and use these new values of switching instants in the succeeding batch operation. Over several batch operations, this can lead to improved cost. Another limitation for this method is the fact that the derivation of analytical expressions for the feedback laws can become rather involved, especially when the degree of singularity is high. For most common batch operations, the number of states is small (3 or 4) and the degree of singularity is usually no more than 2. However, for large-scale problems it may be necessary to use symbolic manipulation software. For systems with a large number of states and high degree of singularity, one has to make a choice between making potentially tedious analytical calculations versus solving a large-order DAE numerically via a classical PMP formulation.

16.4 Application

Consider a semibatch reactor in which the following elementary reactions are taking place:

$$A + B \rightarrow C \rightarrow D. \quad (16.26)$$

The objective is to maximize the production of C at a fixed final time of 0.5 h by manipulating the reactor temperature, u, and the feed rate, v, of species B. There are constraints on the reactor volume, feed rate, and reactor temperature as well as the maximum heat that can be removed by the cooling system. The following endpoint optimization can be formulated:

$$\min J = -C_C \cdot V \tag{16.27}$$

subject to the dynamics

$$\begin{aligned}
\dot{C}_A &= -k_{10}\exp\left(-\frac{E_1}{Ru}\right)C_A C_B - \frac{v}{V}C_A; \qquad C_A(0) = C_{A0};\\
\dot{C}_B &= -k_{10}\exp\left(-\frac{E_1}{Ru}\right)C_A C_B + \frac{v}{V}(C_{B_{in}} - C_B); \qquad C_B(0) = C_{B0};\\
\dot{C}_C &= k_{10}\exp\left(-\frac{E_1}{Ru}\right)C_A C_B - k_{20}\exp\left(-\frac{E_2}{Ru}\right)C_C - \frac{v}{V}C_C; \qquad C_C(0) = C_{C0};\\
\dot{V} &= v; \qquad V(0) = V_0
\end{aligned} \tag{16.28}$$

and constraints

$$\begin{aligned}
V &\le V_{max},\\
T_{min} &\le u \le T_{max},\\
v_{min} &\le v \le v_{max},\\
(-\Delta H_1)k_{10}\exp\left(-\frac{E_1}{Ru}\right)C_A C_B V &+ (-\Delta H_2)k_{20}\exp\left(-\frac{E_2}{Ru}\right)C_C V \le q_{max},
\end{aligned} \tag{16.29}$$

where C_A, C_B, and C_C are the concentrations of species A, B, and C, respectively, V is the reactor volume, u is the reactor temperature (nonaffine input), v is the flow rate of species B (affine input), and ΔH_1, ΔH_2, E_1, E_2, R, k_{10}, k_{20} are constant parameters of the dynamic system. The initial conditions of the system states are given by C_{A0}, C_{B0}, C_{C0}, and V_0 and the inlet concentration of the feed v is $C_{B_{in}}$.

The batch is initially charged with species A and a suitable amount of B such that the maximum heat removal constraint is satisfied as an equality. Since the dynamic system represented by (16.28) is linearly dependent, one of the dynamic state equations can be removed using the following algebraic expression:

$$C_B = \frac{1}{V}\left(C_{B_{in}}V + C_A V + V_0(C_{B0} - C_{A0} - C_{B_{in}})\right). \tag{16.30}$$

It can be shown that the optimal solution consists of u_{max} and $u_{feasible}$ for the nonaffine input and v_{path} and v_{min} for the affine input as described below:

- Since the initial condition of B is such that the maximum heat removal constraint is satisfied as an equality, the input feed rate of B is applied to keep this path constraint active. This is obtained by differentiating the path constraint regarding the heat production rate once with respect to time:

$$\begin{aligned}
v_{path} &= V\frac{(-\Delta H_1)k_1^2 C_A C_B(C_A + C_B) - (-\Delta H_2)k_2(k_1 C_A C_B - k_2 C_C)}{(-\Delta H_1)k_1 C_A(C_{B_{in}} - C_B)}\\
&\quad - \frac{\dot{u}V}{Ru^2}\frac{((-\Delta H_1)E_1 k_1 C_A C_B + (-\Delta H_2)E_2 k_2 C_C)}{(-\Delta H_1)k_1 C_A(C_{B_{in}} - C_B)},
\end{aligned} \tag{16.31}$$

where $k_1 = k_{10}\exp\left(-\frac{E_1}{Ru}\right)$ and $k_2 = k_{20}\exp\left(-\frac{E_2}{Ru}\right)$. It can be seen that the relative degree for this constraint with respect to v_{path} is 1 while the DAE index is 2.

- Once the reactor reaches its maximum volume, V_{max}, the flow rate of B is set to the lower constraint value of zero ($v = v_{min}$).
- The temperature, u, starts at its upper bound value of T_{max} to favor the desired reaction ($u = u_{max}$). Later, the temperature switches to $u_{feasible}$ to take advantage of the temperature-dependent compromise between production and consumption of C. This is computed from the following dynamic state feedback:

$$\dot{u} = -\frac{Ru^2 k_1 C_A C_B}{E_2 C_C} - \frac{Ru^2(C_{B_{in}} - C_B)}{C_B(E_1 - E_2)}\frac{v}{V}. \tag{16.32}$$

For numerical values presented in Table 16.1, an objective function of −2.02 is obtained, which implies that a maximum of 2.02 moles of C can be produced in this batch operation. The temperature switches from the path constraint to feasible region at 0.05 h and the flow rate switches from the path constraint to its minimum value at 0.3165 h. The small number of switching instants shown in this example is fairly typical for a large class of industrially relevant batch reactor operations.

Table 16.1. Numerical values of system parameters and initial conditions.

k_{10}	4	$l/mol\ h$
k_{20}	800	$l/mol\ h$
E_1	6×10^3	J/mol
E_2	20×10^3	J/mol
R	8.31	$J/mol\ K$
ΔH_1	-3×10^4	J/mol
ΔH_2	-1×10^4	J/mol
v_{min}	0	l/h
v_{max}	1	l/h
T_{min}	20	^oC
T_{max}	50	^oC
V_{max}	1.1	l
q_{max}	1.5×10^5	J/h
C_{A0}	10	mol/l
C_{B0}	1.1685	mol/l
C_{C0}	0	mol/l
V_0	1	l
$C_{B_{in}}$	20	mol/l
t_f	0.5	h

16.5 Further Reading

A general exposition of the application of PMP to endpoint optimization of finite-time systems can be found in Bryson and Ho [3]. A review of differential geometric concepts for nonlinear controller design as well as the definition of relative degree can be found in Marino and Tomei [6]. Details of the connections between DAE index and relative degree can be found in Krishnan and McClamroch [4]. The synthesis of state feedback laws for batch processes with affine scalar input can be found in Palanki, Kravaris, and Wang [7, 8]. A generalization of this approach to batch processes with multiple inputs and constraints can be found in Srinivasan, Palanki, and Bonvin [9]. The application of this methodology in a measurement-based optimization scheme can be found in Srinivasan, Bonvin, Visser, and Palanki [10]. A numerical approach to this class of problems that converts the endpoint optimization problem into a nonlinear programming problem can be found in Biegler [1] and Biegler, Cervantes, and Wachter [2]. A numerical approach to endpoint optimization that involves iterative dynamic programming can be found in Luus [5].

Bibliography

[1] L. T. BIEGLER, *Solution of dynamic optimization problems by successive quadratic programming and orthogonal collocation*, Comput. Chem. Eng., 8 (1984), pp. 243–248.

[2] L. T. BIEGLER, A. M. CERVANTES, AND M. A. WACHTER, *Advances in simultaneous strategies for dynamic process optimization*, Chem. Eng. Sci., 57 (2002), pp. 575–593.

[3] A. E. BRYSON, AND Y. C. HO, *Applied Optimal Control*, Hemisphere, Mineola, NY, 1975.

[4] H. KRISHNAN AND N. H. MCCLAMROCH, *Tracking in nonlinear differential-algebraic control systems with applications to constrained robot systems*, Automatica, 30 (1994), pp. 1885–1897.

[5] R. LUUS, *Optimal control of batch reactors by iterative dynamic programming*, J. Process Control, 4 (1994), pp. 218–226.

[6] R. MARINO AND P. TOMEI, *Nonlinear Control Design*, Prentice–Hall, London, 1995.

[7] S. PALANKI, C. KRAVARIS, AND H. Y. WANG, *Synthesis of state feedback laws for end-point optimization in batch processes*, Chem. Eng. Sci., 48 (1993), pp. 135–152.

[8] S. PALANKI, C. KRAVARIS, AND H. Y. WANG, *Optimal feedback control of batch reactors with a state inequality constraint and free terminal time*, Chem. Eng. Sci., 49 (1994), pp. 85–97.

[9] B. SRINIVASAN, S. PALANKI, AND D. BONVIN, *Dynamic optimization of batch processes:* I. *Characterization of the optimal solution*, Comput. Chem. Eng., 27 (2003), pp. 1–26.

[10] B. SRINIVASAN, D. BONVIN, E. VISSER, AND S. PALANKI, *Dynamic optimization of batch processes:* II. *Measurement based optimization*, Comput. Chem. Eng., 27 (2003), pp. 27–44.

Index